FLORA ZAMBESIACA

Flora terrarum Zambesii aquis conjunctarum

VOLUME ELEVEN: PART ONE

FLORA ZAMBESIACA

MOZAMBIQUE

MALAWI, ZAMBIA, ZIMBABWE

BOTSWANA

VOLUME ELEVEN: PART ONE

Edited by
G.V. POPE

on behalf of the Editorial Board:

G.Ll. LUCAS
Royal Botanic Gardens, Kew

I. MOREIRA
*Centro de Botânica, Instituto de Investigação
Científica Tropical, Lisboa*

G.V. POPE
Royal Botanic Gardens, Kew

Published by the Managing Committee on behalf of
the contributors to Flora Zambesiaca
1995

Typeset at the Royal Botanic Gardens, Kew, by
Christine Beard and Dominica Costello

Printed in Great Britain by
Whitstable Litho Printers Ltd., Whitstable, Kent.

ISBN 0 947643 89 3

CONTENTS

FAMILY INCLUDED IN VOLUME 11, PART 1

163. Orchidaceae

Acknowledgements
The Flora Zambesiaca Managing Committee thanks M.A. Diniz and E. Martins of
the Centro de Botânica, Lisbon, for their valuable help in reading and
commenting on the text.

163. ORCHIDACEAE

By I. la Croix and P.J. Cribb[1]

Perennial, terrestrial, saprophytic or epiphytic herbs or rarely scrambling climbers, with rhizomes, root-stem tuberoids or rootstocks with mycorrhizal fungi in the roots and often elsewhere. Growth either sympodial or less commonly monopodial. Stems usually leafy, but leaves often reduced to bract-like scales, one or more internodes at the base often swollen to form a "pseudobulb"; aerial, assimilating adventitious roots, often bearing one or more layers of dead cells called a velamen, are borne in epiphytic species. Leaves glabrous or occasionally hairy, entire except at the apex in some cases, alternate or occasionally opposite, often distichous, frequently fleshy and often terete or canaliculate, almost always with a basal sheath which frequently sheaths the stem, sometimes articulated at the base of the lamina and sometimes with a false petiole. Inflorescences erect to pendent, spicate, racemose or paniculate, one to many-flowered, basal, lateral or terminal, the flowers rarely secund or distichously arranged. Flowers small to large, often quite showy, hermaphrodite or rarely monoecious and polymorphic, sessile or variously pedicellate, most often twisted through 180 degrees, occasionally not twisted or twisted through 360 degrees. Ovary inferior, unilocular and the placentation parietal, or rarely trilocular and the placentation axile. Perianth epigynous, of two whorls of three segments; outer perianth whorl (sepals) usually free but sometimes variously adnate, the median (dorsal) often dissimilar to the laterals, the laterals sometimes adnate to the column foot to form a saccate, conical or spur-like mentum; inner whorl comprising two lateral petals and a median lip; petals free or rarely partly adnate to sepals, similar to sepals or not, often showy; lip entire, variously lobed or two or three-partite, ornamented or not with calli, ridges, hair cushions or crests, with or without a basal spur or nectary, margins entire to laciniate. Stylar and filamentous tissue forming a long or short column, with or without a basal foot, occasionally winged or with lobes or arms at apex or ventrally; anther one (or rarely two or three in extra African taxa), terminal or ventral on column, with a concave anther cap or opening by longitudinal slits; pollen in tetrads, agglutinated into discrete masses called pollinia; pollinia mealy, waxy or horny, sectile or not, 2, 4, 6 or 8, sessile or attached by caudicles, a stipes or stipites to one or two sticky viscidia; stigma 3-lobed, the mid-lobe often modified to form a rostellum, the other lobes either sunken on the ventral surface of the column behind the anther or with two lobes porrect. Fruit a capsule, opening laterally by 3 or 6 slits; seeds numerous, dust-like, lacking endosperm, sometimes markedly winged.

The orchids comprise one of the largest families of flowering plants, with an estimated 800 genera and upwards of 17,000 species, some estimates suggesting as many as 30,000 species. They are distributed in all continents except for Antarctica, but are most numerous in the humid tropics and subtropics. Some 600 species in 70 genera are found in the Flora Zambesiaca region. They are extensively grown around the world as ornamentals but other economic uses are few. Two or possibly three species of *Vanilla* Miller are grown commercially to produce the flavouring vanillin, the tubers of several species are used for food and as aphrodisiacs, and a number of species are used in folk medicine, particularly in China.

The classification of the family is currently the subject of some debate, particularly the number of subfamilies that should be recognised and the placement in those of certain tribes, subtribes and genera. The classification of the Orchidaceae outlined by Summerhayes (1968) in his account of the family for the Flora of Tropical East Africa differs in only relatively minor points from the most widely accepted of recent classifications (Dressler 1981, 1993). Summerhayes accepted three subfamilies; Apostasioideae, Cypripedioideae and Orchidoideae, dividing the last into four tribes; Orchideae, Neottieae, Epidendreae and Vandeae. All African orchids are placed in Summerhayes's Orchidoideae, neither the Apostasioideae nor the

---

[1]*Disa, Brownleea, Herschelianthe, Monadenia* & *Oligophyton* by P. Linder, *Nervilia* by Börge Pettersson.

Cypripedioideae being represented in the continent. Dressler (1993) accepts five subfamilies: Apostasioideae, Cypripedioideae, Spiranthoideae, Orchidoideae and Epidendroideae, the last two equivalent more or less to the Orchideae and a combined Epidendreae and Vandeae of Summerhayes. The Neottieae of Summerhayes largely fall into Dressler's Spiranthoideae but *Vanilla, Didymoplexis, Epipogium, Epipactis* and *Nervilia*, all found in tropical Africa, are considered by Dressler to be primitive Epidendroideae. Dressler's 1993 classification is followed but the genera, except for the Spiranthoideae which precede the Orchidoideae here, are arranged more or less in the same order as in Summerhayes in the Flora of Tropical East Africa to facilitate comparison.

Key to the subfamilies of ORCHIDACEAE

1. Anthers two or three · 2
 – Anther solitary, although the anther loculi can be well separated · · · · · · · · · · · · · · · 3
2. Anther filaments fused to the column near the base only; staminode absent; lip petaloid or nearly so, never saccate · · · · · · · · · · · · · · · · APOSTASIOIDEAE (not found in Africa)
 – Anther filaments fused to the column stalk; staminode present, usually terminal and shield-shaped; lip deeply saccate · · · · · · · · · · · · · · CYPRIPEDIOIDEAE (not found in Africa)
3. Anther attached to the column by the base, the loculi adnate to the column, persistent; pollinia granular or sectile with caudicles and two (rarely one) viscidia; mostly terrestrial with erect leafy stem, terminal inflorescence and root-stem tuberoids · · · · · ORCHIDOIDEAE
 – Anther attached to the column by the apex, usually at the back of the column and either operculate or erect and persistent, caudicle and viscidia then being at the top · · · · · · · 4
4. Pollinia granular, occasionally mealy, or much divided into small masses attached to a common axis; anther opening lengthwise; plants usually terrestrial · · SPIRANTHOIDEAE
 – Pollinia waxy, entire, 2, 4 or 8, free or adhering at one end by caudicles or attached to 1 or 2 viscidia; anther operculate; plants terrestrial, epiphytic, lithophytic or scrambling lianas · EPIDENDROIDEAE

Key to the genera of SPIRANTHOIDEAE

1. Plants with woody stems and rhizomes; inflorescences 1–several, branching, glabrous; flowers more than 3 cm long · **1. Corymborkis**
 – Plants up to c. 50 cm; stems and rhizomes fleshy; inflorescence solitary, unbranched, often pubescent or glandular; flowers less than 1 cm long · 2
2. Lip tapering to a blunt apex; sepals and petals 7–9 mm long · · · · · · · · · · · **2. Platylepis**
 – Lip with a transversely oblong apical lobe; sepals and petals 6 mm long or less · · · · · · · 3
3. Lateral sepals united for about half their length · · · · · · · · · · · · · · · · · · **3. Cheirostylis**
 – Lateral sepals free to the base · **4. Zeuxine**

Key to the genera of ORCHIDOIDEAE

1. Spur formed by the dorsal sepal, not formed by the lip or lateral sepals · · · · · · · · · · · 2
 – Spur or spurs or sacs, if present, formed by the lip or lateral sepals · · · · · · · · · · · · · 5
2. Lip with lobed or deeply divided margins, appearing bearded; flowers blue, purplish-blue or rarely white; leaves grass-like, tufted · · · · · · · · · · · · · · · · · · **21. Herschelianthe**
 – Lip entire; flowers not blue; leaves lanceolate, elliptic or ovate · · · · · · · · · · · · · · · · 3
3. Lateral sepals appressed to the ovary; pollinia attached to a single viscidium · · · · · · · · ·
 · **22. Monadenia**
 – Lateral sepals spreading; each pollinium attached to its own viscidium · · · · · · · · · · · 4
4. Petals free from the dorsal sepal; anther erect or horizontal · · · · · · · · · · · · · · **19. Disa**
 – Petals united to the dorsal sepal; anther horizontal · · · · · · · · · · · · · · · · **20. Brownleea**
5. Lip with 1 or 2 distinct, sometimes small, spurs · 6
 – Lip not spurred, sometimes boat-shaped at the base · 18
6. Spurs 2; flowers resupinate with the lip uppermost · · · · · · · · · · · · · · · · · **23. Satyrium**
 – Spur single; flowers not resupinate · 7
7. Lip more or less united to the column · 8
 – Lip free from the column · 11
8. Leaves 1 or 2, radical, orbicular or ovate · **10. Holothrix**
 – Leaves usually several, borne on the stem, not orbicular · 9
9. Stigmas sessile · **9. Schizochilus**
 – Stigmas borne on variously lengthened stalks · 10

10. Stigmatic arms divided into 2 elongated branches, the lower fertile and capitate, the upper sterile and horn-like · **13. Centrostigma**
   − Stigmatic arms unbranched · **16. Habenaria** (in part)
11. Stigmas sessile · 12
   − Stigmas borne on, or forming, club-shaped processes, projecting from the front of the column, free or partly united with the lateral lobes of the rostellum · · · · · · · · · · · · 13
12. Flowers usually purple, mauve, white or rarely yellow; sepals and petals more than 3 mm long · **6. Brachycorythis** (in part)
   − Flowers tiny, yellow with a green lip; sepals and petals up to 2 mm long · · **15. Oligophyton**
13. Lateral sepals and front lobe of the petals joined to the lip and to the stigmatic arms in their lower parts; tooth in mouth of spur · **12. Bonatea**
   − Lateral sepals and front lobe of petals not united at the same time; tooth absent in mouth of spur · 14
14. Stigmas 2-lobed, the lobes projecting in different directions, upwards and downwards · **18. Roeperocharis**
   − Stigmas simple, not branched · 15
15. Stigmatic processes partly united to the side lobes of the rostellum · · · · · · · · · · · · 16
   − Stigmatic processes free from the rostellum · 17
16. Flowers pink or mauve; side lobes of lip entire · · · · · · · · · · · · · · · · · · · **14. Cynorkis**
   − Flowers green; side lobes of lip fimbriate · · · · · · · · · · · · · · · **16. Habenaria** (in part)
17. Flowers yellow or orange; mid-lobe of rostellum concave, placed in front of the anther · **17. Platycoryne**
   − Flowers green or white; mid-lobe of rostellum usually flat or subulate · **16. Habenaria** (in part)
18. Lip adnate to front of the column, bearing incurled lateral appendages · · · · · · · · · 19
   − Lip not adnate to the front of the column, lateral appendages absent · · · · · · · · · · · 20
19. Lateral sepals not spurred; petals much wider than long · · · · · · · · · · · · · **24. Corycium**
   − Lateral sepals spurred or saccate; petals longer than wide · · · · · · · · · · · · · **25. Disperis**
20. Plant dwarf, leafless · **7. Schwartzkopffia**
   − Plant small to large, leafy · 21
21. Margins of lip divided in the apical part · · · · · · · · · · · · · · · · · · · **11. Stenoglottis**
   − Margins of lip entire · 22
22. Lip with a basal callus, not boat-shaped at the base · · · · · · · · · · · · · · · **8. Neobolusia**
   − Lip without a basal callus, boat-shaped at the base · · · · · · · · · **6. Brachycorythis** (in part)

Key to the genera of EPIDENDROIDEAE

1. Plants terrestrial saprophytes, lacking leaves and chlorophyll · · · · · · · · · · · · · · · · · 2
   − Plants terrestrial, epiphytic or lithophytic, usually with green leaves, rarely epiphytic and lacking green leaves but then with green roots · 4
2. Lip lacking a spur; pedicels elongating rapidly after fertilzation · · · · · · **28. Didymoplexis**
   − Lip with a spur; pedicels not elongating after fertilization · · · · · · · · · · · · · · · · · · · 3
3. Lip entire, lacking longitudinal crests on lip; pollinia soft, sectile · · · · · · · **29. Epipogium**
   − Lip 3-lobed, bearing longitudinal crests in basal half; pollinia hard, attached to a noticeable viscidium · **Eulophia** (in part)
4. Plants terrestrial, scrambling or climbing · 5
   − Plants epiphytic · 16
5. Plants scrambling or climbing, liana-like · **26. Vanilla**
   − Plants terrestrial · 6
6. Inflorescence appearing well before the solitary leaf emerges · · · · · · · · · · · **27. Nervilia**
   − Inflorescence appearing with the leaves, or sometimes after the leaves have fallen · · · · · 7
7. Pollinia 8, clavate; flowers turning blue if bruised · 8
   − Pollinia 2 or 4; flowers not turning blue when bruised · 9
8. Column short, adnate to the lip base; spur as long as or longer than the lip · · **30. Calanthe**
   − Column elongate, not adnate to the column except at the base; spur much shorter than the lip · **31. Phaius**
9. Inflorescence terminal · 10
   − Inflorescence lateral or basal · 13
10. Flowers pubescent; lip bipartite, the hypochile saccate and articulated to a fleshy epichile · **5. Epipactis**
   − Flowers glabrous; lip entire or lobed · 11

11. Inflorescence branching; lip spurred · · · · · · · · · · · · · · · **Acrolophia** (see vol. 11 part 2)
  – Inflorescence simple; lip lacking a spur · 12
12. Column very short, with basal auricles of lip on either side of the column · · · **33. Malaxis**
  – Column elongate, lacking basal auricles around the lip · · · · · · · · · · · · · · **32. Liparis**
13. Plants with above-ground pseudobulbs · 14
  – Plants with underground tubers or swollen rhizomes · 15
14. Pseudobulbs of one node · **Oeceoclades** (see vol. 11 part 2)
  – Pseudobulbs of several nodes, or stems caulescent · · · · · · · · · · · · · · **Eulophia** (in part)
15. Lip lacking a spur; column short, lacking a foot · · · · · **Pteroglossaspis** (see vol. 11 part 2)
  – Lip with a spur or saccate base; column with a short to long foot · · · · · · · · · · · · · · · · ·
  · **Eulophia** (in part, see vol. 11 part 2)
16. Plants with monopodial growth, always lacking pseudobulbs; leaves when present distichous, fleshy, conduplicate, usually unequally 2-lobed at the apex; inflorescences always axillary; pollinia with stipes and viscidium · · · · · · · · · · · · **Tribe Vandeae** (see key below)
  – Plants with sympodial growth, with or without pseudobulbs; leaves plicate to conduplicate or iridiform; pollinia with or without stipes and viscidium · · · · · · · · · · · · · · · · · · · 17
17. Leaves iridiform, bilaterally compressed, distichous; flowers minute, less than 2 mm across, in whorls in a tapering densely-flowered inflorescence · · · · · · · · · · · · · · · **34. Oberonia**
  – Leaves flat, needle-like or terete · 18
18. Flowers with a more or less distinct conical to cylindrical mentum formed by the lateral sepals and column-foot · 19
  – Flowers lacking a mentum; column-foot absent · 22
19. Pollinia 8 · **35. Stolzia**
  – Pollinia 2 or 4 · 20
20. Inflorescence terminal · **Polystachya** (see vol. 11 part 2)
  – Inflorescences lateral, basal · 21
21. Leaves 1 or 2 on pseudobulb · **37. Bulbophyllum**
  – Leaves several, pine-needle-like · **36. Chaseella**
22. Lip lacking a spur; flowers large, yellow, distinctively spotted with maroon on sepals and petals; pseudobulbs elongate-fusiform · · · · · · · · · · · · · · · **Ansellia** (see vol. 11 part 2)
  – Lip spurred at the base; flowers small, yellow, flushed but not spotted with maroon-brown on sepals and petals · **Graphorkis** (see vol. 11 part 2)

Key to the genera of VANDEAE

1. Leaves reduced to achlorophyllous scales, or absent · 2
  – Leaves present, with chlorophyll · 4
2. Stems elongate; roots scattered along the stem · · · **Solenangis** (in part, see vol. 11 part 2)
  – Stems very short; roots clustered · 3
3. Flowers orange; sepals and petals united into a tube in lower part; lip with a hooded apex; pollinia 4 · **Taeniophyllum** (see vol. 11 part 2)
  – Flowers usually white; sepals and petals free; lip apex not hooded; pollinia 2 · · · · · · · · ·
  · **Microcoelia** (see vol. 11 part 2)
4. Sepals and petals pale yellow, blotched or marked with reddish-brown; lip fleshy; pollinia 4 in 2 pairs · **Acampe** (see vol. 11 part 2)
  – Sepals and petals unicoloured; pollinia 2 · 5
5. Flowers borne in a capitate head; rostellum distinctively reflexed in apical half to lie parallel to the basal part · **Ancistrorhynchus** (see vol. 11 part 2)
  – Flowers solitary or in a lax to dense raceme; rostellum not as above · · · · · · · · · · · · · 6
6. Leaves iridiform · **Bolusiella** (see vol. 11 part 2)
  – Leaves flat, semi-terete or terete · 7
7. Rostellum retuse, not projecting beyond the column apex; stipites obscure or apparently absent · 8
  – Rostellum projecting beyond the column apex, often pendent; stipes 1 or 2, elongate, attached to 1 or 2 viscidia · 10
8. Column-foot saccate, spurred, the lip articulated below the spur at the apex of the column-foot · **Aeranthes** (see vol. 11 part 2)
  – Column lacking a foot; lip spurred · 9
9. Lip ovate to cordate with the base enfolding the column · · **Angraecum** (see vol. 11 part 2)
  – Lip lanceolate or subpandurate, the base not enfolding the column · · · · · · · · · · · · · · · ·
  · **Jumellea** (see vol. 11 part 2)

10. Flowers borne in whorls at the nodes of the inflorescence · · · · · · · · · · · · · · · · · · 11
  − Flowers borne singly at the inflorescence nodes · 12
11. Lip obscurely 3-lobed, 10 mm or more long, fringed ·
  · **Diaphananthe** (in part, see vol.11 part 2)
  − Lip entire; 4 mm long or less, margins entire · · · · · · · · **Chamaeangis** (see vol. 11 part 2)
12. Spur abruptly geniculate in middle and dilated at the apex; rhachis markedly zigzag · · · ·
  · **Calyptrochilum** (see vol. 11 part 2)
  − Spur saccate to cylindrical, not as above · 13
13. Pollinia attached by a single stipes to a viscidium · 14
  − Pollinia attached by two stipites to one or two viscidia · · · · · · · · · · · · · · · · · · 22
14. Lip cordate · **Cardiochilos** (see vol. 11 part 2)
  − Lip not cordate · 15
15. Lip uppermost in flower; viscidium shoe-shaped · · · · **Summerhayesia** (see vol. 11 part 2)
  − Lip lowermost in flower · 16
16. Stipes Y-shaped · **Ypsilopus** (see vol. 11 part 2)
  − Stipes linear or oblanceolate · 17
17. Lip 2- or 3-lobed in apical half, sometimes obscurely so · · · · · · · · · · · · · · · · · · 18
  − Lip entire, not lobed in apical half · 19
18. Lip 3-lobed in apical part, lobes triangular, tapering or fimbriate · · · · · · · · · · · · · · ·
  · **Tridactyle** (in part, see vol. 11 part 2)
  − Lip broadly 2-lobed in apical half · · · · · · · · · · · · · · · · **Nephrangis** (see vol. 11 part 2)
19. Flowers stellate, white, sometimes variously tinged with green or pink on spur; stem usually
  short; leaves oblanceolate or obovate, flat · · · · · · · · · · · · · · **Aerangis** (see vol. 11 part 2)
  − Flowers white, pale green, yellow or brown; lip with basal auricles; stem usually elongate;
  leaves linear or oblong, flat or terete · 20
20. Leaf with a distinct ligule borne at top of sheath opposite the lamina · · · · · · · · · · · · ·
  · **Eggelingia** (see vol. 11 part 2)
  − Leaf lacking a ligule · 21
21. Leaves awl-shaped, up to 13 × 1 mm; flowers not opening widely, white; lip lacking basal
  auricles · **Solenangis** (in part, see vol. 11 part 2)
  − Leaves linear to oblong, more than 25 mm long; flowers opening widely, not white; lip with
  basal auricles · **Tridactyle** (in part)
22. Pollinia attached by separate stipes to a separate viscidium · · · · · · · · · · · · · · · · · 23
  − Pollinia with stipites attached to a single viscidium · 26
23. Lip strongly 3-lobed in the middle; rostellum of 3 equal lobes · · · · · · · · · · · · · · · · ·
  · **Angraecopsis** (in part, see vol. 11 part 2)
  − Lip entire or obscurely lobed · 24
24. Lip with a small tooth in the mouth of the spur; mid-lobe of rostellum fleshy and shortly
  clavate, the side lobes obscure · **Diaphananthe** (in part)
  − Lip lacking a tooth in the mouth of the spur; rostellum thin-textured, 3-lobed · · · · · · 25
25. Rostellar lobes papillate, the outer lobes exceeding the mid-lobe · · · · · · · · · · · · · · ·
  · **Mystacidium** (see vol. 11 part 2)
  − Rostellar lobes glabrous, more or less equal · · · **Angraecopsis** (in part, see vol. 11 part 2)
26. Flowers pale translucent yellow or green · 27
  − Flowers white, rarely marked with green on lip · 29
27. Lip with a small tooth or fleshy swelling in the mouth of the spur · · **Diaphananthe** (in part)
  − Lip lacking any tooth or swelling in mouth of spur · 28
28. Inflorescences erect, laxly many-flowered; leaves usually more than 8 cm long · · · · · · · ·
  · **Cribbia** (see vol. 11 part 2)
  − Inflorescences pendent, of many closely-spaced secund pale green flowers; leaves very
  small, less than 5 cm long · **Angraecopsis** (in part)
29. Lip lanceolate, similar to sepals and petals; rostellar side arms longer than the column;
  viscidium linear, saddle-shaped, hyaline or bipartite · · · · · **Cyrtorchis** (see vol. 11 part 2)
  − Lip obscurely 3-lobed, differently shaped from sepals and petals; rostellar side arms shorter
  than the column; viscidium suboblong, never bipartite · · · **Rangaeris** (see vol. 11 part 2)

1. CORYMBORKIS Thouars

Corymborkis Thouars in Nouv. Bull. Sci. Soc. Philom. Paris **1**: 318 (1809).
—Rasmussen in Bot. Tidsskr. **71**: 161–192 (1977); *Corymborchis* Thouars, Hist.
Orchid. [Fl. Iles Austr. Afr.]: tab. gen.; figs. 37, 38 (1822), orth. var. *Corymbis*
Thouars, Hist. Orchid. [Fl. Iles Austr. Afr.]: prem. tab. espèc.; figs. 37, 38 (1822).

Rhizomatous terrestrial herb with tall, erect leafy stems. Leaves distichous, sessile
or shortly petiolate, plicate. Inflorescences axillary panicles, few to many-flowered.
Flowers white or greenish-white. Sepals and petals subequal, long, linear, basally
connivent; lip similar, but with ovate apex. Column long, straight, slender but
dilated at apex, with 2 lateral auricles. Anther erect, narrow, acuminate, almost as
long as column; pollinia 2, narrow, sectile, with a long slender caudicle attached to
a peltate viscidium descending behind column. Stigma broad, deeply 2-lobed;
rostellum erect, bifid.

A pan-tropical genus with two African species.

Corymborkis corymbis Thouars, Hist. Orchid. [Fl. Iles Austr. Afr.]: prem. tab. espèc.; figs. 37, 38
(1822). —Rasmussen in Bot. Tidsskr. **71**: 169 (1977). —P.F. Hunt in F.T.E.A., Orchidaceae:
243 (1984). —la Croix et al., Orch. Malawi: 24 (1991). TAB. **1**. Type from Réunion.
    Corymbis thouarsii Rchb.f. in Bot. Zeit. **7**: 868 (1849), nom. superfl., based on *Corymborkis
corymbis*.
    Corymbis welwitschii Rchb.f. in Flora **48**: 183 (1865). —Rolfe in F.T.A. **7**: 180 (1897). Type
from Angola.
    Corymbis corymbosa Ridl. in J. Linn. Soc. **21**: 498 (1885). —Rolfe in F.T.A. **7**: 180 (1897),
nom. superfl. based on *Corymborkis corymbis*.
    Corymborkis corymbosa (Ridl.) Kuntze, Revis. Gen. Pl., 2: 658 (1891). —Summerhayes in
Kew Bull. **11**: 224 (1956); in F.W.T.A. ed. 2, **3**: 211 (1968). —Piers, Orch. E. Afr.: 141
(1959), as "*Corymborchis*". —Grosvenor in Excelsa **6**: 79 (1976).
    Corymborkis welwitschii (Rchb.f.) Kuntze, Revis. Gen. Pl., 2: 658 (1891). —Summerhayes
in F.W.T.A. **2**: 423 (1936). —Robyns & Tournay, Fl. Parc Nat. Alb. **3**: 459 (1955).

Terrestrial herb 0.5–1.5 m tall. Roots elongate, c. 2 mm in diameter. Stem leafy
along its length. Leaves dark green, distichous, plicate, up to 24 × 8.5 cm, lanceolate
to elliptic, acuminate, the margins slightly undulate, the base clasping the stem.
Inflorescences terminal and axillary, the rhachis zig-zag, often branched, several-
flowered; ovary and pedicel 10–15 mm long; bracts 7–15 mm long, lanceolate to
ovate. Flowers non-resupinate; sepals green, petals and lip green, turning white
towards apices. Sepals 70–85 × 2 mm, linear, curling back at tips. Petals 70–80 mm
long, linear-spathulate, 2 mm wide at base, 3 mm at apex. Lip 60–80 mm long, the
basal part 2 mm wide, linear, the apex 6–7 mm wide, ovate. Column elongated, lying
along the lip and slightly shorter than it, c. 1 mm wide.

    Zimbabwe. E: Chipinge, Chirinda Forest Reserve, fl. i.1975, *Goldsmith* 1/75 (K; SRGH).
Malawi. S: Mulanje Distr., Ruo Tea Estate, fl. immat. 30.xii.1982, *la Croix & Spurrier* 382 (K).
Mozambique. MS: Mt. Bandula, 600 m, st. 1.iii.1958, *Chase* 6844 (K; SRGH; PRE). GI: Massinga,
fl. 11.ii.1898, *Schlechter* 12129 (BM; K).
    Widely distributed in tropical Africa, also in South Africa, Madagascar and Mascarene
Islands. Evergreen, usually lowland, forest, in densely shady undergrowth, often forming large
colonies; 100–1120 m.

2. PLATYLEPIS A. Rich.

Platylepis A. Rich. in Mém. Soc. Hist. Nat. Paris **4**: 34 (1828), nom. conserv.
Erporkis Thouars in Nouv. Bull. Sci. Soc. Philom. Paris **1**: 317 (1809); *Erporchis*
Thouars, Hist. Orchid. [Fl. Iles Austr. Afr.]: tab. gen., fig. 28 (1822), orth. var.
Notiophrys Lindl. in J. Linn. Soc., Bot. **1**: 189 (1857). *Diplogastra* Welw. ex
Rchb.f. in Flora **48**: 183 (1865).

Terrestrial herb, stems creeping at the base, roots tomentose. Leaves ovate,
petiolate, sheathing at the base. Upper stem with leaves reduced to sheaths;
inflorescence terminal, racemose. Bracts conspicuous, broad, glandular-hairy, longer

Tab. 1. CORYMBORKIS CORYMBIS. 1, part of upper stem and inflorescence (×1), from *la Croix & Spurrier* 382; 2, stem base and roots (×1), from *Chapman* 445; 3, dorsal sepal (×1); 4, lateral sepal (×1); 5, petal (×1); 6, lip (×1); 7, column (×1); 8, column apex, viewed from above (×4); 9, column apex with anther removed, viewed from above (×4); 10, column apex, viewed from below (×4); 11, anther cap, back view (×4); 12, anther cap, front view (×4), 3–12 from *Drummond & Hemsley* 3202 (in K spirit coll. 20631). Drawn by Judi Stone.

than the ovaries. Sepals free; petals partly joined to dorsal sepal; lip adnate to the column for part of its length, reflexed at the apex, saccate at the base, with calli of various shapes. Column elongate; clinandrium and anther erect, oblong; rostellum erect, 2-lobed, set in front of the anther. Stigma oblong.

A genus of 10 species in tropical Africa, South Africa, Madagascar and the Mascarene Islands.

Platylepis glandulosa (Lindl.) Rchb.f. in Linnaea **41**: 62 (1877). —Rolfe in F.T.A. **7**: 184 (1897). —Summerhayes in Kew Bull. **11**: 223 (1956); in F.W.T.A. ed. 2, **3**: 208 (1968). —Grosvenor in Excelsa **6**: 84 (1976). —Williamson, Orch. S. Centr. Africa: 106, t.53 (1977). —Stewart et al., Wild Orch. South. Africa: 212 (1982). —P.F. Hunt in F.T.E.A., Orchidaceae: 245 (1984). —Geerinck in Fl. Afr. Centr., Orchidaceae pt. 1: 13 (1984). —la Croix et al., Orch. Malawi: 25 (1991). TAB. **2**. Type from Principe.
   Notiophrys glandulosa Lindl. in J. Linn. Soc., Bot. **6**: 138 (1862).
   Diplogastra angolensis Welw. ex Rchb.f. in Flora **48**: 183 (1865). Type from Angola.
   Platylepis angolensis (Welw. ex Rchb.f.) T. Durand & Schinz, Consp. Fl. Afric. **5**: 58 (1895).
   Platylepis australis Rolfe in Bull. Misc. Inform., Kew **1906**: 378 (1906). Type from South Africa.
   Platylepis nyassana Schltr. in Bot. Jahrb. Syst. **53**: 557 (1915). —Mansfeld in Fedde, Repert. Spec. Nov. Regni Veg., Beih. 68, t. 51, fig. 201 (1932). Type from Tanzania.

Terrestrial herb, 20–36 cm tall; stems creeping at the base, erect in the upper part. Roots arising from the lower nodes, fleshy, villous, c. 2 mm in diameter. Leaves spirally arranged, 4.8–8 × 1.2–4 cm, ovate, slightly asymmetrical, glabrous, margins often undulate, lamina with 3 prominent longitudinal veins; petiole 2–4 cm long, flattened, widening to a sheath at the base; upper 5–6 leaves reduced to ovate sheaths up to about 2 cm long. Inflorescence terminal, densely many-flowered; rhachis, bracts, ovary, pedicel and sepals all glandular-hairy; ovary and pedicel c. 7 mm long; bracts up to 15 × 7 mm at the base of the inflorescence, ovate. Sepals green, the lateral sepals sometimes pink-brown along the margin; petals green or pink-brown; lip white, pink-brown at the base. Dorsal sepal 6–7 × 3 mm, ovate, convex; lateral sepals joined to the dorsal sepal for 3 mm, then spreading, the free part c. 4 × 2.5 mm. Petals 5–8 × 1 mm, oblanceolate or spathulate, lying inside the dorsal sepal. Lip 5.5–8.5 × 2.2–2.5 mm, adnate to the column for half its length, fleshy, deeply concave with 2 rounded calli at the base, the apex rounded and recurved. Column 4–6.5 mm long; rostellum 2-lobed, the lobes acuminate, as long as the anther; stigma papillose.

Zambia. W: Chingola, fl. immat. 16.i.1964, *Fanshawe* 8217 (K). C: Lusaka Distr., Mowashi R., 30 km east of Rufunza, 940 m, fl. 4.iii.1973, *Kornaś* 3417 (K). **Zimbabwe**. E: Nyanga, fl. 16.i.1960, *Ball* 866 (K; SRGH). **Malawi**. N: Mzimba Distr., Lunyangwa, Mzuzu, 1330 m, fl. 23.i.1987, *la Croix* 943 (K; MAL). S: Mulanje Distr., Mimosa, 700 m, fl. 16.i.1984, *la Croix & Spurrier* 523 (K). **Mozambique**. GI: Inhambane, "insilva masuku", 30 m, fl. 10.ii.1898, *Schlechter* 12110 (BM).
   Also in West Africa, Uganda, Tanzania and South Africa (Natal). In deep shade in marshy forest; 30–1330 m.

3. CHEIROSTYLIS Blume

Cheirostylis Blume, Bijdr.: 413, t. 16 (1825).
Mariarisqueta Guinea, Ensayo Geobot. Guin. Continent. Espan.: 268 (1946);
in Anales Jard. Bot. Madrid **6**(2): 470 (1946).

Small terrestrial herb with erect leafy stems arising from fleshy rhizomes. Leaves thin-textured, petiolate with a sheathing base. Inflorescence terminal, erect; upper stem with few sheaths. Flowers small, mostly white. Sepals joined for about half their length; petals adnate to the dorsal sepal. Lip equalling or longer than the tepals, erect, joined to base of the column, saccate at the base with 2 calli, lobed at the apex; the lobes divergent, entire or toothed. Column short, erect, with 2 apical appendages parallel to the elongate rostellum and 2 lateral stigmas. Anther dorsal, acuminate; pollinia 2, sectile; caudicle short; viscidium oblong.

A genus of more than 20 species, mostly in tropical Asia and Australasia. Three species are known in Africa, one of which also occurs in Madagascar and the Comoro Islands; one species occurs in the Flora Zambesiaca area.

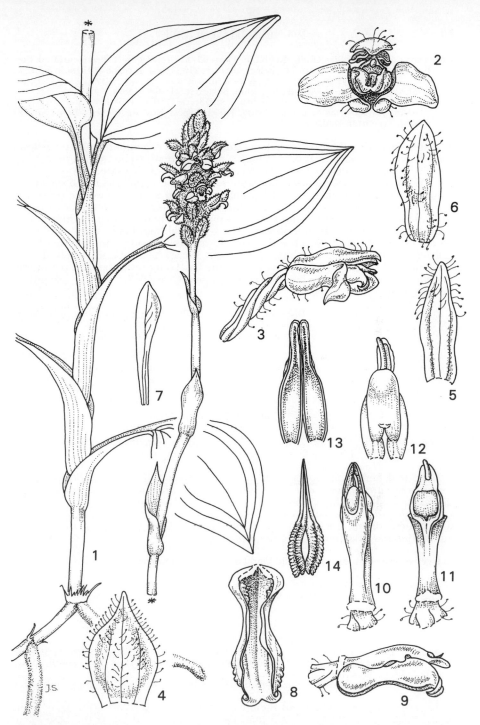

Tab. 2. PLATYLEPIS GLANDULOSA. 1, habit (×1), from *la Croix & Spurrier* 523; 2, flower, front view (×4); 3, flower, side view (×3); 4, bract (×3); 5, dorsal sepal (×4); 6, lateral sepal (×4); 7, petal (×4); 8, lip (×4); 9, column and lip, side view (×4); 10, column, back view (×4); 11, column, front view (×4); 12, anther cap, back view (×8); 13, anther cap, front view (×8); 14, pollinarium (×8), 2–14 from *Cribb & Grey-Wilson* 10683 (in K spirit coll. 25694.245). Drawn by Judi Stone.

Cheirostylis gymnochiloides (Ridl.) Rchb.f. in Flora **68**: 537 (1885). —H. Perrier in Bull. Soc.
    Bot. France **83**: 23 (1936); in Fl. Madag., Orch. 1: 216 (1939). —Cribb & Hunt in F.T.E.A.,
    Orchidaceae: 249 (1984). TAB. **3**. Type from Madagascar.
    Monochilus gymnochiloides Ridl. in J. Linn. Soc., Bot. **21**: 499 (1885).
    Cheirostylis sarcopus Schltr. in Bot. Jahrb. Syst. **53**: 558 (1915). Type from Tanzania.
    Cheirostylis micrantha Schltr. in Fedde, Repert. Spec. Nov. Regni Veg., Beih. **33**: 127
    (1925). Type from Madagascar.

Terrestrial herb with a creeping fleshy rhizome; roots c. 3 mm in diameter, villous.
Leaves several, crowded near base of the stem, petiolate with a broad sheathing base;
petiole c. 2 cm long, blade 4–6 × 2–2.5 cm, slightly obliquely ovate. Flowering stem
to c. 35 cm high, usually sparsely hairy but sometimes glabrous, with 5–6 sheaths to
15 mm long. Inflorescence 2–6 × 1.5–2 cm, fairly densely many-flowered; ovary and
bracts usually sparsely glandular-hairy but sometimes glabrous; bracts membranous,
to 6 mm long. Sepals buff to old-rose coloured; petals and lip hyaline-white. Sepals
3–3.5 × 1.5 mm long, oblong, obtuse, joined in basal half. Petals 2.5–3 × 1.5 mm,
oblong, obtuse. Lip 2.5–3 mm long, concave at base, usually with variously lobed
calli, constricted at about two thirds of its length, the apical third transversely oblong,
1.5–2 mm wide. Glabrous plants do not have the basal outgrowths on the lip. Column
1.7 mm long with 2 clavate rostellar arms at apex. Ovary 6–7 mm long, starting to
swell as soon as flowers open.

Zimbabwe. E: Chimanimani Distr., near Nyabamba Falls, 760 m, fl. 14.ix.1954, *Ball* 360 (K;
SRGH); Lusito riverine forest, 300 m, fl. 20.ix.1959, *Ball* 818 (K; SRGH).
    Also in Tanzania, South Africa (Natal) and Madagascar. Deep shade in leaf mould in riverine
forest; 300–760 m.
    Glandular-hairy plants with basal calli on the lip and glabrous plants without calli (*Ball* 818)
apparently grow in the same area but in separate colonies. Apart from the above characters the
two forms seem identical.

4. ZEUXINE Lindl.

Zeuxine Lindl., Coll. Bot., Append.: [1] (1826), as "*Zeuxina*"; corr. Roeper in
            Linnaea **2**: 528 (1827), orth. et nom. conserv.

Terrestrial herb with creeping rhizomes and erect leafy stems. Inflorescence a
terminal raceme; flowers small or very small, not opening widely. Petals adnate to the
dorsal sepal and forming a hood with it; lateral sepals enclosing the base of the lip.
Lip saccate at the base with 2 calli. Column very short, sometimes winged in front,
with 2 lateral stigma lobes. Rostellum deeply divided; pollinia pyriform, the caudicle
attached to the rostellum by an oblong viscidium.

A genus of 70–80 species in the Old World tropics and subtropics.

1. Leaves linear (less than 10 mm wide), sheathing at the base · · · · · · · · · · · · · · 1. *africana*
– Leaves ovate (more than 10 mm wide), petiolate · 2
2. Sepals 5–6 mm long, lip 6–7 mm long · 2. *ballii*
– Sepals and lip 2–3 mm long · 3. *elongata*

1. **Zeuxine africana** Rchb.f. in Flora **50**: 103 (1867). —Rolfe in F.T.A. **7**: 181 (1897). —
            Summerhayes in F.W.T.A. ed. 2, **3**: 208 (1968). —Stewart et al., Wild Orch. South. Africa:
            212 (1982). Type from Angola.
            Zeuxine cochlearis Schltr. in Bot. Jahrb. Syst. **20**, Beibl. 50: 11 (1895). Type from South Africa.

Terrestrial herb 7.5–35 cm tall; stem base creeping, upper part erect, leafy; roots
short, thick. Leaves 20–47 × 2.5–5 mm, linear, acute, sheathing at the base. Stem
pinkish-white. Inflorescence terminal; rhachis 1–4 cm long, densely few to many-
flowered; ovary and pedicel 5 mm long; bracts longer than the flowers, up to 18 mm
long at the base of the inflorescence, linear, acuminate. Flowers small, whitish with a
yellowish lip; not opening fully. Dorsal sepal 3 × 1.25 mm, convex; lateral sepals 3.5
× 1 mm, ligulate, obtuse. Petals hyaline, 2.75–3 × 0.75 mm, oblong, acute, forming a
hood with the dorsal sepal. Lip c. 3 mm long, unguiculate, the base triangular-

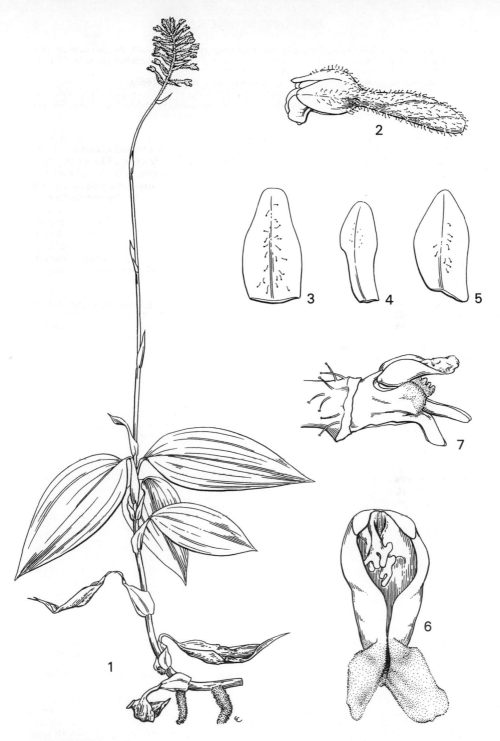

Tab. 3. CHEIROSTYLIS GYMNOCHILOIDES. 1, habit (×⅔); 2, flower (×5); 3, dorsal sepal (×8); 4, petal (×8); 5, lateral sepal (×8); 6, lip (×16); 7, column , side view (×16), 1–7 from *Ball* 360. Drawn by Eleanor Catherine.

hastate, the apex linear-ligulate, obtuse. Column very short, with a pair of ligulate, retuse arms at the base.

Botswana. N: near Xobega Lediba, fl. & fr. 18.viii.1979, *P.A. Smith* 2809, 2809A (K; SRGH).
Also in Nigeria, Angola and South Africa (Natal). Moist sand in grassland; in mud in shade.
The species closely resembles the Asiatic *Zeuxine strateumatica* (L.) Schltr., which has become naturalized in Florida and Cuba, but the lips differ sufficiently in shape for the species to be kept separate.

2. **Zeuxine ballii** P.J. Cribb in Kew Bull. **32**: 150 (1977). —la Croix et al., Orch. Malawi: 26 (1991). TAB. **4**. Type: Zimbabwe, Chimanimani (Melsetter), *Ball* 815 (K, holotype; SRGH). *Zeuxine sp. no. 1*, Grosvenor in Excelsa **6**: 86 (1976).

Terrestrial herb 12–24 cm tall; stem creeping at the base and rooting at the nodes, the upper part erect, leafy, covered with white spots, 4–5 mm in diameter. Leaves c. 8, spirally arranged, light green with reticulate veining; lamina 2.6–6 × 1.3–2.6 cm, ovate, acute, glabrous; petiole 1–2 cm long, widening into a sheathing base. Inflorescence terminal; rhachis 3–7 cm long, laxly 3–8-flowered with flowers more or less secund; ovary and pedicel 8–9 mm long; bracts up to 8 mm long, ovate, acuminate, margins ciliate. Peduncle, rhachis, ovary, bracts and sepals densely pubescent; sepals green, petals brownish, lip white, orange or yellow at the base. Dorsal sepal 6–7 × 3–3.5 mm, ovate, convex; lateral sepals 5–7 × 1.8–2.5 mm, obliquely lanceolate or ovate. Petals 4–6 × 2–2.2 mm, adnate to the dorsal sepal and forming a hood with it. Lip 6–8 mm long; hypochile narrow and channelled; epichile 3–3.5 mm long and 5.5–7 mm wide, transversely oblong, slightly emarginate. Column 3–3.5 mm long, including rostellum.

Zimbabwe. E: Chimanimani, 300 m, fl. 20.ix.1959, *Ball* 815 (K; SRGH). **Malawi**. N: Mzimba Distr., Lunyangwa, Mzuzu, 1330 m, fl. 15.ix.1986, *la Croix* 852 (K; MAL).
Also in south Tanzania. Riverine forest, in deep shade in leaf litter, rooting near the surface; 300–1330 m.
Plants in the Malawi locality produce flower buds during the late rains and these develop very slowly. In the dry season the colony dies back completely. Presumably flowering takes place in the wild only when the rains are unusually prolonged and the habitat does not dry out completely.

3. **Zeuxine elongata** Rolfe in Bol. Soc. Brot. **9**: 142 (1892); in F.T.A. **7**: 181 (1897). —Summerhayes in F.W.T.A. ed. 2, **3**: 208 (1968). —Grosvenor in Excelsa **6**: 86 (1976). —Williamson, Orch. S. Centr. Africa: 106 (1977). —Cribb & Bowden in F.T.E.A., Orchidaceae: 251 (1984). —Geerinck, Fl. Afr. Centr., Orchidaceae pt. 1: 20, pl. 3A (1984). Type from Principe.

Terrestrial herb 17–25 cm tall; stem creeping at the base, erect in the upper part, leafy. Leaves petiolate, the base sheathing; lamina 3–3.5 × 1.7 cm, ovate. Inflorescence terminal, fairly densely many-flowered; rhachis 3.5–13 cm long; ovary and pedicel erect, 5 mm long; bracts 5–7 mm long, lanceolate, acuminate. Flowers green and white with a yellowish throat, very small, c. 2.5 mm long. Sepals 2–2.5 × 0.75–2 mm, ovate, obtuse, the dorsal sepal convex, the lateral sepals somewhat oblique. Petals 1.75 × 1 mm, linear, falcate, adnate to lip, the apices free. Lip 2.5 mm long; hypochile narrowly saccate with sides incurved, 1.5 mm wide when flattened, with 2 hooked calli in the base; epichile 1.25–2 mm wide, transversely oblong or reniform, entire. Column 1.5 mm long, rostellum arms elongate, triangular.

Zambia. W: Salujinga, c. 70 km north of Mwinilunga, fl. vii-viii.1969, *G. Williamson* 2058 (K).
Also in W Africa, Central African Republic, Zaire, Uganda, Kenya, Tanzania. Evergreen rainforest, in deep shade.

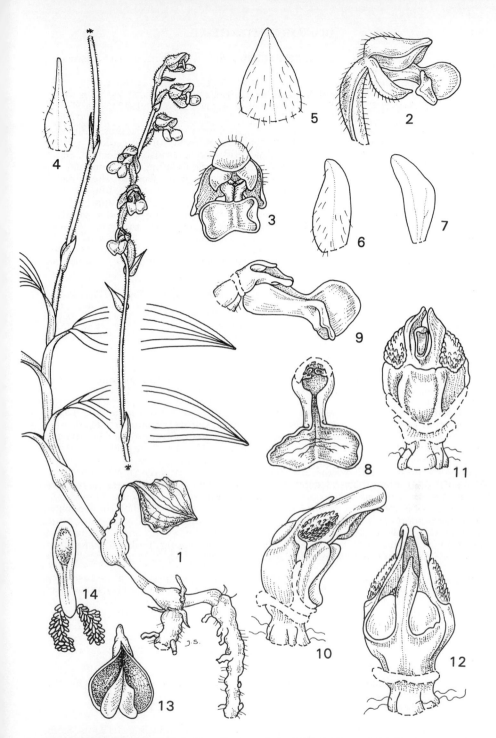

Tab. 4. ZEUXINE BALLII. 1, habit (×1), from *la Croix* 852; 2, flower, side view (×3); 3, flower, front view (×3); 4, bract (×3); 5, dorsal sepal (×4); 6, lateral sepal (×4); 7, petal (×4); 8, lip (×4); 9, column and lip, side view (×4); 10, column, side view (×8); 11, column, front view (×8); 12, column, back view (×8); 13, anther cap (×8); 14, pollinarium (×8), 2–14 from *la Croix* 852 (in K spirit coll. 50948). Drawn by Judi Stone.

5. EPIPACTIS Zinn

Epipactis Zinn, Cat. Pl. Hort. Gott.: 85 (1757) nom. conserv.
Helleborine Mill., Gard. Dict. Abr., ed. 4 (1754).
Amesia A. Nelson & J.F. Macbr. in Bot. Gaz. **56**: 472 (1913).

Terrestrial herb, occasionally saprophytic, with very short rhizomes, numerous fleshy roots and simple erect leafy stems. Leaves ovate or lanceolate, plicate. Inflorescence racemose; flowers usually dull reddish or greenish, pedicelled. Sepals and petals free. Lip with no spur, but with the hypochile forming a nectar-containing cup; hypochile articulated to the epichile by a narrow joint. Column short, flat or concave in front, with a shallow cup at the apex; anther free, 2-celled, hinged at the back of the column apex, behind the stigma and rostellum; pollinia 2, attached near the rostellum apex, each longitudinally divided, caudicles absent; pollen grains in friable masses loosely bound by fine threads. Stigma broad, with the large, globular rostellum placed centrally above it.

A genus of c. 25 species, most in north temperate regions, but with 3 species in tropical Africa.

Epipactis africana Rendle in J. Bot. **33**: 252 (1895). —Rolfe in F.T.A. **7**: 189 (1897). —Robyns & Tournay, Fl. Parc Nat. Alb. **3**: 455, fig. 64 (1955). —Brummitt in Wye College 1972 Malawi Project Report: 70 (1973). —Williamson, Orch. S. Centr. Africa: 107, fig. 55 (1977). —Cribb & Hunt in F.T.E.A., Orchidaceae: 241 (1984). —Geerinck in Fl. Afr. Centr., Orchidaceae pt. 1: 26 (1984). —la Croix et al., Orch. Malawi: 27 (1991). TAB. **5**. Type from Zaire.
   Serapias africana (Rendle) Eaton in Proc. Biol. Soc. Wash. **21**: 66 (1908).
   Helleborine africana (Rendle) Druce in Bull. Torrey Bot. Club **36**: 546 (1909).
   Epipactis excelsa Kraenzl. in Bot. Jahrb. Syst. **43**: 332 (1909). Type from Zaire.
   Amesia africana (Rendle) A.Nelson & J.F.Macbr. in Bot. Gaz. **56**: 472 (1913).

Terrestrial herb 1–2.5 m tall; stem purplish, leafy throughout its length. Leaves dark green, 16–21 × 3.5–10 cm, lanceolate to ovate, plicate, acuminate, clasping the stem at the base. Inflorescence an elongated few- to many-flowered raceme; rhachis, bracts, ovary and pedicel with a rusty pubescence; pedicel 1.8–3 cm long; ovary 1.3–2 cm long; bracts 4–5.5 cm long, leaf-like. Flowers pendent, yellow-green with purple veining on the sepals; lip maroon-brown to brick red. Dorsal sepal 20–27 × 4.5–7 mm, lanceolate, acuminate; lateral sepals 20–27 × 7–9 mm, obliquely ovate, acuminate. Petals 16–24 × 2.5–6 mm, lanceolate, acuminate. Lip 16–20 mm long, 3-lobed near the base; side lobes 9 × 2 mm, oblanceolate, subacute, parallel to hypochile; hypochile channelled, 7–10 mm long, 3 mm wide; epichile 9–10 × 4.5 mm, ovate, acute, with a fleshy area at the tip, bent upwards almost at a right angle. Column 5–9 mm long. Capsules purplish, 40 × 13 mm, narrowly oblong, 6-ribbed, pendulous, with the remains of the flowers at the apex.

Malawi. N: Nyika Plateau, 8 km east of Nganda, by tributary of Wovwe River, 1980 m, fl. 2.viii.1972, *Brummitt, Munthali & Synge*, WC 126 (K; MAL); S Viphya, Chikangawa, below Kasitu Resthouse, 1750 m, fl. 15.viii.1987, *la Croix* 1047 (K).
Also in Ethiopia, Zaire, Uganda, Kenya and Tanzania. Riverine forest, in deep shade in wet soil by streams; 1300–1980 m.

6. BRACHYCORYTHIS Lindl.

Brachycorythis Lindl., Gen. Sp. Orchid. Pl.: 363 (1838).
Gyaladenia Schltr. in Beih. Bot. Centralbl. **38**, abt. 2: 124 (1921).
Diplacorchis Schltr. in Beih. Bot. Centralbl. **38**, abt. 2: 127 (1921).

Terrestrial, rarely epiphytic, herb with fusiform or ellipsoid tuberous roots. Stems leafy. Leaves often numerous and overlapping. Bracts leaf-like and longer than the flowers at least at the base of the inflorescence. Flowers few to many in a terminal inflorescence, white, yellow, pink, mauve or purple often with darker spots. Sepals free, the lateral sepals spreading and oblique. Petals usually adnate at the base to the side of the column. Basal part of lip (hypochile) spurred, saccate and boat-shaped;

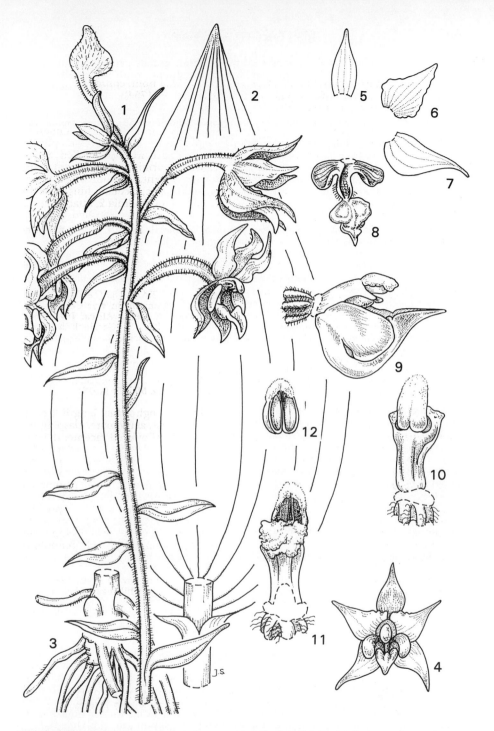

Tab. 5. EPIPACTIS AFRICANA. 1, inflorescence (×1); 2, leaf (×1), 1 & 2 from *Richards* 24048; 3, stem base and roots (×1), from *Wye College* 126; 4, flower, front view (×1); 5, dorsal sepal (×1); 6, petal (×1); 7, lateral sepal (×1); 8, lip (×1); 9, column and lip, side view (×2); 10, column, from above (×3); 11, column, from below (×3); 12, anther cap (×3), 4–12 from *de Leyser* 157 (in K spirit coll. 52342). Drawn by Judi Stone.

upper part (epichile) flattened, entire and 2–3-lobed, usually projecting forwards. Column erect, slender; anther loculi parallel, canals absent; caudicles usually short; viscidia naked. Stigma hollowed out; rostellum mid-lobes small, erect, folded; side lobes fleshy, surrounding the viscidia.

A genus of 30–35 species, occurring throughout tropical and South Africa, and tropical Asia to Taiwan.

1. Leaves glabrous · 2
– Leaves densely pubescent · 12
2. Lip hypochile forming a distinct spur at least 2 mm long · · · · · · · · · · · · · · · · · 3
– Lip hypochile saccate or boat-shaped, but not forming a spur · · · · · · · · · · · · · · · · 8
3. Spur over 5.5 mm long · 4
– Spur up to 5 mm long · 5
4. Spur slightly wider towards apex, 5.5–10.5 mm long; lip epichile 3-lobed · · · · · 4. *tenuior*
– Spur conical, tapering towards apex, 6–7 mm long; lip epichile entire · · · · · · · · 1. *conica*
5. Spur 1.5–2 mm long; dorsal sepal 3–4 mm long · · · · · · · · · · · · · · · · 2. *rhodostachys*
– Spur over 2 mm long; dorsal sepal 4–9 mm long · 6
6. Spur slender with a narrow mouth; epichile with a large, round mid-lobe and much smaller narrower basal lobes · 3. *friesii*
– Spur broad and rounded with a wide mouth; epichile 3-lobed towards the apex · · · · · · 7
7. Leaves up to 28 cm long, petiolate, borne on a separate sterile shoot; mid-lobe of the epichile longer than the side lobes · 6. *angolensis*
– Leaves up to 8 cm long, borne on the flowering stem only; all lobes of the lip equal · 5. *congoensis*
8. Dorsal sepal 4–5.5 mm long · 9
– Dorsal sepal over 5.5 mm long · 10
9. Lip with a conical callus in front of the junction of the epichile and hypochile; flowers mauve-purple · 8. *buchananii*
– Lip lacking a callus; flowers yellow-green with brown spots · · · · · · · · · · · 7. *inhambanensis*
10. Epiphytic plants; dorsal sepal 15–16 mm long; epichile 17–20 mm long · · · · 11. *kalbreyeri*
– Terrestrial plants; dorsal sepal less than 9 mm long; epichile less than 15 mm long · · · 11
11. Leaves tapering to a very long, fine point; hypochile 1.5–2.5 mm long · · · · 9. *pleistophylla*
– Leaves acuminate but not with a long, fine point; hypochile 3.5–6 mm long · · · · 10. *ovata*
12. Lip hypochile forming a spur 2–3 mm long · 12. *pilosa*
– Lip hypochile saccate, but not spurred · 13
13. Lip with a knee-like bend at the junction of the epichile and hypochile; flowers mauve-pink · 13. *pubescens*
– Lip projecting horizontally; flowers cream or greenish and yellow · · · · · · · · · · · · · · 14
14. Epichile slightly broader than long, the mid-lobe shorter than or equalling the side lobes · 15. *mixta*
– Epichile longer than broad, the mid-lobe longer than the side lobes · · · · · · · 14. *velutina*

1. **Brachycorythis conica** (Summerh.) Summerh. in Kew Bull. **10**: 244 (1955). —Williamson, Orch. S. Centr. Africa: 31, t. 28 (1977). —Geerinck in Fl. Afr. Centr., Orchidaceae pt. 1: 40 (1984). Type from Gabon.
    Diplacorchis conica Summerh. in Bull. Misc. Inform., Kew **1938**: 141 (1938).

Terrestrial herb 20–50 cm tall, with fusiform tuberous roots. Stem leafy. Leaves 4.5–6.5 × 1.2–2 cm, lanceolate, acute, glabrous. Inflorescence ± densely many-flowered, 14–16 × 5–6 cm; pedicel 11–12 mm long; bracts leaf-like, longer than the flowers. Flowers purple-pink to whitish-mauve, the lip with darker marks. Dorsal sepal 8–11 × 2.8–5 mm, ovate, obtuse, very convex; lateral sepals 6.5–12.5 × 2.3–5 mm, obliquely lanceolate. Petals up to 11.8 × 4.5 mm, ovate with a dorsal keel, obtuse, joined to column at the base. Lip hypochile 3.5–6 mm long; epichile 9–10 × 4.8–6.2 mm, ± orbicular to oblong; spur 6–7 mm long, conical, obtuse. Column 3.5–6 mm long.

Subsp. **longilabris** Summerh. in Kew Bull. **10**: 244 (1955). Type: Zambia, Mwinilunga Distr., *Milne-Redhead* 3941 (K, holotype).

Distinguished by having a robust habit and densely imbricate leaves, a dense many-flowered inflorescence with bracts much longer than the flowers, and a ligulate-oblong or obovate epichile about twice as long as wide, rather fleshy with crisped margins and a ± truncate apex.

Zambia. W: Mwinilunga Distr., SW of Dobeka Bridge, fl. 3.i.1938, *Milne-Redhead* 3941 (K); Mwinilunga Distr., 34 km W on Matonchi Road, c. 1380 m, fl. 24.i.1975, *Brummitt, Chisumpa & Polhill* 14084 (K).
Also in Zaire. Grassy savanna, and dry sandy dambos; 1300–1400 m.
Subsp. *conica* occurs in Nigeria, Gabon and Zaire, while subsp. *transvaalensis* occurs in South Africa.

2. **Brachycorythis rhodostachys** (Schltr.) Summerh. in Kew Bull. **10**: 246 (1955); in F.T.E.A., Orchidaceae: 18 (1968). —Williamson, Orch. S. Centr. Africa: 29 (1977). —Geerinck in Fl. Afr. Centr., Orchidaceae pt. 1: 39 (1984). —la Croix et al., Orch. Malawi: 33 (1991). Type from Angola.
  Platanthera rhodostachys Schltr. in Warb., Kunene-Sambesi-Exped. Baum: 203 (1903).
  Gyaladenia rhodostachys (Schltr.) Schltr. in Beih. Bot. Centralbl. **38**, abt. 2: 126 (1921).

A slender terrestrial herb 20–40 cm tall, with villous cylindrical tuberous roots. Stem leafy. Leaves 2.5–5.5 × 0.8–2 cm, ovate, acute, sheathing at the base. Inflorescence densely many-flowered, 3.5–12 × 1.5–2 cm; pedicel 3 mm long; ovary c. 5 mm long; bracts up to 15 × 4 mm, lanceolate, acuminate. Flowers greenish-mauve to pink-mauve with purple marks on lip. Sepals 3–5 × 1.5–3 mm, the dorsal sepal elliptic, convex; lateral sepals oblique, projecting forwards. Petals 2–3.5 × 1–2 mm, oblong, attached to column at the base. Lip 4–5 mm long; epichile 2.7–3.5 mm long, ± obovate, shortly 3-lobed near the apex, the mid-lobe up to 1 mm long; spur 1.5–3 mm long, slightly incurved, shortly 2-lobed at the apex. Column 2–2.5 mm long.

Zambia. N: Kawambwa Distr., Ntumbachushi (M'tunatusha) R., c. 1290 m, fl. 28.xi.1961, *Richards* 15415 (K). W: Mwinilunga Distr., Sinkabolo Swamp, 1200 m, fl. & fr. 20.xi.1962, *Richards* 17435 (K). C: Kundalila Falls, *G. Williamson* 631 (K). **Malawi**. N: S Viphya, Lichelemu (Luchilemu) Dambo, 1300 m, fl. 4.i.1983, *la Croix* 393 (K).
Also in Zaire, Tanzania and Angola. Marshy grassland and dambos; 600–2285 m.

3. **Brachycorythis friesii** (Schltr.) Summerh. in Kew Bull. **10**: 246 (1955); in F.T.E.A., Orchidaceae: 18 (1968). —Williamson, Orch. S. Centr. Africa: 30, t. 26 (1977). —Geerinck in Fl. Afr. Centr., Orchidaceae pt. 1: 32 (1984). Type from Burundi.
  Platanthera friesii Schltr. in Fries, Wiss. Ergebn. Schwed. Rhod.-Kongo-Exped. **1**: 240, fig. 22 (1916).
  Gyaladenia friesii (Schltr.) Schltr. in Beih. Bot. Centralbl. **38**, abt. 2: 126 (1921).

A slender terrestrial herb 8–27 cm tall, with ovoid to cylindrical tuberous roots up to 5 cm long. Stem leafy. Leaves up to 10, 2.5–6 × 2–3.5 cm, linear-lanceolate. Inflorescence laxly (1)3–15-flowered, up to 9 cm long; ovary and pedicel arched, 5–8 mm long; bracts 13–15 mm long, lanceolate, acuminate. Flowers almost white, pale to deep pink or purple with darker marks on lip. Dorsal sepal 4–6.7 × 1.6–2.8 mm, ovate, obtuse; lateral sepals 4.5–7.6 × 2.3–2.6 mm, rather oblique. Petals 3–6.5 × 2–3.3 mm, obliquely ovate, obtuse, attached to column at the base. Lip 4–7 mm long; epichile 5–6.3 × 5.7–7.3 mm, 3-lobed near the base, the mid-lobe quadrate, 2–3 mm long, the side lobes 1–2 mm long, narrow and spreading; spur 2.5–4.5 mm long, slender, blunt, curved. Column 2.5–3.3 mm long.

Zambia. N: Mbala Distr., top of Kambole Escarpment, 1650 m, fl. & fr. 1.ii.1959, *Richards* 10831 (K). W: Mwinilunga Distr., 6 km N of Kalene Hill, fl. 12.xii.1963, *E.A. Robinson* 5899 (K). C: Kundalila Falls, near Serenje, fl. 17.i.1972, *Kornaś* 0876 (K).
Also in Zaire, Burundi, Tanzania and Angola. Shallow soil and seepage areas on granite outcrops, and in open areas in woodland; 1200–1650 m.

4. **Brachycorythis tenuior** Rchb.f. in Flora **48**: 183 (1865). —Summerhayes in Kew Bull. **10**: 247 (1955). —Goodier & Phipps in Kirkia **1**: 53 (1961). —Summerhayes in F.W.T.A. ed. 2, **3**: 187 (1968); in F.T.E.A., Orchidaceae: 19 (1968). —Grosvenor in Excelsa **6**: 78 (1976). —Williamson, Orch. S. Centr. Africa: 31, t. 27 (1977). —Stewart et al., Wild Orch. South.

Africa: 73 (1982). —Geerinck in Fl. Afr. Centr., Orchidaceae pt. 1: 41 (1984). —la Croix et al., Orch. Malawi: 33 (1991). TAB. **6**. Type from South Africa (Natal).
 Habenaria tenuior (Rchb.f.) N.E. Br. in Gard. Chron., ser. 2, **24**: 307 (1885). Type as above.
 Platanthera tenuior (Rchb.f.) Schltr. in Bot. Jahbr. Syst. **20**, Beibl. 50: 12 (1895). —Rolfe in F.T.A. **7**: 205 (1898). Type as above.
 Diplacorchis engleriana (Kraenzl.) Schltr. in Beih. Bot. Centralbl. **38**, abt. 2: 129 (1921). Type from Cameroon.
 Diplacorchis tenuior (Rchb.f.) Schltr. in Beih. Bot. Centralb. **38**, 2: 128 (1921). —Summerhayes in Bot. Not. **1937**: 182 (1937). Type as above.
 Brachycorythis crassicornis Kraenzl. in Vierteljahrsschr. Naturf. Ges. Zürich **74**: 103 (1929). Type from Uganda.

Terrestrial herb 20–50 cm tall with ovoid tuberous roots. Stem leafy throughout. Leaves dark green, numerous, 30–50 × 9–15 mm, lanceolate, acute, glabrous. Inflorescence 7–16 cm long, densely several- to many-flowered; ovary and pedicel green with purple spots, 6–13 mm long; bracts leaf-like, up to 30 mm long. Flowers pale to deep purple with darker spots. Dorsal sepal 6–10 × 2.3–5 mm, elliptic, convex; lateral sepals 6.5–12 × 2.3–4.5 mm, oblique, spreading upwards. Petals 6–9.5 × 2.3–3.5 mm, oblong-falcate, attached to column at the base. Lip 10–13 mm long, with 2 keels at the junction of epichile and hypochile; hypochile forming a spur 5–9 mm long with a blunt, green bulbous tip; epichile 6.5–9.5 × 3.8–7.5 mm, 3-lobed at about halfway, the mid-lobe fleshy and convex. Column slender, 4–6 mm long.

 Zambia. N: Mbala District, Zombe Plain, 1500 m, fl. 20.i.1968, *Richards* 22950 (K). C: c. 11 km E of Lusaka, fl. 4.ii.1958, *King* 416 (K). **Zimbabwe.** N: Makonde (Lomagundi), 1200 m, fl. 20.i.1963, *Jacobsen* 2087 (PRE). C: Rusape, fl. 4.ii.1953, *Dehn* in *GHS* 42107 (K; SRGH). E: Chimanimani Mts., fl. 10.ii.1954, *Ball* 218 (K; SRGH). **Malawi.** N: Nyika National Park, Mbuzinandi Road, 1800 m, fl. 9.i.1983, *la Croix* 416 (K). S: Blantyre, fl. 7.i.1888, *Scott* s.n. (K).
 Widely distributed in tropical Africa and South Africa. Open grassy areas, dambo margins, and in long grass in woodland; 1050–1800 m.

5. **Brachycorythis congoensis** Kraenzl., Orch. Gen. Sp. **1**: 544 (1898). —Summerhayes in Kew Bull. **10**: 250 (1955); in F.T.E.A., Orchidaceae: 19 (1968). —Grosvenor in Excelsa **6**: 78 (1976). —Williamson, Orch. S. Centr. Africa 31, t. 29 (1977). —Geerinck in Fl. Afr. Centr., Orchidaceae pt. 1: 39 (1984). —la Croix et al., Orch. Malawi: 30 (1991). Type from Zaire (Katanga).
 Brachycorythis hirschbergii Braid in Bull. Misc. Inform., Kew **1925**: 358 (1925). Type from Zaire.

Robust terrestrial herb 18–70 cm tall with cylindrical tuberous roots. Leaves numerous, up to 7.5 × 3 cm, lanceolate, acuminate, decreasing in size up the stem and becoming narrower. Inflorescence up to 23 cm long, 5 cm wide, densely many-flowered; ovary and pedicel 13–20 mm long; bracts leaf-like, up to 34 × 5 mm. Flowers pink-purple, spotted, inner parts often whitish. Sepals elliptic or oblong, rounded; dorsal sepal 5–7 × 2–3.5 mm; lateral sepals 5.5–7.5 × 2–3.5 mm, rather oblique. Petals 4–7.5 × 1.5–3 mm, narrowly oblong. Lip 8–12 mm long, the hypochile forming a conical spur 2–3.5 mm long; epichile 5.2–7 × 4.5–7.5 mm, obovate or oblong, 3-lobed in apical third, all lobes truncate, the mid-lobe larger than the side lobes. Column 3.5–5 mm long.

 Zambia. N: Mbala Distr., Zambia State Ranch, Saisi Valley, 1500 m, fl. 13.i.1968, *Richards* 22907 (K). W: Solwezi Distr., near source of Chifubwa, fl. 5.i.1962, *Holmes* 0318 (K; SRGH). C: 11 km E of Lusaka, 1200 m, fl. 2.i.1956, *King* 251 (K). S: Batoka Distr., c. 20 km N of Choma, 1240 m, fl. 14.i.1954, *E.A. Robinson* 459 (K). **Zimbabwe.** N: Umvukwes Distr., Mazowe, Ruorka Ranch, fl. 17.xii.1952, *Wild* 3996 (K; SRGH). W: Matobo Distr., Farm Besna Kobila, 1450 m, fl. i.1957, *Miller* 4037 (K; SRGH). E: Chimanimani Distr., 'Fairfield', 1360 m, fl. 19.i.1955, *Ball* 472 (K; SRGH). S: Masvingo, 1909–12, *Monro* 1878 (BM). **Malawi.** C: Mchinji Distr., near Tamanda Mission, 1400 m, fl. 8.i.1959, *Robson* 1104 (K).
 Also in Zaire, Tanzania and Burundi. Seepage areas at foot of slopes, dambo and wet grassland; 1000–1500 m.

6. **Brachycorythis angolensis** (Schltr.) Schltr. in Beih. Bot. Centralbl. **38**, abt. 2: 113 (1921) pro parte. —Summerhayes in Kew Bull. **10**: 250 (1955); in F.T.E.A., Orchidaceae: 21 (1968). —Williamson,

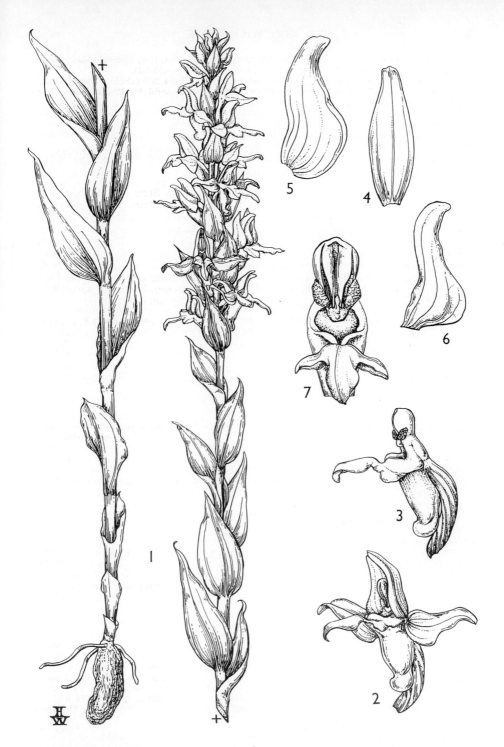

Tab. 6. BRACHYCORYTHIS TENUIOR. 1, habit (×1); 2, flower (×2); 3, flower, tepals removed to show lip and column (×2); 4, dorsal sepal (×4); 5, lateral sepal (×4); 6, petal (×4); 7, column and lip, front view (×5), 1–7 from *Milne-Redhead & Taylor* 8829. Drawn by Heather Wood. From F.T.E.A.

Orch. S. Centr. Africa: 31, t. 30, 31 (1977). —Geerinck in Fl. Afr. Centr., Orchidaceae pt. 1: 38 (1984). Type from Angola.
    Platanthera angolensis Schltr. in Warb., Kunene-Sambesi-Exped. Baum: 203 (1903).
    Brachycorythis oligophylla Kraenzl. in Warb., Kunene-Sambesi-Exped. Baum: 208 (1903). Type from Angola.

Terrestrial herb 30–65 cm tall with long, thick cylindrical roots. Leaves on flowering stem 5–10, well spaced out along the stem, up to 10 × 2.5 cm, lanceolate; leaves on sterile shoot 2, up to 28 × 3.5 cm, oblanceolate, petiolate. Inflorescence dense, many-flowered, 5.5–12.5 × 3–5 cm; ovary and pedicel 11–15 mm long; bracts up to 26 × 4.5 mm, lanceolate. Flowers white or mauve, the lip darker and spotted. Sepals elliptic; dorsal sepal 7–10.5 × 2.5–4.5 mm, lateral sepals usually slightly longer and oblique. Petals 6–10 × 1.5–4.5 mm, obliquely oblanceolate or elliptic, attached to column at the base. Lip 7–10 mm long, the hypochile forming a sac-like spur 2.5–4.5 mm long; the epichile 5.5–8 × 7–9.2 mm, 3-lobed in apical half, the lobes rounded, ± equal in length. Column slender, 4.5–5.5 mm long.

Zambia. N: Kasama Distr., Mungwi, fl. 19.xi.1960, *E.A. Robinson* 4082 (K). W: Mwinilunga Distr., Sinkabolo Dambo, fl. 12.xi.1937, *Milne-Redhead* 3201 (K). C: Serenje Distr., M'Kushi R., 1350 m, fl. 27.xi.1962, *Richards* 17316 (K).
    Also in Zaire, Tanzania and Angola. Swampy grassland and permanently wet dambo; 1200–1440 m.
    Material referred to *Brachycorythis oligophylla* Kraenzl. has very large flowers and Summerhayes has suggested that it might be a tetraploid with sepals up to 13 mm long, petals up to 11.5 mm and the lip up to 12 mm. It otherwise resembles *B. angolensis*.

7. **Brachycorythis inhambanensis** (Schltr.) Schltr. in Beih. Bot. Centralbl. **38**, abt. 2: 112 (1921). —Mansfeld in Fedde, Repert. Spec. Nov. Regni Veg., Beih. 68, tafel. 3, nr. 9 (1932). —Summerhayes in Kew Bull. **10**: 252 (1955). —Goodier & Phipps in Kirkia **1**: 53 (1961). —Grosvenor in Excelsa **6**: 78 (1976). —Williamson, Orch. S. Centr. Africa: 28 (1977). —Stewart et al., Wild Orch. South. Africa: 73 (1982). —Geerinck in Fl. Afr. Centr., Orchidaceae pt. 1: 34 (1984). —la Croix et al., Orch. Malawi: 30 (1991). Type: Mozambique, *Schlechter* 12091 (B†, holotype).
    Platanthera inhambanensis Schltr. in Bot. Jahrb. Syst. **26**: 330 (1899).

Slender terrestrial herb 20–35 cm tall with fleshy roots. Leaves c. 8, up to 8 × 3 cm, bright green, lanceolate or ovate, acute, the 2 lowest sheath-like. Inflorescence 6–9.5 × 2.5 cm, laxly or densely several to many-flowered; ovary and pedicel 8–10 mm long; bracts leaf-like, up to 25 mm long, longer than the lowest flowers. Flowers whitish- or yellowish-green, the inside of the petals and lip spotted with purple-brown. Sepals 4.5–5.6 × 2.5–3 mm, ovate, obtuse or apiculate, projecting forwards. Petals 4.5 × 1.5–2.7 mm, lanceolate, obtuse. Lip 5–7 mm long, the hypochile saccate but not forming a spur; epichile 4.8–6 × 3.5–5.5 mm, 3-lobed near the apex, the lobes triangular, c. 1 mm long, the side lobes broader than the mid-lobe. Column 2–2.5 mm long, reddish.

Zambia. N: Mbala Distr., Ndundu Dambo, 1680 m, fl. 8.ii.1957, *Richards* 8111 (K). W: Mwinilunga Distr., SW of Dobeka Bridge, fl. 14.xii.1937, *Milne-Redhead* 3655 (K). **Zimbabwe**. E: Chimanimani Mts., 1980 m, fl. 30.i.1954, *Ball* 190 (K; SRGH). **Malawi**. N: Nyika National Park, near Lake Kaulime, fl. 17.xii.1969, *Ball* 1193 (SRGH). **Mozambique**. GI: near Inhambane, in marsh, fl. ii.1898, *Schlechter* 12091 (B†).

8. **Brachycorythis buchananii** (Schltr.) Rolfe in F.T.A. **7**: 570 (1898). —Schlechter in Beih. Bot. Centralbl. **38**, abt. 2: 113 (1921). —Summerhayes in Bot. Not. **1937**: 182 (1937). —Suessenguth & Merxmüller, Contrib. Fl. Marandellas Distr.: 82 (1951). —Summerhayes in Kew Bull. **10**: 252 (1955). —Goodier & Phipps in Kirkia **1**: 53 (1961). —Summerhayes in F.T.E.A., Orchidaceae: 21 (1968). —Grosvenor in Excelsa **6**: 78 (1976). —Williamson, Orch. S. Centr. Africa: 32, t. 32 (1977). —Geerinck in Fl. Afr. Centr., Orchidaceae pt. 1: 33 (1984). —la Croix et al., Orch. Malawi: 29 (1991). Type: Malawi, no locality, *Buchanan* s.n. (B†, holotype).
    Platanthera buchananii Schltr. in Bot. Jahrb. Syst. **24**: 420 (1897).
    Brachycorythis parviflora Rolfe in F.T.A. **7**: 202 (1898). Type: Zambia, Mbala Distr., Fwambo, *Nutt* s.n. (K, holotype).

Slender terrestrial herb 20–50 cm tall with fusiform, tuberous roots. Leaves numerous, up to 3 × 1 cm, lanceolate, acute or acuminate, gradually decreasing in size up the stem. Inflorescence 4–12 × 1.5–2 cm, densely many-flowered; ovary and pedicel 5–6 mm long; bracts lanceolate, acuminate, larger than the flowers. Flowers mauve to purple. Sepals 4–5 × 1.5–2.5 mm, ovate, obtuse, projecting forwards, the lateral sepals rather oblique. Petals 4.8–6 × 3–3.7 mm, obovate, rounded. Lip 4–5.5 mm long, with a small callus in front of the hypochile; hypochile boat-shaped; epichile vertical, 2.4–3.5 × 4–6 mm, transversely elliptical, 3-lobed near the apex, the side lobes larger than the mid-lobe. Column 2 mm long.

Zambia. N: Mbala Distr., Nkali (Kali) Dambo, 1500 m, fl. 26.i.1952, *Richards* 784 (K). W: Mwinilunga Distr., Sinkabolo Swamp, 1200 m, fl. 20.xi.1962, *Richards* 17430 (K). C: Kabwe, Chibwe P.F.A., fl. 18.i.1961, *Morze* 48 (K). **Zimbabwe.** W: Matobo, 1450 m, fl. i.1958, *Miller* 4995 (K; SRGH). C: Makoni, Chimbe Farm, 1575 m, fl. 14.ii.1960, *Chase* 7265 (K; SRGH). E: Chimanimani Mts., Bundi Valley, 1500 m, fl. 2.ii.1957, *Phipps* 429 (K; SRGH). **Malawi.** N: S Viphya, Lichelemu (Luchilemu) Dambo, 1250 m, fl. 21.ii.1984, *la Croix* 559 (K). C: Nkhota Kota, 515 m, fl. 16.ii.1944, *Benson* 296 (K; PRE). S: Zomba Mt., Chitinji Marsh, fl. January 1976, *Welsh* 118 (K).
Also in Nigeria, Central African Republic, Zaire, Ethiopia, Uganda, Kenya, Tanzania and Angola. Bogs, swampy grassland; 500–1800 m.

9. **Brachycorythis pleistophylla** Rchb.f., Otia Bot. Hamburg: 104 (1881). —Rolfe in F.T.A. **7**: 202 (1898). —Kraenzlin, Orch. Gen. Sp. **1**: 540 (1898). —Schlechter in Beih. Bot. Centralbl. **38**, abt. 2: 117 (1921). —Summerhayes in Bot. Not. **1937**: 182 (1937). —H. Perrier in Fl. Madag., Orch. 1: 8 (1939). —Summerhayes in Kew Bull. **10**: 253 (1955). —Goodier & Phipps in Kirkia **1**: 53 (1961). —Summerhayes in F.T.E.A., Orchidaceae: 22 (1968). —Moriarty, Wild Fls. Malawi: t. 27, 4 (1975). —Grosvenor in Excelsa **6**: 78 (1976). —Williamson, Orch. S. Centr. Africa: 32, t 33 (1977). —Geerinck in Fl. Afr. Centr., Orchidaceae pt. 1: 36 (1984). —la Croix et al., Orch. Malawi: 31 (1991). Type: Mozambique, Morrumbala Mt., *Kirk* s.n. (K, holotype).
Platanthera pleistophylla (Rchb.f.) Schltr., Westafr. Kautschuk-Exped.: 274 (1900).
Brachycorythis briartiana Kraenzl. [apud De Wild. & Duv.] in Bull. Soc. Bot. Belg. **38**: 219 (1900). Type from Zaire.
Brachycorythis pulchra Schltr. in Bot. Jahrb. Syst. **53**: 485 (1915) pro parte. —Suesseguth & Merxmüller, Contrib. Fl. Marandellas Distr.: 82 (1951). Type from Tanzania.
Brachycorythis perrieri Schltr. in Beih. Bot. Centralbl. **34**, abt. 2: 296 (1916); op. cit. **38**, abt. 2: 118 (1921). Type from Madagascar.
Brachycorythis macclouniei Braid in Bull. Misc. Inform., Kew **1925**: 357 (1925). Type: Malawi, Mt. Mulanje, Oct. 1895, *McClounie* s.n. (K, holotype).

Terrestrial herb 36–90 cm tall with fleshy, woolly, cylindrical roots. Leaves numerous, overlapping, up to 10 × 1.5 cm, lanceolate, acuminate, the largest in the middle of the stem. Inflorescence 12–25 × 2.5–4.5 cm, usually densely many-flowered; ovary and pedicel up to 18 mm long; bracts leaf-like, up to 3 cm long. Flowers mauve-purple to purple. Sepals ovate or elliptic; dorsal sepal 5.5–8.3 × 2–4.6 mm; lateral sepals (5)6–9 × 3.5–5.5 mm, rather oblique. Petals 6–8.4 × 3.7–8 mm, semi-orbicular to obliquely orbicular, standing forwards. Lip 11–15 mm long; hypochile boat-shaped, 2–3 mm long; epichile 8–13 × 6.6–14.4, horizontal, upcurved at apex, 3-lobed with the mid-lobe very small and tooth-like, the lateral lobes folded down. Column stout, c. 2 mm long.

Key to subspecies

Epichile longer than broad, or as long as broad; flowers drying dark blackish-brown · · · · · · ·
· subsp. *pleistophylla*
Epichile broader than long; flowers drying pale brown or mid-brown · · · · · · · · subsp. *leopoldi*

Subsp. **pleistophylla**

Flowers usually mauve-purple, drying dark brown. Petals 6–7.3 × 3.7–4.6 mm, epichile 8–11 × 6.6–9.5 mm.

Zambia. N: Mbala Distr., Kawimbe, 1680 m, fl. 21.xi.1956, *Richards* 7322 (K). W: Mwinilunga Distr., c. 6 km from Mwinilunga on road to Kalene Hill, fl. 28.xi.1972, *Strid* 2518 (K). C: Mkushi Distr., Chinshinshi Dambo, 1400 m, fl. 9.i.1958, *E.A. Robinson* 2708 (K). E: Nyika National Park, fl. xii.1966, *G. Williamson* 128 (K; SRGH). **Zimbabwe.** W: Matobo, 1450 m, fl. xii.1955, *Miller* 2601 (PRE; SRGH). E: Chimanimani, 'The Corner', 1500 m, fl. xii.1954, *Ball* 423 (K; SRGH). **Malawi.** N: Nyika Plateau, Vitumbi Road, 2000 m, fl. 28.xii.1975, *Phillips* 676 (K; MAL); Nkhata Bay Distr., Mzuzu-Chikwina road near Chipunga, 1350 m, fl. 22.xii.1986, *la Croix* 907 (K; MAL). S: Thyolo Distr., Mindale Estate, 1100 m, fl. 9.xii.1981, *la Croix* 228 (K). **Mozambique.** Z: Morrumbala, 900 m, fl. xii.1958, *Kirk* (K).

Widespread in tropical Africa, east of Nigeria. Grassland, open woodland and dambo margins; 1100–2200 m.

Subsp. **leopoldi** (Kraenzl.) Summerh. in Kew Bull. **10**: 254 (1955). —la Croix et al., Orch. Malawi: 32 (1991). Type from Zaire.

    Brachycorythis leopoldi Kraenzl., Orch. Gen. Sp. **1**: 542 (1898). —Schlechter in Beih. Bot. Centralbl. **38**, abt. 2: 117 (1921).

    Brachycorythis pulchra Schltr. in Bot. Jahrb. Syst. **53**: 485 (1915) pro parte. —Suessenguth & Merxmüller, Contrib. Fl. Marandellas Distr.: 82 (1951). Type from Tanzania.

Flowers usually rich purple, yellow in the throat, drying light to medium brown. Petals 6–8.5 × 5.3–8 mm; epichile 9–13 × 9.5–14.5 mm.

Malawi. N: Nkhata Bay Distr., Mzuzu-Chikwina road near Chipunga, 1350 m, fl. 22.xii.1986, *la Croix* 906 (K).

Also in Gabon, Zaire, Tanzania and Angola. Grassland.

10. **Brachycorythis ovata** Lindl., Gen. Sp. Orchid. Pl.: 363 (1838). —Bolus, Icon. Orchid. Austro-Afric. **1**: tab. 62 (1896). —Kraenzlin, Orch. Gen. Sp. **1**: 541 (1898). —Rolfe in Dyer, F.C. **5**,3: 85 (1912). —Schlechter in Beih. Bot. Centralbl. **38**, abt. 2: 115 (1921). —Summerhayes in Kew Bull. **10**: 256 (1955); in F.T.E.A., Orchidaceae: 23 (1968). —Williamson, Orch. S. Centr. Africa: 32 (1977). —Stewart et al., Wild. Orch. South. Africa: 73 (1982). —Geerinck in Fl. Afr. Centr., Orchidaceae pt. 1: 35 (1984). Type from South Africa.

    Platanthera ovata (Lindl.) Schltr. in Bot. Jahrb. Syst. **20**, 1: 12 (1895).

Terrestrial herb 25–55 cm tall with woolly, cylindrical roots. Leaves numerous, overlapping, up to 7 × 2.5 cm, lanceolate or ovate, acute or acuminate. Inflorescence 13–25 × 4–6 cm, fairly laxly to densely many-flowered; ovary and pedicel 1.5–1.6 cm long; bracts up to 4 cm long. Flowers mauve and white. Dorsal sepal 6.8–7.5 × 3–4.2 mm, elliptic or ovate, convex. Lateral sepals 7–9.3 × 2–4.7 mm, obliquely ovate. Petals 6–10 × 2.8–7 mm, obliquely oblong or ovate with a rounded auricle in front, standing forwards beside the column. Lip 11.5–14.5 mm long; hypochile boat-shaped 3.5–6 mm long; epichile 8.5–10.5 × 6.5–10.5 mm, broadly obovate, 3-lobed near the apex or in the upper part, with a central ridge running into the mid-lobe; the side lobes reflexed. Column stout, 3–4 mm long.

Subsp. **welwitschii** (Rchb.f.) Summerh. in Kew Bull. **10**: 257 (1955). —Goodier & Phipps in Kirkia **1**: 53 (1961). —Grosvenor in Excelsa **6**: 78 (1976). —la Croix et al., Orch. Malawi: 31 (1991). Type from Angola.

    Brachycorythis welwitschii Rchb.f. in Flora **50**: 99 (1867).

    Brachycorythis acutiloba Rendle in J. Linn. Soc., Bot. **40**: 208 (1911). —Eyles in Trans. Roy. Soc. S. Afr. **5**: 334 (1916). Type: Zimbabwe, near Chirinda, *Swynnerton* 6632 (BM, holotype).

Differs from the typical subspecies in the stem being less densely leafy, the inflorescence laxer and the flowers slightly larger. The lip is lobed at the apex only (as in subsp. *ovata*) but the lobes are more distinct, with the mid-lobe usually shorter than the acute or subacute lateral lobes. In subsp. *schweinfurthii* (Rchb.f.) Summerh. the epichile is lobed in the upper part.

Zambia. N: Mporokosa Distr., 40 km E of boma, fl. 10.i.1961, *Holmes* 0281 (K; SRGH). W: Mwinilunga Distr., 100 m NE of Dobeka Bridge, fl. 11.xii.1937, *Milne-Redhead* 3616 (K). E: Nyika Plateau, 2200 m, fl. 3.i.1964, *Benson* NR 427 (K). **Zimbabwe.** E: Mutare Distr., Vumba Mts., 1650

m, fl. 13.xii.1950, *Chase* 4213 (K; SRGH). **Malawi**. N: Nyika Plateau, track to Rukuru Falls, 1800 m, fl. 6.i.1959, *Richards* 10520 (K; SRGH).
Also in Zaire, Burundi, Tanzania and Angola. Montane grassland and dry dambo; 1650–2200 m.
Subsp. *ovata* is found only in South Africa and Swaziland, while subsp. *schweinfurthii* occurs in W Africa, Zaire, Sudan, Uganda, Kenya and Tanzania.

11. **Brachycorythis kalbreyeri** Rchb.f. in Flora **61**: 77 (1878). —Rolfe in F.T.A. **7**: 201 (1898). —Kraenzlin, Orch. Gen. Sp. **1**: 540 (1898). —Schlechter in Beih. Bot. Centralbl. **38**, abt. 2: 120 (1921). —Summerhayes in Kew Bull. **10**: 259 (1955); in F.W.T.A. ed. 2, **3**: 187 (1968); in F.T.E.A., Orchidaceae: 24 (1968). —Williamson, Orch. S. Centr. Africa: 28 (1977). —Geerinck in Fl. Afr. Centr., Orchidaceae pt. 1: 33 (1984). Type from Cameroon.

Epiphytic herb 15–40 cm tall with fleshy, woolly, cylindrical roots. Leaves up to 15, c. 8 × 2–2.5 cm, lanceolate, acute. Inflorescence lax, up to c. 20-flowered, usually fewer; ovary and pedicel 2–2.5 cm long; bracts leaf-like, up to 5.5 cm long. Flowers violet to mauve-purple. Dorsal sepal 15–16 × 6–10 mm, elliptic, subacute; lateral sepals 16–17 × 6–10 mm, obliquely ovate, obtuse. Petals 10–14 × 7.5–11 mm, obliquely ovate, standing forward beside the column. Lip 22–26 mm long; hypochile boat-shaped, c. 6 mm long; epichile 17.5–20 × 13–19.5 mm, broadly obovate, 3-lobed in the upper half, the mid-lobe triangular, curved upwards, much smaller than lateral lobes, lateral lobes slightly dentate. Column stout, c. 4 mm long.

Zambia. W: Upper Kabompo R., *Holmes* 0329 (SRGH).
Also in Guinea Republic, Sierra Leone, Liberia, Cameroon, Zaire, Kenya, Uganda and Tanzania. Riverine forest and rainforest; 1800–2100 m.
Only one specimen is known from the Flora Zambesiaca area.

12. **Brachycorythis pilosa** Summerh. in Kew Bull. **10**: 259 (1955); in F.T.E.A., Orchidaceae: 25 (1968). —Williamson, Orch. S. Centr. Africa: 33 (1977). —Geerinck in Fl. Afr. Centr., Orchidaceae pt. 1: 42 (1984). Type from Tanzania.

Terrestrial herb 25–70 cm tall with fleshy, cylindrical, woolly roots. Leaves numerous, up to 5 × 1 cm, lanceolate to ovate, acute or acuminate, velvety-hairy. Inflorescence up to 21 × 5 cm, densely many-flowered; ovary and pedicel 20 mm long; bracts leaf-like, c. 30 mm long. Flowers faintly scented, pale greenish-cream, brown-spotted inside; lip pinkish. Sepals slightly pubescent on the outside; dorsal sepal 5.7–6.5 × 1.5–2.8 mm, elliptic, rounded; lateral sepals 6–7 × 2.8–3.5 mm, obliquely ovate. Petals 4.5–5.8 × 2–2.5 mm, obliquely lanceolate or ovate. Lip hypochile forming a rounded spur 2–3 mm long; epichile flat, 4–5 × 4–6 mm long, hanging down, broadly obovate to ± orbicular, ± equally 3-lobed at the apex. Column stout, 3–4 mm long.

Zambia. N: Mbala Distr., Kawimbe, 1680 m, fl. 21.xii.1956, *Richards* 7318 (K). W: Mwinilunga Distr., source of W Lungu R., fl. 18.xi.1961, *Holmes* 0307 (K; SRGH).
Also in Zaire and Tanzania. Scrub woodland, grassland and swamp; 1650–1700 m.

13. **Brachycorythis pubescens** Harv. in Thes. Cap. **1**: 35, t. 54 (1859). —Rolfe in F.T.A. **7**: 201 (1898). —Kraenzlin, Orch. Gen. Sp. **1**: 542 (1898). —Schlechter in Beih. Bot. Centralbl. **38**, abt. 2: 108 (1921). —Summerhayes in Kew Bull. **10**: 260 (1955); in F.T.E.A., Orchidaceae: 25 (1968). —Williamson, Orch. S. Centr. Africa: 34, pl. 34 (1977). —Stewart et al., Wild Orch. South. Africa: 75 (1982). —Geerinck in Fl. Afr. Centr., Orchidaceae pt. 1: 44 (1984). —la Croix et al., Orch. Malawi: 32 (1991). TAB. **7**. Type from South Africa.
Peristylus hispidulus Rendle in J. Linn. Soc., Bot. **30**: 398 (1895), as "*hispidula*". Type: Malawi, 1891, *Buchanan* 572 (BM, holotype).
Platanthera hispidula (Rendle) Gilg in Pflanzenw. Ost-Afr. **C**: 151 (1895).
Brachycorythis goetzeana Kraenzl. in Bot. Jahrb. Syst. **28**: 176 (1900). Type from Tanzania.
Brachycorythis kassneriana Kraenzl. in Bot. Jahrb. Syst. **51**: 378 (1914). Type: Zambia, Kamibinga Spruit, 24.xii.1907, *Kassner* 2117 (K).
Brachycorythis hispidula (Rendle) Schltr. in Beih. Bot. Centralbl. **38**, abt. 2: 108 (1921).
Brachycorythis stolzii Schltr. in Beih. Bot. Centralbl. **38**, abt. 2: 110 (1921). Type from Tanzania.

Tab. 7. BRACHYCORYTHIS PUBESCENS. 1, upper part of stem and inflorescence ($\times\frac{2}{3}$); 2, lower stem and roots ($\times\frac{2}{3}$); 3, flower, tepals removed ($\times$3); 4, lip ($\times$5); 5, lateral sepal ($\times$5); 6, column and petals, lip removed ($\times$5); 7, pollinarium, much enlarged. Drawn by W.E. Trevithick. From F.W.T.A.

Terrestrial herb 15–75 cm tall with fleshy, cylindrical, woolly, tuberous roots. Leaves numerous, overlapping, up to 6 × 2 cm in the middle of the stem, lanceolate, acute or acuminate, densely pubescent. Inflorescence 4–35 × 2.5–3.5 cm, densely or sometimes laxly many-flowered; ovary and pedicel 1.5–2.5 cm long, pubescent, purple; bracts leaf-like, up to 25 × 6 mm at base of inflorescence, pubescent. Flowers mauve-pink to purple-pink, orange or yellow in the centre, the sepals often darker at the tips. Sepals 5–8 × 2–4.5 mm, pubescent; dorsal sepal ovate, convex, forming a hood with the petals; lateral sepals spreading, obliquely ovate. Petals adnate to column at the base, 4.5–7 × 2–4 mm, obliquely ovate, sparsely pubescent on outside. Lip hypochile saccate, 2–3 mm long, with a knee-bend at the junction with the epichile, which hangs down at right angles to it; epichile 6–9 × 6–9 mm, cuneate from a narrow base, 3-lobed near the apex, the side lobes broadly rounded, the mid-lobe narrow, shorter than or equal to the side lobes. Column 2.5–4 mm long, stout.

Zambia. N: Mbala, Sandpits, 1500 m, fl. 4.i.1961, *Richards* 13771 (K). W: Mwinilunga, fl. 28.xi.1958, *Holmes* 0104 (K; SRGH). C: Kabwe Distr., Mpundo Mission, Kelongwe R., 1130 m, fl. 20.i.1973, *Kornaś* 3034 (K). E: Nyika Plateau, fl. 5.i.1959, *E.A. Robinson* 3065. **Malawi**. N: S Viphya, Luwawa Link Road near Mzimba R. Bridge, 1500 m, fl. 24.ii.1984, *la Croix* 562 (K). C: Dedza Distr., Kachere, near Mphemzi, 1550 m, fl. 22.i.1959, *Robson & Jackson* 1299 (K). S: Zomba Distr., Malosa Mt., above Domasi Church, c. 1300 m, fl. 29.xii.1979, *Morris* 551 (K; MAL). **Mozambique**. N: near Lake Malawi (Nyasa), *Johnson* (K).
Throughout tropical Africa and in South Africa. Grassland, edge of dambo and open woodland; 500–2250 m.

14. **Brachycorythis velutina** Schltr. in Bot. Jahrb. Syst. **53**: 483 (1915); in Beih. Bot. Centralbl. **38**, abt. 2: 107 (1921). —Mansfeld in Fedde, Repert. Spec. Nov. Regni Veg., Beih. 68, t. 4 no. 13 (1932). —Summerhayes in Kew Bull. **10**: 263 (1955); in F.T.E.A., Orchidaceae: 27 (1968). —Grosvenor in Excelsa **6**: 78 (1976). —Geerinck in Fl. Afr. Centr., Orchidaceae pt. 1: 45 (1984). —la Croix et al., Orch. Malawi: 33 (1991). Type from Tanzania.
Brachycorythis hispidula sensu Eyles in Trans. Roy. Soc. S. Afr.: 334 (1916) quoad *Rand* 266.

Terrestrial herb 25–70 cm tall with fleshy, cylindrical, tuberous, villous roots. Leaves numerous, overlapping, up to 5 × 1.5 cm in the middle of the stem, lanceolate or ovate, acuminate, densely pubescent. Inflorescence 8–13 × 2.5–4 cm, densely many-flowered; ovary and pedicel 1.2–2 cm long; bracts leaf-like, up to 25 × 7 mm, densely pubescent. Flowers cream-coloured or greenish-brown, lip yellow. Sepals 5–6.5 × 2.5–4 mm, pubescent outside; dorsal sepals elliptic, obtuse; lateral sepals obliquely ovate. Petals 4.5–5 × 1.7–2.5 mm, oblong, obtuse, glabrous. Lip hypochile 2–2.5 mm long and deep, cup-shaped, rounded; epichile projecting forwards horizontally, 4.5–6 × 3.8–5.8 mm, obovate, 3-lobed towards the apex, the mid-lobe longer than the side lobes. Column stout, 2.5–3.5 mm long.

Zimbabwe. C: Harare Distr., Highlands, c. 1500 m, fl. 3.xi.1948, *Greatrex* 104 (K; SRGH). E: Nyanga (Inyanga), 8 km N of Mudzora R., fl. 4.i.1961, *Ball* 971 (K; SRGH). **Malawi**. N: Chitipa Distr., Misuku Hills, near Itera, c. 1800 m, fl. 27.xii.1977, *Pawek* 13406 (K; MAL; MO; SRGH; UC). C: Dedza Distr., Chongoni Forest Reserve, fl. 13.xii.1968, *Salubeni* 1243 (K; MAL; SRGH).
Also in Tanzania. Grassland and open woodland; 1500–1800 m.

15. **Brachycorythis mixta** Summerh. in Kew Bull. **10**: 263 (1955). —Williamson, Orch. S. Centr. Africa 35, t. 35 (1977). Type from Angola.

Terrestrial herb 30–50 cm tall. Leaves numerous, up to 5 × 1 cm, lanceolate or linear-lanceolate, acuminate, densely pubescent. Inflorescence 10–15 × 3 cm, densely many-flowered; ovary and pedicel c. 15 mm long; bracts leaf-like, up to 20 × 3 mm, pubescent. Sepals and petals yellow-green, purplish on the outside; lip bright yellow, mauve-tinged in throat. Sepals 6.5–7 × 3.7–4.5 mm; dorsal sepal elliptic, obtuse; lateral sepals obliquely ovate. Petals c. 5.5 × 2.5 mm, oblong, obtuse, forming a hood with the dorsal sepal. Lip hypochile 2–3 mm long, cup-shaped; epichile projecting horizontally, rather fleshy, c. 7.3 × 7.5 mm, obovate, 3-lobed towards the apex, the sides reflexed, lateral lobes larger than the mid-lobe. Mid-lobe often upturned. Column short and stout.

Zambia. W: Mwinilunga Distr., fl. 20.x.1958, *Holmes* 059 (K).
Also in Angola. Dambo and seasonally wet upland grassland.
This species is close to *B. pubescens*, but is separated on flower colour and differences in the lip.

7. SCHWARTZKOPFFIA Kraenzl.

Schwartzkopffia Kraenzl. in Bot. Jahrb. Syst. **28**: 177 (1900).

Terrestrial herb only a few centimeters tall, leafless and lacking in chlorophyll.
Roots thick, fleshy. Stem short, covered with imbricate sheaths. Flowers 1–5, relatively
large, pedicellate. Sepals free, petals joined to the column in lower part. Lip with a
cup-like hypochile, and an epichile 3-lobed near the apex. Column erect; anther
loculi parallel, canals scarcely developed; pollinia caudicles short, viscidia naked;
stigma cushion-like; rostellum mid-lobe small, erect, cucullate, the side lobes
auriculate, surrounding the column.

A genus of 2 species in tropical Africa.
Schwartzkopffia is very close to *Brachycorythis*, but differs in that it is a leafless saprophyte with
fewer (usually 2–3), relatively large flowers.

Schwartzkopffia lastii (Rolfe) Schltr., Die Orchideen: 63 (1914). —Summerhayes in Kew Bull.
   14: 130 (1960); in F.T.E.A., Orchidaceae: 15 (1968). —Morris, Epiphyt. Orch. Malawi: 30
   (1970). —Moriarty, Wild Fls. Malawi: t. 28, 4 (1975). —Grosvenor in Excelsa **6**: 86 (1976).
   —Williamson, Orch. S. Centr. Africa: 26 (1977). —la Croix et al., Orch. Malawi: 34 (1991).
   TAB. **8**. Type from Malawi, Shire Highlands, Blantyre, *Last* s.n. (K, holotype).
      Brachycorythis lastii Rolfe in F.T.A. **7**: 203 (1898). —Geerinck, Fl. Afr. Centr., Orchidaceae
   pt. 1: 46 (1984).
      Schwartzkopffia angolensis Schltr. in Beih. Bot. Centralbl. **38**, abt. 2: 123 (1921). Type
   from Angola.

Dwarf plant 3–12 cm tall; roots few to many, spreading, up to 25 × 6 mm,
thickened, villous. Stem with whitish imbricate sheaths up to 4.5 cm long, often
recurved at the tips. Inflorescence 2–3-flowered; ovary and pedicel 3–4 cm long;
bracts c. 2.7 cm long. Flowers with a strong lemon scent; sepals and petals white or
pink; lip white or magenta with a yellow spot inside. Dorsal sepal 12–17 × 3–6 mm,
lanceolate, acute; lateral sepals 13–23 × 3.5–8.5 mm, obliquely lanceolate, acute,
keeled. Petals 10–16 × 3–7 mm, lanceolate, acute. Lip 17–22 mm long, cuneate, ±
horizontal; hypochile 2–3 mm long, rounded; epichile 15–20 mm long, 3-lobed, with
a ridge from the base of the column to the apex, c. 16 mm wide across side lobes, 2–3
mm across mid-lobe. Column 5–6 mm long.
   Considerable variation is seen in flower-size, shape of the lip mid-lobe and its
length relative to the side lobes, and the shape of the sides of the hypochile.

Zambia. N: Mbala Distr., near middle Lunzua Falls, c. 1300 m, fl. 12.i.1975, *Brummitt & Polhill*
13751 (K). W: Kitwe, fl. 27.xi.1962, *Fanshawe* 7167 (K; NDO). **Zimbabwe**. E: Nyanga, Honde
Valley, fl. 18.xi.1960, *Wild* 5280 (K; SRGH). **Malawi**. N: Nkhata Bay Distr., Mzenga Estate, 650 m,
fl. 20.xi.1986, *la Croix* 884 (K). C: Dedza Distr., between Chiwao and Kanjoli, fl. 12.i.1967, *Jeke*
52 (K; SRGH). S: Mulanje Mt., Lichenya Path, c. 950 m, fl. 28.xi.1982, *la Croix* 372 (K).
Mozambique. N: near Lake Malawi (Nyasa), 1902, *Johnson* s.n. (K).
   Also in Tanzania, Zaire and Angola. *Brachystegia* woodland; 650–1800 m.

8. NEOBOLUSIA Schltr.

Neobolusia Schltr. in Bot. Jahrb. Syst. **20**, Beibl. 50: 5 (1895).

Slender terrestrial herb; stem erect, leafy, from small ovoid, ellipsoid or fusiform
tubers. Leaves lanceolate or linear, 1 or 2 basal, the rest spaced out along the stem.
Inflorescences spicate; flowers laxly arranged, ± sessile. Sepals and petals of similar
length; sepals free; petals very oblique, attached to the column along the lower part

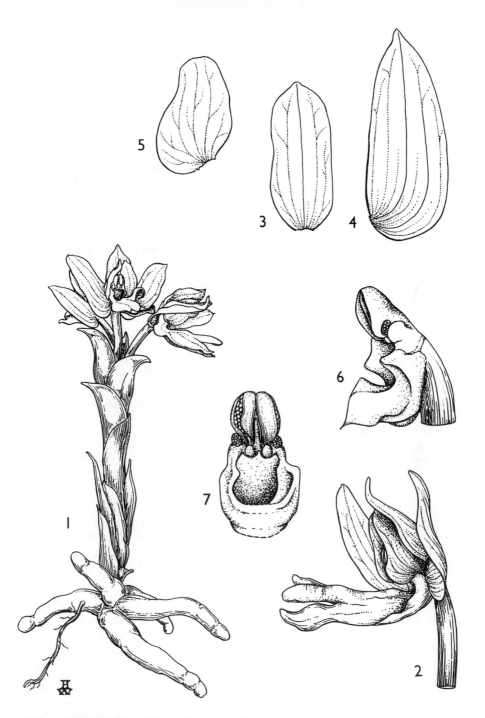

Tab. 8. SCHWARTZKOPFFIA LASTII. 1, habit (×1); 2, flower (×2); 3, dorsal sepal (×3); 4, lateral sepal (×3); 5, petal (×3); 6, column, side view (×4); 7, column, front view (×4), 1–7 from *Milne-Redhead & Taylor* 7917. Drawn by Heather Wood. From F.T.E.A.

of their dorsal margin. Lip entire or 3-lobed, neither saccate nor spurred, but with a callus towards the base. Column erect, slender; anther loculi parallel; pollinia granular, caudicles short, viscidia rather large, naked. Stigma cushion-like, 2-lobed; rostellum mid-lobe small, erect, cucullate.

A genus of 4 species in east tropical Africa and South Africa.

1. Lip entire, margin glabrous · 1. *stolzii*
– Lip 3-lobed, margin ciliate · 2. *ciliata*

1. **Neobolusia stolzii** Schltr. in Bot. Jahrb. Syst. **53**: 482 (1915). —Summerhayes in Bot. Not. **1937**: 182 (1937); in F.T.E.A., Orchidaceae: 13 (1968). —Grosvenor in Excelsa **6**: 84 (1976). —Williamson, Orch. S. Centr. Africa: 26, t. 25 (1977). —la Croix et al., Orch. Malawi: 35 (1991). Type from Tanzania.

Terrestrial herb 18–65 cm tall; tubers c. 10 mm long, spherical or ellipsoid. Leaves 3–6, the largest near the base, 5–13.5 × 0.5–1.4 cm, lanceolate, acute, decreasing in size up the stem, the upper ones bract-like. Inflorescence laxly 2–12-flowered; ovary 5–9 mm long; bracts leaf-like, 10–17 × 3.5–5 mm, acute. Sepals and petals yellow-green, lip dull maroon with a white or greenish margin. Sepals 6–11 × 2.5–3.5 mm, lanceolate to ovate, acuminate; dorsal sepal projecting over the column; lateral sepals spreading, oblique, slightly longer than the dorsal sepal. Petals 4–9 × 2.2–5 mm, lanceolate, acuminate, ciliolate. Lip 5–9 × 3–7 mm, oblong-spathulate, apiculate, entire, convex with a large hairy callus in the lower centre, margins glabrous. Column 2.5–3 mm long.

Var. **stolzii**

Petals ciliolate.

Zambia. E: Nyika Plateau, near Rest House, 2150 m, fl. 5.ii.1982, *Dowsett-Lemaire* 198 (K). **Malawi**. N: Nyika National Park, just S of Chelinda Camp, 2200–2300 m, fl. 10.iii.1977, *Grosvenor & Renz* 1110 (K; SRGH). S: Zomba Plateau, edge of Chitinji Marsh, c. 1800 m, fl. 6.iii.1983, *la Croix* 466 (K); Mulanje Mt., Lichenya Path, c. 1800 m, fl. 19.ii.1983, *la Croix, Jenkins & Killick* 459 (K).
Also in Tanzania. Damp montane grassland, seepage slopes amongst rocks; 1800–2300 m.

Var. **glabripetala** Summerh. in Kew Bull. **16**: 255 (1962). Type: Zimbabwe, Nyanga, foot of Mt. Inyangani, *Norlindh & Weimarck* 5067 (K, holotype).

Petals not ciliolate on margin, otherwise similar to the typical variety.

Zimbabwe. E: Nyanga, foot of Mt. Inyangani, c. 2000 m, fl. 15.ii.1931, *Norlindh & Weimarck* 5067 (K).
So far known only from Mt. Inyangani in Zimbabwe. Damp montane grassland amongst rocks on seepage slopes; 1800–2300 m

2. **Neobolusia ciliata** Summerh. in Kew Bull. **11**: 217 (1956). —Goodier & Phipps in Kirkia **1**: 53 (1961). —Grosvenor in Excelsa **6**: 84 (1976). TAB. **9**. Type: Zimbabwe, Makoni Distr., Rusape, *Eyles* 8358 (K, holotype; SRGH).

Terrestrial herb, glabrous except for lip and petal margins and the tuber; tuber c. 20 × 3 mm, fusiform, villose. Leaves usually 3, largest near the stem base, 4.2–6.5 × 1–1.2 cm, lanceolate, apiculate. Inflorescence laxly 3–10-flowered, flowers suberect; ovary 5–7 mm long; bracts 10–12 × 5 mm. Sepals greenish to straw-coloured, purple-veined; petals pale green and rimmed with pale purple, pellucid-punctate; lip deep glossy-purple, pale or greenish at tip. Dorsal sepal erect, 9.2–9.6 × 3–4 mm, lanceolate, acute; lateral sepals spreading, 9.7–10.5 × 3–3.7 mm, obliquely lanceolate, acute. Petals 5.9–8.2 × 2.7–3.7 mm, lanceolate, slightly ciliolate on margin. Lip 11–13.5 × 9–10.5 mm, broadly rhomboid-hastate, 3-lobed in the basal half, ciliate on

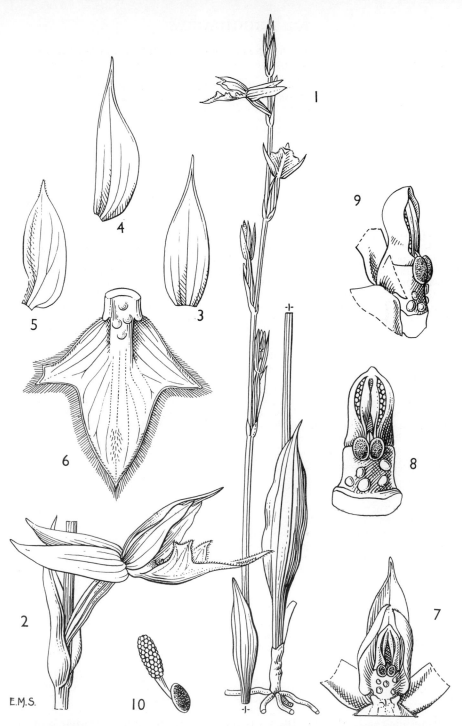

Tab. 9. NEOBOLUSIA CILIATA. 1, habit (×1), from *Eyles* 8358; 2, flower and bract, side view
(×2); 3, dorsal sepal (×4); 4, lateral sepal (×4); 5, petal (×4); 6, lip, spread out (×4); 7,
centre of flower, showing column, dorsal sepal and petals (×4); 8, column, front view (×8);
9, column, side view (×8); 10, pollinarium (×8), 2–10 from *Ball* 200. Drawn by Margaret
Stones. From Hooker's Icon. Pl.

the margin; mid-lobe 6.2–7.5 × 5.3–7 mm; lateral lobes 1–2.5 mm long along the front margin. Column 4.3–5 mm high.

Zimbabwe. C: Makoni Distr., Rusape, fl. & fr. 1.i.1935, *Eyles* 8385 (K; SRGH). E: Chimanimani Mts., c. 2000 m, fl. & fr. 31.i.1954, *Ball* 200 (K; SRGH). **Mozambique**. MS: Chimanimani Mts., between Skeleton Pass and Namadima, 1540 m, fl. 30.xii.1959, *Goodier & Phipps* 344 (K; SRGH).

Only known from the eastern border area of Zimbabwe and adjacent Mozambique. High, grassy plateau and montane grassland; 1540–2000 m.

9. SCHIZOCHILUS Sond.

Schizochilus Sond. in Linnaea **19**: 78 (1846). —Linder in J. S. African
Bot. **46**: 379–434 (1980).

Terrestrial herb with perennating tubers. Tubers 1–5 cm long, testicular. Stems slender, mostly flexuose, 5–80 cm tall. Basal sheath(s) hyaline or white, less than 5 cm long, obtuse or acute. Leaves 5–30, linear, rarely elliptic, narrowly oblanceolate or ovate; lower leaves with free blades up to 15 cm long, acute, midrib prominent below; upper leaves lanceolate to narrowly lanceolate, acuminate, erect, grading into the floral bracts. Inflorescence mostly nodding, lax to dense; bracts narrowly ovate to narrowly lanceolate, usually acuminate, green, about as long as the ovary. Flowers white, white and yellow, yellow or white with a mauve tint, 1.5–10 mm in diameter. Sepals subequal, usually 3-nerved; lateral sepals suboblique, to 14 mm long, lanceolate to ovate, acute; dorsal sepal usually shorter than the lateral sepals, narrowly to broadly elliptic, or elliptic-oblong to rotund, often apiculate, shallowly galeate. Petals one third to two thirds as long as the sepals, single-veined, oblique, ± rhomboid and acute. Lip about as long as the sepals, epichile ± 3-lobed with the central lobe longer than the lateral lobes, hypochile concave, leading into the spur, frequently with calli between hypochile and epichile; spur terete to bifid, slender to clavate, almost always shorter than the lip, generally straight. Rostellum 3-lobed, the central lobe a fold between the anther cells, the lateral lobes square, carrying the naked viscidia. Anther erect or at an angle of 45°, 0.5–1.5 mm long. Stigma flat, single, borne on the column below the rostellum, above the entrance to the spur. Ovary 4–14 mm long, flowers not resupinate.

A genus of 11 species in southern and East Africa, with 4 in the Flora Zambesiaca area.

1. Spur 11–12 mm long · 4. *calcaratus*
 – Spur less than 5 mm long · 2
2. Sepals less than 4 mm long; flowers white or yellow · 1. *cecilii*
 – Sepals more than 4 mm long; flowers yellow · 3
3. Spur 0.8–1.2 mm long, curved · 2. *sulphureus*
 – Spur 1.2–2.2 mm long, straight · 3. *lepidus*

1. **Schizochilus cecilii** Rolfe in Bull. Misc. Inform., Kew **1906**: 168 (1906). —Schlechter in Beih. Bot. Centralbl. **38**, abt. 2: 93 (1921). —Summerhayes in Bot. Not. **1937**: 183 (1937). —Grosvenor in Excelsa **6**: 86 (1976). —Linder in J. S. African Bot. **46**: 392 (1980). —Stewart et al., Wild Orch. S. Afr.: 76 (1982). Type: Zimbabwe, Nyanga (Inyanga) Mts., *E. Cecil* 202 (K, holotype).

Subsp. **cecilii** Rolfe in Bull. Misc. Inform., Kew **1906**: 168 (1906). —Linder in J. S. African Bot. **46**: 394 (1980). Type as above.
    sp. no. 1, Grosvenor in Excelsa **6**: 86 (1976) quoad *Ball* 358.

Plants subflexuous, slender, 10–30 cm tall. Basal sheaths hyaline or white, 1–3 cm long, obtuse to acute. Leaves 6–15, the lower 3–6 clustered near base of stem, semi-erect, up to 10 cm long, narrowly oblanceolate to very narrowly elliptic, acute, the remaining leaves scattered on the stem, lanceolate, acute to subacuminate, smaller towards the apex and grading into the floral bracts. Inflorescences dense, up to 7 × 1 cm, c. 40-flowered, nodding; bracts at least 3 mm long, narrowly ovate to ovate, subacuminate. Flowers bright yellow or white, 2–3 mm in diameter. Sepals subequal, 1 or 3 veined; dorsal sepal 2–2.7 mm long, elliptic or rotund, obtuse; lateral sepals 2.7–3.7 mm long, subobliquely lanceolate to ovate, subacute. Petals 1-nerved, 1.3–3 mm long, rhomboid-rotund to ovate, acute, curved over the anther. Lip 3–3.6 × 1.5–1.8 mm, epichile 3-lobed, central lobe c. 1 mm long, lateral lobes less than half as long as the central lobe, lobes subacute; hypochile deeply concave, somewhat smaller than the epichile; disk between hypochile and epichile with 3 fleshy calli; spur cylindrical, straight, 0.1–0.5 mm long, anther curved, c. 1 mm long. Ovary c. 4 mm long.

Zimbabwe. E: Nyanga, Inyanga Fort, fl. 31.i.1948, *Fisher* 1451 (K; NU; SRGH).
Known only from the Nyanga area. It appears to grow in shallow soil in mountain grassland above 1500 m.
The flower colour of *S. cecilii* subsp. *cecilii* is variable, and both white and yellow forms have been reported from the same population.
Subsp. *transvaalensis* and subsp. *culveri*, distinguished from the typical subspecies by their generally larger dorsal sepals, both occur in the eastern Transvaal mountains of South Africa.

2. **Schizochilus sulphureus** Schltr. in Bot. Jahrb. Syst. **53**: 486 (1915). —Schlechter in Beih. Bot. Centralb. **38**, abt. 2: 90 (1921). —Summerhayes in F.T.E.A., Orchidaceae: 27 (1968). —Williamson, Orch. S. Centr. Africa: 35, 213 (1977). —Linder in J. S. African Bot. **46**: 420 (1980). —la Croix et al., Orch. Malawi: 36 (1991)). TAB. **10**. Type from Tanzania.

Plants 8–20(30) cm tall. Basal sheaths usually 2, the outer hyaline, obtuse; the inner white, obtuse to acute, to 2 cm long. Leaves 6–14, scattered on the stem, generally 2–5 clustered at the base, semi-erect to deflexed, 2–5 cm long, narrowly lanceolate to linear, acute, the upper leaves grading into the floral bracts. Inflorescences semi-dense, 2–6 cm long, 15–40-flowered, nodding; bracts about 6 mm long, slightly larger than the ovary, narrowly ovate, acuminate. Flowers c. 4 mm in diameter, butter-yellow, the petals slightly paler than the sepals. Sepals subequal, 3-nerved; dorsal sepal shallowly concave, 4.5–5.5 mm long, elliptic to narrowly elliptic, obtuse, apiculate; lateral sepals shallowly boat-shaped, sub-oblique, 4.6–6.5 mm long, lanceolate, acute. Petals 1-nerved, oblique, 3–4 mm long, rhombic-lanceolate, acute. Lip 3.5–5.5 × 1.5–2.5 mm; epichile 3-lobed with the central lobe subacute, c. 1.5 mm long, lateral lobes obtuse, subobsolete; disk with 3 small calli; spur 0.8–1.2 mm long, curved towards the ovary. Anther c. 1.5 mm long. Ovary c. 5 mm long.

Malawi. N: Nyika Plateau, fl. 10.ii.1960, *Holmes* 0203 (K; SRGH). S: Mt. Mulanje, fl. 12.ii.1979, *Blackmore et al.* 366 (K). **Mozambique**. Z: Gurué, fl. 5.i.1968, *Torre & Correia* 16928 (LISC).
Also known from the Southern Highlands of Tanzania. This is a montane species, occurring above 1800 m. It is often found in moss in seepages over rock and in damp grassland, often after fire.

3. **Schizochilus lepidus** Summerh. in Kew Bull. **14**: 130 (1960). —Grosvenor in Excelsa **6**: 86 (1976). —Linder in J. S. African Bot. **46**: 419 (1980). Type: Mozambique, Mt. Tsetserra, *Wild* 4471 (K, holotype; PRE; SRGH).

Plants slender, 15–25 cm tall. Basal sheaths 2, white to hyaline, up to 3 cm long, obtuse, apiculate. Leaves 8–12, scattered on the stem, the lower 2–5 semi-erect, 3–8 cm long, narrowly oblanceolate to narrowly elliptic, acute, the remaining leaves more or less sheathing, c. 1 cm long, lanceolate, acute. Inflorescence semi-dense, 2–8 cm long, 20–30-flowered, nodding; bracts 5–7 mm long, narrowly acute or acuminate. Flowers c. 5 mm in diameter, yellow. Sepals subequal, 3-nerved, 4.5–6.5 mm long, acute to apiculate; dorsal sepal elliptic, concave; lateral sepals somewhat longer than the dorsal sepal, oblique, lanceolate to narrowly ovate, shallowly concave. Petals 1-nerved, 2.5–4 mm long, obliquely ovate-rhomboid, acute. Lip 4–6.5 mm long; epichile 3-lobed, central lobe 1–2.5 mm long, lateral lobes c. half as long; disk with 3 obscure calli; spur 1.5–2 mm long, subclavate. Anther c. 1.5 mm long, Ovary 5–7 mm long.

Zimbabwe. E: Mt. Binga, fl. 29.i.1966, *Pereira* 39 (K). **Mozambique**. MS: Mt. Tsetserra, 2140 m, 7.ii.1955, *Exell, Mendonça & Wild* 232 (LISC).

Tab. 10. SCHIZOCHILUS SULPHUREUS. 1, habit (×1), from *Stolz* 1075; 2, flower (×4); 3, flower, tepals removed to show lip and column (×4); 4, dorsal sepal (×5); 5, lateral sepal (×5); 6, petal (×5); 7, lip (×5); 8, column, front view (×15), 2–8 from *Ball* 795. Drawn by Heather Wood. From F.T.E.A.

Tab. 11. SCHIZOCHILUS CALCARATUS. 1, habit (×⅔); 2, flower (×6); 3, dorsal sepal (×8); 4, petal (×8); 5, lateral sepal (×8); 6, lip (×8); 7, column, side view (×12); 8, column, front view (×12), 1–8 from *Whellan* 2211b. Drawn by Eleanor Catherine. From Kew Bull.

Eastern border mountains of Zimbabwe and adjacent Mozambique. Montane grassland; usually above 2000 m.

This species is very closely allied to *S. zeyheri* from South Africa, a very variable species, and its separation may not be warranted. It can be distinguished by the short, sub-clavate spur.

4. **Schizochilus calcaratus** P.J. Cribb & la Croix in Kew Bull. **48**, 2: 364 (1993). TAB. **11**. Type: Zimbabwe, Chimanimani Mts., *Whellan* 2211B (MAL, holotype).

Glabrous terrestrial herb 18–25 cm tall. Leaves c. 7, 5–7 × 0.4–1 cm, linear or oblong, the lowermost 3–4 semi-erect, the upper more or less sheathing and bract-like. Inflorescence 2–3.5 × 1.5 cm, densely several-flowered; pedicel and ovary 6–8 mm long; bracts 9 mm long, lanceolate, acuminate. Flowers bright yellow. Dorsal sepal erect, 3–3.5 mm long, oblong and cucullate at apex, forming a hood with the petals. Lateral sepals ± deflexed, slightly longer and wider than the dorsal sepal. Petals c. 3 × 2 mm, obliquely ovate. Lip 3 × 2.5 mm, wedge-shaped, 3-lobed in the apical third; mid-lobe c. 0.7 mm long, cucullate and rounded or truncate at the apex; side lobes shorter, cucullate at the apex; spur 11–12 mm long, slender, slightly curved. Column very short, less than 2 mm long; anther erect; rostellum obscure.

Zimbabwe. E: Chimanimani Mts., fl. xii.1964, *Whellan* 2211 (MAL). Not known elsewhere. Montane grassland; 1550–1600 m.

10. HOLOTHRIX Lindl.

Holothrix Lindl., Gen. Sp. Orchid. Pl.: 257, 283 (1835), nom. conserv.
Deroemera Rchb.f., De Poll. Orch.: 29 (1852).

Terrestrial herb with small ovoid or ellipsoid tubers. Leaves radical, 1–2, appressed to ground, sessile, ovate or orbicular, sometimes withered by flowering time. Scape erect, unbranched, with or without sheaths. Flowers sessile or shortly stalked, often secund. Sepals subequal, free, often hairy, usually green. Petals usually longer (often much longer) than sepals, entire or lobed, often rather fleshy. Lip lobed or toothed, rarely entire, spurred; adnate to column at base. Column very short; anther loculi parallel, pollinia granular; caudicles very short, viscidia small, naked; stigma sessile.

A genus of 50–60 species in tropical Africa, South Africa and tropical Arabia.

1. Scape with several acute sheaths · 2
   – Scape without sheaths · 4
2. Petals and lip 3-lobed, less than 8 mm long · · · · · · · · · · · · · · · · · · · 11. *tridactylites*
   – Petals and lip 5–11-lobed, over 10 mm long · 3
3. Leaves, scape and bracts hairy · 4. *longiflora*
   – Leaves, scape and bracts glabrous · 10. *randii*
4. Petals entire · 5
   – Petals 3-lobed · 11
5. Lip entire or with 2–3 lobes or teeth · 6
   – Lip with 5–10 lobes or teeth · 7
6. Lip entire, or minutely 3-lobed or notched, flowers lilac · · · · · · · · · · · · · · · · 1. *puberula*
   – Lip 3-lobed, the lobes 0.5–1.2 mm long; flowers buff or straw-yellow · · · · · · · · 12. *villosa*
7. Petals over 10 mm long · 3. *johnstonii*
   – Petals less than 6 mm long · 8
8. Lip 9–13 mm long with 1 pair of lobes near base, the mid-lobe 3-toothed · · · 8. *orthoceras*
   – Lip 2.5–9.5 mm long; lobed in apical half · 9
9. Lip lobed to about half-way; flowers buff · 7. *micrantha*
   – Lip with short apical teeth; flowers white · 10
10. Ovary densely hairy · 5. *buchananii*
    – Ovary glabrous · 6. *macowaniana*
11. Flowers densely papillose; lip 5–7-lobed · 9. *papillosa*
    – Flowers not papillose; lip 8–10-lobed · 2. *pleistodactyla*

1. **Holothrix puberula** Rendle in J. Bot. **33**: 278 (1895). —Rolfe in F.T.A. **7**: 191 (1897). —Summerhayes in F.T.E.A., Orchidaceae: 6 (1968). —Williamson, Orch. S. Centr. Africa: 24, t.24 (1977). —la Croix et al., Orch. Malawi: 41 (1991). Type from Kenya.

Terrestrial herb, 5–11 cm tall. Tubers one or two, 1–2 cm long, ovoid, hairy. Leaves 2, appressed to ground, 7–15 × 6–15 mm, orbicular or heart-shaped, apiculate, glabrous. Scape hairy, without sheaths. Inflorescence up to 5 cm long, 2–5-flowered, flowers secund; pedicels very short; ovary 5 mm long, arched; bracts up to 3 mm long, lanceolate, acute, hairy. Flowers lilac-purple, sepals and spur green. Sepals 2.7–5 × 2–2.2 mm, ovate, acute, hairy, the lateral sepals oblique. Petals 2–6 × 1–2 mm, lanceolate, acute, entire. Lip 6–11 × 3–4 mm, elliptic, minutely 2-lobed or 3-lobed at tip; spur c. 4 mm long, conical, somewhat incurved. Ovary hairy.

Zambia. C: Kundalila Falls, fl. i.1969, *G. Williamson* 409 (K). **Malawi**. N: Nyika National Park, c. 4 km N of Thazima, fl. 13.ii.1987, *la Croix* 962 (K).
Also in Ethiopia, Kenya and Tanzania. *Brachystegia* woodland, in sparse grass on shallow soil overlying rock, open rocky slopes; c. 1660 m.

2. **Holothrix pleistodactyla** Kraenzl. in Pflanzenw. Ost-Afr. **C**: 181 (1895). —Rolfe in F.T.A. **7**: 193 (1898). —Summerhayes in F.T.E.A., Orchidaceae: 8 (1968). —Williamson, Orch. S. Centr. Africa: 23 (1968). —la Croix et al., Orch. Malawi: 41 (1991). Type from Tanzania (Kilimanjaro).
Holothrix rorida G. Will. in Kirkia **13**, 2: 245 (1990). Type: Malawi, Nyika Plateau, *G. Williamson* 395 (SRGH, holotype; K) synon. nov.

Terrestrial herb 10–18 cm tall. Tubers 2, c. 15 × 18 mm, ovoid. Leaves 2, appressed to ground, usually withered at flowering time, c. 15 mm long and wide, orbicular, apiculate. Scape without sheaths, pink with white spreading hairs. Inflorescence fairly laxly 3–10-flowered, flowers ± secund; ovary 4–6 mm long, erect; bracts 2–5 mm long, ovate, acute, hairy. Flowers non-resupinate, c. 7 mm in diameter, cream to pale yellow, sepals green. Sepals 2–4 × 1–1.5 mm, lanceolate to ovate, acute, glabrous. Petals 5–7 mm long, with 3 slender lobes 2–2.5 mm long. Lip similar but with 8–10 lobes. Spur 1.7–3 mm long, obtuse, more or less straight then upcurved at apex. Column very short. Ovary glabrous.

Malawi. N: Nyika National Park, Chelinda Bridge, fl. 29.ix.1969, *Pawek* 2827 (K; MO); Chelinda Bridge, fl. 27.ix.1987, *la Croix* 1054 (K; MAL).
Also in Tanzania. Dry montane grassland, after burning; 2130–2300 m.
Plants from Malawi have shorter bracts and slightly larger flowers than the Tanzanian specimens seen.

3. **Holothrix johnstonii** Rolfe in Bull. Misc. Inform., Kew **1896**: 47 (1896); in F.T.A. **7**: 194 (1898). —Summerhayes in Mem. N.Y. Bot. Gard. **9**, 1 80 (1954). —Morris, Epiphyt. Orch. Malawi: 28 (1970). —la Croix et al., Orch. Malawi: 39 (1991). Type: Malawi, Mt. Mulanje, *Whyte* s.n. (K, holotype).

Terrestrial herb (6)16–30 cm tall. Tubers two or three, c. 15 × 10 mm, ovoid. Leaves 2, appressed to ground, fleshy, 3–6 × 2.3–6 cm, broadly ovate to heart-shaped, dark glossy-green with long, weak white hairs. Scape without sheaths, with spreading hairs. Inflorescence up to 10(15)-flowered, flowers secund; ovary 3–6 mm long; bracts 5–7 × 2.5–4 mm, ovate, hairy. Flowers c. 2 cm in diameter, white to rose-pink, usually opening pale and darkening with age; sepals greenish-purple. Sepals 5–7 × 2–3 mm, ovate, hairy. Petals entire, 12–17 × 2 mm, linear-oblong, obtuse, with a slightly raised median nerve. Lip 12–17 mm long, 5–7-lobed, the lobes 2–4 mm long, obtuse, with raised median nerves. Spur 2–3 mm long, conical, obtuse, upturned at apex. Ovary ovoid, constricted at top just below the flower, hairy.

Malawi. S: Chiradzulu Mt., southern end, fl. 11.v.1980, *Brummitt & Patel* 15619 (K; MAL); Mulanje Mt., Chambe Plateau, fl. 2.v.1965, *Morris* 165 (K); Mulanje Mt., Thuchila Plateau, fl. 18.viii.1983, *Johnston-Stewart* 78 (K).
Endemic. Rock slabs and seepage slopes, exposed or in shade of rocks; 1300–2310 m.

4. **Holothrix longiflora** Rolfe in Bol. Soc. Brot. **7**: 237 (1890). —Summerhayes in F.T.E.A., Orchidaceae: 13 (1968). —Morris, Epiphyt. Orch. Malawi: 29 (1970). —Williamson, Orch. S. Centr. Africa: 24, t.23 (1977). —la Croix et al., Orch. Malawi: 39 (1991). TAB. **12**. Type from Angola.

    Holothrix lastii Rolfe in F.T.A. **7**: 195 (1898). Type: Malawi, Shire Highlands near Blantyre, *Last* s.n. (K, holotype).

Terrestrial herb 23–53 cm tall. Tubers 2, up to 3.5 cm long, ovoid, villous. Leaves 2, one larger than the other, fleshy, appressed to ground, 1.5–7 × 1.5–8.5 cm, heart-shaped to reniform, glossy light green, sparsely hairy. Scape densely hairy, with several bract-like leaves c. 7 mm long mostly in the upper half. Inflorescence 6–16 cm long, laxly to fairly densely 6 to many-flowered, flowers ± secund; ovary 3–9 mm long; bracts 4–8 mm long, hairy. Flowers white, strongly and sweetly scented; sepals green. Sepals 6–9 × 3.6–5 mm, ovate, acute, hairy. Petals 16–35 mm long, divided to more than halfway into 9 slender lobes, tapering to become more or less filiform. Lip similar, 9–11-lobed. Spur 3–4 mm long, conical, tightly incurled. Anther 2 mm long. Ovary hairy.

    Zambia. N: Mbala Distr., Sumbawanga Road, fl. 1.iii.1957, *Richards* 8413 (K). **Malawi**. C: Dedza Distr., Chencherere Hill, fl. 11.iii.1967, *Salubeni* 583 (K; MAL). S: Zomba Mt., Chingwe's Hole, fl. & fr. 23.ii.1978, *Taylor* 16 (K; MAL); Mulanje Mt., Sombani Plateau, *Johnston-Stewart* 363.
    Also in Tanzania and Angola. Open *Brachystegia* woodland, rock crevices, montane grassland overlying rock slabs, seepage slopes; 1650–1900 m.

5. **Holothrix buchananii** Schltr. in Ost. Bot. Zeitschr. **48**: 446, 447 (1898); in Bot. Jahrb. Syst. **26**: 330 (1899). —Summerhayes in F.T.E.A., Orchidaceae: 6 (1968). —la Croix et al., Orch. Malawi: 37 (1991). Type: Malawi, without precise locality, *Buchanan* s.n. (B†, holotype).

Terrestrial herb 5–19 cm tall. Tubers one or two, 1–2 cm long, ellipsoid, hairy. Leaves 2, appressed to ground, 8–25 × 6–22 mm, broadly ovate to reniform-circular, dark green with silvery-white venation. Scape purplish, without sheaths, with some spreading or retrorse hairs. Inflorescence up to 5 cm long, fairly laxly 3–7-flowered, flowers secund; ovary 4–7 mm long; bracts 3 mm long, hairy. Flowers white, the lip with 3–5 purple lines, sepals green with purplish tips. Sepals 2.4–3 × 1–1.5 mm, ovate to ovate lanceolate, acute, pubescent, the lateral sepals oblique. Petals 3–5 × 1.3–2 mm, lanceolate, acuminate, glabrous. Lip 6–9.5 × 4.5–5.5 mm, cuneate from a narrow base, very shortly 5–7-lobed at the apex; spur 2–3 mm long, narrowly conical. Column c. 1 mm long. Anther loculi parallel; pollinaria 2. Stigma sessile; rostellum shallowly 3-lobed. Ovary hairy.

    Malawi. N: Masuku Plateau, fl. vii.1896, *Whyte* s.n. (K); Nyika Plateau, on road to Kasaramba, fl. 14.v.1970, *Brummitt* 10684 (K; MAL). S: without precise locality, *Buchanan* s.n. (B†).
    Also in Tanzania. Montane grassland; 1900–2300 m.

6. **Holothrix macowaniana** Rchb.f., Otia Bot. Hamburg.: 108 (1881). —Bolus, Icon. Orchid. Austro-Afric. **3**: t.19a (1913). —Goodier & Phipps in Kirkia **1**: 52 (1961). —Grosvenor in Excelsa **6**: 83 (1976). —Stewart et al., Wild Orch. South. Africa: 61 (1982). Type from South Africa (E Cape).

Terrestrial herb 5–8 cm tall. Scape without sheaths, sparsely hairy. Leaves 2, appressed to ground, c. 13 × 10 mm, broadly ovate. Inflorescence laxly several-flowered, flowers secund; ovary 4 mm long. Flowers white. Dorsal sepal 1.75–2.2 × 1.2 mm, lateral sepals 2.25–2.75 × 1.2–1.5 mm. Petals 2.75–4.2 × 1–1.4 mm; lanceolate, acuminate, entire, united to lip at the base. Lip 5.3–7 × 3.5–5.2 mm, obovate or cuneate, with 6–9 shallow teeth. Spur 2.5–3 mm long, slender, tapering. Anther 1.25 mm tall. Ovary glabrous.

    Zimbabwe. E: Chimanimani Mts., 1800 m, fl. 17.ix.1976, *Ball* 579 (K; SRGH); Mt. Nuzi, 1850 m, fl. ix.1934, *Gilliland* 835 (K).
    Also in South Africa (E Cape). Shallow soil near granite boulders; 1800–1850 m.
    Known in the Flora Zambesiaca area from only 2 collections. This species closely resembles *Holothrix buchananii*, differing in the glabrous ovary and slightly smaller flowers.

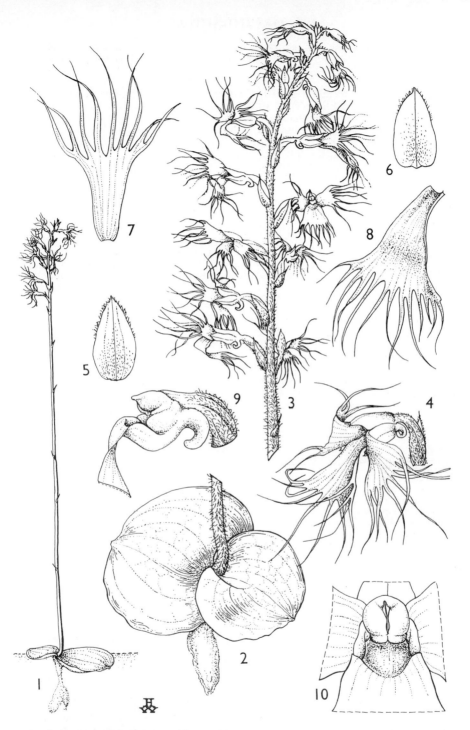

Tab. 12. HOLOTHRIX LONGIFLORA. 1, habit (×⅓); 2, base of plant (×1); 3, inflorescence (×1);
4, flower (×2); 5, dorsal sepal (×3); 6, lateral sepal (×3); 7, petal (×3); 8, lip (×3); 9, ovary,
column and base of lip, side view (×5); 10, column and base of lip, front view (×10), 1–10
from *Richards* 12140. Drawn by Heather Wood. From F.T.E.A.

7. **Holothrix micrantha** Schltr. in Bot. Jahrb. Syst. **20**: 31 (1895). —Grosvenor in Excelsa **6**: 83 (1976). —Stewart et al., Wild Orch. South. Africa: 61 (1982). Type from South Africa (S Transvaal).

Terrestrial herb c. 16 cm tall. Leaf 1, ± withered at flowering time. Scape without sheaths, with long spreading hairs. Inflorescence 4–7 cm long, several- to many-flowered; ovary 2–5 mm long; bracts long ciliate. Flowers buff. Sepals 1.5–1.7 × 1.3 mm, ovate, hairy. Petals 3.8 × 1 mm, narrowly lanceolate, obtuse, entire, opaque and fleshy towards the tip. Lip 3.5 × 3.9 mm, divided to about halfway into 5 lobes, the entire part 1.7 mm long, the lobes 1.4–1.8 mm long. Spur 1.7 mm long, conical, obtuse. Anther 2 mm long. Ovary glabrous.

Zimbabwe. E: Nyanga (Inyanga), Danakay Hotel, fl. 5.x.1949, *James* in *GHS* 25171 (K; SRGH). Also in South Africa (S Transvaal). Submontane grassland, usually following annual burn.

8. **Holothrix orthoceras** (Harv.) Rchb.f. in Otia Bot. Hamburg.: 119 (1881). —Goodier & Phipps in Kirkia **1**: 52 (1961). —Dyer (ed.), Fl. Pl. Afr. **36**: t.1469 (1964). —Morris, Epiphyt. Orch. Malawi: 29 (1970). —Grosvenor in Excelsa **6**: 83 (1976). —Stewart et al., Wild Orch. South. Africa: 62 (1982). —la Croix et al., Orch. Malawi: 40 (1991). Type from South Africa.
   Tryphia orthoceras Harv. in Thes. Cap. **2**: 4, t.105 (1863).

Terrestrial herb 10–26 cm tall. Tubers up to 20 × 15 mm, ovoid. Leaves 2, appressed to ground, 3–6.5 × 1.5–4.2 cm, ovate, glabrous, dark green with whitish veining. Scape sparsely hairy, without sheaths. Inflorescence 6–11 cm long, up to 20-flowered, flowers secund; ovary 6 mm long; bracts 4–5 mm long, lanceolate, acuminate. Flowers white with 5 purple lines on lip, one running to each lobe. Sepals green, sparsely hairy, 2.5–3 × 1 mm, ovate. Petals 3–5 × 1 mm, lanceolate, acute, erect. Lip 7–12 × 3.5–5 mm, with 1 pair of lobes 1–2 mm long near base; mid-lobe 4–6 mm long, cuneate with 3 shallow teeth. Spur 2–2.8 mm long, conical.

Zimbabwe. E: Nyanga (Inyanga), lip of Pungwe Falls, fl. 15.x.1953, *Chase* in *GHS* 44155 (K; SRGH). **Malawi**. C: Dedza Mt., near radio mast, fl. & fr. 24.iv.1970, *Brummitt* 10125 (K). S: Mulanje Mt., Lichenya Path, fl. & fr. 5.vi.1962, *Richards* 16531 (K).
Also in South Africa. Submontane grassland or amongst mossy rocks, sometimes in heavy shade; 1500–2050 m.

9. **Holothrix papillosa** Summerh. in Kew Bull. **14**: 128 (1960); in F.T.E.A., Orchidaceae: 7 (1968). —Morris, Epiphyt. Orch. Malawi: 30 (1970). —Williamson, Orch. S. Centr. Africa: 23, t.22 (1977). —la Croix et al., Orch. Malawi: 40 (1991). Type from Tanzania.
   Holothrix brachycheira Summerh. in Kew Bull. **14**: 127 (1960). Type: Zambia, Mwinilunga Distr., between Mwinilunga and Matonchi Farm, 26.i.1938, *Milne-Redhead* 4367 (K, holotype).

Terrestrial herb 9–32 cm tall. Tubers 2, up to 13 × 8 mm, ovoid. Leaves 2, appressed to ground, 10–23 × 7–33 mm, orbicular to reniform, pilose above. Scape without sheaths, with long spreading hairs. Inflorescence 2–8 cm long, fairly densely 2–many-flowered, flowers secund; ovary 3 mm long; bracts 1–4 mm long, ovate, pubescent. Flowers white, non-resupinate, c. 6 mm long including ovary. Sepals green or purplish, 1.7–3 × 1.3–1.4 mm, ovate, long ciliate. Petals 2.5–6 mm long, densely papillose, 3-lobed to about one third of the length. Lip 2.5–5 mm long, similar to petals but with 5–7 lobes. Spur 1.5–3 mm long, conical, obscure, straight. Anther 1 mm high. Ovary pubescent.

Zambia. W: Mwinilunga Distr., between Mwinilunga and Matonchi Farm, fl. 26.i.1938, *Milne-Redhead* 4367 (K). E: Nyika National Park, c. 0.5 km SW of Zambian Rest-house, fl. & fr. 17.iv.1986, *Philcox, Pope & Chisumpa* 9975 (K). **Malawi**. N: Nyika Plateau, near Nganda Peak, fl. 16.iv.1975, *Pawek* 9277 (K; MO; UC); S Viphya, Mtangatanga Forest Reserve, fl. 1.iii.1987, *Cornelius* in *la Croix* 992 (K; MAL).
Also in Tanzania. Montane grassland, *Brachystegia* woodland; 1650–2300 m.
Holothrix brachycheira from the Mwinilunga area has small flowers (petals 2–2.7 mm long, lip 2.5–2.9 mm long) but falls within the range of variation of *H. papillosa*, which varies considerably in size depending largely on the habitat.

Tab. 13. HOLOTHRIX TRIDACTYLITES. 1, habit (×1); 2, flower, side view, with one petal turned back (×8); 3, dorsal sepal (×6); 4, lateral sepal (×6); 5, petal (×6); 6, column (×14), 1–6 from *McLoughlin* 9. Drawn by Mary Grierson. From F.T.E.A.

10. **Holothrix randii** Rendle in J. Bot. **37**: 208 (1899). —Eyles in Trans. Roy. Soc. S. Afr. **5**: 334
(1916). —Summerhayes in Bot. Not. **1937**: 182 (1937); in Kew Bull. **14** (1): 129 (1960);
in F.T.E.A., Orchidaceae: 11 (1968). —Grosvenor in Excelsa **6**: 83 (1976). —Stewart et
al., Wild Orch. South. Africa: 65 (1982). Type: Zimbabwe, Harare (Salisbury), *Rand* 596
(BM, holotype).
   Holothrix reckii Bolus, Icon. Orchid. Austro-Afric. **3**: t. 21 (1913). Type from South Africa.

Terrestrial herb 23–35.5 cm tall, glabrous except for the roots. Tubers c. 25 × 17
mm, ovoid, villous. Leaves 2, appressed to ground, up to 3 × 4 cm, reniform,
glabrous, fleshy, usually withered at flowering time. Scape glabrous, pinkish, with
several bract-like leaves up to 1 cm long. Inflorescence 5–18 cm long, laxly 5–18-
flowered, flowers secund; ovary 7–10 mm long, arched; bracts 6 × 2 mm, ovate. Sepals
green, 3.2–4.2 × 1.9–2.2 mm, ovate, acute. Petals white, 12.5–15 mm long, fimbriately
divided to over halfway into 9–13 very slender, tapering lobes. Lip 15–18 mm long,
fimbriately c. 15-lobed. Spur 4.5–6.5 mm long, lightly incurved. Anther 2.3 mm high.

Zimbabwe. C: Marondera (Marandellas), fl. & fr. 4.x.1964, *Plowes* 2500 (K; SRGH); Harare, *Rand*
596 (BM). E: Nyanga (Inyanga), c. 1650 m, fl. & fr. 3.xi.1930, *Fries, Norlindh & Weimarck* 2578 (K).
   Also in Kenya, Tanzania and South Africa (Transvaal and Cape Province). Dry open
grassland and shady woodland; c. 1650 m.

11. **Holothrix tridactylites** Summerh. in Kew Bull. **16**: 253 (1962); in F.T.E.A., Orchidaceae: 9
(1968). —Williamson, Orch. S. Centr. Africa: 23 (1979). —la Croix et al., Orch. Malawi: 42
(1991). TAB. **13**. Type from Tanzania.

Terrestrial herb 5–11 cm tall. Tubers 10–20 × 5–10 mm, ovoid. Leaf 1, appressed
to ground, 7–15 × 7–15 mm, orbicular to reniform, glabrous, more or less withered
at flowering time. Scape with retrorse hairs at the base, glabrous towards the apex
and with 2–3 bract-like leaves. Inflorescence laxly 2–10-flowered, flowers ± secund;
ovary 5–7 mm long; bracts 2.5 mm long. Flowers white, sometimes tinged with lilac.
Sepals green, 2–2.5 × 1.25–2 mm, ovate, acute or apiculate, the lateral sepals oblique.
Petals 4–5 × 2–6 mm, 3-lobed; lobes 2–3.5 mm long, acute, very shortly papillose. Lip
4–6 mm long, cuneate or oblong, adnate to column at base, 3-lobed, the lobes similar
to the petal lobes but slightly shorter. Spur 3–4 mm long, slender, tapering, straight
or slightly incurved. Anther c. 2 mm tall.

Zambia. E: Nyika National Park, near Kasoma Forest, fl. 11.ix.1981, *Dowsett-Lemaire* 346 (K).
Malawi. N: Nyika National Park, Chelinda Hill, fl. 29.viii.1987, *la Croix* 1049 (K; MAL).
   Also in Tanzania. Dry montane grassland, usually recently burnt; 2050–2300 m.

12. **Holothrix villosa** Lindl. in Companion Bot. Mag. **2**: 207 (1837). —Goodier & Phipps in
Kirkia **1**: 52 (1961). —Grosvenor in Excelsa **6**: 83 (1976). —Stewart et al., Wild Orch.
South. Africa: 61 (1982). Type from South Africa (Cape Province).
   Holothrix villosa subsp. *chimanimaniensis* G. Will. in Kirkia **13**: 247 (1990). Type:
Zimbabwe, *Grosvenor* 190 (SRGH, holotype) synon. nov.

Terrestrial herb 18–30 cm tall. Tubers c. 17 × 10 mm, ovoid, villous. Leaves 2,
appressed to ground, 16–25 × 15–27 mm, heart-shaped, pilose. Scape without
sheaths, with spreading hairs. Inflorescence 4–7 cm long, fairly densely many-
flowered, flowers secund; ovary 3 mm long; bracts 2.5 mm long, sparsely hairy.
Flowers buff to straw-yellow. Sepals 2–2.5 × 1.4–1.7 mm, ovate, obtuse, glabrous, the
lateral sepals oblique. Petals 3.3–3.4 × 1–1.5 mm, lanceolate-ligulate, obtuse,
thickened towards the tip, entire. Lip 2.8–3.5 × 2.2–2.8 mm, 3-lobed; lobes 0.6–1.2
mm long, obtuse, somewhat papillose. Spur c. 2 mm long, the apex obtuse and
slightly upturned. Anther 1 mm high. Ovary hairy.

Zimbabwe. E: Chimanimani Mts., Mt. Peza, fl. 15.x.1950, *Wild* 3593 (K; SRGH); Chimanimani
Mts., fl. 26.ix.1956, *Ball* 583 (K; SRGH); Chimanimani Mts., near Higher Valley, fl. 25.ix.1966,
Grosvenor 190 (K; SRGH).
   Also in South Africa (Cape Province). This species would also be expected to occur in the
Mozambique Chimanimani. Grassy slopes; 1800–2270 m.

11. STENOGLOTTIS Lindl.

Stenoglottis Lindl. in Companion Bot. Mag. **2**: 209 (1837). —Stewart in Kew
Mag. **6**, 1: 9–22 (1989).

Terrestrial, lithophytic or epiphytic herb with swollen fleshy roots. Leaves several
to many, in a basal rosette. Scape erect with a few scattered sheaths. Inflorescence an
erect lax or dense raceme, flowers sometimes sub-secund; bracts small, shorter than
ovary. Flowers numerous, white, pink or lilac, usually with darker spots. Sepals briefly
united to base of column and lip, otherwise free. Petals erect, oblique. Lip spurred
or unspurred, longer than tepals, 3–5-lobed. Column very short and broad; anther-
loculi parallel, canals absent; pollinia 2, with short caudicles and round viscidia;
staminodes present, longer or shorter than the anther; stigmas ± club-shaped, erect;
rostellum very short. Capsules erect, ellipsoid or cylindrical.

A genus of 4 species in tropical and South Africa.

Lip with a small spur at the base · 1. *woodii*
Lip without a spur · 2. *zambesiaca*

1. **Stenoglottis woodii** Schltr. in Ann. Transvaal Mus. **10**, 4: 242 (1924). —Stewart et al., Wild
    Orch. South. Africa: 82 fig. 8,3 (1982). —Stewart in Kew Mag. **6**, 1: pl. 118 (1989). Type
    from South Africa.
      Cynorchis macloughlinii L. Bolus in Ann. Bol. Herb. **4**, 4: 139 (1928). Type from South Africa.

Terrestrial or lithophytic herb, 10–20 cm tall. Tubers several, cylindrical, up to 8
mm in diameter. Leaves 5–20 in a basal rosette, erect or spreading, 5–15 × 1–3 cm,
linear, lanceolate-elliptic or ovate-lanceolate, glabrous, often somewhat glaucous
green, apex acute, margins not undulate. Scape with several sheaths, ovate, long-
acuminate, decreasing in size up the scape. Inflorescence to 20 cm high, 5–40-
flowered. Flowers white, pink or rosy-crimson, the lip usually spotted. Sepals 4–6 mm
long, ovate, obtuse or rounded, the lateral sepals oblique. Petals 3–5 mm long, ovate,
enfolding the column, the margins entire. Lip 10–14 mm long, obovate-cuneate in
outline, 3-lobed in apical half; side lobes rounded or truncate, wider and usually
slightly longer than mid-lobe; spur 1.5–3 mm long, slender, acute, straight or slightly
curved. Column less than 2 mm long; staminodes erect, narrow, club-shaped,
obscurely lobed at apex; stigmatic arms small, curved upwards then downwards in
front of the staminodes. Capsule 1–1.5 cm long, ribbed.

Zimbabwe. E: Vumba Mts., Mutare (Umtali), fl. i.1932, *Woodburn* 30b (K).
Also in South Africa (Natal) and Transkei. Rocky situations; 0–1500 m.

2. **Stenoglottis zambesiaca** Rolfe in F.T.A. **7**: 190 (1897). —Stewart in Kew Mag. **6**, 1: 17 (1989).
    TAB. **14**. Type: Malawi, Shire Highlands, *Buchanan* 385 (K, holotype).
      Stenoglottis fimbriata sensu Summerhayes in F.T.E.A., Orchidaceae: 29 (1968). —Goodier
    & Phipps in Kirkia **1**: 52 (1961). —Morris, Epiphyt. Orch. Malawi: 21 (1970). —Moriarty,
    Wild Fls. Malawi: t. 27, 5 (1975). —Grosvenor in Excelsa **6**: 86 (1976). —la Croix et al.,
    Orch. Malawi: 107 (1991), non Lindley (1837).

An erect epiphytic or lithophytic herb 10–35 cm tall; tubers to 4 × 1 cm, ellipsoid,
woolly. Leaves 6–12 in a basal rosette, spreading or arched, up to 12 × 2 cm, lanceolate
or oblanceolate, acute, often spotted with dark brown, the margins undulate. Scape
with several lightly spotted sheaths, lanceolate, acuminate, decreasing in size up the
scape. Inflorescence erect, rather laxly 5–many-flowered, flowers subsecund; pedicel
and ovary 9–10 mm long, the bracts shorter. Flowers pink or lilac, the lip with darker
spots. Dorsal sepal 3–5.5 × c. 2 mm, elliptic; lateral sepals slightly longer, obliquely
ovate. Petals 5–6 × 3 mm, enfolding the column; sepals and petals with papillose
margins. Lip 5–12 × 3–6 mm, narrowly wedge-shaped in outline, 3-lobed in apical
third; mid-lobe acute, narrower and usually longer than the truncate side lobes.
Column 1–2 mm tall; staminodes fist-shaped, the apex ± tuberculate; stigmatic arms
slender, erect or slightly curved. Capsule ribbed, 12–18 mm long.

Tab. 14. STENOGLOTTIS ZAMBESIACA. 1, habit (×1), from *Morris* 113; 2, flower, front view (×4); 3, flower, side view (×3); 4, dorsal sepal (×4); 5, lateral sepal (×4); 6, petal (×4); 7, lip (×4); 8, column, side view (×8); 9, column, front view (×8); 10, pollinarium (×8), 2–10 from *J. Stewart*, K spirit coll. 53491. Drawn by Judi Stone.

Zimbabwe. E: Chimanimani Mts., behind Mountain Hut, 1800 m, fl. 6.iv.1969, *Simon & Kelly* 1852 (K; SRGH). S: Bikita Distr., S face of Mt. Horzi, 1180 m, fl. 11.v.1969, *Biegel* (K; SRGH). **Malawi**. N: S Viphya, Kawandama Forest, 1750 m, fl. 15.v.1983, *Dowsett-Lemaire* 720 (K). C: Dedza Mt., near radio relay station, 2000 m, fl. 3.iii.1977, *Grosvenor & Renz* 1005 (K; SRGH). S: Chiradzulu Mt., 1450 m, fl. 6.iv.1985, *la Croix* 690 (K). **Mozambique**. M: Namaacha, M'Ponduine Mt. (Monte Ponduine), fl. 12.iii.1967, *Carvalho* 889 (K).

Also in Tanzania and South Africa (Transvaal). Submontane evergreen forest, occasionally woodland, on rotting logs or mossy rocks, or epiphytic on tree trunks; 1300–2150 m.

12. BONATEA Willd.

Bonatea Willd., Sp. Pl. **4**: 43 (1805).

Terrestrial herb with elongated fleshy and tuberous roots. Stem unbranched, leafy, leaves sometimes withered by flowering time. Inflorescence terminal, one- to many-flowered. Flowers resupinate, green or yellow and white. Dorsal sepal free, usually forming hood with upper petal lobes; lateral sepals united for some distance to base of lip, the lower petal lobes and the stigmatic arms. Petals 2-lobed, the upper lobe generally adnate to dorsal sepal, lower lobe adnate to stigmatic arms and lip. Free part of lip 3-lobed, usually with a tooth in the mouth of the spur; spur cylindrical, long or short. Anther erect, loculi adjacent and parallel; canals usually ± elongated, adnate to side lobes of rostellum; auricles entire, rugulose; pollinaria 2, each with a sectile pollinium, a long slender caudicle and a small naked viscidium. Stigmatic processes elongated, the lower part adnate to lip, free part club-shaped. Rostellum standing out in front of anther, convex, 3-lobed, with relatively short mid-lobe and often long, slender side lobes.

A genus of about 20 species in mainland Africa, with 1 species in Yemen.

1. Spur over 10 cm long · 1. *steudneri*
   – Spur less than 8 cm long · 2
2. Spur 1–2.5 cm long; lower petal lobes and side lobes of lip oblanceolate, acute · 4. *cassidea*
   – Spur 3–7 cm long; lower petal lobes and side lobes of lip linear · · · · · · · · · · · · · · · · · 3
3. Leaves ± withered at flowering time · 4
   – Leaves not withered at flowering time · 5
4. Sepals to 12 mm long · 5. *porrecta*
   – Sepals over 15 mm long · 3. *antennifera*
5. Leaves in a basal rosette; spur 5–7 cm long · 6. *pulchella*
   – Leaves borne along the stem, not in a basal rosette; spur 3–5 cm long · · · · · · · · 2. *speciosa*

1. **Bonatea steudneri** (Rchb.f.) T. Durand & Schinz, Consp. Fl. Afric. **5**: 90 (1895). —Rolfe in F.T.A. **7**: 253 (1898). —Summerhayes in Kew Bull. **17**: 531 (1964); in F.T.E.A., Orchidaceae: 137 (1968). —Grosvenor in Excelsa **6**: 78 (1976). —Williamson, Orch. S. Centr. Africa: 68 (1977). —Geerinck, Fl. Afr. Centr., Orchidaceae pt. 1: 147 (1984). TAB. **15**. Type from Eritrea.
   Habenaria steudneri Rchb.f., Otia Bot. Hamburg.: 101 (1881). Type as above.
   Habenaria kayseri Kraenzl. in Bot. Jahrb. Syst. **19**: 246 (1894). Type from Tanzania.
   Habenaria ecaudata Kraenzl. in Pflanzenw. Ost-Afr. **C**: 152 (1895). Type from Tanzania.
   Bonatea kayseri (Kraenzl.) Rolfe in F.T.A. **7**: 255 (1898).
   Bonatea ugandae Summerh. in Bull. Misc. Inform., Kew **1931**: 383 (1931). Type from Uganda.

Terrestrial herb up to c. 110 cm tall; roots villous, thick, fleshy and tuberous. Stem erect, leafy; leaves 7–12.5 × 3–4.5 cm, ovate, acute or apiculate, not overlapping, the base clasping the stem. Inflorescence fairly laxly 3–30-flowered; ovary and pedicel 5–6 cm long; bracts ± scarious, c. 4 × 1 cm. Flowers spreading, mainly green, white in centre. Dorsal sepal erect, 2.2–2.5 × 1.4–1.6 cm, ovate, acute, very convex; lateral sepals deflexed, 2.3–2.6 × 1.1–1.3 cm, broadly lanceolate, apiculate. Petals 2-lobed almost to base; posterior lobe erect, adnate to dorsal sepal, 21–24 × 1–2 mm; anterior lobe 50–70 × c. 1 mm, adnate to claw of lip for 6–20 mm, then curving down and outwards; both lobes linear, the anterior becoming ± filiform. Lip with basal claw 15–30 mm long, then 3-lobed; mid-lobe 20–37 × 1–3 mm, linear, tapering, decurved; side lobes 25–85 × 1–2 mm, spreading, linear, tapering, longer and narrower than

Tab. 15. BONATEA STEUDNERI. 1, upper stem and inflorescence (×⅔); 2, dorsal sepal (×1); 3, lateral sepal (×1); 4, petal (×1); 5, column, with part of lip, anterior petal-lobe, spur and ovary (×1), 1–5 from *Drummond & Hemsley* 2523. Drawn by Heather Wood. From F.T.E.A.

mid-lobe. Spur 10–21 cm long, pendent, slender but slightly swollen at apex. Anther 10–18 mm high, loculi parallel; canals slender, 18–25 mm long. Stigmatic arms projecting forwards, 22–32 mm long, the basal third to half adnate to the lip, the free part club-shaped. Rostellum placed in front of column, mid-lobe 5–8 mm long, side lobes slender, 10–20 mm long.

Zambia. S: 8 km S of Kalomo, fl. iii.1969, *G. Williamson* 420 (K; SRGH); Machili, fl. 11.v.1962, *Fanshawe* 6831 (K; NDO). **Zimbabwe**. E: Mutare, Vumba Mts., 900 m, fl. 22.v.1956, *Ball* 443 (K; SRGH).
Also in Ethiopia, Somalia, S Arabia, Yemen, Zaire, Rwanda, Uganda, Kenya and Tanzania. Dry woodland, scrub; 900–1100 m.

2. **Bonatea speciosa** (L.f.) Willd., Sp. Pl. ed. 4, **1**: 43 (1805). —Rolfe in F.C. **5**, 3: 141 (1912). —Grosvenor in Excelsa **6**: 78 (1976). —Stewart et al., Wild Orch. South. Africa: 99 (1982). Type from South Africa.
  Orchis speciosa L.f., Suppl. Pl.: 401 (1781).
  Bonatea boltonii Harv., Thes. Cap. **1**: 55, t. 88 (1859) as "*boltoni*". Type from South Africa.
  Habenaria bonatea Rchb.f., Otia Bot. Hamburg.: 101 (1881). Type as for *Orchis speciosa*.

Robust terrestrial herb 40–100 cm tall; tubers cylindrical, villous. Stem erect, leafy; leaves 7–10 × 2–3 cm, lanceolate to ovate, acute, sheathing at the base. Inflorescence ± densely to c. 15-flowered; ovary and pedicel 40–45 mm long; bracts leafy, 30–45 × 10–12 mm, ovate, acuminate. Flowers green and white, with a slightly spicy scent. Dorsal sepal 18–20 × 12 mm, erect, ovate, acute, very convex. Lateral sepals 19–22 mm long, obliquely ovate, apiculate, slightly rolled lengthwise. Petals 2-lobed, upper lobe 15–16 × 2–3 mm, linear, ± adnate to dorsal sepal; lower lobe 20–25 × 1 mm long, linear or oblanceolate, acuminate. Lip 26–32 mm long, joined to the stigma for 7–8 mm at the base, 3-lobed from a basal claw 9–10 mm long. Mid-lobe 20–23 × 1.5–2 mm long, linear, bent back about the middle; side lobes 20–23 × 1.5–2 mm, linear, falcate, acuminate. Spur 3.5–4 cm long, slender, swollen at apex. Column 9 mm long. Stigmatic arms up to 16 mm long; anther canals 7–8 mm long.

Zimbabwe. C: Marondera (Marandellas), 1500 m, fl. 9.iv.1952, *Greatrex* in *GHS* 36091 (K; SRGH). E: Chimanimani, 8 km from Chimanimani (Melsetter) on Orals Kranz Road, *Ball* 984 (K; SRGH).
Also in South Africa. Open deciduous woodland; c. 1500 m.

3. **Bonatea antennifera** Rolfe in Gard. Chron. ser. 3, **38**: 450 (1905). —Grosvenor in Excelsa **6**: 78 (1976). Type from South Africa.
  Bonatea speciosa var. *antennifera* (Rolfe) Sommerville in Contrib. Bolus Herb. No. 10: 157 (1982). —Stewart et al., Wild Orch. South Africa: 99 (1982).

Robust terrestrial herb 45–70 cm tall; tubers cylindrical, villous. Stem leafy, the lowest 2 ± sheathing, the remainder 7.5–9.5 × 2–3 cm, ovate, acute, usually ± withered by flowering time. Inflorescence densely many-flowered. Pedicel and ovary 3.5–5.5 cm long; bracts 3–3.5 × 1 cm, ovate, acuminate. Flowers green and white. Dorsal sepal 16–18 × 12–13 mm, ovate, acute, very convex; lateral sepals 18–24 mm long, obliquely ovate, rolled lengthways. Petals 2-lobed, both lobes narrowly linear; upper lobe 16–18 mm long, lower lobe 20–25 mm long. Lip 38–45 mm long over all, joined to the stigmatic arms for c. 8 mm at base. Mid-lobe 24–35 mm long; side lobes 18–30 mm long, all lobes filiform. Spur 3–5 cm long, slender, swollen at apex. Stigmatic arms 15–20 mm long; anther canals 8 mm long.

Botswana. SE: 6 km W of Kanye, with much *Euclea undulata*, 1300 m, fl. & fr. 15.iv.1978, *Hansen* 3406 (BM; C; GAB; K; PRE; SRGH). **Zimbabwe**. W: Bulawayo, fl. 1930, *Cheeseman* 205 (K). **Mozambique**. M: Matutuíne (Bela Vista), arredores do Rio Futi, fl. 28.v.1963, *Carvalho* 617 (K; LISC).
Also in South Africa (N Cape and Transvaal) and Lesotho. Woodland, often in drier areas.
This species is similar to *B. speciosa* and may be a variety of it. However, it can be distinguished by the leaves being withered at flowering time, by the more numerous flowers, the narrower petal lobes and longer lip lobes and so here we are keeping it separate. It also tends to grow in drier areas than *B. speciosa* and at higher altitudes.

4. **Bonatea cassidea** Sond. in Linnaea **19**: 81 (1847). —Rolfe in F.C. **5**, 3: 138 (1913). —Grosvenor in Excelsa **6**: 78 (1976). —Stewart et al., Wild Orch. South. Africa: 101 (1982). Type from South Africa.

Slender plant 25–50 cm tall. Stem erect, leafy. Leaves 10–12, sometimes starting to wither at flowering time, particularly towards the base of the plant, 10–40 × 1–3 cm, linear-lanceolate, acute, slightly falcate. Inflorescence 5–20 cm long, several- to many-flowered; ovary and pedicel 1.8–3.2 cm long, semi-erect; bracts 2–3 cm long, ovate, acuminate. Flowers green, the lower petal lobes and lip lateral lobes white. Sepals 8.5–11 × 3.5–8 mm; dorsal sepal erect, ovate, acute, very convex; lateral sepals obliquely ovate, acute, ± rolled lengthwise. Petals 2-lobed; upper (posterior) lobe 10–20 mm long, linear, adnate to dorsal sepal; lower (anterior) lobe 9.5–15 × 3.5–6.5 mm, obovate or semi-orbicular from a narrow base, spreading. Lip 1.6–2 cm long, 3-lobed, with claw c. 8 mm long. Mid-lobe 7–10 × 0.5–0.8 mm, linear; side lobes similar to lower petal lobes. Column 4–6 mm long; stigmatic arms 7–9 mm long; anther canals 4–5 mm long.

Zimbabwe. E: Mutare (Umtali), Murahwa's (Murahivas) Hill below Christmas Pass, 1200 m, fl. 17.viii.1966, *Chase* 8429 (K; SRGH); Chimanimani, Lusitu Reserve, fl. 17.ix.1958, *Ball* 749 (K; SRGH).
Also in South Africa. In shade, amongst boulders; 1200–1500 m.

5. **Bonatea porrecta** (Bolus) Summerh. in Kew Bull. **4**: 430 (1949). —Stewart et al., Wild Orch. South. Africa: 100 (1982). Type from South Africa.
    Habenaria porrecta Bolus in J. Linn. Soc., Bot. **25**: 167 fig. 5 (1889). Type as above.

A fairly robust plant 20–60 cm tall; tubers cylindrical, woolly. Stem erect, leafy; leaves up to c. 7 cm long, ± withered at flowering time. Inflorescence several- to many-flowered; ovary and pedicel 2.5–3 cm long; bracts up to 3 cm long, ovate, acuminate. Flowers green and white. Sepals 12 × 8–10 mm, the dorsal sepal erect, ovate, acute, very convex; lateral sepals obliquely semi-orbicular, apiculate. Petals 2-lobed; upper lobe 1–1.2 cm long, adnate to dorsal sepal; lateral sepals up to 2.2 cm long; both lobes linear. Lip 25 mm long, 3-lobed from a claw 9 mm long; mid-lobe up to 20 mm long, linear, bent back about the middle; side lobes up to 22 mm long, narrowly linear, slightly curled towards the apex. Spur 3–3.5 cm long, slightly swollen towards the apex. Stigmatic arms 10 mm long; anther canals 6 mm long.

Mozambique. M: Maputo Province, Namaacha, Monte Ponduine, 780 m, fl. 25.xii.1980, *Pettersson* 137 (K; LMU).
Also in South Africa. Open grassy places.

6. **Bonatea pulchella** Summerh. in Kew Bull. **17**: 529 (1964). —Stewart et al., Wild Orch. South. Africa: 102 (1982). TAB. **16**. Type: Mozambique, Inhaca Island, Ponta Rasa, *Barbosa* 7646 (K, holotype).

A slender plant 20–35 cm tall; tubers fleshy, cylindrical, villous. Leaves in basal rosette, 5–6 cm long, lanceolate to ovate, acute, with petiole-like base up to 2.5 cm long. Stem with about 3 sheathing leaves 20–25 mm long. Inflorescence 10–12 cm long, laxly 3–5-flowered; ovary and pedicel 3.5–4 cm long; bracts c. 1.5 cm long. Flowers green and white, mostly white. Sepals 11–13 × 6–8 mm; dorsal sepal erect, ovate, acute; lateral sepals adnate to lip for 4–5 mm, obliquely ovate, with an apiculus c. 3 mm long. Petals 2-lobed, lobes filiform; upper lobe 10.5–12.5 mm long, adnate to dorsal sepal; lower lobe up to 47 mm long, adnate to lip for 2.5–4 mm. Lip 3-lobed with all lobes filiform; claw 4.5–6 mm long; mid-lobe 24–31 mm long, side lobes 45–55 mm long. Spur 5–7 cm long, slender, slightly swollen towards apex. Stigmatic arms 8–9.5 mm long, free part 6–6.5 mm long. Anther 5–5.5 mm long, canals 6–8 mm long.

Mozambique. M: Goba, foot of Lebombo Mts., fl. 23.v.1957, *Carvalho* 258 (K); Inhaca Island, Ponta Rasa, fl. 10.vii.1957, *Barbosa* 7646 (K, holotype).
Also in South Africa. In coarse damp, basic sandy soil amongst rocks; 0–c. 600 m.

Tab. 16. BONATEA PULCHELLA. 1, habit (×1), from *Carvalho* 258; 2, dorsal sepal (×2); 3, lateral sepal (×2); 4, petal (×2); 5, column, with base of anterior petal-lobe and of lip, lateral view (×2), 2–5 from *Barbosa* 7646. Drawn by Mary Grierson. From Kew Bull.

163. ORCHIDACEAE

13. CENTROSTIGMA Schltr.

Centrostigma Schltr. in Bot. Jahrb. Syst. **53**: 522 (1915).

Terrestrial herb with tuberous roots. Stems unbranched, leafy, with the inflorescence terminal. Flowers resupinate, white, yellowish or green. Sepals and petals free. Lip joined to column at the base, 3-lobed; side lobes entire or with pectinate teeth; spur cylindrical. Column erect, anther loculi adjacent, parallel, canals slender and upcurved; auricles small, entire; pollinaria 2, each with a sectile pollinium, caudicle and small naked viscidium. Stigmatic processes 2-lobed above the base, the lower lobe receptive, porrect, clavate or capitate; upper lobe sterile, projecting upwards in front of the anther, often slender and horn-like. Rostellum 3-lobed, the mid-lobe narrowly triangular, erect between the anther loculi but shorter than the anther, the side lobes porrect or upcurved.

A genus of 3 species in tropical Africa.

1. Spur 15 cm long; side lobes of lip reflexed · 2. *occultans*
 – Spur less than 10 cm long; side lobes of lip projecting forwards · · · · · · · · · · · · · · · · 2
2. Spur 5.5–7.5 cm long; side lobes of lip over 25 mm long · · · · · · · · · · · · · · · 1. *clavatum*
 – Spur 2–2.5 cm long; side lobes of lip less than 15 mm long · · · · · · · · · · · · 3. *papillosum*

1. **Centrostigma clavatum** Summerh. in Kew Bull. **11**: 221 (1956). —Williamson, Orch. S. Centr. Africa: 78 (1977). —Geerinck in Fl. Afr. Centr., Orchidaceae pt. 1: 152 (1984). —la Croix et al., Orch. Malawi: 43 (1991). TAB. **17**. Type: Zambia, Mbala Distr., between Mbala and Kambole, fl. 31.xii.1949, *Bullock* 2160 (K, holotype).

A robust terrestrial herb, drying dark brown; tubers ellipsoid. Stem leafy; leaves 6–7, the lowest 2–3 sheathing, the largest in the middle, semi-erect, up to c. 14 × 2 cm, lanceolate, acute, ribbed. Inflorescence 5–32 cm long, ± lax, up to 18-flowered; ovary c. 35 mm long, pedicel 25 mm long; bracts leaf-like, up to 9.5 cm long. Flowers white, greenish-cream or greenish-yellow. Dorsal sepal 16.5–20 × 11–13 mm, convex, ovate, erect; lateral sepals 20–22.5 × 5–6.5 mm, spreading; apical parts of sepals puberulous. Petals 17–20 × 5–6 mm, ciliolate in upper half, lying ± inside dorsal sepal. Lip 3-lobed, claw 4–4.5 mm long, attached to base of column and lateral sepals; mid-lobe 20–30 × 1.5–3 mm, linear, deflexed; lateral lobes swept forwards, 25–35 × 2–4 mm, with pectinations c. 5 mm long in apical half. Spur 5.5–7.5 cm long, swollen at apex, usually tucked into the bract. Anther 10–11 mm high; canals 18 mm long. Upper branch of stigma 12.5–14 mm long, papillose at apex, lower branch 8–9 mm long. Auricles 1.5–2 mm long.

Zambia. N: Mbala Distr., between Mbala and Kambole, swampy grassland near Lunzua R., fl. 31.xii.1949, *Bullock* 2160 (K). C: Serenje Distr., Kundalila Falls, fl. i.1969, *G. Williamson* 1606 (K). **Malawi**. N: S Viphya, Elephant Rock Dambo, 1550 m, fl. 7.i.1981, *la Croix* 85 (K).
Also in Zaire and Tanzania. Dambo, swampy grassland; 1200–1680 m.

2. **Centrostigma occultans** (Rchb.f.) Schltr. in Bot. Jahrb. Syst. **53**: 523 (1515) in obs. — Summerhayes in Kew Bull. **11**: 219 (1956); in F.T.E.A., Orchidaceae: 151 (1968). —Grosvenor in Excelsa **6**: 79 (1976). —Williamson, Orch. S. Centr. Africa: 77 (1977). —Stewart et al., Wild Orch. South. Africa: 103 (1982). —la Croix et al., Orch. Malawi: 44 (1991). Type from Angola.
 Habenaria occultans Rchb.f. in Flora **48**: 178 (1865). Type as above.
 Habenaria schlechteri Kraenzl. in Xen. Orch. **3**: 148, t. 286/5–9 (1896). Type from South Africa.
 Centrostigma schlechteri (Kraenzl.) Schltr. in Bot. Jahrb. Syst. **53**: 523 (1915), in obs. —Summerhayes in Bot. Not. **1937**: 187 (1937). Type from South Africa.
 Centrostigma nyassanum Schltr. in Bot. Jahrb. Syst. **53**: 523 (1915). Type from Tanzania.

Terrestrial herb 55–75 cm tall, drying dark brown; tubers c. 2–3 cm long, globose or ellipsoid. Stem leafy; the lowermost leaves sheathing, the remainder semi-erect, up to 15.5 × 2 cm, lanceolate, acute. Inflorescence up to 16 cm long, 1–10-flowered; ovary 3 cm long, pedicel 3 cm long, ± erect; bracts up to 6.5 cm long, lanceolate,

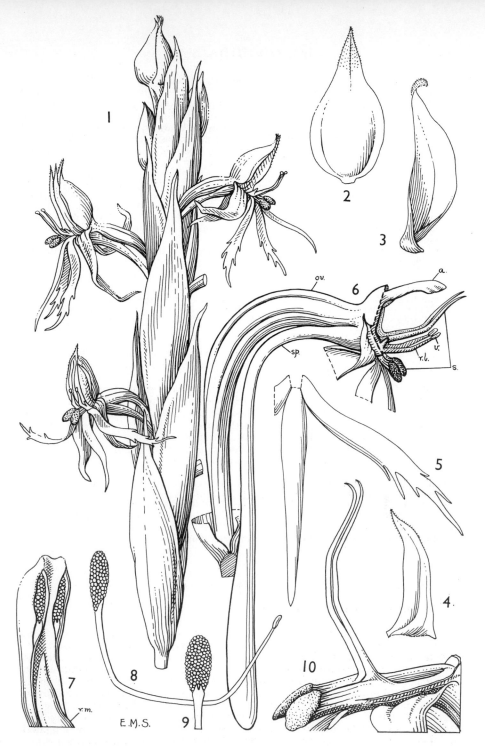

Tab. 17. CENTROSTIGMA CLAVATUM. 1, inflorescence (×1); 2, dorsal sepal (×2); 3, lateral sepal (×2); 4, petal (×2); 5, part of lip, spread out (×2); 6, flower, side view, with perianth removed to show ovary, column and spur (×2); 7, anther and rostellum mid-lobe (×4); 8, pollinarium (×4); 9, apex of pollinarium (×4); 10, stigmata, side view (×4), 1–10 from *Bullock* 2160. a, anther; ov, ovary; rl, rostellum lateral lobe; rm, rostellum mid-lobe; s, stigmata; sp, spur; v, viscidium. Drawn by Margaret Stones. From Hooker's Icon. Pl.

acuminate. Flowers pale green to yellow-green. Dorsal sepal erect, 13–20 × 10–15 mm, ovate, convex; lateral sepals deflexed, 14–25 × 3–6 mm, obliquely lanceolate, acuminate. Petals 12–18 × 3–3.5 mm, obliquely lanceolate, acute, lying inside the dorsal sepal. Lip 3-lobed, claw 1.5–3 mm long, attached to base of column; mid-lobe deflexed, 18–25 × 1–4 mm, linear; side lobes swept back, 18–25 × 2–4 mm, the last c.10 mm pectinate. Spur slender, slightly swollen at apex, 10–15 cm long. Anther 8–11 mm high, canals c. 10 mm long. Upper and lower lobes of stigma 8–9 mm long. Auricles 1–1.5 mm long.

Zambia. N: Kawambwa, Mbereshi Dambo, fl. 11.i.1961, *Holmes* 0287 (K; SRGH). W: Mwinilunga Distr., 16 km on Matonchi Road, fl. 20.xi.1958, *Holmes* 095 (K). C: Serenje Distr., Kundalila Falls, fl. i.1972, *G. Williamson* 2148 (K; SRGH). **Zimbabwe**. C: Harare, Borrowdale, c. 1500 m, fl. 5.xii.1945, *Wild* 560 (K; SRGH). E: Nyanga (Inyanga), c. 1700 m, fl. 21.i.1931, *Norlindh & Weimarck* 4497 (K). **Malawi**. N: S Viphya, Elephant Rock Dambo, 1550 m, fl. 8.i.1981, *la Croix* 86 (K).
Also in Tanzania, Angola and South Africa (Transvaal). Dambo and marshy localities; 1250–1700 m.

3. **Centrostigma papillosum** Summerh. in Kew Bull. **11**: 219 (1956). —Grosvenor in Excelsa **6**: 79 (1976). —Williamson, Orch. S. Centr. Africa: 76 (1977). —la Croix et al., Orch. Malawi: 44 (1991). Type: Zimbabwe, Marondera Distr., fl. 22.xii.1959, *Greatrex* in *GHS* 26451 (K, holotype; SRGH).

A ± slender herb 40–65 cm tall, drying dark brown or almost black; tubers ellipsoid. Stem leafy; leaves 6–8, the two lowest sheathing, the rest semi-erect, up to c. 9 × 1.5 cm, lanceolate, acute. Inflorescence 11–20 × 3.5–4 cm, laxly 2–12-flowered; ovary and pedicel c. 35 mm long; bracts of similar length. Flowers green. Dorsal sepal erect, 10.5–12 × 4–8 mm, ovate, convex; lateral sepals sharply reflexed and curled, 10–11 × 2.5–4 mm, obliquely lanceolate. Petals 7–12 × 1–3 mm, lanceolate, obtuse, usually with a small lateral lobe near the base, lying inside dorsal sepal. Lip 3-lobed; claw 2–4 mm long; mid-lobe 10–13 × 1–2.5 mm, linear, reflexed; side lobes 10–13 × 2–3 mm, projecting forwards, the apex truncate and dentate. Spur 20–25 mm long, tucked into bract. Anther 2.8–3 mm high; canals 3–4 mm long. Stigmatic arms 3–5 mm long.

Zambia. N: Chinsali Distr., Lake Young, Ishiba Ngandu (Shiwa Ngandu), 1350 m, fl. 15.i.1959, *Richards* 10662 (K). W: Mwinilunga, fl. 3.xii.1958, *Holmes* 0114 (K; SRGH). C: Serenje Distr., Kundalila Falls, fl. xii.1967, *G. Williamson* 712 (K). **Zimbabwe**. C: Marondera Distr., Digglefold, 1500 m, fl. 22.xii.1949, *Greatrex* in *GHS* 26451 (K; SRGH). **Malawi**. N: S Viphya, Lichelemu (Luchilemu) Dambo, c. 1300 m, fl. 4.i.1983, *la Croix* 394 (K).
Also in Angola. Dambo and marshy areas; 1300–1500 m.

14. CYNORKIS Thouars

Cynorkis Thouars in Nouv. Bull. Sci. Soc. Philom. Paris **1**: 317 (1809).

Terrestrial, or occasionally epiphytic, herb with fleshy or tuberous roots. Stem, ovary and calyx often with glandular hairs. Leaves 1 to several, radical, with a few sheath-like and cauline. Flowers few to many in a terminal raceme, usually resupinate, pink, mauve or purple, occasionally white. Sepals free or slightly adnate to the lip, the dorsal often forming a hood with the petals, the lateral sepals spreading. Lip entire or 3–5-lobed, spurred at the base. Column short and broad; androclinium erect or sloping; anther loculi parallel, the canals short or long and slender; viscidia 2, rarely 1, auricles distinct; stigmatic processes oblong, papillose, usually joined to the lobes of the rostellum. Side lobes of rostellum usually elongated; mid-lobe often large, projecting forwards. Capsules oblong or fusiform, often ripe at the base of an inflorescence which is still flowering at the apex.

A genus of about 125 species, mostly native to Madagascar and the Mascarene Islands, with up to 20 species on mainland Africa.

1. Spur absent · 1. *anacamptoides* var. *ecalcarata*
 - Spur present · 2
2. Spur over 15 mm long; lip 4-lobed · 4. *kirkii*
 - Spur less than 15 mm long; lip entire, 3-lobed or 5-lobed · · · · · · · · · · · · · · · · · 3
3. Spur less than 2 mm long · 8. *brevicalcar*
 - Spur over 2 mm long · 4
4. Lip entire, or obscurely 3-lobed at apex · 5
 - Lip distinctly 3-lobed or 5-lobed · 6
5. Lip linear to ligulate; ovary densely glandular-hairy · · · 1. *anacamptoides* var. *anacamptoides*
 - Lip cuneate, usually obscurely 3-lobed at apex; ovary sparsely glandular- hairy · · · · · · · ·
 · 2. *buchananii*
6. Lip 3-lobed · 7
 - Lip 5-lobed · 8
7. Side lobes of lip smaller than mid-lobe · 6. *kassneriana*
 - Side lobes of lip much larger than mid-lobe · 7. *symoensii*
8. Side lobes of lip arising in basal half and almost as wide as mid-lobe · · · · 5. *hanningtonii*
 - Side lobes of lip arising about halfway and much narrower than mid-lobe · · · 3. *anisoloba*

1. **Cynorkis anacamptoides** Kraenzl. in Pflanzenw. Ost-Afr. **C**: 151 (July 1895), as "*Cynosorchis*". —Summerhayes in Kew Bull. **12**: 107 (1957); in F.T.E.A., Orchidaceae: 32 (1968). —Morris, Epiphyt. Orch. Malawi: 23 (1970). —Grosvenor in Excelsa **6**: 79 (1976). —Williamson, Orch. S. Centr. Africa: 35 (1977). —Geerinck in Fl. Afr. Centr., Orchidaceae pt. 1: 170 (1984). —la Croix et al., Orch. Malawi: 45 (1991). Type from Zaire.

 Stenoglottis calcarata Rchb.f. in Flora **48**: 180 (1865). Type from Angola. Non *Cynosorchis calcarata* (Thouars) Dur. & Schinz, Consp. Fl. Afr. **5**: 90 (1892).
 Barlaea calcarata (Rchb.f.) Rchb.f. in Linnaea **41**: 54 (1877).
 Habenaria calcarata (Rchb.f.) Benth. in J. Linn. Soc., Bot. **18**: 355 (1881).
 Cynorkis platylepoides Kraenzl. in Bot. Jahrb. Syst. **28**: 367 (1900), sphalm. "*Cynosorchis platyclinoides*" (see Summerhayes in Kew Bull. **12**: 107 (1957)). Type from Tanzania.
 Cynorkis micrantha sensu Schlechter in Bot. Jahrb. Syst. **53**: 488 (Oct. 1915), non (Frapp. ex Cardem.) Schltr. (Aug. 1915).
 Cynorkis gymnadenoides Schltr. in Bot. Jahrb. Syst. **53**: 489 (1915). —Summerhayes in Bot. Not. **1937**: 183 (1937) as "*Cynorchis*". Type from Tanzania.
 Cynorkis barlaea Schltr. Bot. Jahrb. Syst. **53**: 489 (1915), in obs.

Slender terrestrial herb, 12–60 cm tall. Roots fleshy, arising at base of stem. Basal leaves 2–several, 1.8–11 × 0.7–1.8 cm, ovate, lanceolate or elliptic; peduncle leaves several, bract-like. Peduncle glandular-hairy, purplish. Inflorescence densely (2)5–many-flowered, pyramidal at first, later elongated and cylindrical, sometimes up to c. 14 × 1.5 cm; bracts 6–8 mm long, lanceolate, acute; ovary and pedicel 8–10 mm long, semi-erect, arched towards apex; densely glandular-hairy and purple. Flowers small, mauve to deep purple, occasionally pink or white. Dorsal sepal 2–4 × 1–2.7 mm, ovate, very convex; lateral sepals spreading 3.5–4.6(7) × 1–2.5 mm, obliquely oblong, obtuse. Petals 3–4 × 1–1.7 mm, falcate, forming a hood with the dorsal sepal. Lip 3.6–4.6(8) × 0.9–1.3(2.5) mm, linear-ligulate, entire. Spur 3.2–5 mm long, usually somewhat incurved, often swollen at apex, or spur absent. Column 1–1.2 mm long; anther 0.7–0.9 mm high; rostellum mid-lobe slightly overlapping anther.

Var. **anacamptoides**

Spur present.

Zambia. N: Chinsali Distr., Ishiba Ngandu (Shiwa Ngandu), fl. & fr. 19.i.1959, *Richards* 10757 (K). E: Nyika Plateau, near Rest House, fl. 2.i.1959, *Richards* 10406 (K). **Zimbabwe**. C: Makoni Distr., Rusape, fl. & fr. 9.xii.1964, *West* 6197 (K). E: Nyanga (Inyanga), fl. 27.xii.1951, *Whellan* 606 (K); Chimanimani (Melsetter), *Chase* 5121 (K). **Malawi**. N: S Viphya, 1st drift SW of Mzuzu, fl. 15.xi.1970, *Pawek* 3985 (K). S: Zomba, near Chingwe's Hole, fl. 26.x.1980, *la Croix* 75 (K); Mulanje Mt. Madzeka Plateau, fl. & fr. 3.iii.1984, *Johnston-Stewart* 276 (K). **Mozambique**. N: Lichinga, fl. 31.xii.1979, *Pettersson* 43 (K).
 Also in Nigeria, Cameroon, Bioko, Zaire, Rwanda, Ethiopia, Uganda, Kenya, Tanzania and Angola. Wet and marshy areas and at edge of dams, usually in colonies; 600–2300 m.
 C. anacamptoides is notable for its long flowering period of almost 6 months, from late

September to mid March. Certain specimens from Zomba Plateau and Mulanje Mt. in Malawi (*la Croix* 95 and *Johnston-Stewart* 276) and the Eastern Highlands of Zimbabwe (*Chase* 5121) have much larger flowers than usual, but do not seem to differ in any other way. These large-flowered specimens seem always to be pale pink or white.

Var. **ecalcarata** P.J. Cribb in Kew Bull. **32** (1): 137 (1977). —la Croix et al., Orch. Malawi: 45 (1991). Type: Malawi, Nyika Plateau, perennial dambo close to Zambian Border, *G. Williamson* 233 (K, holotype; SRGH).
    Cynorkis sp., Williamson, Orch. S. Cent. Afr.: 37 (1977).

Spur absent.

Malawi. N: Nyika Plateau, perennial dambo close to the Zambian Border, fl. *G. Williamson* 233 (K; ?SRGH); Nyika Plateau, 3 km N of Lake Kaulime, fl. 22.xii.1969, *Ball* 1213 (K; SRGH).
    Known from 2 collections on the Nyika Plateau. It resembles typical variety in every way except for being spurless. The flowers are pale pink.

2. **Cynorkis buchananii** Rolfe in F.T.A. **7**: 260 (1898). —la Cròix et al., Orch. Malawi: 46 (1991). Type: Malawi, top of Zomba Plateau, *Buchanan* 308 (K, holotype).
    Cynorkis nyassana Schltr. in Bot. Jahrb. Syst. **38**: 145 (1906). Type: Malawi, *Buchanan* (B†).
    Cynorkis buchwaldiana subsp. *nyassana* (Schltr.) Summerh. in Kew Bull. **16**: 258 (1962).
    —Morris, Epiphyt. Orch. Malawi: 25 (1970).

Terrestrial herb 10–44 cm tall. Tubers usually 2, c. 1.5 cm long, ovoid. Basal leaves 1 or 2, pale green, 2–16 × 1.5–3.5 cm, lanceolate to elliptic, narrowing to a petiole-like base; peduncle leaves several, reduced, the lowest usually similar to but smaller than the basal leaves, the remainder sheath-like, up to 23 mm long. Inflorescence 2.5–11 cm long, laxly to densely several- to many-flowered, flowers often ± secund; rhachis sparsely glandular-hairy; bracts 8–12 mm long, lanceolate, acute; ovary and pedicel 8–12 mm long, arching, sparsely glandular-hairy. Flowers pale pink or lilac, whitish at base of lip and inside hood. Dorsal sepal 3.5–5.5 × 2–3.2 mm, ovate, acute, convex, the apex slightly recurved. Lateral sepals deflexed, 6–8 × 2–3 mm, obliquely lanceolate, acute. Petals 3–5.5 × 1.5–2.5 mm, falcate, adnate to column and lip at base and forming a hood with the dorsal sepal. Lip 6–7.5 × 2–3.5 mm, wedge-shaped, entire but with slight indentations at the lip, and with 2 small lobules at the base. Spur 3–5 mm long, somewhat incurved. Column short and stout, the rostellum overtopping the anther, up to 1.5 mm tall.

Malawi. S: Zomba Mt., fl. 21.iii.1981, *la Croix* 124 (K); Mulanje Mt., Lichenya Plateau, fl. & fr. 19.ii.1983, *la Croix* 451 (K); Mulanje Mt., Chambe Plateau, fl. & fr. 23.iii.1958, *Jackson* 2191 (K).
    Only known from Malawi, on Zomba Plateau and Mulanje Mt. Pine plantations, seepage slopes and damp grassland; 1200–1900 m.

3. **Cynorkis anisoloba** Summerh. in Kew Bull. **12**: 108 (1957). —Goodier & Phipps in Kirkia **1**: 52 (1961). —Grosvenor in Excelsa **6**: 79 (1976). Type: Zimbabwe, Chimanimani Mts., 1650 m, *Ball* 244 (K, holotype; SRGH).

Terrestrial herb 11–35 cm tall. Tubers usually 2, c. 2 × 0.8 cm, ellipsoid, villous. Leaves 1–3, basal, 3–13 × 1.4–2.7 cm, ovate to lanceolate, acute or apiculate; peduncle hairy with 2–3 sheathing leaves. Inflorescence lax, up to c. 20-flowered; rhachis hairy; bracts c. 7 mm long, lanceolate, acute; ovary and pedicels 9–10 mm long, arched, glandular-hairy. Flowers pale pink to purple, the lip with red spots. Sepals with scattered glandular hairs; dorsal sepal 3–4.2 × 2.5–3.3 mm, ovate, convex; lateral sepals 5.8–8 × 2.5–4 mm, obliquely ovate, obtuse. Petals 3–4 × 1.75–2.5 mm, subfalcate-oblong, attached to the column at the base and forming a hood with the dorsal sepal. Lip 7–10 mm long, 5-lobed, the posterior lateral lobes very small, 0.6–0.9 mm long and less than 1 mm wide; anterior lateral lobes arising about half way along, 2–2.8 × 0.7–1.1 mm. Mid-lobe 2.5–5 × 1–1.8 mm. Spur 5–12 mm long, straight or C-shaped, tapering. Anther 1–1.4 mm high.

Zimbabwe. E: Chimanimani Mts., fl. 21.ii.1954, *Ball* 244 (K; SRGH); Nyanga (Inyanga), fl. 5.iii.1969, *Jacobsen* 3753 (K; SRGH).
    Not known elsewhere. In damp or wet ground, in rock crevasses, or amongst short grasses,

sometimes in deep shade; 1350–1700 m.
This species, apparently endemic to the Eastern Highlands of Zimbabwe, is closely related to the widespread *C. hanningtonii.*

4. **Cynorkis kirkii** Rolfe in F.T.A. **7**: 261 (1898). —Goodier & Phipps in Kirkia **1**: 52 (1961). —Summerhayes in F.T.E.A., Orchidaceae: 37 (1968). —Morris, Epiphyt. Orch. Malawi: 26 (1970). —Moriarty, Wild Fls. Malawi: t. 27, 1 (1975). —Grosvenor in Excelsa **6**: 79 (1976). —la Croix et al., Orch. Malawi: 48 (1991). Type: Mozambique, Morrumbala Mt., *Kirk* s.n. (K, holotype).
   Cynorkis oblonga Schltr. in Bot. Jahrb. Syst. **26**: 333 (1899). Type: Mozambique, near Beira, iv.1898, *Schlechter* s.n. (B†).

Terrestrial herb 9–46 cm tall, glabrous except for the tuber and roots. Tubers usually 2, 3–3.5 cm long, ovoid, villous. Basal leaf usually 1, occasionally 2, pale green, 4–21 × 1.8–5 cm, ovate to elliptic, acute or apiculate, the base clasping the stem. Peduncle usually with 1 sheathing leaf. Inflorescence 1–10-flowered, the flowers set close together, often held with lip pointing upwards; bracts 9–10 mm long, ovate, acute; ovary and pedicel erect, 3–4 × 3 mm, narrowing to a "neck" 3 mm long below the flowers. Flowers lilac, yellowish or white in the centre, the lip paler than the sepals and petals, but darker at the tips of the lobes. Dorsal sepal 5–6 × 3–5 mm, ovate, lobes convex; lateral sepals 6–7 × 2.5–3 mm, obliquely ovate. Petals 5–5.5 × 2–2.5 mm, falcate, forming hood with dorsal sepal. Lip 8–20 mm long, 4-lobed, the lobes oblong or spathulate, rounded or truncate at the ends. Side lobes 4–7 × 2–4 mm; mid-lobe 10 mm long, bifid, the lobes resembling the side lobes, but slightly narrower. Spur slender, more or less straight, 15–30 mm long. Column stout, 2.5 mm long.

Zimbabwe. E: Chimanimani, near Rusitu (Lusito) Mission, fl. 15.xi.1955, *Ball* 462 (K; SRGH). **Malawi**. N: S Viphya, near Nkhalapya, fl. 17.ii.1987, *la Croix* 981 (K; MAL). C: Dedza Distr., main road from Dedza to Ntcheu, fl. 7.iii.1971, *Westwood* 535 (K). S: Zomba Mt., fl. 25.i.1959, *Robson* 1323 (K); Mulanje Mt., Lichenya Path, fl. & fr. 18.ii.1982, *Brummitt & Polhill* 15951 (K). **Mozambique**. Z: Morrumbala Mt., fl. 18.i.1863, *Kirk* s.n. (K). MS: Beira, iv. 1898, *Schlechter* s.n. (B†).
Also in Tanzania (including Pemba and Zanzibar Islands). Seepage slopes, usually in grass tufts overlying rock slabs, often in protected shade of rocks; 0–1800 m.
Cynorkis oblonga Schltr. is probably a small form of *Cynorkis kirkii*, with many colonies including plants with 1–3 flowers.

5. **Cynorkis hanningtonii** Rolfe in F.T.A. **7**: 261 (1898). —Summerhayes in Kew Bull. **12**: 109 (1957); in F.T.E.A., Orchidaceae: 37 (1968). —Morris, Epiphyt. Orch. Malawi: 25 (1970). —Grosvenor in Excelsa **6**: 79 (1976). —Williamson, Orch. S. Centr. Africa: 38, t.37 (1977). —la Croix et al., Orch. Malawi: 47 (1991). TAB. **18**. Type from Tanzania.
   Cynorkis johnsonii Rolfe in F.T.A. **7**: 261 (1898). —Summerhayes in Bot. Not. **1937**: 183 (1937). Type from Tanzania.

Terrestrial herb 8–42 cm tall. Tubers usually 2, 1.5–2 cm long, ovoid, villous. Basal leaves 1–2, usually grey-green, 4–15 × 1.5–5 cm, elliptic, acute; peduncle leaves 1–2, sheathing. Peduncle usually glabrous, occasionally sparsely glandular-hairy. Inflorescence 4–24 cm long, fairly densely few- to many-flowered, flowers ± secund; rhachis usually glabrous; bracts 6–7 mm long, lanceolate, acute; ovary and pedicel 7–10 mm long, glabrous to very sparsely hairy. Flowers white, to pale or deep pink, with purple spots on lip. Sepals more or less glabrous. Dorsal sepal 3–5 × 2–3 mm, ovate, convex; lateral sepals spreading, 4–9 × 2.5–6 mm, obliquely ovate. Petals 3–5 × 1.5–2.5 mm, adnate to column at base and forming a hood with the dorsal sepal. Lip 5–10 mm long, 5-lobed in the basal half. Posterior lateral lobes 0.5–1.3 × 0.4–0.7 mm; anterior lateral lobes 1.5–3 × 0.5–1.2 mm; mid-lobe 2.3–5 × 1–3 mm. Spur (2)3–11 mm long, slender, tapering, ± straight. Anther 1.3–1.4 mm high; mid-lobe of rostellum very small.

Zambia. N: Kasama Distr., Malde Rocks, fl. & fr. 2.iii.1960, *Richards* 21704 (K). W: Kitwe, fl. 12.ii.1967, *Fanshawe* F9916 (K). C: Serenje Distr., Kundalila Falls, fl. 4.ii.1973, *Strid* 2890 (K). **Zimbabwe**. C: Harare, fl. & fr. 10.v.1948, *Wild* 4803 (K; SRGH). E: Nyanga (Inyanga), fl. 3.iv.1962, *Wild* 5686 (K; SRGH). **Malawi**. N: S Viphya, Lichelemu (Luchilemu) Valley, fl. & fr. 11.iii.1986, *la Croix* 817 (K; MAL). C: Lilongwe Distr., Dzalanyama Forest Reserve, fl. 24.ii.1982, *Brummitt, Polhill & Banda* 16088 (K; MAL). S: Blantyre Distr., Ndirande Mt., fl. & fr. 12.iii.1970,

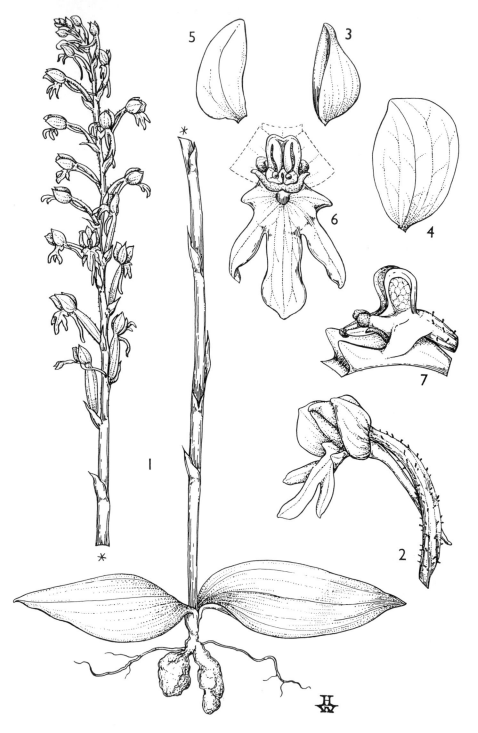

Tab. 18. CYNORKIS HANNINGTONII. 1, habit (×1), from *Stolz* 1245; 2, flower (×4); 3, dorsal sepal, side view (×6); 4, lateral sepal (×6); 5, petal (×6); 6, column and lip, front view (×5); 7, column, side view (×10), 2–7 from *Ball* 757. Drawn by Heather Wood. From F.T.E.A.

Brummitt 9041 (K; MAL).

Also in Zaire, Tanzania and Angola. In high rainfall *Brachystegia* woodland, and amongst damp grass near streams, occasionally in evergreen forest, usually forming large colonies; 900–1900 m.

On the Zomba Plateau in Malawi, *Cynorkis hanningtonii* has colonised pine plantations, growing with and flowering at the same time as *Cynorkis kassneriana* and *Cynorkis buchananii*. These plants are darker in colour than is usual elsewhere, and have very small basal lobes of the lip. It seems possible that some introgression has occurred.

6. **Cynorkis kassneriana** Kraenzl. in Bot. Jahrb. Syst. **51**: 377 (1914). —Summerhayes in Kew Bull. **11**: 218 (1956). —Goodier & Phipps in Kirkia **1**: 52 (1961). —Summerhayes in F.T.E.A., Orchidaceae: 33 (1968). —Morris, Epiphyt. Orch. Malawi: 24 (1970). —Moriarty, Wild Fls. Malawi: t. 27, 2 (1975). —Grosvenor in Excelsa **6**: 79 (1976). —Williamson, Orch. S. Centr. Africa: 36 (1977). —Stewart et al., Wild Orch. South. Africa: 80 (1982). —Geerinck in Fl. Afr. Centr., Orchidaceae pt. 1: 167 (1984). —la Croix et al., Orch. Malawi: 48 (1991). Type from Zaire.

Terrestrial, rarely epiphytic, herb 15–44 cm tall. Tubers usually 2, c. 1.5 cm long, ovoid, villous. Basal leaf usually 1, sometimes 2, dark olive-green above, purple below, 5–20 × 2–5 cm, elliptic, acute, narrowing to a petiole up to 4.5 cm long; peduncle leaves usually 3, sheath-like, 2–2.5 cm long. Peduncle sparsely hairy. Inflorescence 3.5–12 cm long, several flowered; rhachis glandular-hairy; bracts 7–12 mm long, lanceolate, acute; pedicel 2–5 mm long; ovary 8–11 mm, semi-erect, glandular-hairy. Flowers pink-purple to deep purple, with darker marks in throat. Dorsal sepal 5–7 × 4 mm, ovate, acute, convex; lateral sepals 6–9 × 2.5–4 mm, obliquely ovate, spreading or deflexed. Petals 4–5 × 2.3 mm, forming hood with dorsal sepal. Lip 7–10 × 5–7 mm, 3-lobed at about the middle, the mid-lobe larger than the side lobes. All lobes triangular, obtuse; mid-lobe c. 3 mm long, side lobes 1–1.5 mm long. Spur 8–10 mm long, straight, parallel to ovary; slightly swollen towards apex. Column 3 mm long.

Zambia. E: Nyika Plateau, fl. & fr. iii.1967, *Williamson & Odgers* 286 (K). **Zimbabwe**. E: Chimanimani Mts., fl. 2.ii.1958, *Hall* 523 (K; SRGH). **Malawi**. N: South Viphya, Bimbya Hill, fl. 6.iii.1987, *Cornelius* 1063 (K). C: Dedza Distr., Chongoni Forestry School, fl. 28.iii.1967, *Salubeni* 622 (K; MAL). S: Zomba Plateau, fl. 15.iii.1970, *Brummitt* 9111 (K; MAL).

Also in Zaire, Ethiopia, Uganda, Kenya, Tanzania, South Africa and Swaziland. In leaf litter or on rotting logs in forest, including pine forest. Occasionally a low-level epiphyte on tree trunks. Also on rocky seepage slopes; 1300–2300 m.

7. **Cynorkis symoensii** Geerinck & Tournay in Bull. Jard. Nat. Bot. Belg. **47**: 484 (1977). —Geerinck in Fl. Afr. Centr., Orchidaceae pt. 1: 165 (1984). —la Croix et al., Orch. Malawi: 49 (1991). Type from Rwanda.

Terrestrial or epiphytic herb 11–15 cm tall. Tubers 2, c. 1 cm long, ovoid. Basal leaves 2, semi-erect, 8–11 × 1.2–2.3 cm, narrowly elliptic, dark glossy-green; peduncle leaves 2–4, sheath-like, up to 13 mm long. Peduncle glabrous. Inflorescence c. 5 cm long, fairly densely flowered, flowers up to 12, secund; bracts c. 7 mm long, lanceolate, acute; ovary and pedicel 8–12 mm long, semi-erect, arched at the top, the ovary slightly papillose. Sepals green, petals magenta, lip bright magenta-pink. Dorsal sepal 3.5–5 × 2.3 mm, ovate, very convex; lateral sepals 4 × 2.5–3 mm, obliquely ovate. Petals similar to lateral sepals, forming a hood with the dorsal sepal. Lip 10 × 12 mm, wedge-shaped, 3-lobed with the mid-lobe reduced to a small point. Side lobes 5–8 mm long and wide, rounded. Spur 7–8 mm long, parallel to the ovary, swollen to 2 mm in diameter at apex. Anther loculi 1.2 mm high, staminodes very short.

Malawi. N: Misuku Hills, Mugesse (Mughesse) Forest Reserve, fl. 11.iv.1984, *la Croix* 613 (K). Also in Rwanda and Tanzania. On mossy rocks, or on lianas up to 5 m above the ground, embedded in moss; c. 1800 m.

8. **Cynorkis brevicalcar** P.J. Cribb in Kew Bull. **40** (2): 399 (1985). —la Croix et al., Orch. Malawi: 46 (1991). TAB. **19**. Type: Malawi, Mt. Mulanje, Chambe-Sapitwa Path, *Blackmore, Brummitt & Banda* 374 (K, holotype; MAL).
Cynorkis sp. in Morris, Epiphyt. Orch. Malawi: 27 (1970).

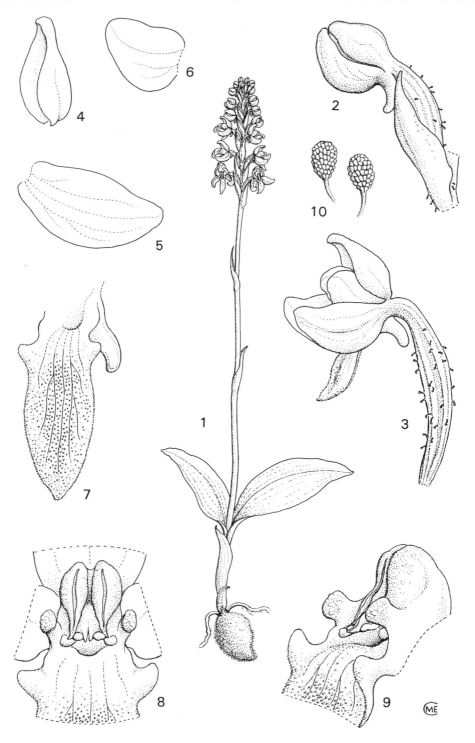

Tab. 19. CYNORKIS BREVICALCAR. 1, habit (×1); 2, flower bud (×8); 1–2 from *Blackmore et al.*
374; 3, flower (×8); 4, dorsal sepal (×10); 5, lateral sepal (×10); 6, petal (×10); 7, lip, front
view (×12); 8, column, front view (×24); 9, column, side view (×30); 10, pollinia (×24),
3–10 from *Brummitt* 9652. Drawn by M.E. Church. From Kew Bull.

Terrestrial herb 5–21 cm tall. Tubers up to 2 cm long, ovoid, villous. Basal leaves 1–2, 25–45 × 7–13 mm, ovate to elliptic, glabrous; peduncle leaves usually 1, sheathing, c. 10 mm long. Peduncle 7–16 cm long, glabrous. Inflorescence 1.5–5 cm long, densely 4–many-flowered, flowers sometimes secund; bracts 4–12 mm long, lanceolate, acute; ovary and pedicel 6–10 mm long, semi-erect, arched at top, purple, sparsely glandular-hairy. Flowers small, white or very pale pink, the lip with purple spots; scented. Dorsal sepal 2–3 × 2 mm, ovate, acute, very convex; lateral sepals 4–6 × 1.8–2.5 mm, obliquely ovate, spreading. Petals 2 × 1 mm, falcate, forming hood with dorsal sepal. Lip 3–6 × 1–2 mm, ligulate, acute, with 2 obscure teeth less than 1 mm long at base. Spur 1–1.5 mm long, straight or slightly incurved. Column 1–1.5 mm long, erect.

Malawi. S: Mt. Mulanje, Lichenya Plateau, fl. 12.ii.1983, *la Croix, Jenkins & Killick* 450 (K; MAL); Sombani Plateau, fl. 17.ii.1985, *Johnston-Stewart* 370 (K).
Known only from Mulanje Mt. where it occurs, often plentifully, on most of the plateaux. Seepage slopes and wet areas in montane grassland, often with underlying rock slabs; 1700–2310 m.

15. OLIGOPHYTON H.P. Linder
by H.P. Linder

Oligophyton H.P. Linder in Kew Bull. **41** (2): 314 (1986).

Terrestrial herb perennating by testicular tubers. Leaves dimorphic, with ovate spreading basal leaves and loosely convoluted cauline leaves. Inflorescence an erect lax spike; bracts half as long as the ovaries, green to somewhat chartaceous. Flowers small, more or less resupinate, with the sepals and petals not spreading; ovary twisted. Sepals free, entire, 1-nerved. Petals more or less enclosed by the sepals, 3-nerved, slightly shorter than the sepals. Lip shorter than the sepals and petals and enclosed by them, 3-lobed and 3-nerved; the side lobes blunt and shorter than the mid-lobe; spur short, straight or curved. Column short; rostellum square with a very small mid-lobe; viscidia separate and held in shallow notches in the rostellum side lobes; staminodes minute; anther erect, 2-celled; stigma 2-lobed, the lobes sessile, slightly convex, diverging from the base.

A monotypic genus endemic to eastern Zimbabwe.

Oligophyton drummondii H.P. Linder & G. Will. in Kew Bull. **41** (2): 314 (1986). TAB. **20**. Type: Zimbabwe, Chimanimani Mts., *Drummond* 8930 (SRGH, holotype).

A terrestrial herb 10–25 cm tall. Basal leaves 2–6 in a basal rosette, 1–4.5 × 0.5–1.5 cm, narrowly elliptic to ovate, acute, narrowed at the base into a pale sheathing petiole; cauline leaves 3–5, lax, 0.5–2 cm long, lanceolate, acute with recurved apices. Inflorescence a laxly 8–25-flowered spike; rhachis 2–11 cm long; bracts half as long as the ovary, green to somewhat chartaceous, acute. Flowers small, 1.5–2 mm long, yellow with green sepals, more or less resupinate; ovary 3–7 mm long. Sepals and petals not spreading. Dorsal sepal concave, 1.5–2 × 1 mm, obovate, cucullate at the obtuse apex; lateral sepals 1.5–2 × 1–1.3 mm, obliquely elliptic, rounded at the apex. Petals more or less enclosed by the sepals, slightly shorter and broader than the sepals, obliquely ovate, obtuse. Lip 1–1.5 × 1.5 mm, 3-lobed; side lobes curved up either side of the column, subquadrate; mid-lobe triangular, 0.5 mm long; spur straight or curved, 0.5–1.2 mm long, subconical. Column with a 0.4 mm long square rostellum; anther 0.5 mm high, erect.

Zimbabwe. E: Chimanimani Mts., ex hort., *James* s.n. (SRGH); fl. 14.ix.1954, *Ball* 358 (SRGH); fl. 29.x.1961, *Ball* 952 (SRGH); between Upper Valley and Point 71, fl. 25.ix.1966, *Drummond* 8952 (SRGH).
Endemic. Known only from Eastern Zimbabwe. Montane zone, in sandy, quartzitic soil.

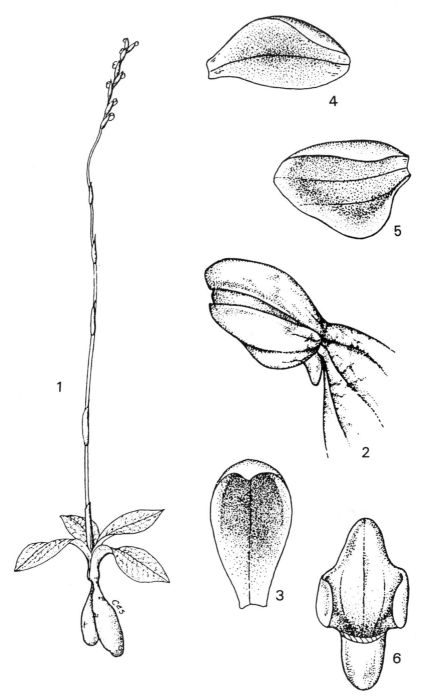

Tab. 20. OLIGOPHYTON DRUMMONDII. 1, habit (×1); 2, flower, side view (×15); 3, dorsal sepal (×20); 4, lateral sepal (×20); 5, petal (×20); 6, lip (×20), 1–6 from *Ball* 952. Drawn by Claire Smith. From Kew Bull.

16. HABENARIA Willd.

Habenaria Willd., Sp. Pl. **4**: 44 (1805).
Podandria Rolfe in F.T.A. **7**: 205 (1898), non Baill. (1890), nom. illegit.
Kryptostoma (Summerh.) Geerinck in Bull. Jard. Bot. Nat. Belgique **52**: 149 (1982).

Terrestrial, rarely epiphytic, herb with tuberoids or long fleshy roots. Stem unbranched. Leaves several–many, arranged along stem or clustered at the base, or with 1–2 basal leaves appressed to the ground and the cauline leaves sheath-like. Inflorescence terminal, 1 to many-flowered. Flowers usually resupinate, in African species green and/or white, rarely yellow. Sepals usually free, the dorsal sepal often forming a hood with the petals; lateral sepals spreading or reflexed. Petals entire, 2-lobed or bifid. Lip entire or 3-lobed, spurred at base; the side lobes sometimes divided; the spur long or short, slender or saccate, often inflated at apex. Column long or short; anther erect or reclinate, the loculi either adjacent or separated by a U-shaped connective; anther canals long or short, almost always adnate to side lobes of rostellum; auricles (staminodes) 2, sometimes 2-lobed; pollinaria 2, each with a sectile pollinium, long or short caudicle and a small, naked viscidium; stigmatic processes 2, long or short, usually free but sometimes joined in lower part to rostellum.

A genus of about 600 species, in tropical and subtropical regions of Old and New World.

Sect. CHLORINAE Kraenzl. in Bot. Jahrb. Syst. **16**: 56, 156 (1892). Type species: *Habenaria chlorina* Par. & Rchb.f. (Burma). *Species 1–6.*

Sect. PSEUDOPERISTYLUS P.F. Hunt in Kew Bull. **22**: 489 (1968). Type species: *Habenaria petitiana* (A. Rich.) T. Durand & Schinz. *Species 7.*

Sect. COMMELYNIFOLIAE Kraenzl. in Bot. Jahrb. Syst. **16**: 56, 136 (1892). Type species: *Habenaria commelynifolia* Lindl. (India, Burma). *Species 8–11.*

Sect. PODANDRIA (Rolfe) P.F. Hunt in Kew Bull. **22**: 489 (1968). Type species: *Habenaria macrandra* Lindl. *Species 12.*

Sect. MULTIPARTITAE Kraenzl. in Bot. Jahrb. Syst. **16**: 56, 189 (1892). Type species: *Habenaria multipartita* Bl. (Java). *Species 13–16.*

Sect. TRACHYPETALAE Summerh. in Bot. Mus. Leafl. Harv. Univ. **10**: 264 (1942). Type species: *Habenaria trachypetala* Kraenzl. *Species 17–18.*

Sect. PENTACERAS (Thouars) Schltr. in Bot. Jahrb. Syst. **53**: 496 (1915). Type species: *Habenaria arachnoides* Thouars (Madagascar). *Species 19–32.*

Sect. REPLICATAE Kraenzl. in Bot. Jahrb. Syst. **16**: 55, 74, 85 (1892). Type species: *Habenaria kilimanjari* Rchb.f. *Species 33–67.*

Sect. PRODUCTAE Summerh. in Kew Bull. **20**: 167 (1966). Type species: *Habenaria singularis* Summerh. *Species 68.*

Sect. MIRANDAE Summerh. in Kew Bull. **16**: 301 (1962). Type species: *Habenaria rautaneniana* Kraenzl. *Species 69–71.*

Sect. CERATOPETALAE Kraenzl. in Bot. Jahrb. Syst. **16**: 55, 64 (1892). Type species: *Habenaria cornuta* Lindl. *Species 72–81.*

Sect. KRYPTOSTOMA Summerh. in Kew Bull. **14**: 135 (1960). Type species: *Habenaria tentaculigera* Rchb.f. *Species 82–83.*

Sect. MACRURAE Kraenzl. in Bot. Jahrb. Syst. **16**: 55, 91 (1892). Type species: *Habenaria perbella* Rchb.f. (Ethiopia). *Species 84–87.*

Sect. DIPHYLLAE Kraenzl. in Bot. Jahrb. Syst. **16**: 56, 147 (1892). Type species: *Habenaria diphylla* Dalz. (India). *Species 88–119.*

Key to artificial groups

1. Basal leaves 1–2, orbicular or heart-shaped, appressed to ground · · · · · · · · · · **Group A**
 – All leaves borne along stem, not appressed to ground · 2
2. Petals entire · **Group B**
 – Petals 2-lobed · 3
3. Spur 3.5 cm long or more · **Group C**
 – Spur less than 3.5 cm long · 4
4. Petals incompletely divided, at least the basal c. 5 mm undivided · · · · · · · · · · **Group D**
 – Petals 2-lobed almost to base · 5
5. Dorsal sepal erect, more or less surrounding the column · · · · · · · · · · · · · · · **Group E**
 – Dorsal sepal reflexed, leaving column exposed · **Group F**

Group A. *Basal leaves 1–2, appressed to ground; cauline leaves sheath-like.*

1. Flowers non-resupinate, clear yellow · 2
 – Flowers resupinate, white, green or yellow-green · 3
2. Spur 9–12 mm long, curving up· 88. *aberrans*
 – Spur 13–16 mm long, pendent · 89. *villosa*
3. Petals entire · 4
 – Petals 2-lobed · 6
4. Basal leaf 1; spur over 2.5 cm long · 90. *odorata*
 – Basal leaves 2; spur less than 2.5 cm long · 5
5. Spur c. 2 cm long; inflorescence 3–8-flowered · · · · · · · · · · · · · · · · · · 91. *nicholsonii*
 – Spur less than 1 cm long; inflorescence to 20-flowered · · · · · · · · · · · · · 92. *debeerstiana*
6. Basal leaf solitary· 7
 – Basal leaves 2· 18
7. Spur over 10 cm long; lower petal lobe and lip side lobes linear to filiform, 5–8 cm long ·
 · 113. *rhopalostigma*
 – Spur less than 4 cm long · 8
8. Sepals hairy · 9
 – Sepals glabrous · 4
9. Petals and lip hairy · 109. *hirsutissima*
 – Petals and lip glabrous, or only the upper petal lobe hairy · · · · · · · · · · · · · · · · · · 10
10. Petals with undivided basal part 3–3.5 mm long· 117. *velutina*
 – Petals 2-lobed to base · 11
11. Dorsal sepal 8–12 mm long; lower (anterior) petal lobe curving down · · · 110. *leucotricha*
 – Dorsal sepal to 7 mm long; lower (anterior) petal lobe upcurved or spreading · · · · · · 12
12. Side lobes of lip to 5.5 mm long, shorter than the mid-lobe · · · · · · · · · · · 114. *holothrix*
 – Side lobes of lip over 7 mm long, longer than the mid-lobe · · · · · · · · · · · · · · · · · 13
13. Lip side lobes 18–20 mm long; spur 8–11 mm long · · · · · · · · · · · · · · 111. *kabompoensis*
 – Lip side lobes 7.5–14 mm long; spur 9–27 mm long · · · · · · · · · · · · · · · · · 112. *pilosa*
14. Dorsal sepal 9–16 mm long · 15
 – Dorsal sepal 3–6.5 mm long · 16
15. Spur 24–30 mm long · 119. *perpulchra*
 – Spur 10–18 mm long · 118. *verdickii*
16. Dorsal sepal 3–3.5 mm long; anther with a sterile stalk so that anther canals and rostellum
 side lobes arise c. half-way up; lower (anterior) petal lobes curving down · · · · · · · · · · · ·
 · 108. *decurvirostris*
 – Dorsal sepal over 4.5 mm long; anther without a sterile stalk; lower petal lobe curving up
 · 17
17. Lip lobes equal, c. 8 mm long; spur 9–10 mm long · · · · · · · · · · · · · · · · · 115. *nyikensis*
 – Side lobes of lip c. 11 mm long, longer than mid-lobe; spur 12–13.5 mm long · · · · · · ·
 · 116. *unifoliata*
18. Side lobes of lip filiform, much longer than mid-lobe · 19
 – Side lobes of lip equal to, or shorter than, mid-lobe · 22
19. Spur less than 4 cm long · 99. *trilobulata*
 – Spur over 5 cm long · 20
20. Inflorescence laxly 2–4-flowered; dorsal sepal over 20 mm long · · · · · · · · · · · 96. *edgarii*
 – Inflorescence fairly densely 10- to many-flowered; dorsal sepal up to 16 mm long · · · · 21

21. Dorsal sepal 7–11 mm long; spur 5–7.5 cm long · · · · · · · · · · · · · · · · · · 98. *subarmata*
  – Dorsal sepal 11–16 mm long; spur 8–18 cm long · · · · · · · · · · · · · · · · · · 97. *armatissima*
22. Spur over 10 cm long · 102. *macrura*
  – Spur less than 8 cm long, usually much less · 23
23. Anther canals 6–9 mm long; stigmatic arms over 7 mm long · · · · · · · · · · · · · · · 24
  – Anther canals less than 3 mm long; stigmatic arms up to 6 mm long · · · · · · · · · · · · 25
24. Dorsal sepal 13–16 mm long; spur 2–2.5 cm long · · · · · · · · · · · · · · · · · · 93. *adolphi*
  – Dorsal sepal 19–26 mm long; spur 3.5–7.5 mm long · · · · · · · · · · · · · · · 95. *mechowii*
25. Dorsal sepal 8–16 mm long; stigmatic arms 4–6 mm long; anther canals over 2 mm long ·
    · 26
  – Dorsal sepal 4–8 mm long; stigmatic arms 1.5–3.5 mm long; anther canals very short, c. 1
    mm long · 28
26. Ovary and dorsal sepal without toothed keels; spur slender, not swollen at apex 103. *stylites*
  – Ovary and dorsal sepal with toothed keels; spur swollen at apex · · · · · · · · · · · · · · · · 27
27. Spur 17–23 mm long; side lobes of the lip longer than the mid-lobe · · · · · · 94. *lindblomii*
  – Spur c. 10 mm long; side lobes of lip slightly shorter than the mid-lobe· 106. *mosambicensis*
28. Inflorescence laxly 5–15-flowered; lower (anterior) petal lobe and lip side lobes
    oblanceolate, 1.5–4 mm wide · 101. *galactantha*
  – Inflorescence 20- to many-flowered; lower (anterior) petal lobe and lip side lobes ligulate,
    linear or filiform, less than 1.5 mm wide · 29
29. Inflorescence pyramidal; flowers pure white · · · · · · · · · · · · · · · · · · 105. *livingstoniana*
  – Inflorescence cylindrical; flowers green or yellow-green· 30
30. Lower (anterior) petal lobe shorter than upper lobe; spur with wide mouth 100. *dregeana*
  – Lower (anterior) petal lobe longer than upper; spur with narrow mouth · · · · · · · · · · · 31
31. Lower (anterior) petal lobe about twice as long as upper lobe; mid-lobe of lip 10–12 mm
    long · 107. *tysonii*
  – Lower (anterior) petal lobe less than twice as long as upper lobe; mid-lobe of lip 6–9.5 mm
    long · 104. *lithophila*

Group B. *Petals entire; leaves borne on stem.*

 1. Lip undivided, narrowly ligulate · 2
  – Lip 3-lobed · 3
 2. Spur c. 1 cm long; leaves linear, less than 1 cm wide · · · · · · · · · · · · · · · · 4. *hologlossa*
  – Spur 2.5–6 cm long; leaves ovate or lanceolate, over 2 cm wide · · · · · · · · · · 8. *zambesina*
 3. Anther connective U-shaped, at least 5 mm wide between loculi and total width at least
    12 mm · 4
  – Anther loculi set close together · 7
 4. Side lobes of lip entire · 13. *falciloba*
  – Side lobe of lip divided into narrow, comb-like segments· 5
 5. Dorsal sepal 30–38 mm long; spur 6.5–8 cm long; base of lip puberulent · · 16. *splendentior*
  – Dorsal sepal to 30 mm long; spur to 5 cm long; base of lip densely pubescent · · · · · · · 6
 6. Dorsal sepal 13–22 mm long; spur 12–22 mm long · · · · · · · · · · · · · · · · 14. *praestans*
  – Dorsal sepal 23–30 mm long; spur 2.5–5 cm long · · · · · · · · · · · · · · · · · · 15. *splendens*
 7. Column tall, over 13 mm high; leaves petiolate, most in a basal cluster · · · · 12. *macrandra*
  – Column short, less than 6 mm high · 8
 8. Petals densely pubescent; side lobes of lip curled at ends · 9
  – Petals glabrous; side lobes of lip not curled at ends · 10
 9. Dorsal sepal 3–4.5 mm long · 17. *pubipetala*
  – Dorsal sepal 10–20 mm long · 18. *trachypetala*
10. Spur over 12 cm long · 9. *gabonensis*
  – Spur less than 7 cm long · 11
11. Dorsal sepal over 8 mm long; petals at least 3 mm wide, lanceolate to orbicular · · · · · 12
  – Dorsal sepal to 7 mm long; petals less than 2 mm wide · 13
12. Dorsal sepal 8–10 mm long; spur over 18 mm long · · · · · · · · · · · · · · 11. *epipactidea*
  – Dorsal sepal 14–15 mm long; spur less than 15 mm long · · · · · · · · · · · · · · 10. *insolita*
13. Spur 1.5–3.5 cm long · 14
  – Spur less than 1 cm long · 15
14. Stem with 2 ovate, petiolate leaves near base; spur slightly inflated near apex, 15–17 mm
    long · 1. *arenaria*
  – Leaves linear or lanceolate, never ovate and petiolate; spur slender, 20–30 mm long · · · ·
    · 3. *filicornis*

15. Spur ± globose, less than 2 mm long, shorter than lip · · · · · · · · · · · · · · · · · 7. *petitiana*
 – Spur not globose, 4–8 mm long, equalling or longer than lip · · · · · · · · · · · · · · · · 16
16. Dorsal sepal 2.5–3 mm long; undivided basal part of lip longer than, or equalling, the mid-
    lobe · 5. *tenuispica*
 – Dorsal sepal 4–7 mm long; undivided basal part of lip shorter than the mid-lobe · · · · 17
17. Dorsal sepal 4–5 mm long; lip divided almost to base · · · · · · · · · · · · · · · · · · · 2. *bicolor*
 – Dorsal sepal 6–7 mm long; lip with an undivided basal part c. 2 mm long · · 6. *xanthochlora*

Group C. *Petals 2-lobed; leaves borne on stem; spur more than 3.5 cm long.*

 1. Lower petal lobes and lip side lobes lanceolate or oblong, thin-textured · · · · · · · · · · · 2
 – Lower petal lobes narrow, either filiform or horn-like and fleshy · · · · · · · · · · · · · · · 3
 2. Spur 13–17 cm long; petals and lip white · 87. *walleri*
 – Spur 5–7.5 cm long; petals and lip yellowish · · · · · · · · · · · · · · · · · · · 85. *macroplectron*
 3. Stigmatic arms up to 2 mm long · 4
 – Stigmatic arms over 5 mm long · 5
 4. Lower (anterior) petal lobes and side lobes of lip 2–3 mm long · · · · · · · · · · 84. *argentea*
 – Lower (anterior) petal lobes and side lobes of lip 11–23 mm long · · · · · · · · · · 20. *elliotii*
 5. Spur up to 8 cm long · 6
 – Spur over 8.5 cm long · 10
 6. Stigmatic arms 5–6 mm long; spur with knee-like bend below lip · · · · · · 75. *gonatosiphon*
 – Stigmatic arms over 8 mm long · 7
 7. Lower (anterior) petal lobe with a wide, folded base on which is borne the thread-like
    upper (posterior) lobe · 76. *harmsiana*
 – Petals 2-lobed ± to base; lower lobe not widened at base · · · · · · · · · · · · · · · · · · · 8
 8. Side lobes of lip lanceolate, much shorter than the mid-lobe and appressed to the stigmatic
    arms and anther canals · 77. *holubii*
 – Side lobes of lip linear, not appressed to the stigmatic arms and anther canals · · · · · · · 9
 9. Dorsal sepal 11–19 mm long; spur 3–5 cm long; stigmatic arms 8–13 mm long · 73. *clavata*
 – Dorsal sepal 20–24 mm long; spur 5.5–8 cm long; stigmatic arms 13–18 mm long
    · 79. *laurentii*
10. Spur 13–22 cm long, less than 2 mm wide at base; anther 7–10 mm high · · · · 72. *cirrhata*
 – Spur 8.5–14 cm long, more than 2 mm wide at base; anther 10–15 mm high · · · · · · · · ·
    · 78. *kassneriana*

Group D. *Petals incompletely 2-lobed; leaves borne on stem.*

Dorsal sepal 16–20 mm long; spur bent forward with a knee-like bend then recurved below
    flower · 83. *tentaculigera*
Dorsal sepal 10–12 mm long; spur forming a loop near base and projecting above flower · · · ·
    · 82. *goetzeana*

Group E. *Petals 2-lobed; leaves borne along stem; spur less than 3 cm long; dorsal sepal erect.*

 1. Mid-lobe of lip lanceolate or oblong-lanceolate, over 4 mm wide · · · · · · · · · · 86. *mirabilis*
 – Mid-lobe of lip linear or narrowly ligulate · 2
 2. Stigmatic arms projecting forwards or upwards, slender but widened and truncate at tips 3
 – Stigmatic arms projecting downwards, often ± appressed to lip, tapering to the tips, not
    suddenly widened · 15
 3. Side lobes of rostellum zigzag, 2–3 mm long, longer than anther canals · · · · 68. *singularis*
 – Side lobes of rostellum not zigzag, of similar length to anther canals · · · · · · · · · · · · · 4
 4. Spur less than 1 cm long · 81. *stenorhynchos*
 – Spur over 1 cm long · 5
 5. Lateral sepals rolled up lengthwise in the open flower · 6
 – Lateral sepals not rolled up lengthwise · 8
 6. Side lobes of lip lacerate on both edges in apical half · · · · · · · · · · · · · · · · · · 80. *mira*
 – Side lobes of lip entire, or toothed on outer margin of basal half · · · · · · · · · · · · · · · 7
 7. Stigmatic arms 3–8 mm long; side lobes of lip widened and flattened some distance above
    base, usually toothed on outer margin · 74. *cornuta*
 – Stigmatic arms 8–13 mm long; side lobes of lip entire, narrow, gradually tapering from base
    · 73. *clavata*

8. Dorsal sepal 9–11 mm long; side lobes of lip wider than the mid-lobe; undivided base of lip adnate to stigmatic arms · 9
 – Dorsal sepal less than 7 mm long; side lobes of lip not wider than the mid-lobe; base of lip not adnate to stigmatic arms · 10
9. Ovary densely papillose; stigmatic arms to 7 mm long · · · · · · · · · · · · · · · 69. *calvilabris*
 – Ovary not papillose; stigmatic arms 11 mm long · · · · · · · · · · · · · · · · · · · 70. *pasmithii*
10. Stigmatic arms 1–2 mm long · 11
 – Stigmatic arms 3–7 mm long · 13
11 Spur 30–37 mm long · 35. *falcicornis* var. *caffra*
 – Spur 4–7 mm long; basal leaf sheaths white with dark green reticulate veining · · · · · · 12
12. Petals and lip densely papillose · 34. *arianae*
 – Petals and lip not papillose · 38. *diselloides*
13. Spur forming a complete loop in apical half · · · · · · · · · · · · · · · · · 33. *anaphysema*
 – Spur not forming a loop · 14
14. Stigmatic arms pubescent; spur 16–23 mm long · · · · · · · · · · · · · · · 44. *hirsutitrunci*
 – Stigmatic arms glabrous; spur 10–16 mm long · · · · · · · · · · · · · · · · · 67. *welwitschii*
15. Anther over 6 mm high; side lobes of lip much wider than mid-lobe, hairy in basal part · 71. *rautaneniana*
 – Anther less than 5 mm high; side lobes of lip the same width as, or narrower than, the mid-lobe · 16
16. Mid-lobe of rostellum large and hooded, standing in front of anther; spur 2–3 cm long · 21. *magnirostris*
 – Mid-lobe of rostellum narrow, set between the anther loculi · · · · · · · · · · · · · · · · 17
17. Spur sharply bent up, the apical half vertical or curved forwards, very swollen at tip · 32. *uncicalcar*
 – Spur not as above · 18
18. Lower (anterior) petal lobes and lip side lobes filiform, 2 cm long or more · 24. *orthocentron*
 – Lower (anterior) petal lobes and lip side lobes less than 2 cm long, usually much less · 19
19. Upper (posterior) petal lobe longer than lower (anterior) lobe; mid-lobe of lip distinctly longer than the side lobes · 20
 – Upper (posterior) petal lobe shorter than lower (anterior) lobe; mid-lobe of lip equal to, or shorter than, the side lobes · 21
20. Dorsal sepal 5–6.5 mm long; spur 6–8 mm long · · · · · · · · · · · · · · · · 27. *silvatica*
 – Dorsal sepal 3–4 mm long; spur 8–10 mm long · · · · · · · · · · · · · · · · · 30. *tridens*
21. Bracts as long as, or longer than flowers, at least at base of inflorescence · · · · · · · · · 22
 – Bracts shorter than flowers at base of inflorescence · · · · · · · · · · · · · · · · · · · 24
22. Lower (anterior) petal lobe about one and a half times as long as upper (posterior) lobe; side lobes of lip one and a half to 2 times as long as the mid-lobe; plants drying yellow-brown or yellow-green · 29. *tetraceras*
 – Lower (anterior) petal lobe only slightly longer than upper (posterior) lobe; side lobes of lip of similar length to the mid-lobe · 23
23. Lower petal lobe and lip side lobes becoming filiform towards tips; plants drying yellow-brown to yellow-green · 23. *njamnjamica*
 – Lower petal lobe and lip side lobes linear but not filiform; plants drying dark brown · 31. *uhehensis*
24. Leaves drying papery, enfolding stem at base; basal sheaths loosely funnel-shaped · 25. *papyracea*
 – Leaves not drying papery; not enfolding stem at base; basal sheaths tight, not funnel-shaped · 25
25. Leaves lanceolate, less than 7.5 cm long and 1.7 cm wide; rhachis slightly zigzag · 28. *supplicans*
 – Largest leaves over 7.5 cm long, usually over 2 cm wide; rhachis not zigzag · · · · · · · · 26
26. Spur horizontal or recurved, projecting above flower, with a hairy tooth in the mouth · 26. *pubidens*
 – Spur pendent or slightly incurved, lacking a hairy tooth in the mouth · · · · · · · · · · · 27
27. Spur 9–15 mm long, inflated in middle; lowest foliage leaf 9–28 cm from base of stem · 22. *malacophylla*
 – Spur 17–22 mm long, slender thoughout or slightly thicker at apex; lowest foliage leaf 3–8 cm from base of stem · 19. *amoena*

Group F. Petals 2-lobed; leaves borne along stem; spur less than 3.5 cm long; dorsal sepal reflexed.

1. Column narrowed towards base, with a sterile stalk 1.5–3 mm long between top of ovary and anther loculi · 2
− Column widest at base; sterile stalk very short or absent · 4
2. Leaves narrowly linear, c. 3 mm wide; lower petal lobe 12–14 mm long; spur 26–28 mm long · 43. *hebes*
− Leaves at least 7 mm wide, linear or lanceolate; lower petal lobe 6–12 mm long; spur 15–25 mm long · 3
3. Lower petal lobe 6–7 mm long, linear, fleshy; mid-lobe of lip 7–8.5 mm long · · · · · · · · · ·
· 51. *macrostele*
− Lower petal lobe 7.5–12 mm long, lanceolate; mid-lobe 8.5–11.5 mm long · · · · 54. *ndiana*
4. Spur forming a complete loop in apical half · 33. *anaphysema*
− Spur often spirally twisted, but not forming a loop · 5
5. Basal leaf sheaths spotted with black · 6
− Basal leaf sheaths not spotted with black · 8
6. Lower petal lobe papillose, slightly pubescent towards base but whole lobe not densely pubescent · 58. *riparia*
− Lower petal lobe densely pubescent · 7
7. Pedicel and ovary 10–20 mm long; spur 9–12.5 mm long, pendent, parallel to ovary and pedicel · 49. *kyimbilae*
− Pedicel and ovary 24–30 mm long; spur 12.5–20 mm long, horizontal or curving up · · · ·
· 56. *petraea*
8. Lip with claw c. 10 mm long, the spur arising 6 mm from base · · · · · · · · · 65. *unguilabris*
− Spur arising at base of lip · 9
9. Leaves tubular with a sheathing base, clustered at base of plant · · · · · · · · · · 64. *tubifolia*
− Leaves not as above · 10
10. Lower (anterior) petal lobe shorter than, or about equalling, upper (posterior) lobe · · 11
− Lower (anterior) petal lobe distinctly longer than upper (posterior) lobe · · · · · · · · · · 13
11. Lower (anterior) petal lobe oblanceolate, much wider than linear upper (posterior) lobe
· 67. *welwitschii*
− Lower (anterior) and upper (posterior) petal lobes of similar shape and width · · · · · 12
12. Petal lobes oblanceolate; mid-lobe of lip 5–9 mm long; spur 13–18 mm long; ends of rostellar arms pincer-like · 48. *kilimanjari*
− Petal lobes lanceolate; mid-lobe of lip 9–10 mm long; spur 10–13 mm long; ends of rostellar arms not pincer-like · 62. *subaequalis*
13. Basal leaf sheaths white or pale green with dark green reticulate veining · · · · · · · · · · 14
− Basal leaf sheaths without noticeable reticulate veining · 19
14. Stigmatic arms diverging at an angle of c. 60° · · · · · · · · · · · · · · · · · · · 39. *disparilis*
− Stigmatic arms ± parallel to each other · 15
15. Leaves narrowly linear, less than 5 mm wide · 16
− Leaves over 5 mm wide · 17
16. Spur less than 1 cm long; lower (anterior) petal lobe oblong, c. 3 mm wide · · · 63. *tortilis*
− Spur over 1 cm long; lower (anterior) petal lobe linear, less than 1 mm wide · · 37. *compta*
17. Lower (anterior) petal lobe 13–19 mm long; spur less than 2 cm long · · · · · 50. *leucoceras*
− Lower (anterior) petal lobe less than 12 mm long; spur over 2.5 cm long · · · · · · · · · · 18
18. Side lobes of lip linear, slightly longer than the mid-lobe; spur incurved, not twisted in the middle · 40. *falcata*
− Side lobes of lip narrowly lanceolate, shorter than the mid-lobe; spur twisted in the middle, parallel to ovary and pedicel · 57. *retinervis*
19. Spur very slender, almost filiform, abruptly widened and almost globose at apex · · · · 20
− Spur not filiform, sometimes swollen at apex but not with the tip almost globose · · · · 21
20. Ovary less than 5 mm long, c. one quarter length of the very slender pedicel, mid-lobe of lip 13–15 mm long · 47. *ichneumonea*
− Ovary almost half as long as pedicel; mid-lobe of lip 6–6.5 mm long · · · · 36. *cataphysema*
21. Ovary bent down at junction with pedicel, forming almost a right angle, so that flowers droop (but straightening in fruit) · 22
− Ovary and pedicel straight or arched, but not forming a distinct angle · · · · · · · · · · · 23
22. Pedicel and ovary 20–40 mm long; lower petal lobe 14–18.5 mm long, ciliate on margin ·
· 59. *schimperiana*
− Pedicel and ovary 16–25 mm long; lower petal lobe 8–16 mm long, not ciliate on margin·
· 42. *genuflexa*

23. Auricles at base of column almost 2 mm long, white, deeply 2-lobed · · · · 52. *macrotidion*
– Auricles at base of column small, sessile, not deeply 2-lobed · · · · · · · · · · · · · · · · · · 24
24. Lower (anterior) petal lobe broadly lanceolate or oblong-lanceolate, usually obtuse, less than twice as long as the upper (posterior) lobe · 25
– Lower (anterior) petal lobe narrowly lanceolate, acute, about twice as long as the upper (posterior) lobe · 26
25. Leaves 5–22 mm wide; inflorescence 3.5–4 (rarely 5) cm wide; pedicel and ovary 15–22 mm long · 46. *humilior*
– Leaves 20–55 mm wide; inflorescence 4–6 cm wide; pedicel and ovary 20–30 mm long · 60. *sochensis*
26. Bracts as long as, or longer than, pedicel and ovary, at least at base of inflorescence · · 27
– Bracts shorter than pedicel and ovary · 28
27. Lower (anterior) petal lobe spreading outwards and upwards; leaves without prominent longitudinal veins · 45. *huillensis*
– Lower (anterior) petal lobe hinged, pendent; leaves with 3 prominent longitudinal veins· 55. *nyikana*
28. Lip divided almost to base · 29
– Lip with undivided basal part 2–3 mm long · 30
29. Leaves lanceolate, 1–3.5 cm wide, mostly clustered in lower part of stem; spur horizontal or ± incurved · 53. *myodes*
– Leaves linear, less than 1 cm wide; spur with tip wrapped round pedicel · · 61. *strangulans*
30. Ovary and pedicel ± arched, the pedicel very slender and about twice as long as the ovary; spur incurved or sigmoid, the apical quarter swollen, obtuse · · · · · · · · · · · · 41. *galpinii*
– Ovary and pedicel slightly upcurved, almost equal in length; spur pendent, very slightly thickened in apical half, the apex acute · 66. *weberiana*

1. **Habenaria arenaria** Lindl., Gen. Sp. Orchid. Pl.: 317 (1835). —Stewart et al., Wild Orch. South. Africa: 83 (1982). Type from South Africa.
    Bonatea micrantha Lindl., Gen. Sp. Orchid. Pl.: 329 (1835). Type from South Africa.

Terrestrial herb 25–40 cm high with villous, ellipsoid tuber up to 25 mm long. Leaves 2–8, with 2 at the base (but not appressed to the ground), petiolate, the lamina 7.5–9.5 × 2.5–3 cm, ovate, the petiole 2.5–3.5 cm long; the remainder sheath-like stem leaves 15–18 mm long. Basal leaves dark green, densely mottled on the upper surface with white or silvery-green. Inflorescence 7–10 cm long, laxly up to c. 12-flowered; bracts 10–15 mm long, lanceolate, acute. Flowers green, very small. Ovary and pedicel arched, 15 mm long. Sepals c. 4 × 2 mm; ovate, dorsal erect, convex; lateral sepals reflexed, oblique. Petals c. 4 × 1 mm, triangular, acute, lying beside dorsal sepal. Lip 3-lobed almost to base; mid-lobe 5–6 mm long, side lobes spreading, 4 mm long; all lobes linear, c. 0.5 mm wide; spur 15–17 mm long, slender but slightly swollen in apical half, straight or slightly incurved. Stigmatic arms very short; rostellum 3-lobed.

Mozambique. M: Missão de Nossa Senhora das Candeias, Marracuene (Vila Luísa), s.d. *Costa* 9 & 10 (K).
    Also in South Africa. In humus in shade, near coast.

2. **Habenaria bicolor** Conrath & Kraenzl. in Vierteljahrsschr. Naturf. Ges. Zürich **51**: 131 (1906). —Stewart et al., Wild Orch. South. Africa: 86 (1982). Type from South Africa.

Slender terrestrial herb up to 75 cm high, with villous tuber c. 5 cm long. Stem leafy; leaves 8–18 cm long, 3–7 mm wide, linear, acute, conduplicate. Inflorescence 10–23 × 1–1.8 cm, densely many-flowered; bracts ovate, acute, longer than the ovary, up to 27 mm long at the base of the inflorescence. Flowers green, the petals and lip drying black, the sepals mid-brown. Ovary and pedicel c. 10 mm long. Dorsal sepal 4–5 × 1–2 mm, lanceolate, acute, erect; lateral sepals 3.5–5.5 × 0.8–1.2 mm, lanceolate, acute, projecting forwards or upwards. Petals 4 mm long, lanceolate, lying beside dorsal sepal. Lip c. 3.5 mm long, 3-lobed; mid-lobe horizontal, 2.5–3 mm long; side lobes deflexed, 2–2.5 mm long; all lobes linear, 0.1–0.3 mm wide; spur 6–8 mm long, slender, parallel to ovary. Stigmatic arms very short.

Zimbabwe. C: Harare, Cranborne, fl. & fr. 1.v.1949, *Greatrex* in *GHS* 23310 (K; SRGH). E: Mutare, fl. 22.iii.1960, *Chase* 7312 (K; SRGH).
Also in South Africa. Open woodland, amongst granite rocks; c. 1800 m.

3. **Habenaria filicornis** Lindl., Gen. Sp. Orchid. Pl.: 318 (1835). —Rolfe in F.T.A. **7**: 216 (1898). —Summerhayes in F.W.T.A. **2**: 410 (1936); in Kew Bull. **8**: 129 (1953); in F.W.T.A. ed. 2, **3**: 193 (1968); in F.T.E.A., Orchidaceae: 54 (1968); in Bot. Not. **1937**: 184 (1937). — Grosvenor in Excelsa **6**: 83 (1976). —Williamson, Orch. S. Centr. Africa: 39 (1977). — Geerinck in Fl. Afr. Centr., Orchidaceae pt. 1: 84 (1984). —la Croix et al., Orch. Malawi: 52 (1991). TAB. **21**. Type from Ghana.
    Habenaria chlorotica Rchb.f. in Flora **48**: 178 (1865). —Rolfe in F.T.A. **7**: 198 (1898). — Summerhayes in F.W.T.A. ed. 2, **3**: 193 (1968); in F.T.E.A., Orchidaceae: 54 (1968). — Grosvenor in Excelsa **6**: 83 (1976). —Williamson, Orch. S. Centr. Africa: 39 (1977). — Stewart et al., Wild Orch. South. Africa: 83 (1982). Type from Angola.
    Habenaria natalensis Rchb.f. & Warm., Otia Bot. Hamburg.: 97 (1881). Type from South Africa.
    Habenaria wilmsiana Kraenzl., Orchid. Gen. Sp. **1**: 464 (1901). Type from South Africa.
    Habenaria filicornis var. *chlorotica* (Rchb.f.) Geerinck, in Bull. Jard. Bot. Belg. **52**: 144 (1982).

Slender terrestrial herb 25–80 cm tall, with villous, ellipsoid tuber 1–2 cm long. Leaves several, spread up stem, 5–18 cm long, 3–15 mm wide, lanceolate or linear. Inflorescence 2–30 cm long, laxly or densely few to many-flowered; bracts up to 12 mm long, shorter than the ovary, lanceolate, acute. Flowers green. Ovary and pedicel 9–13 mm long. Dorsal sepal 3–4.5 × 2–3 mm, ovate, erect, convex; lateral sepals 5–6 × 1.8–2.2 mm, obliquely ovate, deflexed. Petals 3–4.5 × 1–1.5 mm, entire, lying beside dorsal sepal. Lip 7–8 mm long, 3-lobed to base, all lobes linear, the mid-lobe slightly longer and broader than the side lobes; spur 20–35 mm long, slender, straight or slightly incurved. Anther c. 2.6 mm high; canals 0.7 mm long. Stigmatic arms 1.8–2.5 mm long, sometimes papillose below. Rostellum mid-lobe 1–1.8 mm high, obtuse.

Botswana. N: "James Camp" area near Kwando R., fl. 2.ii.1978, *P.A. Smith* 2342 (K; SRGH). **Zambia**. N: Mbala Distr., shores of Lake Chila, 1630 m, fl. 4.i.1952, *Richards* 262 (K). W: Mwinilunga, fl. 15.xi.1958, *Holmes* 091 (K; SRGH). C: 43 km east of Lusaka, fl. i.1968, *G. Williamson* 768 (K; SRGH). S: 42 km north of Choma, c. 1100 m, fl. 5.i.1957, *E.A. Robinson* 2015 (K; SRGH). **Zimbabwe**. N: Makonde Distr., Shinje, SE of Guruve (Sipolilo), fl. 10.ii.1982, *Brummitt & Drummond* 15848 (K; SRGH). W: Matobo Distr., Besna Kobila, c. 1450 m, fl. i.1957, *Miller* 4092 (K; SRGH). C: Harare Distr., Delport Road near Ruvoa R., fl. 6.ii.1977, *Grosvenor & Renz* 881 (K; SRGH). E: Nyanga (Inyanga), c. 1300 m, fl. 15.i.1931, *Norlindh & Weimarck* 4421 (K). **Malawi**. N: Mzimba Distr., South Viphya, c. 60 km SW of Mzuzu, 1700 m, fl. & fr. 3.ii.1974, *Pawek* 8036 (K; MAL; MO; SRGH; UC). C: Dedza Distr., Chongoni Forest Reserve, fl. 24.iii.1969, *Salubeni* 1283 (K; MAL; SRGH). S: Thyolo Distr., Bvumbwe, 1150 m, fl. & fr. 22.ii.1981, *la Croix* 108 (K). **Mozambique**. Z: Mlanje Distr., Serra Tumbine, fl. 16.i.1971, *Hilliard & Burtt* 6281 (E).
    Throughout tropical Africa and South Africa (Natal). Grassland, often poorly drained marshy ground, peat on gravel; 1150–2300 m.

In the F.T.E.A. Orchidaceae key to *Habenaria* Summerhayes differentiated *H. filicornis* and *H. chlorotica* as follows:-
    "Leaves linear, 10–20 times as long as broad, stigmas smooth underneath · · *chlorotica*
    Leaves oblong or narrowly lanceolate, ± 5–6 times as long as broad; stigmas papillose
    underneath · *filicornis*"

Most specimens key out easily to one species or the other, but others, including most of those from the Nyika Plateau in Malawi which have short but narrow leaves, do not. All specimens in the Kew Herbarium of these 2 species from the Flora Zambesiaca area were measured. When length and breadth of the longest leaf of each specimen were plotted against each other for plants with papillose or smooth stigmas, no disjunction was apparent. A trend was visible, in that most plants with a low length to breadth ratio had papillose stigmas (*H. filicornis*) while most of those with a high length to breadth ratio had smooth stigmas (*H. chlorotica*). However, there was a definite overlap and furthermore, the leaf ratios formed a continuous range. Other characters were also measured, eg. length of inflorescence, number of flowers and spur length and again, no disjunctions were found. It would appear that these names are synonymous and that the earlier name, *H. filicornis* Lindl., is therefore the correct name for this species. D. Geerinck has come to the same conclusion in his treatment for the Flore d'Afrique Centrale, but retains the names at varietal level.

Tab. 21. HABENARIA FILICORNIS. 1, habit (×¼); 2, habit (×1); 3, flower (×3); 4, dorsal sepal (×10); 5, lateral sepal (×10); 6, petal (×10); 7, column, with part of lip, spur and ovary (×10); 8, rostellum (×10), 1–8 from *Milne-Redhead & Taylor* 9337. Drawn by Heather Wood. From F.T.E.A.

4. **Habenaria hologlossa** Summerh. in Kew Bull. **13**: 57 (1958); in F.T.E.A., Orchidaceae: 53 (1968). —la Croix et al., Orch. Malawi: 54 (1991). Type from Kenya.

Slender terrestrial herb 20–40 cm high with ellipsoid tuber 1–1.5 cm long. Leaves 4–7, spread along the stem, 5–12 cm long, 3–5 mm wide, linear, the two lowest sheath-like, whitish, with dark green reticulate veining. Inflorescence laxly 2–10-flowered; bracts up to 12 mm long, lanceolate, acute. Flowers green. Pedicel 6–8 mm long, erect; ovary arched, 10–12 mm long. Dorsal sepal 3–4 × 2–2.5 mm, ovate, erect, convex; lateral sepals 4–5.5 × 2–3 mm, obliquely elliptic, spreading. Petals 3.5–4 × 1.2–2 mm, obliquely lanceolate, entire, forming a hood with the dorsal sepal. Lip entire, 5–6 × 1.5–1.7 mm, rather fleshy, ligulate; spur (3)9–12 mm long, slender, sometimes slightly inflated in apical half, slightly incurved. Column c. 3 mm high; stigmatic arms clavate, 1.5–2 mm long; anther erect, c. 1.5 mm long.

Malawi. N: Nyika National Park, Katizi Valley, c. 1900 m, fl. 21.ii.1982, *Elias* 10 (K); Nyika National Park, Zovochipolo, c. 2200 m, fl. 3.iii.1986, *la Croix* 998 (K; MAL).
Also in Kenya and Angola. Wet montane grassland; 1900–2200 m.
The type specimen of *H. hologlossa* (*Tweedie* 324) has a dense, many-flowered inflorescence, but other specimens from Kenya and Angola, (eg. *Milne-Redhead* 4523) have a lax, few-flowered inflorescence, similar to those of the Nyika Plateau plants. In some instances, specimens with dense, many-flowered and lax, few-flowered inflorescences are mounted on the same herbarium sheet.

5. **Habenaria tenuispica** Rendle in J. Bot. **33**: 293 (1895). —Rolfe in F.T.A. **7**: 215 (1898). — Summerhayes in F.T.E.A., Orchidaceae: 51 (1968). —Williamson, Orch. S. Centr. Africa: 41 (1977). —Geerinck in Fl. Afr. Centr., Orchidaceae pt. 1: 86 (1984). —la Croix et al., Orch. Malawi: 55 (1991). Type from Zaire.

Terrestrial herb 40–80 cm high with an ellipsoid tuber c. 1 cm long. Leaves 6–11, spaced along stem, the lowest 2 sheathing, the remainder erect, 12–30 × 1–2 cm, lanceolate or linear. Inflorescence 5–22 cm long, fairly densely many-flowered; bracts c. 20 mm long. Flowers green. Ovary and pedicel arched, 8–10 mm long. Dorsal sepal erect, 2.5–3 × 1.5–2.5 mm, ovate, convex; lateral sepals 2.5–3 × 1.5mm, obliquely ovate, deflexed. Petals entire, 2–2.5 × 1.2–1.7 mm, lying beside dorsal sepal. Lip 2.5–3 mm long, 3-lobed with a basal undivided part 1.2–2 mm long; all lobes linear, subacute, the mid-lobe longer and slightly broader than the side lobes; spur 4–6 mm long. Anther 1 mm high, canals 0.3 mm long. Stigmatic arms c. 1 mm long, clavate, stout.

Malawi. N: Nyika National Park, fl. xii.1966, *G. Williamson* 235 (K).
Also in Zaire, Uganda, Kenya and Tanzania. Montane bog; 2200–2300 m.

6. **Habenaria xanthochlora** Schltr. in Bot. Jahrb. Syst. **53**: 496 (1915). —Summerhayes in F.T.E.A., Orchidaceae: 56 (1968). —Williamson, Orch. S. Centr. Africa: 40 (1977). —la Croix et al., Orch. Malawi: 56 (1991). Type from Tanzania.

Terrestrial herb 40–50 cm high. Stem leafy; leaves 6–10 × 1–3.5 cm, lanceolate. Inflorescence 10–20 cm long, fairly densely many-flowered; bracts up to 27 mm long at base of inflorescence, overtopping the flower, lanceolate, acuminate. Flowers green outside, yellow inside, turning very dark brown (darker than leaves and bracts) when dry. Ovary and pedicel 8 mm long. Dorsal sepal 6–7 × 2–3 mm, broadly lanceolate, acute, erect; lateral sepals spreading, a little longer and narrower than dorsal sepal, oblong or oblong-elliptic, obtuse to subacute. Petals entire, c. 5.5 × 1 mm, narrowly lanceolate, curved. Lip 5.5–7 mm long, 3-lobed with undivided base c. 2 mm long; all lobes narrowly lanceolate, acute, the mid-lobe longer and broader than the side lobes; spur 5–6 mm long. Stigmatic arms clavate, c. 2 mm long.

Zambia. E: Lundazi Distr., Nyika Plateau near Resthouse, fl. iii.1967, *Williamson & Odgers* 266 (K; SRGH). **Malawi.** N: Nyika National Park, south of Chelinda Camp on road to Chosi, c. 2250 m, fl. 10.iii.1977, *Grosvenor & Renz* 1115 (K; SRGH).
Also in Tanzania. Montane grassland; 2130–2350 m.

7. **Habenaria petitiana** (A. Rich.) T. Durand & Schinz, Consp. Fl. Afr. **5**: 83 (1895). — Summerhayes in F.T.E.A., Orchidaceae: 56 (1968). —Grosvenor in Excelsa **6**: 83 (1976). — Williamson, Orch. S. Centr. Africa: 41 (1977). —Stewart et al., Wild Orch. South. Africa: 84 (1982). —Geerinck in Fl. Afr. Centr., Orchidaceae pt. 1: 82 (1984). —la Croix et al., Orch. Malawi: 56 (1991). TAB. **22**. Type from Ethiopia.

Peristylus petitianus A. Rich. in Ann. Sci. Nat., sér. 2, **14**: 266 (1840). —Rolfe in F.T.A. **7**: 199 (1898).

Platanthera petitiana (A. Rich.) Engl. in Abh. Preuss. Akad. Wiss. **1891**: 179 (1892).

Platanthera volkensiana Kraenzl. in Pflanzenw. Ost-Afr. **C**: 151 (1895). Type from Tanzania.

Peristylus volkensianus (Kraenzl.) Rolfe in F.T.A. **7**: 198 (1898).

Peristylus snowdenii Rolfe in Bull. Misc. Inform., Kew **1918**: 237 (1918). Type from Kenya.

Peristylus ugandensis Rolfe in Bull. Misc. Inform., Kew **1918**: 237 (1918). Type from Kenya.

Terrestrial herb 25–55 cm high with an ellipsoid tuber 1–2.5 cm long. Leaves 5–9, spaced along stem, the 2 lowest sheathing, the rest 4–9 × 2–6.5 cm, ovate, acute, the edge often undulate, clasping the stem at the base. Inflorescence 5–20 cm long, laxly to fairly densely many-flowered; bracts 7–15 mm long, lanceolate, acute, longer than the flowers. Flowers green. Ovary and pedicel c. 5 mm long, arched at apex. Sepals 2–3 × 1 mm; dorsal erect, ovate, convex; lateral sepals spreading, elliptic. Petals entire, 2–3.5 × 1–1.5 mm, obliquely oblong, obtuse. Lip 2.5–4 mm long and wide, 3-lobed in the apical half; side lobes slightly longer than mid-lobe, all lobes ± triangular, obtuse; spur 1–1.5 mm long, thick and blunt. Anther c. 1 mm tall, canals absent. Stigmatic arms c. 1 mm long.

Zambia. W: Ichimpi, Kitwe, fr. 13.v.1962, *Mutimushi* (K). C: Serenje, fl. iii.1967, *Williamson & Odgers* 288 (K). **Zimbabwe**. E: Mutare, Himalaya Range, fl. 2.iii.1954, *Wild* 4435 (K; SRGH). **Malawi**. N: Nyika National Park, forest patch on road to Kasaramba, c. 2300 m, fl. 4.iv.1987, *la Croix & Meredith* 1028 (K; MAL). S: Thyolo Mt., c. 1400 m, fl. 23.iii.1980, *la Croix* 11 (K).

Also in Cameroon, Ethiopia, Zaire, Uganda, Kenya, Tanzania and South Africa (E Transvaal). Evergreen forest, in dense shade; 1250–2300 m.

Plants from the Flora Zambesiaca area tend to have laxer inflorescences with fewer flowers than those from East Africa.

8. **Habenaria zambesina** Rchb.f., Otia Bot. Hamburg.: 96 (1881). —Rolfe in F.T.A. **7**: 211 (1898). —Summerhayes in F.W.T.A. ed. 2, **3**: 193 (1968); in F.T.E.A., Orchidaceae: 58 (1968). —Grosvenor in Excelsa **6**: 83 (1976). —Williamson, Orch. S. Centr. Africa: 42 (1977). —Geerinck in Fl. Afr. Centr., Orchidaceae pt. 1: 94 (1984). —la Croix et al., Orch. Malawi: 57 (1991). Type: Mozambique, without precise locality, *Kirk* (K, holotype).

Habenaria myriantha Kraenzl. in Notizbl. Bot. Gart. Berlin **3**: 237 (1903). Type from Tanzania.

Robust terrestrial herb up to c. 1 m tall; roots several, fleshy, up to 12 × 1 cm, cylindrical, sometimes branched. Leaves 8–9, spaced along stem, the lowest 2–3 loosely sheathing, the remainder up to 22 × 5 cm, ovate, acute, light green, loosely clasping the stem at the base, the edge sometimes undulate, decreasing in size towards the stem apex. Inflorescence 11–25 × 4.5–5 cm, densely many-flowered; bracts 17–22 mm long, lanceolate, acute. Flowers white, the sepals turning green towards the apex. Ovary and pedicel 18–20 mm long. Dorsal sepal 5–6 × 3–5 mm, ovate, reflexed; lateral sepals 6–10 × 3–4 mm, ovate, deflexed or spreading, twisted. Petals 5–10 × 3–4 mm, ovate, acute, entire, pointing forwards and more or less forming a hood. Lip 8–13 × 1.5–2.5 mm, ligulate entire; spur (1.5)5.5–9 cm long, slender, pendent. Anther c. 2 mm tall, canals very short. Stigmatic arms 2–2.5 mm long, clavate, usually joined near the tip.

Zambia. N: Kasama Distr., Mungwi, fl. 18.xii.1960, *E.A.Robinson* 4200 (K). W: Mwinilunga Distr., Kalenda Ridge, west of Matonchi Farm, fl. 1.i.1938, *Milne-Redhead* 3905 (K). C: Kabwe (Broken Hill), Chibwe P.F.A., fl. 16.i.1961, *Morze* 44 (K). S: Muckle Neuk, 19 km north of Choma, c. 1250 m, fl. 14.i.1954, *E.A. Robinson* 463 (K). E: Chipata, fl. i.1952, *Benson* N.R. 26 (BM). **Zimbabwe**. C: Harare, Chakoma Farm, c. 24 km north of Harare, fl. 24.i.1977, *Grosvenor* 874 (PRE; SRGH). **Malawi**. N: Nkhata Bay Distr., near Mpamba, c. 600 m, fl. 4.ii.1973, *Pawek* 6422 (K; MAL; MO). C: Lilongwe Distr., Dzalanyama Forest Reserve, near Chiungiza, 1550 m, fl. 9.ii.1959, *Robson* 1523 (K). S: Thyolo Distr., Bvumbe, 1150 m, fl. 17.i.1981, *la Croix* 94 (K). **Mozambique**. *Kirk* s.n. (K).

Throughout most of tropical Africa. Wet grassland, dambos; 600–1550 m.

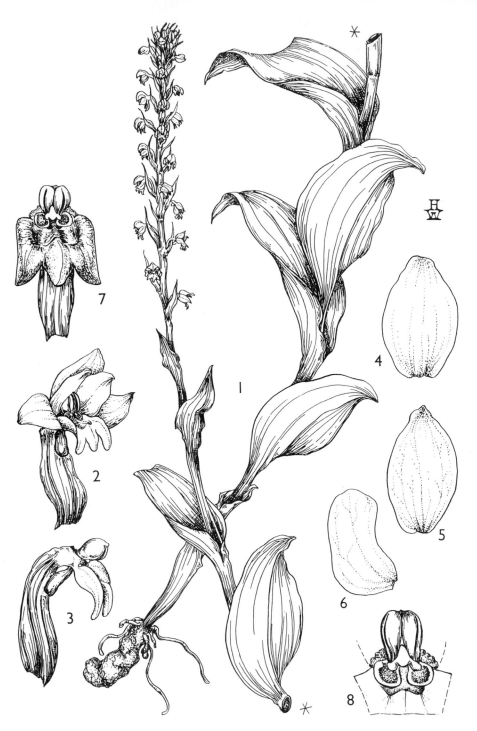

Tab. 22. HABENARIA PETITIANA. 1, habit (×1); 2, flower (×5); 3, flower, tepals removed to show lip and column (×5); 4, dorsal sepal (×10); 5, lateral sepal (×10); 6, petal (×10); 7, column and lip, front view (×5); 8, column, front view (×10), 1–8 from *Drummond & Hemsley* 2423. Drawn by Heather Wood. From F.T.E.A.

9. **Habenaria gabonensis** Rchb.f. in Bot. Zeitung (Berlin) **10**: 934 (1852). —Rolfe in F.T.A. **7**: 220 (1898). —Summerhayes in Bull. Misc. Inform., Kew **1938**: 144 (1938); in F.W.T.A. ed. 2, **3**: 194 (1968). Type from Gabon.

Var. **psiloceras** (Welw. ex Rchb.f.) Summerh. in Bot. Mus. Leafl. Harv. Univ. **10**, 9: 263 (1942). —Williamson, Orch. S. Centr. Africa: 41 (1977). Type from Angola.
    Habenaria psiloceras Welw. ex Rchb.f. in Flora **50**: 99 (1867).

Terrestrial herb up to 75 cm high with many long, fleshy roots. Leaves c. 8, the lowest 2 ± sheath-like, the rest 1–20 × 3–5.5 cm, oblong, acute. Inflorescence rather laxly 10 to many-flowered; bracts 18–28 mm long, ovate. Flowers white. Ovary and pedicel 4.5–5.5 cm long. Dorsal sepal 9–10.5 × 6 mm, ovate, acute, erect, convex; lateral sepals 10–12 × 5–8 mm, deflexed. Petals entire, 8–11 × 3.5–4.5 mm, ovate, projecting forwards. Lip 3-lobed, the undivided basal part 4–6 mm long. Mid-lobe 15–17 × 1–1.5 mm, side lobes 16–23 × 0.5 mm, all lobes linear; spur 13–16 cm long, pendent. Anther 4–5 mm high, canals 6–7 mm long, the last 2 mm upturned. Stigmatic arms 4–7 mm long.

Zambia. W: Mwinilunga Distr., 6 km north of Kalene Hill, fl. 12.xii.1963, *E.A. Robinson* 5960 (K). C: Kabwe, fl. xii.1969, *G. Williamson* 1866 (K).
    Also in Angola. Damp shady areas by rivers.
    Var. *psiloceras* is distinguished from the typical variety by its longer spur (13–16 cm long). Var. *gabonensis*, restricted to West Africa, has a spur up to 9 cm long.

10. **Habenaria insolita** Summerh. in Kew Bull. **16**: 259 (1962); in F.T.E.A., Orchidaceae: 60 (1968). —la Croix et al., Orch. Malawi: 56 (1991). TAB. **23**. Type from Tanzania.

Robust terrestrial herb 80–90 cm tall. Leaves c. 10, the lowest 2–3 sheathing, the upper 3 bract-like, the remainder erect, up to 20 × 4 cm, broadly lanceolate, clasping the stem at the base; the base of the stem and the 3 lowermost leaf sheaths with large, black spots. Inflorescence 12–18 × 3–4 cm, densely many-flowered; bracts prominent, longer than the flower, up to 55 mm long at base of inflorescence, lanceolate, acute. Flowers yellow-green. Ovary and pedicel 12–13 mm long. Dorsal sepal 14–15 × 6–7 mm, lanceolate, acute, erect, convex; lateral sepals 14–15 × 3.5–5 mm, spreading and curving upwards. Petals adnate to dorsal sepal along inner edge, 13–14 × 3–4 mm, curved-lanceolate from a narrow base, irregularly toothed on outer edge. Lip 3-lobed, with undivided basal part 3.5–4 mm long; mid-lobe 12–17 mm long, side lobes 9–15 mm long, all lobes linear, c. 1 mm wide; spur c. 14 mm long, inflated for most of its length to c. 2 mm wide, straight or curving forwards, obtuse. Anther c. 3 mm tall, canals 1.5–2 mm long. Stigmatic arms 2–3 mm long.

Malawi. N: Nyika National Park, c. 8 km north of Thazima, c. 1800 m, fl. 5.iv.1987, *la Croix* 1029 (K). C: Ntcheu Distr., Tsangano Hill, fl. & fr. iii.1956, *Adlard* 238F (FHO; K).
    Also in Tanzania. Long grass at edge of woodland; 1800–1900 m.
    Apparently a rare species.

11. **Habenaria epipactidea** Rchb.f. in Flora **50**: 100 (1867). —Rolfe in F.T.A. **7**: 219 (1898). —Summerhayes in Bull. Misc. Inform., Kew **1939**: 490 (1939); in F.T.E.A., Orchidaceae: 61 (1968). —Grosvenor in Excelsa **6**: 83 (1976). —Stewart et al., Wild Orch. South. Africa: 84 (1982). —Geerinck in Fl. Afr. Centr., Orchidaceae pt. 1: 89 (1984). Type from Angola.
    Orchis foliosa Sw. in Kongl. Vetensk. Acad. Nya Handl. **21**: 206 (1800). Type from South Africa.
    Habenaria foliosa (Sw.) Rchb.f. in Flora **48**: 180 (1865), non A. Rich.
    Habenaria hircina Rchb.f. in Flora **50**: 100 (1867). Type from Angola.
    Habenaria polyphylla Kraenzl. in Bot. Jahrb. Syst. **16**: 214 (1892). Type from South Africa.
    Habenaria schinzii Rolfe in F.T.A. **7**: 219 (1898). Type from Namibia.
    Habenaria perfoliata Kraenzl. in Bull. Herb. Boissier, sér. 2, **2**: 942 (1902). Type from Namibia.
    Habenaria rautanenii Kraenzl. in Bull. Herb. Boissier, sér. 2, **4**: 1008 (1904). Type from Namibia.

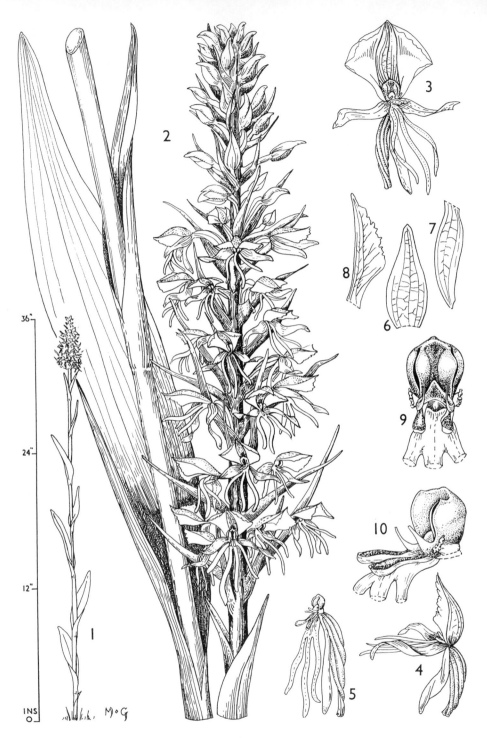

Tab. 23. HABENARIA INSOLITA. 1, habit ($\times\frac{1}{9}$); 2, part of leafy stem and inflorescence ($\times$1);
3, single flower, front view ($\times$2); 4, single flower, side view ($\times$2); 5, flower, tepals removed
($\times$2); 6, dorsal sepal ($\times$2); 7, lateral sepal ($\times$2); 8, petal ($\times$2); 9, column, front view ($\times$6);
10, column, side view ($\times$6), 1–10 from *McLoughlin* 85. Drawn by Mary Grierson. From
Kew Bull.

Terrestrial herb up to 50 cm high; tubers c. 4 × 1.5 cm, ovoid, villous. Stem with 8–15 overlapping leaves, the lowermost 1–2 sheath-like, the upper ones bract-like, the largest in the middle, 5–12 × 1–2.5 cm, lanceolate, acute, the edge often undulate, ± distichously arranged. Inflorescence densely 5 to many-flowered, up to 16 × c. 4 cm; bracts up to 25 mm long, lanceolate, acute. Flowers white or greenish-white. Ovary and pedicel 22–27 mm long. Dorsal sepal 8–10 × 4–6 mm, ovate, erect, convex; lateral sepals 9–12 × 3–4 mm, spreading, curved-lanceolate, subacute. Petals entire, 8–10 × 6.5–8 mm, orbicular to elliptic, rounded. Lip 10–14 mm long, 3-lobed to c. 1 mm from base; mid-lobe c. 2 mm wide, side lobes much shorter and narrower, only 1–4 mm long, thread-like; spur 18–30 mm long, straight or sigmoid, swollen towards the tip. Anther 3.5 mm high, canals very short. Stigmatic arms 2–3 mm long, clavate, truncate. Rostellum mid-lobe almost covering anther.

Botswana. SE: Musi Ranch Guest House, 1250 m, fl. 3.ii.1978, *O.J.Hansen* 3345 (C; GAB; K; PRE; SRGH). **Zimbabwe**. C: Marondera (Marandellas), c. 1500 m, fl. 13.ii.1950, *Corby* 640 (K; SRGH). E: Nyanga (Inyanga), near Rhodes Hotel, c. 1700 m, fl. 16.ii.1958, *Beasley* 60 (K; SRGH).
Also in Ethiopia, Rwanda, Uganda, Kenya, Tanzania, Angola and South Africa. Grassland and open woodland; 1000–1700 m.

12. **Habenaria macrandra** Lindl. in J. Linn. Soc., Bot. **6**: 139 (1862). —Summerhayes in F.W.T.A. ed. 2, **3**: 193 (1968); in F.T.E.A., Orchidaceae: 63 (1968). —Grosvenor in Excelsa **6**: 83 (1976). —Geerinck in Fl. Afr. Centr., Orchidaceae pt. 1: 80 (1984). —la Croix et al., Orch. Malawi: 58 (1991). TAB. **24**. Type from Nigeria.
   Podandria macrandra (Lindl.) Rolfe in F.T.A. **7**: 206 (1898).
   Habenaria stenorhynchus Kraenzl. in Bot. Jahrb. Syst. **33**: 55 (1902), non Schltr. Type from Tanzania.

Terrestrial herb 25–55 cm high with fleshy, cylindrical, woolly roots. Leaves mostly clustered at the stem base, with 3–5 bract-like leaves spaced along the stem. Basal leaves c. 6, lanceolate or ovate, acute, petiolate; lamina 7–13 × 3–5 cm; petiole 2.5–5.5 cm long. Inflorescence laxly 2–11-flowered; bracts 25–27 mm long. Flowers suberect, white with greenish apices to the tepals, the spur green. Ovary and pedicel 23–25 mm long. Dorsal sepal 16–20 × 3–4 mm, lanceolate, acute, erect; lateral sepals 17–22 × 3–4 mm, obliquely lanceolate, deflexed. Petals 17–25 mm long, narrowly linear or filiform, entire, erect. Lip 3-lobed ± to the base; mid-lobe 26–30 mm long, the side lobes 45–50 mm long, all lobes filiform but mid-lobe slightly broader than the rest; spur 6–7.5 cm long, straight or slightly curved, slender but slightly swollen towards tip. Anther 13–22 mm long, slender with parallel loculi; canals 1.5 mm long, free from rostellum side lobes. Stigmas cushion-like, c. 1.5 mm long.

Zimbabwe. E: Chimanimani Mts., Haroni Valley, 1060 m, fl. 11.iv.1954, *Ball* 314 (K; SRGH). **Malawi**. S: Mulanje Distr., Ruo Gorge, 730 m, fl. 27.iii.1967, *Berrie* s.n. (MAL). **Mozambique**. MS: lower SE slopes of Chimanimani Mts., c. 20 km west of Dombe, between Muerera and Mevumesi Rivers, c. 350 m, fl. 23.iv.1974, *Pope & Müller* 1273 (K; SRGH).
Also in West Africa, Zaire, Uganda and Tanzania. Evergreen forest, dense riverine forest; 350–1060 m.

13. **Habenaria falciloba** Summerh. in Bot. Mus. Leafl. Harv. Univ. **14**: 215 (1951). —Williamson, Orch. S. Centr. Africa: 43 (1977). —Geerinck in Fl. Afr. Centr., Orchidaceae pt. 1: 95 (1984). Type from Zaire.

Robust terrestrial herb up to 75 cm high. Stem leafy; leaves c. 17 × 5 cm, broadly lanceolate. Inflorescence c. 15 × 6.5 cm, densely many-flowered; bracts leafy, longer than the flower. Flowers semi-erect, greenish. Ovary and pedicel 20–25 mm long. Dorsal sepal 23–28 × 8 mm, erect, convex; lateral sepals 25–27 × 9–10 mm, deflexed; all sepals with a keeled midrib. Petals 25 × 7 mm, falcate-lanceolate from a narrow base, adnate to dorsal sepal on inner edge, entire. Lip c. 27 mm long from base to apex of mid-lobe, 3-lobed with a claw 5–8 × c. 5 mm; mid-lobe c. 16 × 1.3–2 mm, ligulate; side lobes c. 13 × 3.5 mm, falcate, acute or occasionally truncate, entire; spur 20 mm long, swollen to a width of 3.5 mm near apex. Anther loculi 3.5 mm high, total width 12 mm, 5 mm wide between loculi; canals 4 mm long. Stigmatic arms 9 mm long; rostellum mid-lobe 1.5 mm high.

Tab. 24. HABENARIA MACRANDRA. 1, habit (×1); 2, dorsal sepal (×2); 3, lateral sepal (×2); 4, petal (×2); 5, column, upper part removed, with part of lip, spur and ovary (×4), 1–5 from *Eggeling* 3443. Drawn by Heather Wood. From F.T.E.A.

Zambia. W: Solwezi Distr., 39 km north of Solwezi, on Katanga border, fl. 19.iii.1961, *Drummond & Rutherford-Smith* 7074 (K; SRGH).

Also in Zaire. Wet *Uapaca* woodland.

Summerhayes commented that while this species agrees with Sect. *Multipartitae* in the general habit, large flowers, veining of sepals, character of petals, long, slender anther canals and broad, horse-shoe-shaped anther connective, it differs from all the other species in the section in having entire side lobes of the lip and in the stigmas not being adnate to the lower side of the rostellar arms. Also, the anther connective, although very broad, is comparatively tall, as in *H. egregia* Summerh., but the loculi are almost horizontal. He concluded that it seems to be an aberrant species which might warrant a distinct section.

14. **Habenaria praestans** Rendle in J. Bot. **33**: 293 (1895). —Rolfe in F.T.A. **7**: 225 (1898). —Summerhayes in Bot. Not. **1937**: 186 (1937); in F.T.E.A., Orchidaceae: 70 (1968). —Grosvenor in Excelsa **6**: 83 (1976). —Williamson, Orch. S. Centr. Africa: 43, pl. 39 (1977). —Geerinck in Fl. Afr. Centr., Orchidaceae pt. 1: 98 (1984). —la Croix et al., Orch. Malawi: 58 (1991). Type from Uganda.

Habenaria ctenophora Schltr. in Bot. Jahrb. Syst. **53**: 500 (1915). Type from Tanzania.

Robust terrestrial herb 30–100 cm high; tubers up to 3.5 cm long, ovoid or ellipsoid, woolly. Leaves 6–12, the lower 2–3 sheath-like, the rest spreading, decreasing in size towards the stem apex, the largest 10–25 × 3–6 cm, lanceolate, acute, light green with 3 or 5 prominent longitudinal veins. Inflorescence 13–25 × 4–8 cm, densely many-flowered; bracts leafy, 28–50 mm long. Flowers green, white in centre. Ovary and pedicel 15–25 mm long. Dorsal sepal 13–22 × 7–10 mm, ovate, erect, convex; lateral sepal 15–22 × 6–11 mm, obliquely lanceolate, acuminate, spreading. Petals 15–21 × 3.5–6 mm, curved-lanceolate, acute or obtuse, adnate to dorsal sepal, entire, erect, glabrous. Lip 30–40 mm long, 3-lobed with a basal claw 4–10 mm long; claw and basal part of lobes densely pubescent; mid-lobe 13.5–22 × 1 mm, linear; side lobes diverging, 16–23 mm long, much divided into comb-like segments on the outer margins; spur 12–22 mm long, straight or sigmoid, swollen to c. 2 mm wide at the tip. Anther c. 3 mm tall, connective 11–16 mm wide; canals 3.5–4 mm long; staminodes globose on slender stalks. Stigmatic arms 8–12.5 mm long, the receptive surface swollen, c. 4.5 mm long.

Var. **praestans**

Zambia. N: Kasama Distr., Chishimba Falls, fl. 12.ii.1961, *E.A.Robinson* 4372 (K). W: Mwinilunga Distr., by R. Kasomba, fl. 15.xii.1937, *Milne-Redhead* 3679 (K). C: 8 km west of Lusaka, fl. iii.1969, *G. Williamson* 1381 (K). E: Lundazi Distr., Nyika Plateau, near Rest-house, fl. 10.i.1964, *Benson* NR 476 (K). **Zimbabwe**. C: Harare, fl. 27.ii.1946, *Greatrex* 49 (K; PRE). E: Nyanga (Inyanga), above Cheshire, c. 1700 m, fl. 4.ii.1931, *Norlindh & Weimarck* 4816 (K). **Malawi**. N: Nyika National Park, Kasaramba road, near Juniper Forest junction, c. 2400 m, fl. & fr. 28.iii.1970, *Pawek* 3396 (K; MAL). C: c. 14 km west of Dedza town, fl. 20.i.1971, *Westwood* 589 (K). S: Blantyre Distr., Ndirande Mt., steep rocky slopes, 1450–1570 m, fl. 3.iii.1970, *Brummitt* 8869 (K; MAL).

Also in Ethiopia, Zaire, Rwanda, Uganda and Tanzania. Montane and upland grassland, *Brachystegia* woodland and forest margins, stream banks and in long grass in rocky areas; 1000–2400 m.

Var. **umbrosa** G. Will. in J. S. African Bot. **48**: 11, fig. 1 (1982). Type: Zambia, c. 6 km west of Lusaka, *G. Williamson* 1867 (SRGH, holotype).

Differs from var. *praestans* in the following features; the inflorescence of var. *umbrosa* is laxer and fewer-flowered, the bracts are more prominent and leaf-like, the flowers are smaller and whiter, the spur is shorter than the mid-lobe of the lip and the staminodes are not always stalked. Var. *umbrosa* occurs in forest shade, usually in swampy ground.

Zambia. C: Rufunsa (Rofunza), fl. ii.1964, *Morze* 160 (K); 6 km west of Lusaka, fl. ii.1970, *G. Williamson* 1867 (SRGH). **Malawi**. N: Chitipa Distr., base of Mafinga Mts. near Chisenga Rest-house, fl. ii.1970, *Williamson & Drummond* 1946 (SRGH). C: Ntchisi Forest Reserve, 1400 m, fl. 21.iii.1984, *Dowsett-Lemaire* 1127 (K).

Not known elsewhere. Deep shade in riverine forest and *Syzygium* forest by streams, usually in swampy ground; c. 1400 m.

15. **Habenaria splendens** Rendle in J. Linn. Soc., Bot. **30**: 395 (1895). —Rolfe in F.T.A. **7**: 224 (1898). —Summerhayes in F.T.E.A., Orchidaceae: 70 (1968). —Williamson, Orch. S. Centr. Africa: 43 (1977). —la Croix et al., Orch. Malawi: 59 (1991). TAB. **25**. Type from Tanzania.

Robust terrestrial herb 35–70 cm high with ovoid or ellipsoid tubers 3–4 cm long. Leaves 6–12, the 2 lowermost sheathing, the mid-cauline largest and up to 20 × 6 cm, ovate, acute, loosely funnel-shaped at base, dark green with 3 prominent longitudinal veins. Inflorescence 10–22 × 7–9 cm, fairly laxly 4–20-flowered; bracts up to 6 cm long. Sepals green, petals and lip white. Ovary and pedicel erect, slightly arched, 25–35 mm long. Dorsal sepal 23–30 × 12–13 mm, ovate, erect; lateral sepals 22–31 × 8–10 mm, obliquely ovate, spreading. Petals 20–30 × 4–8 mm, falcate, entire, adnate to dorsal sepal. Lip 25–40 mm long, 3-lobed with a basal claw 7–13 mm long, the claw and base of lobes densely pubescent; mid-lobe 15–23 × 1 mm, linear; side lobes 18–24 mm long, fimbriate towards the apex with 8–12 branches c. 7 mm long; spur 25–50 mm long, sigmoid, slightly swollen at apex, pendent. Anther connective 12–18 mm wide, 3–4 mm high; canals 4–6 mm long, erect. Staminodes stalked, c. 2 mm long. Stigmatic arms 16–20 mm long, joined to the side lobes of the rostellum in the basal half.

Zambia. N: Mbala Distr., Kawimbe, 1680 m, fl. 24.i.1957, *Richards* 7957 (K). C: 11 km west of Lusaka, fl. ii.1969, *G. Williamson* 1380 (K). **Malawi**. N: Mzimba Distr., Mzuzu, Lunyangwa, 1330 m, fl. 19.iii.1987, *la Croix* 1018 (K). S: Zomba Distr., Namasi, fl. v.1899, *Cameron* 10 (K).
Also in Ethiopia, Uganda, Kenya and Tanzania. Junction of forest and woodland or forest and grassland, *Syzygium* thicket; 1270–2100 m.

16. **Habenaria splendentior** Summerh. in Bull. Misc. Inform., Kew **1949**: 427 (1949). —Williamson, Orch. S. Centr. Africa: 43, fig. 16 (1977). —Geerinck in Fl. Afr. Centr., Orchidaceae pt. 1: 99 (1984). Type: Zambia, Mwinilunga Distr., *Milne-Redhead* 3789 (K, holotype).

Terrestrial herb 40–80 cm high with ellipsoid tubers. Stem leafy; lowermost 2 leaves ± sheathing, the rest ovate, the largest 10–15 × 5.5–8 cm. Inflorescence 15–20 cm long, fairly laxly 5–10-flowered; bracts leafy, up to 65 mm long. Flowers pale green and white. Ovary and pedicel 35 mm long. Dorsal sepal 30–38 × 13–17 mm, erect, convex; lateral sepals 30–38 × 11–16 mm, obliquely ovate, spreading. Petals 30–38 × 6–7 mm, falcate. Lip 43–45 mm long, 3-lobed with a claw 12–16 mm long, puberulous; mid-lobe 26–31 × 1–2 mm, linear; side lobes 25–30 mm long, with 10–13 fine branches; spur 6.5–7.8 cm long. Anther connective 14–18 mm wide; canals 6–7 mm long, turned up at the ends; staminodes sessile. Stigmatic arms 25–27 mm long, free part 11–14 mm long.

Zambia. W: Mwinilunga Distr., just north of Matonchi Farm, fl. 22.xii.1937, *Milne-Redhead* 3789 (K); Ndola, fl. 18.i.1954, *Fanshawe* 692 (K).
Also in Zaire. *Brachystegia* woodland and pine plantations.

17. **Habenaria pubipetala** Summerh. in Bot. Mus. Leafl. Harv. Univ. **10**: 266 (1942). —Moriarty, Wild Fls. Malawi: t. 26, 5 (1975). —la Croix et al., Orch. Malawi: 60 (1991). Type: Malawi, Zomba Mt., *Lawrence* 341 (K, holotype).

Terrestrial herb 30–70 cm high, with ellipsoid tubers c. 2 cm long. Leaves 8–10, spaced along stem, the 2 lowermost sheathing, the mid-cauline largest and up to 22 × 5 cm, broadly lanceolate, acute. Inflorescence 16–30 × 5–6 cm, rather laxly many-flowered; bracts 17–18 mm long. Sepals and lip green, petals white, densely pubescent. Pedicel c. 15 mm long, spreading; ovary arched, 12–13 mm long. Dorsal sepal 3–4.5 × 4–4.6 mm, erect; lateral sepals 10–11 × 3.5 mm, reflexed, curled. Petals 8–12 × 3 mm, erect, the basal part rounded, adnate to dorsal sepal; the upper part narrowing abruptly to an acuminate apex. Lip 3-lobed with a basal claw 5–8 mm long; mid-lobe 8–12 mm long; side lobes 12–17 mm long, curled at the ends; all lobes linear, less than 1 mm wide; spur 20–25 mm long, with a spiral twist. Anther 2.8–3.8 mm high; canals 2–3 mm long. Stigmatic arms c. 3.5 mm long.

Tab. 25. HABENARIA SPLENDENS. 1, habit (×⅔), from *Williamson & Gassner* 2362; 2, dorsal sepal (×1); 3, lateral sepal (×1); 4, petal (×1); 5, lip, spur removed (×1); 6, close-up of basal region (×8); 7, close-up of lateral lobe (×8); 8, column, front view (×2); 9, column, back view (×2); 10, anther cap (×3); 11, pollinium (×3), 2–11 from *Polhill & Paulo* 1597 (in K spirit coll. 6683). Drawn by Judi Stone.

Malawi. N: Khondowe to Karonga, fl. vii.1896, *Whyte* s.n. (K). C: Ntchisi Mt. Forest Reserve, fl. 11.v.1984, *Banda & Kaunda* 2164 (K; MAL). S: Zomba Mt., Mulunguzi stream, 1350 m, fl. 28.iii.1937, *Lawrence* 341 (K); Mangochi Distr., Namwera, Mt. Uzuzu, c. 1250 m, fl. 29.iii.1986, *Jenkins* 13 (K).

Although fairly widespread in Malawi, this species has not as yet been found elsewhere. Evergreen forest, sparse grass at edge of *Brachystegia* woodland, seepage slopes; 1250–2000 m.

18. **Habenaria trachypetala** Kraenzl. in Bot. Jahrb. Syst. **30**: 281 (1901). —Summerhayes, F.T.E.A., Orchidaceae: 71 (1968). —Grosvenor in Excelsa **6**: 83 (1976). —Williamson, Orch. S. Centr. Africa: 45, fig. 17 (1977). —Geerinck in Fl. Afr. Centr., Orchidaceae pt. 1: 83 (1984). —la Croix et al., Orch. Malawi: 61 (1991). TAB. **26**. Type from Tanzania.
    Habenaria rhombocorys Schltr. in Bot. Jahrb. Syst. **53**: 501 (1915). Type from Tanzania.

Robust terrestrial herb to c. 1 m tall. Tubers 3–4 × 1.5–3 cm, ovoid or ellipsoid. Stem leafy; leaves 6–9, the lowermost 1–2 sheath-like, the middle ones suberect or spreading, 10–15 × 2.5–7 cm, broadly lanceolate, loosely clasping stem at base; upper leaves grading into bracts. Inflorescence 10–24 × 4–6 cm, densely many-flowered; bracts leafy, shorter than flowers. Flowers green or green and white, suberect or spreading. Pedicel and ovary 2–2.5 cm long. Sepals glabrous; dorsal sepal erect, 10–18 × 6–7 mm, lanceolate, acuminate, convex; lateral sepals deflexed, obliquely lanceolate, acuminate, slightly longer and narrower than dorsal sepal. Petals densely hairy, adnate to dorsal sepal, 3.5–7 mm wide in centre, obliquely lanceolate, angled on the outer edge. Lip glabrous, pendent, 3-lobed with a basal claw 5–7.5 mm long; mid-lobe 12–18 mm long, side lobes 13–23 mm long, curled up at the ends; all lobes linear, c. 1 mm wide; spur 1.5–2.2 cm long, ± parallel to pedicel and ovary, swollen at apex. Anther c. 3 mm high, the loculi parallel; canals 4–6 mm long, projecting up; stigmas 3–5 mm long, clavate, projecting down. Rostellum mid-lobe 4 mm long, ligulate, truncate at the apex, overtopping anther.

Zambia. E: below Nyika, fl. iii.1966, *Williamson & Odgers* 300 (SRGH). **Zimbabwe**. E: Mutare Distr., Vumba, fl. 17.iii.1957, *Chase* 6348 (K; SRGH). **Malawi**. N: South Viphya, Chikangawa, Wozi Hill, 1780 m, fl. 20.iii.1987, *Cornelius in la Croix* 1020 (K). C: Dedza Mt., between radio relay station and summit beacon, c. 2100 m, fl. 3.iii.1977, *Grosvenor & Renz* 1017 (K; SRGH). S: Zomba Mt., near Mandala Cottage, fl. iii.1976, *Welsh* 513 (K; MAL).

Also in Zaire and Tanzania. Evergreen forest margins and montane grassland, *Brachystegia* woodland and tall grass at edge of dambos; 1330–2100 m.

This species is variable in the size and number of the flowers, with plants from Malawi having much larger flowers than those from Zimbabwe.

19. **Habenaria amoena** Summerh. in Kew Bull. **11**: 218 (1956); in F.T.E.A., Orchidaceae: 75 (1968). —Grosvenor in Excelsa **6**: 82 (1976). —Williamson, Orch. S. Centr. Africa: 45 (1977). —Geerinck in Fl. Afr. Centr., Orchidaceae pt. 1: 135 (1984). —la Croix et al., Orch. Malawi: 61 (1991). Type: Zambia, Mbala Distr., Chilongowelo, *Richards* 1000 (K, holotype).

Terrestrial herb 25–55 cm high with globose tubers c. 15 mm in diameter, sometimes with additional tubers produced at the end of c. 13 cm long rhizomes. Stem leafy thoughout, the lowermost foliage leaf c. 8 cm from base. Leaves 9–13, the lowermost 2–4 sheath-like, the largest leaves in the middle, 7.5–12.5 × 1–3.5 cm, oblanceolate (rarely almost linear), narrowing at the base, suberect or spreading, light green. Inflorescence 5–21 × 2.5–4.5 cm, rather laxly many-flowered; bracts 9–15 mm long, lanceolate, acuminate. Flowers yellow-green, spreading, sometimes described as fragrant. Ovary and pedicel arched, 12–18 mm long. Dorsal sepal 4–5.5 × 2.5–4 mm, ovate, erect, convex; lateral sepals 5–7 × 2–3.5 mm, obliquely curved-lanceolate, deflexed. Petals 2-lobed ± to base; posterior (upper) lobe 4–5.5 × c. 1 mm, ± adnate to dorsal sepal; anterior (lower) lobe 5–9.5 × c. 0.5 mm, curving upwards. Lip 3-lobed almost to base; mid-lobe 5.5–9 × 1 mm, linear; side lobes 7–12 mm long, c. 0.5 mm wide, curving up like lower petal lobes; spur 17–22 mm long, slender, parallel to ovary. Anther c. 2 mm high; canals very short, c. 0.5 mm long. Stigmas 1.5–2.5 mm long, clavate, ± appressed to base of lip.

Zambia. N: Mbala Distr., Chilongowelo, 1450 m, fl. 1.iii.1952, *Richards* 1000 (K). S: Choma Distr., Mapanza, 1060 m, fl. 16.iii.1958, *E.A. Robinson* 2793 (K; SRGH). **Zimbabwe**. N: Makonde

Tab. 26. HABENARIA TRACHYPETALA. 1, habit (×⅔), from *Richards* 13800; 2, flower, side view (×⅔); 3, flower, front view (×⅔); 4, leaf bract (×1); 5, dorsal sepal (×1); 6, lateral sepal (×1); 7, petal (×1); 8, column, side view (×2); 9, column, front view (×2); 10, pollinia (×2), 2–10 from *Richards* 13800 (in K spirit coll. 5968). Drawn by Judi Stone.

(Lomagundi) Distr., Umvukwe Range, 11 km south of Mutorashanga (Mtoroshanga), fl. 7.iii.1976, *Grosvenor* 857 (K; SRGH). C: Harare, Twentydales, fl. 22.ii.1970, *Linley* 460 (K; SRGH). E: Mutare Distr., below 'Rhodes Viewpoint', 1100 m, fl. 24.i.1963, *Chase* 7940 (K; SRGH). **Malawi**. S: Chowe Mountain, Mr. Arnal's Estate, *Patel & Banda* 396 (K; MAL); Mulanje Distr., Likhubula Valley, 800 m, fl. 19.iii.1981, *la Croix* 130 (K).

Also in Zaire and Tanzania. Woodland, rocky hillsides; 800–1450 m.

20. **Habenaria elliotii** Rolfe ex Kraenzl. in Bot. Jahrb. Syst. **16**: 70 (1893). —Rolfe in J. Linn. Soc., Bot. **29**: 57, t. 12 (1891). —Grosvenor in Excelsa **6**: 83 (1976). —la Croix et al., Orch. Malawi: 62 (1991). Type from Madagascar.

Terrestrial herb 55–80 cm high. Stem leafy. Leaves 11–13, the lowermost 2–3 sheath-like, the remainder spreading with the mid-cauline largest and 11.5–20 × 3.5–4.5 cm, lanceolate or oblanceolate, acute, the underside silvery with some black dots. Inflorescence 13–17 × 5–7 cm, laxly to fairly densely 16- to many-flowered; bracts c. 20 mm long. Flowers spreading, green and white. Pedicel 10 mm long, ovary 15 mm long, slightly arched. Dorsal sepal 5.3–8 × 4 mm, ovate, erect, convex; lateral sepals 6.5–7 × 3 mm, obliquely lanceolate, spreading. Petals 2-lobed to c. 2 mm from base; posterior lobe 4.5–5 × c. 1 mm, adnate to dorsal sepal; anterior lobe 13–15(23) mm long, filiform, spreading. Lip 3-lobed, with undivided base c. 2 mm long; mid-lobe 7–10 × 1 mm, linear; side lobes 11.5–20 mm long, filiform, deflexed; spur 40–50 mm long, incurved, slender but slightly swollen at apex. Anther c. 2.2 mm high, canals c. 0.8 mm long. Auricles 1 mm long. Stigmatic arms 1.5–2 mm long, truncate.

Zimbabwe. E: Chimanimani, 900 m, fl. vi.1958, *Ball* 735 (K; SRGH). **Malawi**. C: Dedza Distr, Mua-Livulezi For. Res., Naminkokwe (Namkokweh), 640 m, fl. immat. 19.iii.1955, *Exell, Mendonça & Wild* 1052 (BM). S: Thyolo Distr., Kumadzi Estate, 650 m, fl. 15.iv.1982, *Johnston-Stewart in la Croix* 309 (K).

Also in Madagascar. Fringe of riverine forest, in fairly heavy shade; 650–900 m.

21. **Habenaria magnirostris** Summerh. in Kew Bull. **14**: 132 (1960); in F.T.E.A., Orchidaceae: 82 (1968). —Grosvenor in Excelsa **6**: 83 (1976). —Williamson, Orch. S. Centr. Africa: 46 (1977). —Geerinck in Fl. Afr. Centr. Orchidaceae pt. 1: 138 (1984). —la Croix et al., Orch. Malawi: 63 (1991). TAB. **27**. Type from Tanzania.

Slender terrestrial herb 30–40 cm high, with a globose tuber c. 13 mm long. Stem leafy. Leaves about 6, the 2 lowermost sheath-like, the uppermost 1–2 bract-like, the rest semi-erect, 6–16 × 1–2.5 cm, lanceolate or linear, acute, rather fleshy. Inflorescence 8–16 × 3.5-4.5 cm, laxly 4–20-flowered. Flowers green or yellow-green. Dorsal sepal 5–7.5 × 2–3 mm, ovate, erect, convex; lateral sepals 5–7 × 2.5–4 mm, falcate, spreading. Petals 2-lobed ± to base; posterior lobe 5–7 × 1 mm, adnate to dorsal sepal; anterior lobe 5–8 mm long, less than 1 mm wide, spreading downwards. Lip 3-lobed, with a basal undivided part 1–3.5 mm long; mid-lobe 7–13 mm long, c. 1 mm wide; side lobes 7–14 × 0.5 mm, the lobes sometimes fleshy and opaque; spur 15–30 mm long, pendent, slender but slightly swollen torwards apex. Anther erect, 3.5–4.5 mm high; canals 2.5–3.3 mm long. Stigmatic arms 2.5–3 mm long. Mid-lobe of rostellum 2–3.5 mm high, standing in front of the anther.

Zambia. N: Chinsali Distr., Ishiba Ngandu (Shiwa Ngandu), fl. 21.xii.1964, *E.A. Robinson* 6320 (K). W: Mwinilunga Distr., Sinkabolo Dambo, fl. 9.xii.1937, *Milne-Redhead* 3568 (K). C: Mkushi, fl. xii.1967, *Williamson & Simon* 619 (K; SRGH). **Zimbabwe**. C: Marondera (Marandellas), 1500 m, fl. 20.xii.1948, *Glen in GHS* 22341 (K; SRGH). E: Nyanga (Inyanga), near Mtarazi turn-off, c. 1660 m, fl. 18.i.1958, *Beasley* 31 (K; SRGH). **Malawi**. N: South Viphya, Elephant Rock Dambo, 1550 m, fl. 24.xii.1985, *la Croix* 762 (K; MAL).

Also in Zaire, Tanzania and Angola. Perennially wet dambo, boggy grassland; 1450–1650 m.

22. **Habenaria malacophylla** Rchb.f., Otia Bot. Hamburg.: 97 (1881). —Rolfe in F.T.A. **7**: 230 (1898). —Eyles in Trans. Roy. Soc. S. Afr.: 334 (1916). —Wild in Clark, Victoria Falls Handb.: 140 (1952). —Summerhayes in F.W.T.A. ed. 2, **3**: 196 (1968); in F.T.E.A., Orchidaceae: 73 (1968). —Grosvenor in Excelsa **6**: 83 (1976). —Williamson, Orch. S. Centr. Africa: 45 (1977). —Stewart et al., Wild Orch. South. Africa: 88 (1982). —Geerinck in Fl. Afr. Centr., Orchidaceae pt. 1: 135 (1984). —la Croix et al., Orch. Malawi: 63 (1991). Type from South Africa.

Tab. 27. HABENARIA MAGNIROSTRIS. 1, habit (×1); 2, flower (×3); 3, dorsal sepal, 2 views (×6); 4, lateral sepal (×6); 5, petal (×6); 6, column, front view (×6); 7, column, side view (×6); 8, rostellum, side view (×9); 9, pollinarium (×9), 1–9 from *Milne-Redhead & Taylor* 7997. Drawn by Margaret Stones. From Hooker's Icon. Pl.

Habenaria holstii Kraenzl. in Bot. Jahrb. Syst. **19**: 246 (1894). Type from Tanzania.
Habenaria polyantha Kraenzl. in Pflanzenw. Ost-Afr. **C**: 152 (1895). Type from Zaire.
Habenaria pulla Schltr. in Bot. Jahrb. Syst. **53**: 497 (1915). Type from Tanzania.

Terrestrial herb 30–90 cm high with ovoid, ± woolly tubers c. 2 cm long. Stem leafy, with c. 7–15 leaves, the 2–3 lowermost sheath-like, the lowermost foliage leaf 15–20 cm from the base. Leaves spreading, the mid-cauline largest and 10–16 × 3–5 cm, oblanceolate, acute, dark green, the base clasping the stem. Inflorescence 11–30 × 3–4 cm, fairly laxly (occasionally densely) many-flowered; bracts 10–20 mm long, usually shorter than the flower. Flowers green. Ovary and pedicel 11–15 mm long, arched so that flower is held ± horizontally. Dorsal sepal 4–7 × 3–5 mm, ovate, erect, convex; lateral sepals 5–7.5 × 2–4 mm, obliquely lanceolate, spreading. Petals 2-lobed ± to base; posterior lobe 4–7 × 1–1.5 mm, linear, ± adnate to dorsal sepal; anterior lobe 5–9 mm long, less than 0.5 mm wide, narrowly linear, curving up. Lip 3-lobed almost to base, mid-lobe 5–8 × c. 1 mm, linear; side lobes slightly longer and narrower, curving upwards like anterior petal lobes; spur 9–15 mm long, inflated in the middle, parallel to ovary. Anther c. 1.5 mm tall; canals less than 1 mm long. Stigmatic arms 1.5–2.5 mm long, clavate, appressed to lip.

Zambia. N: Mporokoso Distr., Lumangwe Falls, Kalungwishi R., c. 900 m, fl. immat. 9.i.1960, *Richards* 12311 (K). W: Kitwe Distr., Ichimpi, fl. 26.iii.1968, *Mutimushi* 2584 (K; NDO). C: Lusaka Distr., 14–16 km east of Lusaka, fl. 2.iii.1975, *Williamson & Gassner* 2482 (K; SRGH). **Zimbabwe**. W: Hwange Distr., Victoria Falls National Park, 880 m, fl. 11.ii.1980, *Kabisa* 13 (K; SRGH). C: Wedza Mt., fl. 27.ii.1964, *Wild* 6353 (K; SRGH). E: Mutare, Hawkdale, c. 1450 m, fl. & fr. 28.iv.1957, *Chase* 6455 (K; SRGH). **Malawi**. N: Nkhata Bay Distr., 8 km SE of Mzuzu, 1200 m, fl. 31.iii.1975, *Pawek* 9178 (K; MAL; MO; SRGH; UC). S: Zomba Distr., Zomba Plateau by Mulunguzi R. near trout ponds, 1520 m, fl. 18.iv.1970, *Brummitt* 9964 (K; MAL). **Mozambique**. Z: Namuli Mt. (SW sector), 970 m, fl. & fr. 25.vii.1962, *Schelpe & Leach* 7008 (K; BOL). MS: Border Farm, 1120 m, fl. 19.iii.1961, *Chase* 7447 (K; SRGH).
Widespread in tropical and South Africa. Deep shade in forest, often near streams; 880–1800 m.

23. **Habenaria njamnjamica** Kraenzl. in Bot. Jahrb. Syst. **16**: 106 (1892). —Rolfe in F.T.A. **7**: 230 (1898). —Summerhayes in F.T.E.A., Orchidaceae: 80 (1968). —Grosvenor in Excelsa **6**: 83 (1976). —Williamson, Orch. S. Centr. Africa: 47 (1977). —Geerinck in Fl. Afr. Centr., Orchidaceae pt. 1: 133 (1984). —la Croix et al., Orch. Malawi: 64 (1991). Type from Sudan.
    Habenaria foliolosa Kraenzl. in Bot. Jahrb. Syst. **51**: 372 (1914). Type from Zaire.
    Habenaria parvifolia Summerh. in Kew Bull. **16**: 265, fig. 6 (1962); in F.T.E.A., Orchidaceae: 80 (1968). —Williamson, Orch. S. Centr. Africa: 47 (1977). —la Croix et al., Orch. Malawi: 65 (1991). Type: Zambia, Mwinilunga Distr., Dobeka Bridge, *Milne-Redhead* 3794 (K, holotype).

Terrestrial herb 30–55 cm high, with tomentose, globose or ellipsoid tubers 1–3 cm long. Stem leafy along its length. Leaves 9–17, imbricate, the lowermost 2–4 sheath-like, the uppermost bract-like, the rest erect or spreading, the largest 2.5–7.5 × 1.5–4 cm, lanceolate, acute. Inflorescence 6–17 × 1.5–4 cm, rather loosely to fairly densely 8- to many-flowered; bracts 20–55 mm long, lanceolate, acuminate, longer than the flowers at the base of, and usually throughout, the inflorescence. Flowers green or yellow-green, suberect, whole plant drying yellow-green to yellow-brown. Ovary and pedicel 13–25 mm long. Dorsal sepal 6–11 × 4–7 mm, ovate, erect, convex; lateral sepals 8–16 × 1.5–5.5 mm, obliquely lanceolate, acute, deflexed. Petals 2-lobed almost to base; posterior lobe erect, 5–11 × 1.5–2.5(4) mm, narrowly lanceolate; anterior lobe 6–14.5 mm long, 0.5–1 mm wide at the base, becoming ± filiform, spreading and curving upwards. Lip 3-lobed to 1.5–2 mm from base; mid-lobe 10–14 × 1–2 mm, linear; side lobes similar to lower petals lobes, 9–14 mm long; spur dependent, 1–2 cm long, much swollen in apical half. Anther erect, 2–5 mm high; canals 1–1.5 mm long. Auricles as long as, or shorter than the anther, sometimes shortly stalked. Stigmatic arms 2–3 mm long, clavate, deflexed on to lip base.

Zambia. N: Mbala Distr., Kanyika Flats, Chinakila, 1200 m, fl. & fr. 12.i.1965, *Richards* 19501 (K). W: Mwinilunga Distr., just NW of Kalalima (Kamwezhi)-Dobeka junction, fl. 17.xii.1937, *Milne-Redhead* 3717 (K); Kasempa Distr., c. 32 km east of Mufumbwe (Chizela), fl. 28.xii.1969, *Williamson & Simon* 1840 (K; SRGH). C: Chakwenga Headwaters, 100–129 km east of Lusaka, fl. 19.i.1964, *E.A. Robinson* 6216 (K). E: Chipata, fl. iii.1952, *Benson* NR 18 (BM). **Zimbabwe**. N:

Makonde Distr., Umvukwes, Aranbira, 1200 m, fl. 25.i.1954, *Pollitt* in *GHS* 46252 (K; SRGH). C: Harare Distr., Denda, 1300 m, fl. 29.i.1954, *Greatrex* in *GHS* 46253 (K; SRGH). **Malawi**. N: Nyika National Park, Thazima, fl. 12.i.1983, *Elias* 3 (K). S: Blantyre Distr., Ndirande Mt., 1250 m, fl. 15.ii.1970, *Brummitt* 8583 (K).

Also in Sudan, Zaire, Kenya and Tanzania. *Brachystegia* woodland and montane grassland; 1180–2250 m.

Summerhayes separated plants with stalked auricles and erect, appressed leaves as *H. parvifolia*, but plants with appressed leaves can have short auricles and the characters do not seem to be consistent. Otherwise, the flowers are identical.

24. **Habenaria orthocentron** P.J. Cribb in Kew Bull. **32**, 1: 145 (1977). Type: Zambia, Kawambwa Distr., *Williamson & Drummond* 1976 (K, holotype; SRGH).
   Habenaria sp., Williamson, Orch. S. Centr. Africa: 45 (1977).

Terrestrial herb 50–60 cm tall; tuber ovoid, c. 22 × 15 mm, tomentose; roots long and slender. Stem leafy throughout its length. Leaves 10–16, the basal 3–4 sheath-like, the lowermost foliage leaf 16 cm from the ground; upper leaves bract-like; the mid-cauline largest and 9–11 × 4–4.5 cm, ovate, acute. Inflorescence 16–20 cm long, fairly laxly many-flowered; bracts 12–15 × 3 mm, lanceolate, acute, shorter than the flower. Flowers light green. Pedicel 12 mm long, ovary 10 mm long, slightly arched. Dorsal sepal 7–8 × 3–4 mm, ovate, acute, erect, convex; lateral sepals deflexed, 8 × 4 mm, obliquely ovate, acute. Petals 2-lobed ± to base; posterior lobe 7 × 0.5 mm, filiform, erect, ± adnate to dorsal sepal; anterior lobes 22 mm long, porrect, filiform. Lip 3-lobed with basal undivided part 1–1.5 mm long; mid-lobe 10 mm long; side lobes 20 mm, all lobes filiform; spur 20–30 mm long, slightly inflated in distal half but with an acute apex, held horizontally. Anther erect, 1.5–2 mm high; canals 2 mm long, crossed. Stigmas decurved, 2 mm long. Rostellum mid-lobe small and triangular; side lobes 2 mm long, crossed.

Zambia. N: Kawambwa Distr., above Lumangwe Falls, fl. ii.1969, *Williamson & Drummond* 1976 (K; SRGH).

Endemic. Very wet swamp forest.

This species resembles *H. elliotii*, differing mainly in the shorter spur, held horizontally rather than incurved, and in the crossed rostellum arms and anther canals.

25. **Habenaria papyracea** Schltr. in Bot. Jahrb. Syst. **53**: 498 (1915). —Summerhayes in F.W.T.A. ed. 2, **3**: 194 (1968); in F.T.E.A., Orchidaceae: 79 (1968). —Grosvenor in Excelsa **6**: 83 (1976). —Williamson, Orch. S. Centr. Africa: 48 (1977). —Geerinck in Fl. Afr. Centr., Orchidaceae pt. 1: 129 (1984). —la Croix et al., Orch. Malawi: 64 (1991). Type from Tanzania.

Terrestrial herb 30–60 cm tall, with densely tomentose, ellipsoid to globose tuber 1.5–3 cm long. Leaves 6–14, ± evenly spread up the stem, the lowermost 2–4 loosely sheathing, the upper ones grading into bracts; those in the middle spreading, the largest 5–14 × 2–4.5 cm, elliptic, oblanceolate or ovate, acute or shortly acuminate, the base enfolding the stem, drying papery. Inflorescence 10–30 × 2.5–4 cm, densely many-flowered; bracts 15–30 mm long, usually about equalling or shorter than the flower. Flowers green, spreading. Ovary and pedicel 12–20 mm long. Dorsal sepal 4.5–7(9) × 3.5–5 mm, ovate, erect, convex; lateral sepals 5.5–10 × 2.5–4 mm, obliquely lanceolate, deflexed. Petals 2-lobed almost to base; upper lobe erect, 4–6.5 × 1–1.5 mm, linear; lower lobe 6–12 mm long, c. 0.5 mm wide, spreading and curving upwards. Lip 3-lobed with an undivided base 1–2 mm long; mid-lobe 5.5–10 × c. 1 mm, linear, obtuse; side lobes 8–14 mm long, c. 0.5 mm wide; spur 7–20 mm long, pendent, slender. Anther erect, 1.5–2 mm tall; canals very short, c. 0.5 mm long. Stigmatic arms 2–2.5 mm long.

Zambia. N: Mbala Distr., Chilongowelo, 1450 m, fl. 26.ii.1955, *Richards* 4703 (K). W: Solwezi Distr., tributary of Lumwana R., fl. 17.i.1962, *Holmes* 0327 (K; SRGH). **Zimbabwe**. N: Makonde Distr., 29 km west of Chinhoyi (Sinoia), fl. 9.ii.1975, *Cannell* 639 (K; SRGH). C: Harare Distr., fl. 6.iii.1946, *Greatrex* 93 (K; SRGH). **Malawi**. N: Nyika Plateau, Vitumbi, 1800 m, fl. & fr. 19.iii.1976, *Phillips* 1498 (K; MO). C: Lilongwe Distr., Dzalanyama Forest Reserve, 1260 m, fl. 22.iii.1970, *Brummitt* 9299a (K). S: Chiradzulu Distr., base of Chiradzulu Mt., c. 1300 m, fl. 26.ii.1983, *la Croix & Jenkins* 464 (K).

Also in Nigeria, Sudan, Zaire and Tanzania. *Brachystegia* woodland and mixed woodland, on rocky hillsides and amongst rocks; 1250–1800 m.

26. **Habenaria pubidens** P.J. Cribb in Kew Bull. **32**, 1: 147 (1977). —la Croix et al., Orch. Malawi: 65 (1991). Type: Zambia, Nyika Plateau, *Williamson & Odgers* 280 (K, holotype; SRGH).
    Habenaria sp., Williamson, Orch. S. Centr. Africa: 46 (1977).

Terrestrial herb up to 80 cm high with ± globose tubers. Stem leafy along its length. Leaves 8–12, the basal 4 or 5 sheath-like, the lowermost foliage leaf about one third up the stem; upper leaves bract-like, those in the middle spreading, 8–15 × 2.5–4.5 cm, oblanceolate or elliptic with noticeable reticulate venation, the edge often undulate, clasping the stem at the base. Inflorescence 11–27 × 3–3.5 cm, laxly 10–25-flowered; bracts c. 10 mm long, lanceolate, acute. Flowers green. Ovary and pedicel c. 15 mm long, arched. Dorsal sepal 4–6 × 3–5 mm, ovate, erect; lateral sepals 6–7 × 2–2.5 mm, elliptic, deflexed. Petals 2-lobed ± to the base; upper lobe 5–6 mm long, less than 1 mm wide, linear, erect, adnate to dorsal sepal; lower lobes 6–7 × 0.5 mm, spreading and curving up. Lip 3-lobed ± to base; mid-lobe 5–8 × 0.5–1 mm, deflexed; side lobes 4–7 × 0.5 mm, parallel to lower petal lobes; spur 15–20 mm long with a hairy tooth in the mouth; slightly swollen in apical half, recurved and projecting above the flower. Anther c. 1.2 mm high; canals 1.5 mm long. Stigmatic arms decurved, 1.2–2 mm long. Rostellum mid-lobe recurved, triangular.

Zambia. E: Nyika Plateau, Chowo Forest, fl. iii.1967, *Williamson & Odgers* 280 (K; SRGH). **Malawi**. N: Chitipa Distr., Misuku Hills, Mugesse (Mughese) Forest, 1700 m, fl. 25.iv.1972, *Pawek* 5206 (K); Nyika National Park, Zovochipolo forest patches, 2225 m, fl. 10.iii.1982, *Dowsett-Lemaire* 339 (K).
    Not known elsewhere. Deep shade in evergreen forest, forming colonies; 1700–2450 m.
    This species closely resembles *H. malacophylla*, differing in the recurved spur with a hairy tooth in the mouth.

27. **Habenaria silvatica** Schltr. in Bot. Jahrb. Syst. **53**: 497 (1915). —Summerhayes in F.T.E.A., Orchidaceae: 76 (1968). Type from Tanzania.
    Habenaria debilis G. Will. in J. S. African Bot.: **46**, 4: 329 (1980), nom. illegit. non Hook.f. (1864). Type: Zambia, Ishiba Ngandu (Shiwa Ngandu), *G. Williamson* 1416 (K; SRGH, holotype).
    Habenaria debiliflora G. Will. in J. S. African Bot. **47**, 2: 133 (1981), type as for *Habenaria debilis* G. Will.
    Habenaria sp., Williamson, Orch. S. Centr. Africa: 47 (1977).

Slender terrestrial herb 13–40 cm high. Stem leafy, the lowermost 2–4 leaves sheath-like, the rest 4–7 × 0.4–1.7 cm, erect, linear. Inflorescence 4–25 cm long, fairly densely several- to many-flowered; bracts c. 15 mm long, lanceolate, acute. Flowers green, spreading. Ovary and pedicel arched, 8–12 mm long. Dorsal sepal 5–6.5 × 3.5 mm, ovate, obtuse, erect; lateral sepals 5.5–6.5 × 2 mm, oblong-ovate, acute, deflexed. Petals 2-lobed ± to base; posterior lobe 4.5–6.5 × 0.8 mm, ± adnate to dorsal sepal; anterior lobe 2.5–3 × 0.5 mm. Lip 3-lobed to c. 1 mm from the base, all lobes linear; mid-lobe 5.5–9 × 1 mm; side lobes 3.5 × 0.5 mm; spur 6–8 mm long, slender. Anther 1.2–2 mm high; canals very short, less than 1 mm long. Auricles less than 1 mm long. Stigmatic arms 1–1.8 mm long.

Zambia. N: Chinsali Distr., Ishiba Ngandu (Shiwa Ngandu), fl. ii.1969, *G. Williamson* 1416 (K; SRGH); Kasama, fl. 7.ii.1961, *Fanshawe* 7172 (K).
    Also in Tanzania. Woodland.
    The specimens cited are the only ones in the Flora Zambesiaca area which match Schlechter's description of *H. silvatica*.

28. **Habenaria supplicans** Summerh. in Bot. Mus. Leafl. Harv. Univ. **10**: 267 (1942). —Williamson, Orch. S. Centr. Africa: 50 (1977). Type: Zambia, Mwinilunga Distr., *Milne-Redhead* 3902 (K; holotype).

Slender terrestrial herb 30–55 cm tall; tuber c. 20 × 15 mm, ellipsoid, villous. Leaves 7–10, fairly evenly spaced up stem, the lowermost 1–3 sheath-like, the rest spreading, the largest in the middle, 4–7.5 × 1–1.7 cm, lanceolate, acute. Inflorescence 12–19 × 2.5–3 cm, laxly 14–20-flowered, the rhachis slightly zigzag. Bracts 12–17 mm long, equalling or shorter than the flower. Flowers green. Ovary and pedicel 11–14 mm long, the ovary arched. Dorsal sepal 4–4.5 × 3–3.6 mm, ovate, erect, convex; lateral sepals 4.5–6 × 1.8–2.5 mm, deflexed. Petals 2-lobed ± to base; posterior lobe 3.5–4 mm long, c. 1 mm wide, rather fleshy, lying beside dorsal sepal; anterior lobe 4–4.7 × 0.6–0.7 mm, spreading and curving up. Lip 3-lobed ± to base, all lobes linear; mid-lobe 4.5–6 × 1 mm, deflexed; side lobes 4–6.5 × 0.5 mm, spreading and curving up; spur 10–13 mm long, slightly incurved, slightly thickened for most of its length but tapering to apex. Anther 1.75 mm tall; canals very short, less than 1 mm long. Stigmatic arms c. 2.3 mm long.

Zambia. W: Mwinilunga Distr., c. 2 km south of Matonchi Farm, fl. 31.xii.1937, *Milne-Redhead* 3902 (K). **Malawi**. N: Mzimba Distr., Champhila (Champhira) Hill, 1650 m, fl. 24.ii.1978, *Pawek* 14096 (K; MO).
Also in Angola and Zaire. *Brachystegia* woodland; up to 1650 m.

29. **Habenaria tetraceras** Summerh. in Kew Bull. **16**: 267, fig. 7 (1962); in F.T.E.A., Orchidaceae: 81 (1968). —Williamson, Orch. S. Centr. Africa: 47 (1977). —Geerinck in Fl. Afr. Centr., Orchidaceae pt. 1: 131 (1984). TAB. **28**. Type from Tanzania.

Terrestrial herb 20–60 cm high; tubers up to 1.5 cm long, ellipsoid or subglobose, sparsely hairy. Stem leafy along its length. Leaves 11–16, the lowermost 3–4 sheath-like, the rest erect, appressed to stem, grading into bracts at the top, the largest 2.5–7 × 1–3.6 cm, broadly lanceolate, acute. Inflorescence 6–15 × 3–5 cm, densely 8- to many-flowered; bracts up to 40 mm long, longer than the flower. Flowers emerging horizontally from the bracts, green or yellow-green; whole plant drying yellow-green or yellow-brown. Ovary and pedicel 15–20 mm long, arched. Dorsal sepal 8.5–14 × 5–8 mm, ovate, erect, convex; lateral sepals 10–15 × 3–4.5 mm, obliquely lanceolate, deflexed. Petals 2-lobed almost to base; posterior lobe 9–16 × 1–1.5 mm, linear, erect; anterior lobe (13)17–26 mm long, 0.5 mm wide at the base, becoming filiform, spreading and curving up. Lip 3-lobed, with basal undivided part 1–2 mm long; mid-lobe 10–17 mm long, c. 1 mm wide, linear, deflexed; side lobes 17–30 mm long, 0.5–1 mm wide at base, becoming filiform and curving up like lower petal lobe; spur 7–13 mm long, very swollen at tip, pendent. Anther erect, 3–4 mm high; canals c. 1 mm long. Auricles c. 2 mm long, shorter than the anther. Stigmatic arms 3–4 mm long, deflexed along the base of the lip.

Zambia. N: Mbala Distr., marsh by Lumi R., 1500 m, fl. 9.ii.1955, *Richards* 4393 (K). C: Kabwe (Broken Hill), fl. ii.1907, *Allen* 486 (K).
Also in Zaire and Tanzania. Marshy grassland and dambo; c. 1500 m.
This species is closely related to *H. njamnjamica*. It differs in the broader inflorescence, the longer and more slender lower petal lobes and lip side lobes, and the more inflated spur apex.

30. **Habenaria tridens** Lindl. in Companion Bot. Mag. **2**: 208 (1837). —Grosvenor in Excelsa **6**: 83 (1976). —Stewart et al., Wild Orch. South. Africa: 87 (1982). Type from South Africa.
Habenaria gerrardii Rchb.f., Otia Bot. Hamburg.: 97 (1881). Type from South Africa
Habenaria barberae Schltr. in Bot. Jahrb. Syst. **20**, Beibl. 50: 7 (1895). Type from South Africa.

Slender terrestrial herb 18–32 cm high. Stem leafy along its length. Leaves 8–10, the lowermost 2–3 sheath-like, the upper 2–3 bract-like, those in the middle 4.5–11 × 0.5–1 cm, narrowly lanceolate to linear. Inflorescence 5–12 × 1 cm fairly laxly up to c. 20-flowered; bracts c. 15 × 4 mm, lanceolate, acute. Flowers pale green. Ovary and pedicel 8–14 mm long. Dorsal sepal 3–3.6 × 2.5–3 mm, ovate, erect, convex; lateral sepals 4.5–5.2 × 2–2.2 mm, obliquely lanceolate, deflexed. Petals 2-lobed almost to base, the lobes more or less opaque; upper lobe 3–3.5 × 1.2 mm; lower lobe c. 2 × 0.4 mm. Lip 3-lobed with basal undivided part 1.3 mm long, all lobes linear; mid-lobe c. 3.5 × 0.5 mm; side lobes c. 2.4 mm long, slightly narrower than mid-lobe; spur 8–10 mm long, slender, pendent. Anther 1.3 mm high; canals very short, 0.4 mm long. Stigmas c. 1 mm long.

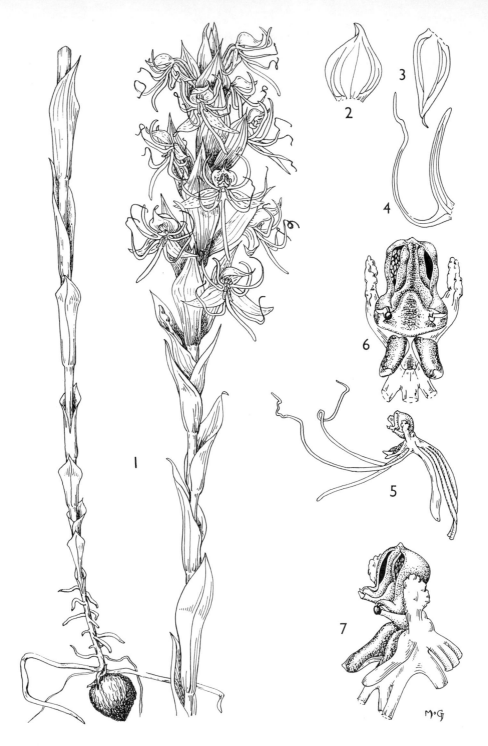

Tab. 28. HABENARIA TETRACERAS. 1, habit (×1), from *Milne-Redhead & Taylor* 8567; 2, dorsal sepal (×2); 3, lateral sepal (×2); 4, petal (×2); 5, flower, tepals removed, showing lip and column, side view (×2); 6, column, front view (×6); 7, column, side view (×6), 2–7 from *Milne-Redhead & Taylor* 8159. Drawn by Mary Grierson. From Kew Bull.

Zimbabwe. E: Chimanimani Distr., Haroni Gorge, 600 m, fl. 11.iv.1954, *Ball* 315 (K; SRGH).
Mozambique. M: Maputo, Namaacha, fl. 30.vi.1960, *Balsinhas* 504 (K; LISC).
Also in South Africa. Wet ground, rocky stream bed, standing in running water for c. 2
months before flowering; 300–600 m.

31. **Habenaria uhehensis** Schltr. in Bot. Jahrb. Syst. **38**: 148 (1906). —Summerhayes in Kew Bull.
    16: 269 (1962); in F.T.E.A., Orchidaceae: 77 (1968). —Williamson, Orch. S. Centr. Africa:
    47 (1977). —la Croix et al., Orch. Malawi: 67 (1991). Type from Tanzania.
    Habenaria lurida Schltr. in Bot. Jahrb. Syst. **53**: 499 (1915). Type from Tanzania.
    Habenaria tenuifolia Summerh. in Bot. Mus. Leafl. Harv. Univ. **10**: 269 (1942); in F.T.E.A.,
    Orchidaceae: 79 (1968). —Williamson, Orch. S. Centr. Africa: 49 (1977). —Geerinck in Fl.
    Afr. Centr., Orchidaceae pt. 1: 132 (1984). —la Croix et al., Orch. Malawi: 66 (1991). Type
    from Tanzania.
    Habenaria modica Summerh. in Kew Bull. **16**: 263 (1962). —Grosvenor in Excelsa **6**: 83
    (1991). —Williamson, Orch. S. Centr. Africa: 47 (1977). —la Croix et al., Orch. Malawi: 63
    (1991). Type: Zimbabwe, Harare Distr., *Greatrex* 107 (K, holotype; SRGH).

Terrestrial herb up to c. 50 cm high, tubers up to c. 2 cm long, ellipsoid to ±
globose. Stem leafy along its length. Leaves 6–12, the lowermost 1–3 sheath-like, the
upper ones grading into bracts, the largest in the middle, 3–13 × (0.4)0.8–2.8 cm,
linear to lanceolate. Inflorescence 3–21 × 1.5–4(5) cm, fairly laxly to fairly densely
few- to many-flowered; bracts usually equal to or longer than the flowers at the base
of the inflorescence. Flowers green, yellow-green or green and white. Ovary and
pedicel arched, 10–20 mm long. Dorsal sepal (4)5–11 × 3–6 mm, ovate, erect,
convex; lateral sepals 5.5–15 × 3–5 mm, obliquely lanceolate, deflexed. Petals 2-
lobed ± to base, both lobes linear, the lower lobe usually slightly longer but
narrower than the upper lobe. Upper lobe 3–10(12) × 0.8–1.5 mm, erect, ± adnate
to dorsal sepal; lower lobe 4–12 × 0.4–0.8 mm, spreading sideways and curving up.
Lip 3-lobed almost to base; mid-lobe 5–12 × 0.5–1 mm, pendent; side lobes similar
in length but slightly narrower, curving up like lower petal lobes; spur 8–20 mm
long, pendent or slightly incurved, swollen at the apex. Anther 1.75–2.5 mm high;
canals short, up to 1 mm long. Stigmatic arms 2–3 mm long, clavate, lying along
base of lip.

Zambia. N: Mbala Distr. c. 1500 m, fl. & fr. 18.ii.1955, *Richards* 4574 (K). C: 13 km west of
Lusaka, fl. 19.ii.1975, *Williamson & Gassner* 2372A (K). **Zimbabwe**. N: Makonde Distr.,
Umvukwes, Aranbira, 1200 m, fl. 25.i.1953, *Pollitt* in *GHS* 46169 (K; SRGH). C: north east of
Harare, fl. 3.ii.1946, *Greatrex* 107 (K; SRGH). **Malawi**. N: Nyika National Park, Jalawe Hill, 2250
m, fl. 9.iv.1984, *la Croix* 607 (K). C: Dowa Distr., Kongwe Forest Reserve, 1500 m, fl. 7.iii.1982,
Brummitt, Polhill & Banda 16380 (K; MAL). S: Zomba Mt., near stables, c. 1200 m, fl. 12.ii.1984,
la Croix 527 (K).
Also in Zaire, Tanzania and Angola. Woodland and grassland; 600–2285 m.
This is a widely distributed species, very variable in size, which has been described under
several names. Although specimens with the largest flowers look different from those with the
smallest, plants can be found showing a complete gradation of flower size and it is impossible
to pick any particular measurements as forming dividing lines. *H. silvatica* and *H. supplicans*
have been kept separate, but it is possible that they also belong here. Detailed studies are
necessary over the complete range of the species or complex of species.

32. **Habenaria uncicalcar** Summerh. in Kew Bull. **16**: 271 (1962); in F.T.E.A., Orchidaceae: 76
    (1968). —Grosvenor in Excelsa **6**: 83 (1976). —Williamson, Orch. S. Centr. Africa: 47
    (1977). —Geerinck in Fl. Afr. Centr., Orchidaceae pt. 1: 131 (1984). —la Croix et al.,
    Orch. Malawi: 67 (1991). Type from Tanzania.

Slender terrestrial herb 25–66 cm high; tubers up to 3 cm long, globose to
ellipsoid, tomentose. Stem leafy along its length. Leaves 8–10, the lowermost 3–4
sheath-like, the upper ones grading into bracts, the rest 3.5–8.5 × 1.2–1.8 cm,
lanceolate, acute. Inflorescence fairly laxly up to c. 20-flowered; bracts leafy, usually
shorter than the flower. Flowers green. Ovary and pedicel (10)15–22 mm long,
arched. Dorsal sepal 6.5–11 × 4.5–8.5 mm, ovate, erect, convex; lateral sepals 8–12 ×
3–4.5 mm, obliquely lanceolate, deflexed. Petals 2-lobed ± to base; posterior lobe
6.5–10.6 × c. 1 mm, erect, linear, adnate to dorsal sepal; anterior lobe 7.5–15.5 mm
long, less than 1 mm wide, deflexed then curving up. Lip 3-lobed with basal

undivided part 2–3.5 mm long; mid-lobe 6.5–12 × c. 1 mm, linear, deflexed; side lobes 5.4–10 × c. 0.5 mm, usually shorter and narrower than mid-lobe; spur 8–11 mm long, the basal half parallel to ovary, then abruptly bent up so that the apex is held above the flower, the apical 3 mm very inflated. Anther 2–3 mm high; canals 1.3–2.2 mm long. Stigmas 2.5–4 mm long, lying along base of lip.

Zambia. N: Mbala Distr., Chilongowelo, 1450 m, fl. 5.ii.1955, *Richards* 4340 (K). W: Mwinilunga Distr., fl. 10.xii.1958, *Holmes* 0121 (K; SRGH). C: 13 km west of Lusaka, fl. 9.ii.1975, *Williamson & Gassner* 2373 (K). **Zimbabwe**. N: Hurungwe National Park, 306 km from Harare on road to Chirundu, 825 m, fl. 20.ii.1981, *Philcox, Leppard & Duri* 8773 (K). C: Harare Distr., Lochinvar, 1360 m, fl. 20.i.1946, *Wild* 715 (K; SRGH). **Malawi**. N: Mzimba Distr., c. 11 km south of Euthini, 1250 m, fl. 31.i.1976, *Pawek* 10777 (K; MAL; MO; SRGH; UC). C: Nkhota Kota (Nkotakota) Escarpment, fl. 28.ii.1953, *Jackson* 1134 (K). S: Blantyre Distr., Michiru Mt., fl. 20.i.1979, *V.A.Taylor* 6 (K; MAL).

Also in Zaire and Tanzania. *Brachystegia* woodland, rocky areas in open woodland; 660–1700 m.

33. **Habenaria anaphysema** Rchb.f. in Flora **50**: 101 (1867). —Rolfe in F.T.A. **7**: 235 (1898). —Suessenguth & Merxmüller, Contrib. Fl. Marandellas Distr.: 83 (1951). —Goodier & Phipps in Kirkia **1**: 52 (1961). —Summerhayes in F.T.E.A., Orchidaceae: 95 (1968). —Grosvenor in Excelsa **6**: 83 (1976). —Williamson, Orch. S. Centr. Africa: 56 (1977). —Geerinck in Fl. Afr. Centr., Orchidaceae pt. 1: 106 (1984). —la Croix et al., Orch. Malawi: 68 (1991). Type from Angola.

Slender terrestrial herb 20–75 cm tall. Tubers 1–2 × 0.5–1 cm, globose or ellipsoid. Stem leafy; leaves 6–12, the lowermost sheath-like, the upper ones grading into bracts, the remainder ± erect, the largest 7–16 cm × 3–10 mm, linear, the edges sometimes inrolled. Inflorescence 6–20 × 3–3.5 cm, rather laxly or densely 6–26-flowered. Flowers green. Pedicel c. 10 mm long, ovary c. 7 mm; bracts 5–15 mm long, shorter than pedicel and ovary, often brown at tips. Dorsal sepal usually reflexed but occasionally erect, 4–6.5 × 4–5 mm, elliptic, convex; lateral sepals deflexed, 6–9 × 4–5 mm, obliquely obovate with lateral apiculum. Petals 2-lobed almost to base; upper lobe 4–5 × 0.5 mm, linear, erect or recurved, ciliolate; lower lobe 8–15 × 1–2 mm, narrowly lanceolate, curving downwards. Lip deflexed, 3-lobed to 1–2 mm from base; mid-lobe 8–15 mm long, side lobes 4–8 mm long, all lobes linear, c. 0.5 mm wide; spur 8–15 mm long, slender but swollen at apex, at first parallel to ovary, then forming a complete loop so that the tip points up. Anther 2.5–4 mm high; anther canals 3.5–6.5 mm long, slightly incurved. Stigmatic arms 3–6 mm long, slender, enlarged and truncate at apex. Rostellum mid-lobe 1.5–3 mm long, triangular, acute.

Zambia. B: Zambezi (Balovale) Distr., Kucheka Valley, fl. 1.xi.1961, *Holmes* 0132 (K; SRGH). N: Kasama Distr., 8 km north of Kasama, fl. 6.xii.1960, *E.A. Robinson* 4153 (K). C: Mkushi Distr., Fiwila, 1200 m, fl. 5.i.1958, *E.A. Robinson* 2646 (K; SRGH). **Zimbabwe**. C: Marondera (Marandellas), fl. 20.i.1952, *Corby* 759 (K; SRGH). E: Chimanimani Mts., 1660 m, fl. 1.ii.1957, *Phipps* 369 (K; SRGH). **Malawi**. N: Nkhata Bay Distr., Mzenga Estate, 600 m, fl. 7.i.1987, *la Croix* 927 (K; MAL; MO).

Also in Angola. Swamp and seasonally flooded grassland; 600–1200 m..

34. **Habenaria arianae** Geerinck in Geerinck & Coutrez, Not. Taxon. Orchid. Afr. Centr., Habenaria: 7, fig. 3–4, (1977); in Bull. Jard. Bot. Belg., **50**: 117 (1980); in Fl. Afr. Centr., Orchidaceae pt. 1: 100 (1984). —la Croix et al., Orch. Malawi: 69 (1991). Type from Burundi.
Habenaria williamsonii P.J. Cribb in Kew Bull. **32**: 143 (1977). Type: Malawi, Nyika Plateau, *G. Williamson* 283 (K, holotype; SRGH).
Habenaria sp., Williamson, Orch. S. Centr. Africa: 63 (1977).

Small terrestrial herb 13–23 cm high; tubers c. 13 × 5 mm, ovoid or elliptic. Leaves 10–20, the lowermost 2–3 sheathing, white with dark green reticulate veining; the middle 5–8 semi-erect, the largest c. 5 cm long, 3–4 mm wide, linear, the base wider and sheathing; the upper few bract-like, appressed to stem. Inflorescence 4–7 × 2 cm, densely several- to many-flowered; bracts to 15 mm long. Flowers suberect, light green. Pedicel and ovary 10–14 mm long. Dorsal sepal erect, 4 × 4 mm; lateral sepals ± spreading, the tips reflexed, 5 × 3 mm, obliquely ovate. Petals papillose, 2-lobed almost to base; upper lobe 4 mm long, erect, lower lobe 5 mm long, both lobes narrowly lanceolate, c. 1 mm wide. Lip deflexed, papillose, 3-lobed almost to base; mid-lobe 4.5 mm long, side lobes c. 5 mm long, all lobes linear, c. 1 mm wide; spur 5–7 mm long, parallel to pedicel and ovary, usually slightly inflated near apex. Anther erect, 2 mm high; canals very short, c. 0.5 mm long. Stigmatic arms porrect, 1 mm long. Rostellum mid-lobe 1.5 mm high, triangular.

Malawi. N: Nyika Plateau, c. 2300 m, fl. iii.1967, *G. Williamson* 283 (K; SRGH); Nyika Plateau, Chelinda Bridge, fl. 13.iii.1977, *Grosvenor & Renz* 1213 (K; SRGH).
Also in Burundi, Zaire, Angola and Tanzania. Wet grassland and seepage areas with shallow soil overlying rock; 600–2300 m.

35. **Habenaria falcicornis** (Lindl.) Bolus in J. Linn. Soc., Bot. **19**: 340 (1882), in error as "*H. falciformis*". —Grosvenor in Excelsa **6**: 83 (1976). Type from South Africa.
Bilabrella falcicornis Lindl. in Bot. Reg. **20**: sub t. 1701 (1835).
Habenaria tetrapetala Kraenzl. ex Rolfe in F.C. **5**, 3: 131 (1912) non Rchb.f. (1865). Type from South Africa.

Var. **caffra** (Schltr.) Renz & Schelpe in S. Afr. Orchid. J. **11**,1:6 (1980).
Habenaria caffra Schltr. in Ann. Transvaal Mus. **10**: 242 (1924). —Grosvenor in Excelsa **6**: 83 (1976). Type from South Africa.
Habenaria sp. No. 2 (*Ball* 236), Goodier & Phipps in Kirkia **1**: 52 (1961).

Terrestrial herb 30–45 cm tall. Stem leafy; leaves 6–10, the lower ones 4–11 cm long and 4–7 mm wide, linear or linear-lanceolate, with fine reticulate veining; upper leaves grading into bracts. Inflorescence 4–11 × 3.5 cm, rather laxly 6–14-flowered; bracts 12–30 × 4–5 mm, finely reticulate, usually equalling or slightly shorter than pedicel and ovary. Flowers green and white. Dorsal sepal 4.5–5.5 × 2–4 mm, ovate, erect, convex; lateral sepals deflexed, 5.5–8 × 4–5.5 mm, obliquely ovate with lateral apiculum. Petals 2-lobed almost to base; upper lobe 4–5.7 × 2–2.5 mm, elliptic to obovate, ciliolate; lower lobe about 7 × 2–3 mm, ovate or lanceolate. Lip 3-lobed almost to base; mid-lobe c. 10 mm long, side lobes 6.5–7 mm long, all linear, c. 1.5 mm wide; spur 30–37 mm long, ± incurved, slender, very slightly inflated at apex. Anther c. 2 mm high. Stigmatic arms c. 2 mm long; rostellum mid-lobe 2 mm long, triangular.

Zimbabwe. E: Chimanimani Mts., 1500 m, fl. 21.ii.1954, *Ball* 236 (K; SRGH).
Also in South Africa and Swaziland. Grassland.

36. **Habenaria cataphysema** Rchb.f. in Flora **50**: 101 (1867). —Rolfe in F.T.A. **7**: 237 (1898). —Williamson, Orch. S. Centr. Africa: 56 (1977). —Geerinck in Fl. Afr. Centr., Orchidaceae pt. 1: 115 (1984). Type from Angola.

Slender terrestrial herb 30–80 cm tall; tubers 1–1.5 × c. 1 cm, globose to ovoid. Stem leafy; leaves 7–9, the lowermost 1–2 sheath-like and the uppermost 3–4 bract-like, the rest 5.5–13.5 cm long and 4–8 mm wide, linear. Inflorescence 11–23 × 3–3.5 cm, many-flowered; bracts 7–10 mm long. Flowers pale green. Pedicel and ovary 16–18 mm long. Dorsal sepal 2.5–4 × 2 mm, ovate, reflexed, convex; lateral sepals deflexed, 5–6 × 3–4 mm, obliquely obovate. Petals shortly papillose, 2-lobed almost to base; upper lobe 3–4 × 0.3–0.4 mm; lower lobe 5–7 × 0.5–0.8 mm, both lobes linear. Lip 3-lobed from an undivided base c. 1 mm long; mid-lobe 6–6.5 mm long, c. 0.5 mm wide; side lobes 3.3–4.2 × 0.2–0.4 mm, all lobes linear; spur 12–13 mm long, very slender but with apical 2–3 mm abruptly inflated, almost globose. Anther 2–3 mm high; canals 2–3 mm long. Stigmas 2.5–3 mm long; rostellum mid-lobe c. 1.5 mm high.

Zambia. N: Kasama Distr., Mungwi, fl. 7.i.1961, *E.A. Robinson* 4239 (K). W: Mwinilunga Distr., Sinkabolo Dambo, fl. 21.xii.1957, *Milne-Redhead* 3757 (K).
Also in Zaire and Angola. Dambo, damp marshy places; c. 1350 m.

37. **Habenaria compta** Summerh. in Kew Bull. **16**: 276 (1962). —Williamson, Orch. S. Centr. Africa: 54 (1977). TAB. **29**. Type: Zambia, Mbala Distr., Kawimbe, *Richards* 5867 (K, holotype).

Terrestrial herb 37–45 cm tall. Tubers 2–2.5 × c. 1 mm, ellipsoid. Stem leafy; leaves 7–9, the basal sheaths reticulate. Lower leaves 8–11.5 cm long and 3–6 mm wide, linear, acute; upper leaves bract-like. Inflorescence 7–13 × 4–5 cm, laxly or densely several- to many-flowered; bracts 15–20 mm long, finely reticulate. Flowers pale green. Ovary and pedicel 18–20 mm long. Dorsal sepal 4.5–5 × 1–1.5 mm, elliptic, reflexed, convex; lateral sepals 5.8–6.7 × 3–3.5 mm, oblique. Petals 2-lobed almost to base; upper lobe 3.9–6 × 0.2–0.7 mm; lower lobe 6.9–9.5 × 0.3–0.6 mm. Lip 3-lobed to less than 1 mm from base; mid-lobe 9–11 mm long, side lobes 7–9.5 mm long, all lobes c. 0.5 mm wide, linear; spur 25–30 mm long, slightly inflated at apex. Anther 3–3.5 mm high, canals c. 2 mm long, somewhat incurved. Stigmatic arms 2–3.3 mm long; rostellum mid-lobe 1.5 mm high.

Zambia. N: Mbala Distr., Kawimbe, marsh by Lumi R., 1500 m, fl. 17.viii.1956, *Richards* 5867 (K). Also in Angola. Marsh; c. 1500 m.

38. **Habenaria diselloides** Schltr. in Bot. Jahrb. Syst. **53**: 513 (1915). —Williamson, Orch. S. Centr. Africa: 56 (1977). —la Croix et al., Orch. Malawi: 69 (1991). Type from Tanzania.

Terrestrial herb 14–30 cm tall. Tubers 11–15 × 11–13 mm, globose or ellipsoid. Stem densely leafy; leaves 7–12, the lowermost 3 sheath-like, white with green reticulate veining, the next 7–8 semi-erect, set close together up to 6.5 × 1 cm, lanceolate or linear, acute, conduplicate; the rest much shorter and narrower. Inflorescence 4.5–9 × 1.5–2.5 cm, densely many-flowered; bracts to 15 mm long, ovate, acute. Flowers yellow-green, papillose. Ovary and pedicel erect, the pedicel 3 mm long, ovary c. 10 mm long. Dorsal sepal 3–4 × 2.5 mm, erect, convex; lateral sepals ± spreading, 4–5 × 2.5–3 mm, obliquely ovate. Petals 2-lobed to c. 1.5 mm from base, both lobes 3–3.5 × 1–2 mm, upper lobe erect, lying inside dorsal sepal; lower lobe spreading. Lip 3-lobed to 1.5–2 mm from base, c. 5 mm long in all; mid-lobe 3–3.5 × 1.5–2 mm, triangular; side lobes similar but slightly narrower and sometimes slightly shorter; spur 4–7 mm long, swollen at apex, parallel to ovary. Anther 2 mm long, the loculi set at an angle; canals 1 mm long. Stigmatic arms 1–1.5 mm long, porrect, short and stout. Rostellum mid-lobe 1.2 mm long, triangular.

Malawi. N: Nyika National Park, Chelinda Bridge, 2250 m, fl. 26.ii.1987, *la Croix* 986 (K; MAL); South Viphya, Luwawa link road, c. 1800 m, fl. 24.iv.1974, *Pawek* 8549 (K; UC). Also in Tanzania. Seepage rocks, old gravel workings; 1700–2250 m.

39. **Habenaria disparilis** Summerh. in Kew Bull. **16**: 277 (1962). —Grosvenor in Excelsa **6**: 83 (1976). —Williamson, Orch. S. Centr. Africa: 53 (1977). —Geerinck in Fl. Afr. Centr., Orchidaceae pt. 1: 102 (1984). —la Croix et al., Orch. Malawi: 70 (1991). TAB. **30**. Type: Malawi, Ntcheu, Nkhande Hill, *Robson* 1345 (K, holotype).

Robust terrestrial herb 50–80 cm high. Tubers 2.5–3 × 1–2 cm, ellipsoid. Stem leafy; leaves 7–13, the lowermost 2–3 loosely sheathing, rather funnel-shaped, white with dark green reticulate veining; next 5–6 spreading, the largest 10–20 × 1.5–2.5 cm, linear-lanceolate; the rest becoming bract-like. Inflorescence 10–30 × 3–4 cm, fairly densely many-flowered, almost all flowers opening together; bracts c. 20 mm long, somewhat scarious. Flowers green, white in centre. Pedicel and ovary arched, 12–22 mm long. Sepals all reflexed; dorsal 5–5.5 × 2 mm, lateral sepals 7–8 × 3–4.5 mm, oblique with lateral apiculum. Petals 2-lobed almost to base, upper lobe 6–9 mm long, less than 1 mm wide, erect but curving back; lower lobes 10–13 × 2.5–3 mm, lanceolate, spreading, acute, often with a secondary tip (a common abnormality in this species). Lip deflexed, 3-lobed almost to base; mid-lobe 10.5–16 mm long, side lobes usually slightly shorter, all lobes c. 1 mm wide, linear; spur 20–30 mm long, curving forwards with 1 twist about the middle, inflated in apical third. Anther 2.5–3 mm long; canals 4.5–6 mm long; auricles 2-lobed, papillose. Stigmatic arms 5–7 mm long, slender but truncate at apex, diverging from each other at an angle of c. 60°. Rostellum mid-lobe c. 2 mm long; side lobes diverging, lying above stigmatic arms.

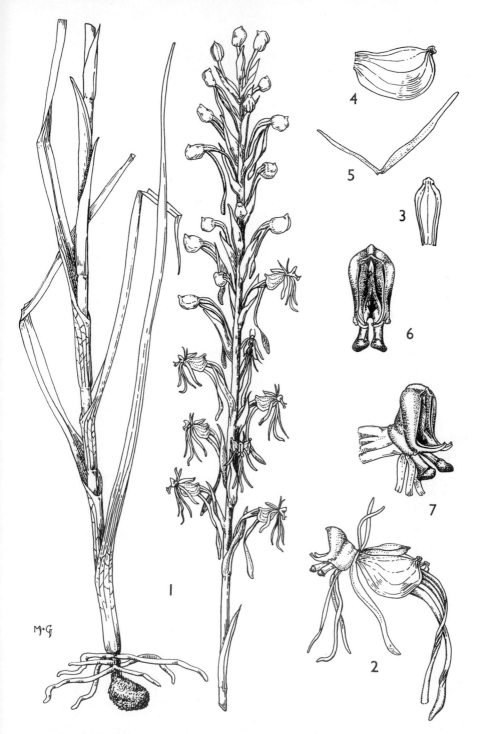

Tab. 29. HABENARIA COMPTA. 1, habit (×1); 2, flower, side view (×3); 3, dorsal sepal (×4); 4, lateral sepal (×4); 5, petal (×4); 6, column, front view (×4); 7, column, three-quarter profile (×4), 1–7 from *Richards* 5867. Drawn by Mary Grierson. From Kew Bull.

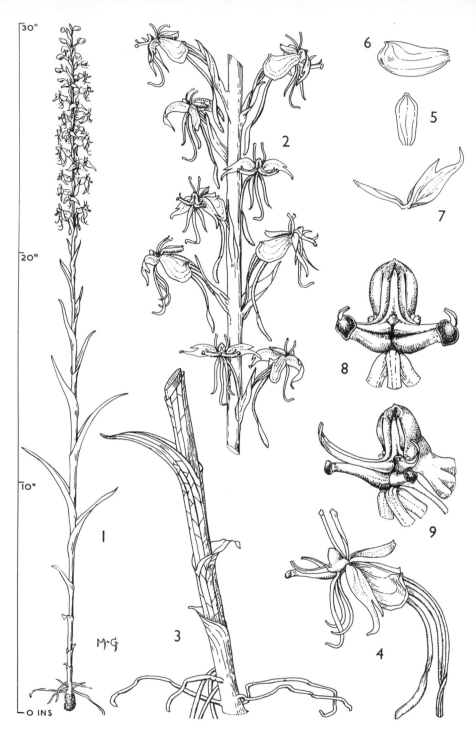

Tab. 30. HABENARIA DISPARILIS. 1, habit (×$\frac{1}{3}$); 2, part of inflorescence (×1); 3, base of stem (×1); 4, flower, side view (×2); 5, dorsal sepal (×2); 6, lateral sepal (×2); 7, petal (×2); 8, column, front view (×6); 9, column, semi-profile view (×6), 1–9 from *Robson* 1345. Drawn by Mary Grierson. From Kew Bull.

Zambia. N: Mbala Distr., Kawimbe, 1680 m, fl. 24.i.1957, *Richards* 7961 (K). C: Serenje Distr., Kundalila Falls, fl. i.1972, *G. Williamson* 2151 (K; SRGH). **Zimbabwe**. N: Makonde Distr., "Alaska Mine", fl. 3.ii.1965, *Wild* 6775 (K; SRGH). C: Harare Distr., Greendale, 1450 m, fl. 8.i.1958, *Leach* 4235 (K; SRGH). E: Nyanga (Inyanga), 1700 m, fl. 20.i.1931, *Norlindh & Weimarck* 4488 (K). **Malawi**. N: South Viphya, c. 1450 m, fl. 18.iv.1982, *la Croix & Johnston-Stewart* 311 (K). C: Dedza Distr., c. 12 km from Dedza on road to Lilongwe, fl. 3.ii.1982, *Gassner & Cribb* 209 (K). S: Mulanje Mt., 1500 m, fl. 13.v.1963, *Wild* 6244 (K; SRGH); Ntcheu Distr., Nkhande (Nkande) Hill, 1370 m, fl. 29.i.1959, *Robson* 1345 (K).

Also in Zaire. *Brachystegia* woodland, grassland or dambo, also in road cuttings; 1050–1500 m.

40. **Habenaria falcata** G. Will. in Pl. Syst. Evol. **134**: 54 (1980). Type: Zambia, Mwinilunga Distr., *Williamson & Simon* 1775 (K; SRGH, holotype).

Terrestrial herb c. 45 cm tall; tubers ovoid or ellipsoid. Stem leafy; leaves about 11, the lowest 3 sheathing, white with green reticulate veining, the next 4 erect, c. 10 × 1 cm, lanceolate, acute; the rest becoming bract-like. Inflorescence c. 15 × 4 cm, fairly laxly 16–30-flowered; bracts shorter than pedicel and ovary. Flowers green. Pedicel and ovary curved, c. 20 mm long. Dorsal sepal ± reflexed, 5 × 2 mm, elliptic, convex; lateral sepals deflexed, c. 8 × 5 mm, obliquely obovate with lateral apiculum. Petals 2-lobed almost to base; upper lobe c. 5 × 0.6 mm, erect; lower lobes reflexed, c. 9.5 mm long, lanceolate, the apex upturned. Lip pendent, 3-lobed almost to base; mid-lobe 8–11 mm long, obtuse; side lobes 9–12 mm long, acute; all lobes linear, c. 0.8 mm wide; spur c. 30 mm long, incurved, slender but swollen at apex, not twisted. Anther erect, slightly stalked, 1.8 mm long, the loculi parallel; canals porrect, c. 5 mm long; auricles c. 1 mm long, ± 3-lobed. Stigmatic arms c. 3 mm long, somewhat papillose, porrect, the apices truncate, lying parallel to each other. Rostellum mid-lobe 1.5 mm high, triangular.

Zambia. W: Mwinilunga Distr., 40 km east of Mwinilunga, fl. xii.1969, *Williamson & Simon* 1775 (K; SRGH).

?Also in Zaire. Seasonally damp, open, sandy grassland.

41. **Habenaria galpinii** Bolus, Icon. Orchid. Austro-Afr. **1**, 1: t.17 (1893). —Grosvenor in Excelsa **6**: 83 (1976). —Stewart et al., Wild Orch. South. Africa: 88 (1982). Type from South Africa.

Terrestrial herb 30–40 cm tall. Stem leafy; leaves 4–7, the upper ones bract-like, the lower ones set close together near base of stem, semi-erect, the largest 10–17 × 0.5–1.7 cm, the margins ± undulate. Inflorescence to 15 × 3–5 cm, densely several- to many-flowered; bracts 12–14 mm long. Flowers semi-erect, green, green and white or yellow-green. Pedicel and ovary 20–25 mm long. Dorsal sepal 5–6 × 1.5–2 mm, reflexed; lateral sepals 7–8 × 4 mm, obliquely obovate with lateral apiculum. Petals 2-lobed almost to base; upper lobe 5.5–6.5 × 0.5 mm, linear, erect; lower lobe 10–11 × 0.8–1.2 mm. Lip 3-lobed, with basal claw c. 2 mm long; mid-lobe 12–13 × 0.5–0.8 mm; side lobes 7.5–8 × 0.4 mm; all lobes linear; spur 18–23 mm long, sigmoid. Anther 3 mm high, canals 4.6–5.7 mm long. Stigmatic arms 5.5 mm long, porrect.

Zimbabwe. W: Matobo Distr., Besna Kobila Farm, 1450 m, fl. iii.1957, *Miller* 4161 (K; SRGH). N: Makonde Distr., Mwami (Miami), fl. iv.1926, *Rand* 44 (BM). E: Nyanga (Inyanga), 1800 m, fl. iii.1954, *Payne* 24 (K; SRGH). S: Mberengwa Distr., Sikanjena (Sikanajena) Hill, c. 1420 m, fl. 4.v.1973, *Pope, Biegel & Simon* 1109 (K; SRGH).

Also in South Africa (Transvaal). Wet soil over rock, *Eucalyptus* plantations; 1400–1800 m.

42. **Habenaria genuflexa** Rendle in J. Bot. **33**: 279 (1895). —Rolfe in F.T.A. **7**: 242 (1898). —Summerhayes in Kew Bull. **16**: 280 (1962); in F.W.T.A. ed. 2, **3**: 196 (1968); in F.T.E.A., Orchidaceae: 89 (1968). Type from Uganda.

Habenaria confusa Rolfe in F.T.A. **7**: 241 (1898). Type from Angola.
Habenaria stenoloba Schltr. in Bot. Jahrb. Syst. **38**: 1 (1905). Type from Gabon.
Habenaria anaphysema sensu Summerh. in F.W.T.A. **2**: 412 (1936), non Rchb.f.

Terrestrial herb 20–100 cm high. Tubers 1–2.5 × 0.5–1 cm, ovoid to ellipsoid. Stem leafy; leaves 7–11, the lowermost sometimes sheathing, the next 4–6 semi-erect or spreading, 7–26 × 0.5–2(2.5) cm, linear to lanceolate, acute; upper leaves becoming bract-like. Inflorescence 5–27 × 3–5.5 cm, several- to many-flowered; bracts 8–25 mm long, usually shorter than pedicel and ovary. Flowers green, white in centre, with an unpleasant smell. Pedicel and ovary 16–25 mm long, straight or more often bent at the junction so that the flowers are drooping. Dorsal sepal 5–7.5 × 2–3.5 mm, elliptic, reflexed; lateral sepals deflexed, 7–10 × 3.5–6 mm, obliquely obovate with a lateral apiculum. Petals 2-lobed almost to base; upper lobe 4.5–7.5 × c. 1 mm, linear, reflexed, papillose and ciliolate; lower lobe spreading forwards or deflexed, 8–16 × 1–1.5 mm, narrowly lanceolate, glabrous or papillose but not ciliate on margin. Lip 3-lobed to 2–3.5 mm from base; mid-lobe 9–13 × c. 0.5 mm, side lobes 6–11.5 mm long, slightly narrower than mid-lobe; all lobes linear; spur 8–10(13) mm long, twisted in the middle and swollen at the apex, usually incurved with a knee-like bend at about halfway. Anther 2–3 mm high, the loculi reclinate; canals 4–6 mm long, porrect, the tips turned up. Stigmatic arms 5–7.5 mm long, porrect, slender but enlarged and truncate at the apex. Rostellum mid-lobe c. 2 mm long, narrowly triangular, acute.

Zambia. N: Mbala Distr., Lumi Marsh below Kawimbe, 1740 m, fl. 26.ii.1959, *Richards* 10996 (K). W: Mwinilunga Distr., below Kalene Hill Mission, fl. 22.ii.1975, *Williamson & Gassner* 2470 (K; SRGH).

Widespread from West Africa south to Angola and east to the Sudan, Uganda and Tanzania. Wet or swampy grassland.

This species closely resembles *H. schimperiana* A. Rich., and it is possible that they form one variable species, in which case *H. schimperiana* would be the correct name. *H. schimperiana* is a larger, more robust plant with bigger flowers and a longer pedicel; the anterior petal lobe is ciliate on the margin.

43. **Habenaria hebes** la Croix & P.J. Cribb in Kew Bull. **48**, 2: 357 (1993). TAB. **31**. Type: Zambia, Mwinilunga Distr., Ikelenge, *E.A. Robinson* 3666 (K, holotype).

Slender terrestrial herb 40–45 cm high, tubers c. 12 × 10 mm, ovoid. Stem leafy; leaves 8–13, the lowermost 1–2 sheathing, with rather obscure reticulate veining; the next 2–5 suberect, to c. 11 × 3 mm, narrowly linear, the base sheathing; the upper 5–6 bract-like and appressed to the stem. Inflorescence 8–10 × 5–7 cm, fairly laxly 6–9-flowered; bracts to 15 mm long, much shorter than the pedicel. Flowers green, semi-erect to semi-spreading. Pedicel 20–22 mm long, ovary 10–13 mm long, straight or slightly arched. Sepals reflexed; dorsal sepal 4–5 × 1.5–2 mm, elliptic; lateral sepals 8 × 3–5 mm, obliquely obovate from a narrow base, with a lateral apiculum. Petals 2-lobed almost to base; upper lobe 4–5 × 0.4 mm, linear, erect but curving back; lower lobe deflexed, 12–14 × 0.5–0.7 mm, narrowly linear-lanceolate, slightly papillose towards the base. Lip 3-lobed almost to base; mid-lobe 13–16 × 0.6 mm, linear or narrowly lanceolate; side lobes similar but shorter; spur 26–28 mm long, straight, ± parallel to ovary and pedicel, very slender but swollen to c. 1.5 mm wide in apical 6 mm. Column 4 mm long in all, including a sterile basal part 2 mm long; anther 2 mm high, canals 5 mm long, porrect, the tips upturned; auricles large, truncate; stigmatic arms 5 mm long, porrect, slender but enlarged and truncate at apex; rostellum mid-lobe c. 1 mm long.

Zambia. W: Mwinilunga Distr., Ikelenge, 16.v.1965, *E.A. Robinson* 6601 (K); 30 km west of Mwinilunga, 17.v.1960, *E.A. Robinson* 3666 (K).

Apparently endemic in the Mwinilunga area. Seasonally damp grassland and sandy plateau grassland.

Habenaria hebes belongs to Sect. *Replicatae* Kraenzl. and is one of the group including *H. ugandensis* and *H. macrostele* which have a distinct stalk at the base of the column. It most closely resembles *H. ugandensis* but differs from it in the much narrower leaves, the shorter and fewer-flowered inflorescence, longer pedicel and ovary and the longer spur which is straight (and not incurved as in *H. ugandensis*), and is more slender for most of its length but more inflated at the apex.

The name is given as it is rather a dull species with no very distinctive features.

44. **Habenaria hirsutitrunci** G. Will. in Pl. Syst. Evol. **134**: 54, fig.1 (1980). —la Croix et al., Orch. Malawi: 70 (1991). Type: Zambia, Luangwa R., c. 50 km south of Mporokoso, *Williamson & Drummond* 1973 (SRGH, holotype).

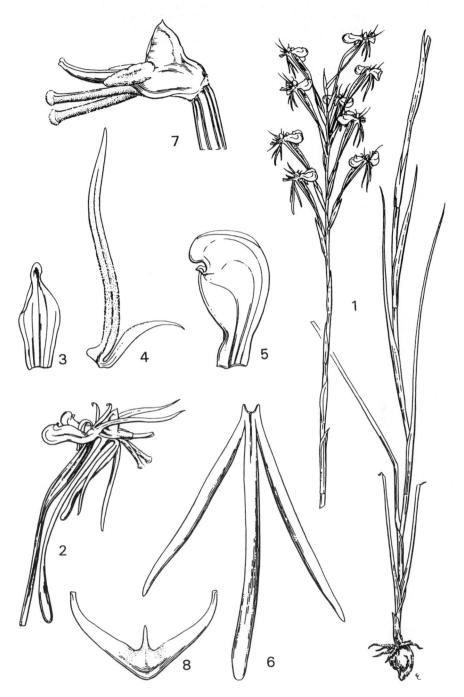

Tab. 31. HABENARIA HEBES. 1, habit (×⅔); 2, flower (×2); 3, dorsal sepal (×5); 4, petal (×5); 5, lateral sepal (×5); 6, lip (×5); 7, column (×5); 8, rostellum (×5), 1–8 from *Robinson* 6601 and *Milne-Redhead* 3769. Drawn by Eleanor Catherine. From Kew Bull.

Slender terrestrial herb 25–40 cm high. Stem leafy; leaves 7–8, the lowest 3 ± sheathing, the upper ones grading into the bracts, the middle ones semi-erect, 5–15 cm × 4–19 mm, linear. Inflorescence 9–13 × 3–5 cm, fairly densely several- to many-flowered; bracts leafy, to 25 × 10 mm. Flowers green. Pedicel and ovary erect, pedicel 15–16 mm long, ovary 8–10 mm. Dorsal sepal 4–5 × 1–3 mm, ovate, erect; lateral sepals reflexed, oblique with a lateral apiculum, 6–8 × 3–4.5 mm, usually rather brown and withered by flowering time. Petals 2-lobed almost to base; upper lobe c. 4 × 0.5 mm, narrowly linear, erect; lower lobes curving sideways and up, 14–18 mm long, c. 1 mm wide at base, becoming filiform. Lip 3-lobed to c. 1 mm from base, all lobes 12–17 mm long, barely 1 mm wide, narrowly linear, acute; spur 16–23 mm long, slender but very swollen in the apical 5 mm, parallel to the pedicel and ovary and tucked into the bracts. Anther 2–4 mm tall; canals 4–5 mm long. Stigmatic arms 4–5.5 mm long, projecting forwards but diverging, truncate at apex, pubescent. Rostellum mid-lobe 3 mm long, side lobes 4–5 mm long, pubescent, lying above stigmatic arms.

Zambia. N: Mbala Distr., road to Uningi Pans, 1500 m, fl. 5.iii.1965, *Richards* 19729 (K).
Malawi. N: Nyika National Park, Jalawe Hill, c. 2100 m, fl. 9.iv.1984, *la Croix* 606 (K).
Not known elsewhere. Montane grassland; 1500–2200 m.

45. **Habenaria huillensis** Rchb.f. in Flora **48**: 179 (1865). —Rolfe in F.T.A. **7**: 240 (1898). —Summerhayes in Bot. Not. **1937**: 184 (1937; in F.W.T.A. ed. 2, **3**: 196 (1968); in F.T.E.A., Orchidaceae: 96 (1968). —Geerinck in Fl. Afr. Centr., Orchidaceae pt. 1: 113 (1984). Type from Angola.
Habenaria humilior sensu Geerinck in Fl. Afr. Centr., Orchidaceae pt. 1: 113 (1984) non Rchb.f. (1881).

Terrestrial herb to 1 m high. Tubers 1–3 × 1 cm, ellipsoid or ovoid. Stem leafy; leaves (4)8–16, the lowermost 1–3 sheathing, the upper ones grading into the bracts, the middle ones ± erect, the largest 10–15(37) × 0.5–1.5 cm, linear. Inflorescence 6–27 × 3–5 cm, fairly densely many-flowered; bracts 10–25 mm long, rather scarious at flowering time, the lower ones longer than the ovary and pedicel. Flowers green, white in centre. Pedicel and ovary 16–21 mm long. Dorsal sepal reflexed, 4.5–7.5 × 2–3 mm, elliptic; lateral sepals deflexed, 6.5–10 × 3–5 mm, obliquely semi-ovate with a lateral apiculum. Petals 2-lobed almost to base; upper lobe 4.5–7.5 × 0.5 mm, linear, recurved; lower lobes 9–14 × 1–2 mm, lanceolate, acute, spreading forwards and upwards, usually glabrous, rarely papillose. Lip 3-lobed almost to base; mid-lobe 9–14 mm long, side lobes 6–12 mm long, all lobes linear, 0.5–1 mm wide; spur 1.5–2.5(3) cm long, somewhat swollen in apical half, ± parallel to pedicel and ovary. Anther 2.5–4 mm high; canals 2.5–5.5 mm long. Stigmatic arms 3–6.5 mm long, porrect, slender but enlarged and truncate at apex. Rostellum mid-lobe c. 1.5 mm high.

Zimbabwe. N: Makonde Distr., c. 22 km from Mvurwi (Umvukwes) to Guruve (Sipolilo), fl. 10.ii.1982, *Brummitt & Drummond* 15855 (K; SRGH). E: Nyanga (Inyanga), Inyangani, 1700 m, fl. 20.i.1931, *Norlindh & Weimarck* 4488 (BM).
Also in Ghana, Nigeria, Central African Republic, Zaire, Sudan, Uganda, Kenya and Angola. Grassland; up to 1700 m.

46. **Habenaria humilior** Rchb.f., Otia Bot. Hamburg.: 100 (1881). —Rolfe in F.T.A. **7**: 236 (1898). —Summerhayes in Kew Bull. **16**: 283 (1962); in F.T.E.A., Orchidaceae: 91 (1968). —Grosvenor in Excelsa **6**: 83 (1976). —Williamson, Orch. S. Centr. Africa: 53 (1977). —la Croix et al., Orch. Malawi: 71 (1991). Type from Ethiopia.
Habenaria replicata A. Rich., Tent. Fl. Abyss. **2**: 296 (1851), non A. Rich. (1850) nom illegit. Type from Ethiopia.
Habenaria hochstetteriana Kraenzl. in Bot. Jahrb. Syst. **16**: 73 (1892). —Rolfe in F.T.A. **7**: 243 (1898). Type as above.
Habenaria culicifera Rendle in J. Linn. Soc., Bot. **33**: 278 (1895). —Rolfe in F.T.A. **7**: 237 (1898). Type from Uganda.
Habenaria kassneriana Kraenzl. in Vierteljahrsschr. Naturf. Ges. Zürich **74**: 104 (1929), non Kraenzl. (1912), nom illegit. Type from Zaire.

Terrestrial herb 15–70 cm tall. Tubers 1–2 cm long, globose, ovoid or ellipsoid. Stem leafy; leaves 4–13, the lowermost 1–2 usually sheath-like, the middle ones suberect or spreading, the largest 5–23 × 0.5–2.2 cm, lanceolate or linear; upper leaves grading into bracts. Inflorescence 5–25 × 3.5–5 cm, few to many-flowered; bracts 11–16 mm long, lanceolate, rather scarious. Flowers spreading, white, green or green and white. Pedicel and ovary 15–22 mm long. Dorsal sepal 4–7 × 1.5–3 mm, elliptic, reflexed; lateral sepals deflexed, 5.5–9.5 × 3–5.5 mm, obliquely obovate with a lateral apiculum. Petals 2-lobed almost to base; upper lobe 4–6 × 0.5–1 mm, linear, ± reflexed; lower lobe 6–11 × 1.5–3 mm, oblong-lanceolate, obtuse, ± spreading. Lip deflexed, 3-lobed almost to base; mid-lobe (6)9–13 mm long, side lobes 5–9 mm long, all lobes linear, less than 1 mm wide; spur 15–25 mm long, usually slightly incurved, swollen towards apex, not or only slightly twisted in middle. Anther 1.5–3 mm tall; canals 3–6 mm long. Stigmatic arms 3–6 mm long, porrect, truncate at apex; rostellum mid-lobe 1.5–2 mm high, triangular, acute.

Zambia. N: Mbala Distr., Mpulungu Marsh, near Nyamkole, 780 m, fl. 20.ii.1964, *Richards* 19054 (K). W: Nkana, fl. 14.i.1936, *R. Brenan* 11 (K). C: Lusaka Distr., 56 km south of Lusaka, fl. ii.1967, *G. Williamson* 143 (K). E: 48 km south of Nyika, fl. ii.1969, *G. Williamson* 1394 (K). S: Choma, 1300 m, fl. 9.i.1957, *E.A. Robinson* 2114 (K). **Zimbabwe**. C: Harare Distr., Ruwa R. tributary, fl. 31.i.1971, *Linley* 610 (K; SRGH). E: Chimanimani, 1500 m, fl. 16.ii.1954, *Ball* 252 (K; SRGH). **Malawi**. C: Lilongwe Distr., Dzalanyama Forest Reserve, valley NW of Kazuzu Hill, 1420 m, fl. 24.ii.1982, *Brummitt, Polhill & Banda* 16079 (K). S: Mulanje Mt., Madzeka Basin and Little Ruo Gorge, 900–2100 m, fl. 8.iv.1974, *Allen* 532 (K; MAL).
Also in Ethiopia, Sudan, Zaire, Uganda and Tanzania. Grassy dambos, damp places in woodland clearings, and montane grassland; 900–2200 m.

47. **Habenaria ichneumonea** (Sw.) Lindl., Gen. Sp. Orchid. Pl.: 313 (1835). —Rolfe in F.T.A. **7**: 240 (1898). —Summerhayes in F.W.T.A. ed. 2, **3**: 196 (1968); in F.T.E.A.,.Orchidaceae: 94 (1968). —Grosvenor in Excelsa **6**: 83 (1976). —Williamson, Orch. S. Centr. Africa: 56 (1977). —Geerinck in Fl. Afr. Centr., Orchidaceae pt. 1: 107 (1984). —la Croix et al., Orch. Malawi: 72 (1991). Type from Sierra Leone.
Orchis ichneumonea Sw. in Schrad., Neues J. **1**: 21 (1805).
Habenaria pedicellaris Rchb.f., Otia Bot. Hamburg.: 100 (1881). —Rolfe in F.T.A. **7**: 244 (1898), pro parte. —Eyles in Trans. Roy. Soc. S. Afr. **5**: 334 (1916). Type from Ethiopia.

Rather slender terrestrial herb 18–85 cm tall. Tubers 1–2.5 cm long, ellipsoid or ovoid. Stem leafy; leaves 5–12, the lowermost usually sheathing, the next few erect or suberect, the largest 7–21 cm long and 4–10 mm wide, narrowly linear. Inflorescence 6–22 × 3–5 cm, laxly or fairly densely few- to many-flowered; bracts 7–18 × 2–3 mm, usually shorter than pedicel and ovary. Flowers suberect, green or green and white. Pedicel and ovary straight or slightly incurved; pedicel 24–30 mm long, ovary 5–6 mm long. Dorsal sepal reflexed, 3–5.5 × 1.5–2.5 mm, elliptic; lateral sepals deflexed, 4–8.5 × 3–5 mm, obliquely obovate with a lateral apiculum. Petals 2-lobed almost to base; upper lobe reflexed, 3–4.5 × 0.3–0.4 mm, linear; lower lobe 6.5–13 × 1–1.5 mm, narrowly lanceolate, acute, curving forwards. Lip 3- lobed with an undivided base 1.5–4 mm long; mid-lobe 13–15 mm long; side lobes 13–17 mm long; all lobes linear, c. 0.5 mm wide; spur 10–25 mm long, usually incurved, the apex very swollen. Anther 1.2–2.5 mm high; canals 3.5–6.5 mm long, porrect, the tips upcurved. Stigmatic arms 4–8 mm long, porrect, slender with the apices enlarged and truncate. Rostellum mid-lobe 1.5–2 mm long, triangular.

Botswana. N: Dabonga Boat Harbour, fl. 21.vii.1973, *P.A. Smith* 675 (K; PRE; SRGH). **Zambia**. N: Mbala Distr., Saisi Valley, 1500 m, fl. 20.vii.1970, *Sanane* 1131 (K). W: Mwinilunga Distr., Sinkabolo Dambo, fl. 9.xii.1937, *Milne-Redhead* 3570 (K). C: Chakwenga Headwaters, 100–129 km east of Lusaka, fl. 25.viii.1963, *E.A. Robinson* 5637 (K; PRE). E: 48 km west of Chipata, fl. 3.vii.1974, *Williamson & Gassner* 2295 (K). **Zimbabwe**. C: Harare Distr., in sandy vleis, c. 1500 m, fl. 24.ii.1945, *Greatrex* 26 (K; PRE). S: Masvingo (Fort Victoria), 1909–12, *Monro* 1786 (BM). **Malawi**. N: Mzimba Distr., 3 km west of Mzuzu, at Katoto, 1360 m, fl. 25.vi.1976, *Pawek* 11491 (K; SRGH; SRGH). C: Kasungu Distr., Lisasadzi Experimental Station, fl. 22.viii.1956, *Jackson* 2031 (K).
Also in Sierra Leone, Senegal, Nigeria, Guinée, Central African Republic, Zaire, Angola, Ethiopia, Uganda and Tanzania. Boggy grassland, sometimes in shallow, standing water; c. 1350–1500 m.
Almost all the specimens from the Flora Zambesiaca area have relatively short, incurved spurs.

48. **Habenaria kilimanjari** Rchb.f., Otia Bot. Hamburg.: 119 (1881). —Rolfe in F.T.A. **7**: 240
(1898). —Summerhayes in Bull. Misc. Inform., Kew **1933**: 249 (1933); in Kew Bull. **16**: 284
(1962); in F.T.E.A., Orchidaceae: 83 (1968). —Williamson, Orch. S. Centr. Africa: 51
(1977). —Geerinck in Fl. Afr. Centr., Orchidaceae pt. 1: 110 (1984). —la Croix et al.,
Orch. Malawi: 72 (1991). Type from Tanzania.
  Habenaria martialis Rchb.f., Otia Bot. Hamburg.: 99 (1881). —Rolfe in F.T.A. **7**: 236
(1898). Type from Tanzania.
  Habenaria uhligii Kraenzl. in Bot. Jahrb. Syst. **43**: 396 (1909). Type from Tanzania.
  Habenaria lutaria Schltr. in Bot. Jahrb. Syst. **53**: 515 (1915). Type from Tanzania.
  Habenaria theodorii Kraenzl. in Vierteljahrsschr. Naturf. Ges. Zürich **74**: 107 (1929). Type
from Zaire.

Terrestrial herb 15–90 cm tall. Tubers 1–2.5 cm long, ovoid or ellipsoid. Stem leafy;
leaves 7–13, the lowermost 1–2 usually sheath-like, the upper ones grading into the
bracts, the middle ones ± spreading, the largest 6.5–21 × 1–2.5 cm, linear or
lanceolate. Inflorescence 4–37 × 3.5–5 cm, laxly or fairly densely several- to many-
flowered; bracts 12–17 mm long. Flowers white, the spur green towards the apex.
Pedicel and ovary 15–26 mm long, ± straight. Dorsal sepal reflexed, 5–7.5 × 2–3 mm,
elliptic; lateral sepals deflexed, 6–8 × 2–4 mm, obliquely semi-orbicular. Petals 2-
lobed almost to base; upper lobe 5–7.5 × 1–2 mm, lower lobe 3.5–6 × 1–1.5 mm, both
lobes oblanceolate, sometimes toothed in upper half. Lip 3-lobed almost to base,
mid-lobe 5–9 mm long, side lobes 3–5 mm long, all lobes linear, c. 0.5 mm wide, the
margins sometimes serrate; spur pendent, 13–18 mm long, very slender, swollen near
apex. Anther erect, 2.5–3 mm high; canals c. 2 mm long, slightly incurved. Stigmatic
arms c. 2 mm long, porrect, the tips enlarged and truncate. Rostellum mid-lobe c.
2.5 mm long, side lobes slightly longer, the tips pincer-like.

Zambia. B: Mongu, fl. 20.i.1966, *E.A. Robinson* 6808 (K). N: Mbala Distr., Mpulungu, 900 m,
fl. 12.ii.1957, *Richards* 8175 (K). W: near Kasempa, Lufupa R., fl. 1921–22, *Foster* s.n. (BM). C:
Kabwe, dambo at Mukobeko road, fl. 24.i.1961, *Morze* 54 (K). S: Mazabuka, near bank of
Miangama Stream, c. 6 km west of Monze African Compound, fl. 23.i.1963, *Linley* 313 (K;
SRGH). **Malawi**. N: Nyika Plateau, Mwenembwe (Mwanemba), 2400 m, fl. Feb.–March 1903,
McClounie 57 (K).
  Also in Zaire, Angola, Kenya and Tanzania. Damp grassland, sandy soil in marsh; 900–2400 m.

49. **Habenaria kyimbilae** Schltr. in Bot. Jahrb. Syst. **53**: 515 (1915). —Summerhayes in Kew
Bull. **15**: 285 (1962); in F.T.E.A., Orchidaceae: 87 (1968). —Williamson, Orch. S. Centr.
Africa: 51 (1977). —Geerinck in Fl. Afr. Centr., Orchidaceae pt. 1: 111 (1984). —la Croix
et al., Orch. Malawi: 73 (1991). Type from Tanzania.
  Habenaria furcipetala Schltr. in Bot. Jahrb. Syst. **53**: 514 (1915). Type from Tanzania.

Terrestrial herb up to 1 m tall. Tubers 1–2.5 cm long, globose or ellipsoid. Stem
leafy; leaves 6–11, the lowermost 1–2 sheathing and spotted with black; the
uppermost 2–4 becoming bract-like; the middle ones suberect, the largest 6.5–22 ×
1–2 cm, linear or lanceolate. Inflorescence 5–25 × 2–3.5 cm, densely several- to many-
flowered; bracts of similar length. Flowers green, the anterior petal lobe usually white
but sometimes greenish. Pedicel and ovary 10–20 mm long, arched. Dorsal sepal
reflexed, 5–6 × 2–3 mm, elliptic; lateral sepals deflexed, 6–10 × 3–5 mm, obliquely
obovate with a lateral apiculum. Petals densely pubescent, 2-lobed almost to base;
upper lobe erect, 5–6 × 0.3–1 mm, linear; lower lobe spreading, 6–10 × 1–2 mm,
lanceolate. Lip 3-lobed almost to base; mid-lobe 8–12 mm long; side lobes 6–8.5 mm
long, all lobes linear, less than 1 mm wide, but the side lobes slightly wider; spur
9–12.5 mm long, ± parallel to the pedicel and ovary, swollen at apex, twisted in the
middle. Anther erect, 2–3 mm long; canals 3.5–5 mm long, slightly incurved.
Stigmatic arms 3–5 mm long, porrect, slender but enlarged and truncate at apex.
Rostellum mid-lobe 1.5 mm high.

Malawi. N: Karonga Distr., Misuku Hills, 1360 m, fl. 9.v.1947, *Benson* 1239 (K); South Viphya,
between Elephant Rock and Chikangawa, 1750 m, fl. 10.iv.1986, *la Croix* 831 (K; MAL).
  Also in Zaire and Tanzania. Montane grassland and *Brachystegia* woodland; 1300–2300 m.
  This species is rather prone to producing abnormal flowers.

50. **Habenaria leucoceras** Schltr. in Bot. Jahrb. Syst. **53**: 518 (1915). —Summerhayes in F.T.E.A., Orchidaceae: 90 (1968). —Williamson, Orch. S. Centr. Africa: 53 (1977). —la Croix et al., Orch. Malawi: 73 (1991). Type from Tanzania.

Terrestrial herb 30–60 cm tall. Tubers 1.5–2 × 1–1.5 cm, ellipsoid or ovoid. Stem leafy. Leaves 7–11, the lowermost 1–3 sheath-like, white with prominent dark green reticulate veining; middle ones suberect, the largest 7–20 × 0.5–1 cm, linear; the uppermost becoming bract-like. Inflorescence 8–15 × 4–6 cm, fairly laxly several- to many-flowered; bracts 16–30 mm long, slightly longer than pedicel and ovary. Flowers greenish-white. Pedicel and ovary 16–25 cm long, spreading and slightly arched. Dorsal sepal reflexed, 6–8 × 2–3 mm, elliptic; lateral sepals reflexed, 8–11 × 4.5–7 mm, obliquely obovate with a lateral apiculum. Petals 2-lobed almost to base; upper lobe 7–9.5 × c. 1 mm, linear, ciliolate, reflexed, the 2 lobes often crossing; lower lobes pendent, hinged, 13–19 × 2–3.5 cm, curved lanceolate, acute, glabrous. Lip deflexed, 3-lobed almost to base; mid-lobe 13–19 × 0.5–1 mm, the tip often curved back; side lobes 8.5–12.5 mm long, slightly narrower than mid-lobe; spur slightly incurved or parallel to pedicel and ovary, 15–18.5 mm long, twisted in middle, the apex swollen. Anther 2–4 mm high; canals 3–5 mm long. Stigmatic arms 3.5–5 mm long, porrect, truncate at apex; rostellum mid-lobe c. 2 mm long, triangular.

Zambia. E: Nyika National Park, c. 0.5 km SW of Zambian Rest-house, fl. 17.iv.1986, *Philcox, Pope & Chisumpa* 9976 (K). **Malawi**. N: Nyika National Park, grassland near Dam 3, c. 2100 m, fl. 7.iv.1984, *la Croix* 594 (K). S: Zomba Plateau, 1800 m, fl. 25.iv.1949, *Wiehe* N/84 (K).

Also in Tanzania. Montane grassland; 1500–2400 m.

51. **Habenaria macrostele** Summerh. in Bot. Not. **1937**: 184 (1937); in Mem. N.Y. Bot. Gard. **9**, 1: 80 (1954). —Goodier & Phipps in Kirkia **1**: 52 (1961). —Moriarty, Wild Fls. Malawi: t. 26, 4 (1975). —Grosvenor in Excelsa **6**: 83 (1976). —Geerinck in Fl. Afr. Centr., Orchidaceae pt. 1: 110 (1984). —la Croix et al., Orch. Malawi: 74 (1991). Type: Zimbabwe, Nyanga, Mt Inyangani summit, *Norlindh & Weimarck* 4989 (K; LD, holotype).

Slender terrestrial herb 30–60 cm tall. Tubers 1–2.5 × 1–1.2 cm, ovoid or ellipsoid. Stem leafy; leaves 3–7, the lowermost sheathing, the largest in the middle, 4–18 cm × 8–17 mm, linear or narrowly lanceolate; the uppermost grading into the bracts. Inflorescence 10–15 × 2.5–4 cm, rather laxly several- to many-flowered; bracts 12–19 × 3–4 mm, glandular, often brown-tipped, often slightly longer than pedicel and ovary. Flowers green, white in centre, smelling very unpleasant. Pedicel and ovary 15–22 mm long. Sepals reflexed; dorsal sepal 4–5 × 1.5–2 cm, elliptic; lateral sepals 5.5–6.5 × 2.7–3 mm, obliquely obovate with a lateral apiculum. Petals 2-lobed almost to base, rather fleshy; upper lobe 4–5 × 0.5–1 mm, linear, erect, the 2 often crossing; lower lobe pendent, hinged, 6–7 × 1 mm, linear, the margin ciliate. Lip 3-lobed to 1–1.5 mm from base; mid-lobe 7–8.5 mm long; side lobes 4.5–5.5 mm long; all lobes linear, c. 1 mm wide; spur 13.5–20 mm long, inflated at apex, not twisted, usually tucked into bracts. Column 4–5 mm long; anther c. 2 mm long with sterile stalk c. 2.5 mm long; canals c. 1 mm long. Stigmatic arms 2.5–3.5 mm long, porrect, truncate at apex; rostellum mid-lobe c. 1 mm long.

Zimbabwe. C: Makoni Distr., Chimbi Farm, 1575 m, fl. 14.ii.1960, *Chase* 7271 (in part, mixed with *H. schimperiana*) (K; SRGH). E: Nyanga, Inyangani, 2450 m, fl. 14.ii.1931, *Norlindh & Weimarck* 4989 (K). **Malawi**. C: Dedza Mt., grassland by summit, 2135 m, fl. 23.ii.1982, *Brummitt, Polhill & Banda* 16051 (K; MAL). S: Zomba Mt., 1800 m, on seepage slopes, fl. 21.iii.1981, *la Croix* 128 (K). **Mozambique**. MS: Tsetserra, 2140 m, fl. 7.ii.1955, *Exell, Mendonça & Wild* 241 (BM).

Also in Zaire, Burundi and Tanzania. Montane grassland, wet soil overlying rocks and on seepage slopes; 1800–1500 m.

Similar to *H. ndiana* Rendle but usually a smaller, more slender plant with a laxer, fewer-flowered inflorescence and shorter pedicel and ovary; anther canals about half as long.

Habenaria macrostele is not included in Summerhayes' (1968) account of the genus in F.T.E.A., but specimens are now known from southern Tanzania. Plants from there, and from Burundi, those from central Malawi and some from south Malawi have the basal leaf sheaths spotted with black, sometimes heavily. Zimbabwe plants, and some from Malawi, have unspotted sheaths. Zimbabwe plants usually have broader leaves. The flowers, however, are uniform throughout the range.

52. **Habenaria macrotidion** Summerh. in Kew Bull. **16**: 285 (1962). —Williamson, Orch. S. Centr. Africa: 53 (1977). TAB. **32**. Type: Zambia: Mbereshi, *Richards* 12318 (K, holotype).

Terrestrial herb 40–55 cm tall; tubers c. 20 × 8 mm, ellipsoid. Leaves 6–8, the lowermost reduced to sheaths, the stem leaves 3–5, suberect, c. 16 × 1 cm, linear-lanceolate, acute, sheathing at the base; uppermost leaves becoming bract-like. Inflorescence 10–16 × 3.5–4 cm, laxly many-flowered; bracts less than half the length of pedicel and ovary. Flowers erect to spreading, pale green. Pedicel and ovary 17–23 mm long, slightly arched. Dorsal sepal reflexed, 5–6 × 3 mm, obovate, convex; lateral sepals deflexed, 7–8.5 × 5 mm, obliquely obovate. Petals 2-lobed almost to base; upper lobe erect, 5–6 × 0.5 mm, linear, densely papillose-ciliolate; lower lobe ± deflexed, 6–7.5 × 2–2.5 mm, lanceolate, acute, slightly fleshy and opaque, glabrous. Lip deflexed, 3-lobed with a basal claw 1.5–2 mm long; mid-lobe 9–10 × 0.5 mm, linear, obtuse; side lobes 5–6 × 0.3 mm, linear, acute; spur 8–10 mm long, pendent, only very slightly swollen in apical half. Anther erect, 3–4 mm high; canals c. 4 mm long, porrect, the tips upturned. Auricles c. 1.7 mm long, deeply 2-lobed, the lobes lanceolate-linear. Stigmatic arms porrect, 4–4.5 mm long; rostellum mid-lobe 2 mm long.

Zambia. N: Kawambwa Distr., Mbereshi, west of Kawambwa, 1050 m, fl. 11.i.1960, *Richards* 12318 (K).
Apparently endemic to the area of the Luapula Province of Zambia. Swampy ground.
Very similar to *H. chirensis* Rchb.f., but the spur is much shorter and does not reach the tip of the bracts, and only the posterior petal lobe is ciliolate.

53. **Habenaria myodes** Summerh. in Kew Bull. **16**: 287, fig. 14 (1962); in F.T.E.A., Orchidaceae: 91 (1968). —Grosvenor in Excelsa **6**: 83 (1976). —Williamson, Orch. S. Centr. Africa: 52 (1977). —la Croix et al., Orch. Malawi: 75 (1991). Type: Zimbabwe, Harare Distr., Cleveland, *Greatrex* in *GHS* 27347 (K, holotype; SRGH).

Terrestrial herb 20–75 cm tall; tubers 1–2 cm long, globose or ovoid. Stem leafy; leaves 6–14, the lowermost sometimes sheath-like, the next 6 or so large, semi-spreading, set fairly close together in lower part of stem, 6–19 × 1–2.5(3.5) cm, lanceolate; uppermost leaves grading into the bracts. Inflorescence 6–20 × 3–6 cm, fairly laxly several- to many-flowered; bracts 10–18 mm long, usually shorter than pedicel and ovary. Flowers green, white in centre, with a strong, sweet, sickly smell, especially at night. Pedicel and ovary 20–32 mm long, somewhat arched. Dorsal sepal reflexed, 5–7.5 × 3–4 mm, elliptic; lateral sepals deflexed, 7.5–10 × 5–6.5 mm, obliquely obovate with a lateral apiculum. Petals ciliolate, 2-lobed almost to base; upper lobe 5–7 × c. 1 mm, linear, erect; lower lobes pendent, hinged, 9–18 × 1.5–2.5 mm, curved lanceolate, acute. Lip 3-lobed to 1–1.5 mm from base; mid-lobe 11–20 mm long; side lobes 6.5–15 mm long; all lobes linear, c. 1 mm wide; spur 14–20 mm long with 1 twist in middle, swollen in apical half, horizontal or ± incurved. Anther c. 3 mm high, canals porrect, 4–6 mm long. Stigmatic arms 3.5–6 mm long, porrect, enlarged and truncate at apex. Rostellum mid-lobe c. 2 mm long.

Zimbabwe. C: Harare Distr., Cleveland, fl. 2.iv.1950, *Greatrex* in *GHS* 27347 (K; SRGH). E: "Himalayas", Butler South, 1500 m, fl. 20.iii.1955, *Ball* 536 (K; SRGH). **Malawi**. N: Misuku Hills, 1500 m, fl. 9.v.1947, *Benson* 1242 (K). S: Blantyre, 900 m, fl. 1.iv.1982, *la Croix* 305 (K); Chiradzulu Mt., 1350–1500 m, fl. 14.iv.1970, *Brummitt & Banda* 9832 (K).
Also in Rwanda, Uganda and Tanzania. Open woodland, grass at edge of forest, wet flush on rocky hillsides; 900–2400 m.
Subsp. *myodes* occurs in the Flora Zambesiaca area. Subsp. *latipetala* Summerh., which differs in its oblong-lanceolate, abruptly tapered anterior petals lobes, occurs in Uganda and Rwanda.

54. **Habenaria ndiana** Rendle in J. Linn. Soc., Bot. **30**: 393 (1895). —Rolfe in F.T.A. **7**: 239 (1898). —Summerhayes in F.T.E.A., Orchidaceae: 101 (1968). —la Croix et al., Orch. Malawi: 76 (1991). Type from Kenya.
   Habenaria ingrata Rendle in J. Bot. **33**: 279 (1895). —Rolfe in F.T.A. **7**: 236 (1898). Type from Uganda.
   Habenaria similis Schltr. in Bot. Jahrb. Syst. **38**: 147 (1906). Type from Tanzania.

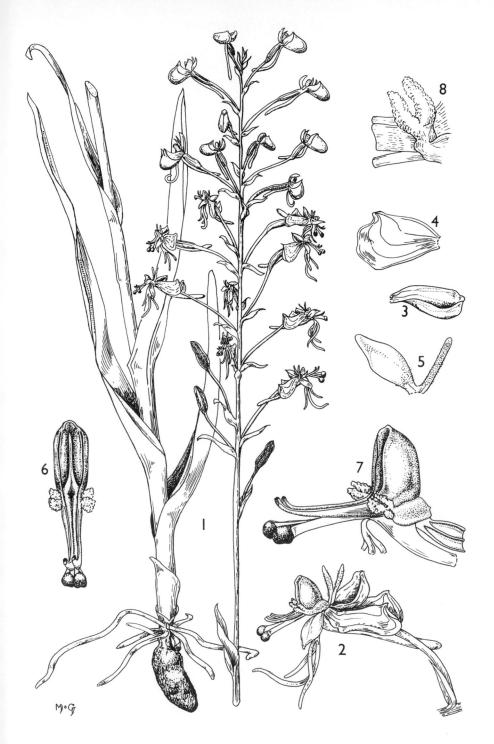

Tab. 32. HABENARIA MACROTIDION. 1, habit (×1); 2, flower, side view (×2); 3, dorsal sepal (×2); 4, lateral sepal (×2); 5, petal (×2); 6, column, front view (×6); 7, column, side view (×6); 8, staminode (×10), 1–8 from *Watmough* 192. Drawn by Mary Grierson. From Kew Bull.

Terrestrial herb to 60 cm tall; tubers 2–3 × 1–1.5 cm, ovoid or ellipsoid. Stem leafy. Leaves 4–16, the lowermost 1–3 sheathing, sometimes white with green reticulate veining; the uppermost grading into the bracts; the middle leaves suberect, the largest 8–20 × 0.7–2 cm, linear or lanceolate. Inflorescence 10–25 × 3–5 cm, densely several- to many-flowered; bracts scarious, 1–2 mm long. Flowers green or green and white. Pedicel 18–20 mm long, ovary 8–12 mm long, together ± straight. Dorsal sepal reflexed, 4–6 × 1–3 mm, elliptic; lateral sepals deflexed, 6–8.5 × 3–4 mm, obliquely obovate with a lateral apiculum. Petals ciliolate, 2-lobed almost to base; upper lobe 4–6.5 × c. 0.5 mm, linear, curving back; lower lobe 7.5–12 × 1–1.5 mm, narrowly lanceolate, spreading downwards. Lip 3-lobed almost to base; mid-lobe 8.5–11.5 mm long; side lobes slightly shorter; all lobes linear, 0.5–1 mm wide; spur 15–25 mm long, twisted in middle, swollen at apex, lying parallel to pedicel and ovary. Column 3–4.5 mm long, including a sterile stalk 1.5–2 mm long; anther canals c. 2 mm long, incurved. Stigmatic arms 2–3 mm long, porrect, fairly stout, truncate at apex. Rostellum mid-lobe c. 2 mm high, triangular.

Malawi. S: Zomba Mt., on "up-road" to plateau, c. 1360 m, fl. 6.iii.1983, *la Croix* 467 (K). Also in Kenya, Uganda and Tanzania. Seepage slope amongst rocks; c. 1360 m.

55. **Habenaria nyikana** Rchb.f., Otia Bot. Hamburg.: 100 (1881). —Rolfe in F.T.A. **7**: 244 (1898). —Summerhayes in F.T.E.A., Orchidaceae: 97 (1968). —Grosvenor in Excelsa **6**: 83 (1976). —la Croix et al., Orch. Malawi: 76 (1991). Type: Mozambique, mouth of R. Zambezi, Nyika Is., Aug. 1862, *Kirk* s.n. (K, holotype).
  Habenaria sp., Eyles in Trans. Roy. Soc. S. Afr. **5**: 334 (1916).

Robust terrestrial herb to 100 cm tall. Tubers 1.5–2.5 × 1 cm, ellipsoid, woolly. Stem leafy. Leaves 10–20, well spaced up the stem, the lowermost usually sheathing, the next 6–10 suberect, the largest 7–23 × 1–2 cm, linear or lanceolate, glossy green with 3 prominent longitudinal veins; the upper leaves appressed to stem and bract-like. Inflorescence 10–35 × 5–6 cm, laxly or fairly densely several- to many-flowered; bracts 15–30 mm long, glandular at the base, scarious at the tips, the lowest about as long as the pedicel and ovary. Flowers green, white in centre. Pedicel and ovary 20–30 mm long, straight or slightly arched, spreading, the ovary usually longer than the pedicel. Sepals reflexed, somewhat scarious after flowers open; dorsal sepal 5–7.5 × 2–3 mm, elliptic; lateral sepals 7–10.5 × 5–7 mm, obliquely obovate with a lateral apiculum. Petals 2-lobed almost to base, slightly ciliolate; upper lobe 5–7.5 × 0.5–1 mm, linear, curving back; lower lobes pendent, hinged, 9.5–14 × 1.5–2.5 mm, lanceolate, acute. Lip 3-lobed almost to base; mid-lobe 11–15 mm long; side lobes 10–13 mm long; all lobes linear, less than 1 mm wide; spur 15–30 mm long, slender, inflated at apex, with 1 twist in the middle, parallel to pedicel and ovary or slightly incurved. Anther 3.5–4 mm high; canals 4–7 mm long, porrect, curving up at tips. Stigmatic arms 4.5–6 mm long, porrect, truncate at apex. Rostellum mid-lobe 2.5–3 mm long, narrowly triangular.

Zimbabwe. W: Matopos Hills, 1500 m, fl. iii.1902, *Eyles* 1037 (BM). E: Nyanga, Rhodes View, 1120 m, fl. 28.ii.1963, *Chase* 7958 (K; PRE; SRGH). S: Masvingo Distr., Hillingdale, near Tokwe crossing on road to Beit Bridge, 1060 m, fl. iii.1958, *Dale* 10 (K; SRGH). **Malawi**. N: South Viphya, Luwawa link road, turn-off to Kawandama, 1750 m, fl. 25.iii.1977, *Pawek* 12525 (K; MAL; MO). S: Mulanje Distr., Likhubula Valley, c. 800 m, fl. 6.iii.1981, *la Croix* 113 (K). **Mozambique**. Z: Expedition Island, fl. viii.1858, *Kirk* (K). GI: Inhambane, Massinga, Pomene, fl. 16.vii.1981, *de Koning & Hiemstra* 9049 (K; LMU).
Also in South Africa (Transvaal) and Tanzania (subsp. *pubipetala*). Grassland, woodland and dambos; 0–1750 m.
Subsp. *nyikana* occurs in the Flora Zambesiaca area. Subsp. *pubipetala* Summerh., recognised by its pubescent anterior petal lobes, occurs in Tanzania.

56. **Habenaria petraea** Renz & Grosvenor in Candollea **34**, 2: 358 (1979). —la Croix et al., Orch. Malawi: 77 (1991). Type: Malawi, Nyika Plateau, Mt. Chosi, *Grosvenor & Renz* 1104 (K; SRGH, holotype).

Terrestrial herb 25–70 cm high; tubers small, ovoid. Leaves 5–9, the lowermost 1–2 sheathing, spotted with black, the next few leaves semi-spreading, 6–14 × 0.5–2 cm, linear-lanceolate, folded, the uppermost grading into the bracts. Inflorescence

5–26 × 3.5–5 cm, laxly or fairly densely several- to many-flowered; bracts 10–22 mm long, lanceolate, acuminate, finely pubescent or glandular. Flowers green, or green with the anterior petal lobes white. Pedicel 16–20 mm long, semi-erect, ovary 8–10 mm long, horizontal, all becoming straight and semi-erect in fruit. Dorsal sepal reflexed, 5.5–8 × 2–3 mm, oblong; lateral sepals deflexed, 8.3–10 × 4–6 mm, obliquely obovate with a lateral apiculum. Petals densely pubescent, 2-lobed almost to base; upper lobe 5–7 × 0.6–0.8 mm, linear, erect or recurved; lower lobes 9–13 × 1.5–2 mm, lanceolate, acute, spreading, hinged. Lip deflexed, 3-lobed almost to base; mid-lobe 12–15 mm long; side lobes 8–10 mm long; all lobes linear, to 1 mm wide; spur (12.5)15–20 mm long, swollen towards apex, twisted in the middle, horizontal or upcurved so that the tip is higher than the ovary. Column 3.5–4 mm high; anther canals porrect, 4–5 mm long. Stigmatic arms porrect, 3.6–5 mm long, (usually slightly shorter than anther canals), slender, truncate at apex. Rostellum mid-lobe c. 2 mm long, triangular, acute.

Zambia. E: Nyika Plateau, Chowo Rocks, fl. iii.1967, *Williamson & Odgers* 273 (K). **Malawi**. N: Nyika Plateau, SW of Chelinda Camp near top of Mt. Chosi, 2380 m, fl. 10.iii.1977, *Grosvenor & Renz* 1104 (K; SRGH); Chitipa Distr., Misuku Hills, Mugesse (Mughese) Forest, c. 1800 m, fl. 14.vii.1970, *Brummitt* 12108 (K).
Not known elsewhere. Damp areas of montane grassland, edge of forest or woodland; 1300–2380 m.
This species is closely related to *H. kyimbilae*, differing in the longer pedicel and ovary, longer stigmatic arms and anther canals, and longer spur, which is horizontal or upcurved. It is also very similar to the East African *H. altior* Rendle, differing only in the spur position.

57. **Habenaria retinervis** Summerh. in Kew Bull. **16**: 291 (1962); in F.T.E.A., Orchidaceae: 98 (1968). —Grosvenor in Excelsa **6**: 83 (1976). —Williamson, Orch. S. Centr. Africa: 54 (1977). —Geerinck in Fl. Afr. Centr., Orchidaceae pt. 1: 117 (1984). —la Croix et al., Orch. Malawi: 77 (1991). TAB. **33**. Type: Zambia, Kalenda Plain, *Milne-Redhead* 3805 (K, holotype).

Terrestrial herb up to 80 cm tall; tubers 1–3 cm long, globose, ellipsoid or ovoid. Leaves 7–11, the lowermost 2–3 sheathing, white with dark green reticulate veining; the next few leaves semi-erect, the largest 5–16 × 1–3 cm, lanceolate; upper leaves grading into the bracts. Inflorescence 10–27 × 4–6 cm, fairly laxly 8–30-flowered; bracts scarious, 10–20 mm long, shorter than the pedicel and ovary. Flowers pale green, curved outwards. Pedicel and ovary slightly arched; pedicel 7–10 mm long, ovary 15–20 mm long. Dorsal sepal reflexed, 5–7 × 2.5–3 mm, elliptic; lateral sepals deflexed, 7–9.5 × 4.5–6.5 mm, obliquely obovate with a lateral apiculum. Petals 2-lobed almost to base; upper lobe 11–14 × 1 mm, recurved, slightly toothed towards tip; lower lobes spreading, 9–11.5 × 1–2.5 mm, lanceolate, acute. Lip deflexed, 3-lobed almost to base; mid-lobe 11–14 × 1 mm, linear; side lobes 7.5–11 × 1.5 mm, narrowly lanceolate; spur 25–32 mm long, with 1 twist in middle, swollen at apex, parallel to pedicel and ovary. Anther 2–3 mm high; canals porrect, c. 4 mm long. Stigmas porrect, c. 4 mm long, enlarged and truncate at apex. Rostellum mid-lobe c. 2 mm long, triangular.

Zambia. N: Mbala Distr., Chenda Farm, 1740 m, fl. 13.i.1965, *Richards* 19562 (K). W: Mwinilunga Distr., just east of Matonchi Farm, fl. 5.i.1938, *Milne-Redhead* 3805B (K). C: Lusaka Distr., c. 10 km east of Lusaka, 1270 m, *King* 412 (K). **Malawi**. N: Mzimba Distr., Lunyangwa Forest Reserve near Mzuzu, 1320 m, fl. 25.i.1986, *la Croix* 786 (K). C: Ntchisi Forest Reserve, 1350–1500 m, fl. 31.i.1983, *Dowsett-Lemaire* 607 (K).
Also in Zaire and Tanzania. *Brachystegia* woodland, often in grassy clearings and on forest margins; 1300–1740 m.

58. **Habenaria riparia** Renz & Grosvenor in Candollea **34**: 362 (1979). —la Croix et al., Orch. Malawi: 77 (1991). Type: Malawi, Nyika Plateau, *Grosvenor & Renz* 1180 (K; SRGH, holotype).

Robust terrestrial herb to 80 cm high; tubers small, ellipsoid. Leaves 7–10, the lowermost 1–2 sheath-like, spotted with black, the next few 11–20 × 1–2 cm, linear or narrowly lanceolate; the uppermost leaves grading into the bracts. Inflorescence 8–16 × 3–4 cm, fairly laxly or fairly densely several- to many-flowered; bracts to 20 mm long, sparsely hairy, much shorter than pedicel and ovary. Flowers green, whitish in

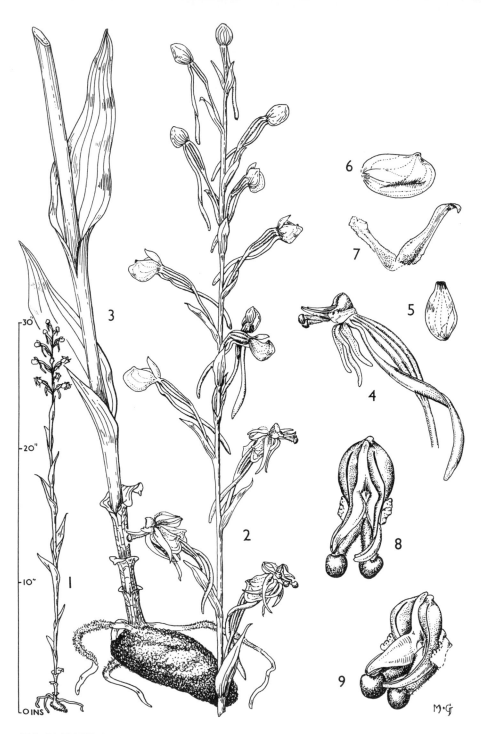

Tab. 33. HABENARIA RETINERVIS. 1, habit ($\times\frac{1}{5}$); 2, inflorescence ($\times$1); 3, base of plant ($\times$1); 4, flower, tepals removed, side view ($\times$2); 5, dorsal sepal ($\times$2); 6, lateral sepal ($\times$2); 7, petal ($\times$2); 8, column, front view ($\times$4); 9, column, three-quarter profile view ($\times$4), 1–9 from *Milne-Redhead* 3805. Drawn by Mary Grierson. From Kew Bull.

centre. Pedicel 20–25 mm long, straight, semi-erect; ovary 7–9 mm long, arched, papillose. Dorsal sepal reflexed, c. 7 × 3 mm, elliptic; lateral sepals deflexed, 10–12 × 6 mm, obliquely obovate with a lateral apiculum. Petals 2-lobed almost to base; upper lobe 6 × 0.5 mm, erect or recurved, papillose-pubescent; lower lobe deflexed, 7.5–9 × 3.5–4 mm, ovate-lanceolate, papillose-pubescent towards base. Lip pendent, 3-lobed almost to base; mid-lobe 12–14 × 1 mm, linear; side lobes 6–7 × 1.5–2 mm, curved linear-lanceolate; spur 13–19 mm long, parallel to pedicel and ovary, sometimes with 1 twist in middle, swollen in apical half, slightly flexuous. Column 3 mm high; anther canals 7 mm long. Stigmatic arms porrect, 6 mm long, papillose-pubescent above. Rostellum mid-lobe c. 1.5 mm long.

Malawi. N: Nyika Plateau, Dembo Bridge, 6 km east of Chelinda Camp, 2150 m, fl. iii.1977, *Renz & Grosvenor* 1180 (K; SRGH); Nyika Plateau, Dembo Bridge, fl. 13.ii.1987, *la Croix* 965 (K; MAL). Not known elsewhere. Wet grass near stream; 2150 m.

Close to the East African *H. thomsonii* Rchb.f., but differing in the longer spur, longer pedicel and ovary, and slightly narrower leaves.

59. **Habenaria schimperiana** A. Rich., Tent. Fl. Abyss. **2**: 295 (1851). —Rolfe in F.T.A. **7**: 241 (1898). —Summerhayes in Bot. Not. **1937**: 186 (1937); in F.T.E.A., Orchidaceae: 89 (1968). —Grosvenor in Excelsa **6**: 83 (1976). —Williamson, Orch. S. Centr. Africa: 51 (1977). —Stewart et al., Wild Orch. South. Africa: 92 (1982). —Geerinck in Fl. Afr. Centr., Orchidaceae pt. 1: 104 (1984). —la Croix et al., Orch. Malawi: 78 (1991). Type from Ethiopia.

Robust terrestrial herb to 1 m high. Tubers 1–4 × 1–2 cm, ellipsoid or ovoid. Stem leafy; leaves 5–10, the basal 1–2 sheath-like; the next few semi-erect, the largest 7–28 × 1–2(3.7) cm, narrowly lanceolate; upper leaves becoming bract-like. Inflorescence 6–35 × 5.5–10 cm, fairly laxly several- to many-flowered; bracts 10–20(30) mm long. Flowers green, white in centre, with an unpleasant smell. Pedicel and ovary 20–40 mm long, the pedicel straight, the ovary bent down at approximately right angles to the pedicel, and about half its length; pedicel and ovary becoming straight and suberect in fruit. Dorsal sepal reflexed, 6–8 × 2.5–4 mm, elliptic; lateral sepals 9–11 × 5–8 mm, obliquely obovate with a lateral apiculum. Petals 2-lobed almost to base, both lobes ciliate; upper lobe curving back, 5–8 × 1 mm, linear; lower lobe deflexed, hinged, 14–18.5 × 2 mm, narrowly lanceolate, acute. Lip deflexed, 3-lobed to 1.5–2 mm from base; mid-lobe 13–17 mm long; side lobes 10–16 mm long; all lobes linear, 0.5–1 mm wide; spur 10–16 mm long, spirally twisted in middle, inflated at apex, pendent or slightly incurved. Anther 3 mm high; canals porrect, 5–6 mm long. Stigmatic arms porrect, 5–7 mm long, slender but enlarged and truncate at apex. Rostellum mid-lobe 1–2 mm long.

Zambia. N: Mbala Distr., Nkali Dambo, 1500 m, fl. 26.i.1952, *Richards* 783 (K). W: Mwinilunga Distr., Kalenda Dambo, fl. 1.i.1938, *Milne-Redhead* 3907 (K). C: Serenje, fl. xii.1966, *Odgers* 187 (K). E: Chipata Distr., 40 km north of Chipata, 1200 m, fl. 14.ii.1959, *King* 463 (K; SRGH). S: Kalomo Distr., between Choma and Masuku, 1060 m, fl. 14.iii.1962, *Astle* 1560 (K). **Zimbabwe**. N: Mrewa Distr., Mazowe (Mazoe) R. Valley, fl. 11.ii.1969, *Mavi* 942 (K; SRGH). W: Hwange Distr., Victoria Falls National Park, 880 m, fl. 7.ii.1980, *Ncube* 74 (K; SRGH). C: Rusape Distr., fl. 25.i.1949, *Munch* 161 (K; SRGH). E: Mutare Distr., bank of Mpanda R., 900 m, fl. 5.iv.1959, *Chase* 7078 (K; SRGH). S: Masvingo (Fort Victoria), fl. 1909, *Monro* 904 (BM). **Malawi**. N: Mzimba Distr., 8 km north of Mpherembe (Mperembe), 1150 m, fl. 2.iii.1987, *la Croix* 994 (K; MAL). C: Lilongwe Distr., Chitedzi Agricultural Station, 1070 m, fl. 8.ii.1959, *Robson* 1512 (K). S: Zomba Mt., marsh near Chingwe's Hole, c. 1800 m, fl. 23.ii.1979, *Taylor* 14 (K; MAL).

Also in Ethiopia, Sudan, Zaire, Kenya, Tanzania and South Africa. Marsh and flooded grassland; 880–2300 m.

60. **Habenaria sochensis** Rchb.f., Otia Bot. Hamburg.: 100 (1881). —Rolfe in F.T.A. **7**: 239 (1898). —Summerhayes in F.T.E.A., Orchidaceae: 98 (1968). —Grosvenor in Excelsa **6**: 83 (1976). —Williamson, Orch. S. Centr. Africa: 55 (1977). —la Croix et al., Orch. Malawi: 78 (1991). Type: Malawi, Soche Hill, *Kirk* (K, holotype).

Robust plants 0.6–2 m tall. Tubers 3–4.5 × 1–2 cm, ovoid or ellipsoid, woolly. Stem leafy. Leaves 11–16, the 2 lowest loosely sheathing, the next few spreading, 12–25 × 2–5 cm, lanceolate; the upper leaves grading into the bracts.

Inflorescence up to 40 × 6 cm, densely many-flowered; bracts 15–35 mm long, scarious by flowering time. Flowers yellow-green, white in centre, smelling unpleasant. Pedicel and ovary 20–30 mm long, arched. Sepals reflexed, turning brown and scarious at flowering time; dorsal sepal 4–6 × 2–2.5 mm, elliptic; lateral sepals 6.5–10 × 4–5.5 mm, obliquely obovate with a lateral apiculum. Petals 2-lobed almost to base; upper lobes somewhat recurved, often crossing, 5–6.5 mm long, less than 1 mm wide, linear; lower lobes spreading, hinged, 8–11 × 2–3.5 mm, lanceolate or oblong-lanceolate, subacute. Lip deflexed, 3-lobed almost to base; mid-lobe 10–17 mm long; side lobes 6–12 mm long; all lobes linear, 0.5–1.5 mm wide; spur 17–25 mm long, twisted in the middle, swollen in apical half, ± parallel to pedicel and ovary. Anther 2.5–4.5 mm high; canals c. 2.5 mm long, upcurved. Stigmatic arms porrect, 3–4 mm long, rather stout, the tips truncate. Rostellum mid-lobe c. 2.5 mm long, triangular, acute.

Zambia. W: Kitwe, fl. 18.ii.1968, *Mutimushi* 2461 (K; NDO). C: c. 13 km east of Lusaka, 1270 m, fl. 15.iii.1956, *King* 358 (K). S: Mazabuka, fl. iii.1960, *G. Williamson* 418 (K). **Zimbabwe**. C: Harare Distr., Cleveland Dam, fl. 19.ii.1933, *Eyles* 7288 (K). E: Mutare Distr., Odzani Heights, 1500 m, fl. 24.iii.1957, *Chase* 6374 (K; SRGH). **Malawi**. N: South Viphya, Luwawa link road, c. 1500 m, fl. 22.iii.1983, *la Croix* 471 (K). S: Thyolo Distr., Bvumbwe, c. 1100 m, fl. 18.iii.1981, *la Croix* 120 (K).
Also in Tanzania. Grassland and woodland, sometimes persisting in *Eucalyptus* and pine plantations; 1050–1700 m.

61. **Habenaria strangulans** Summerh. in Kew Bull. **16**: 295, fig. 16 (1962). —Grosvenor in Excelsa **6**: 83 (1976). —Williamson, Orch. S. Centr. Africa: 55 (1977). —la Croix et al., Orch. Malawi: 79 (1991). TAB. **34**. Type: Zimbabwe, Umvukwes Distr., Aranbira Estates, *Pollitt* in *GHS* 42189 (K, holotype; SRGH).

Slender terrestrial herb 15–60 cm tall; tubers c. 2 × 1 cm, ovoid, woolly. Stem leafy. Leaves to c. 14, the lowest 1–2 sheath-like; the intermediate 4–6 suberect, 6–16 cm × 4–7 mm, linear; upper leaves bract-like. Inflorescence 8–20 × 4–5 cm, fairly laxly 10–30-flowered; bracts 12–15 mm long. Flowers green. Pedicel and ovary semi-erect, 20–23 mm long. Dorsal sepal reflexed, 5–6 × 2–3 mm, elliptic; lateral sepals deflexed, 7–8 × 5–5.5 mm, oblique with a lateral apiculum. Petals 2-lobed almost to base; upper lobe 4.5–6 × 0.6–0.7 mm, linear, erect; lower lobe 10–12.5 × c. 2 mm, lanceolate. Lip 3-lobed almost to base; mid-lobe 11.5–12.5 mm long; side lobes 8.5–9.5 mm long; all lobes linear, c. 0.5 mm wide; spur 19–22 mm long, slender, swollen at apex, the apical third wrapped round the pedicel. Anther 2–2.5 mm high; canals porrect, 4–5 mm long. Stigmatic arms porrect, 4–5 mm long. Rostellum mid-lobe 1.5–2 mm high.

Zimbabwe. N: Umvukwes, Aranbira Estates, 1200 m, fl. *Pollitt* in *GHS* 42189 (K; SRGH). C: Harare, NE edges, fl. iii.1945, *Greatrex* 94 (K; PRE; SRGH). E: Mutare Distr., "Rhodes View", 1100 m, fl. 15.iii.1960, *Chase* 7293 (K; SRGH). **Malawi**. C: Kasungu Distr., 48 km north of Kasungu, by M1, 1250 m, fl. 7.iv.1978, *Pawek* 14347 (K; MO).
Also in Angola. Open woodland; 1100–1200 m.

62. **Habenaria subaequalis** Summerh. in Kew Bull. **16**: 297 (1962). —Grosvenor in Excelsa **6**: 83 (1976). TAB. **35**. Type: Zimbabwe, Mutare Distr., Himalaya Range, Engwa, *Wild* 4455 (K, holotype; SRGH).

Slender terrestrial herb 15–35 cm tall; tubers 12–16 × 4–7 mm, ovoid. Stem leafy. Leaves 7–10, the largest 6–11 × 0.4–1 cm, linear-lanceolate; upper leaves grading into the bracts. Inflorescence 4–8 × 2.5–3 cm, fairly densely 10–30-flowered; bracts 6–8 mm long. Flowers white, scented (occasionally described as green and white). Pedicel and ovary 10–17 mm long. Sepals reflexed; dorsal sepal 5.4–6 × 3–3.7 mm, elliptic; lateral sepals 6–7 × 3.5–5 mm, oblique with a lateral apiculum. Petals 2-lobed almost to base; upper lobes 5.5–6 × 1.2–1.6 mm; lower lobes 5.6–7.5 × 1.6–2 mm, both lobes lanceolate. Lip white, 3-lobed almost to base; mid-lobe 9–10 × 0.6–0.7 mm; side lobes 6.5–7.8 × 0.5–0.8 mm; all lobes linear; spur 10–13 mm long. Anther c. 3.5 mm high; canals 2–3 mm long. Stigmatic arms c. 3 mm long. Rostellum mid-lobe 1.5–2 mm high.

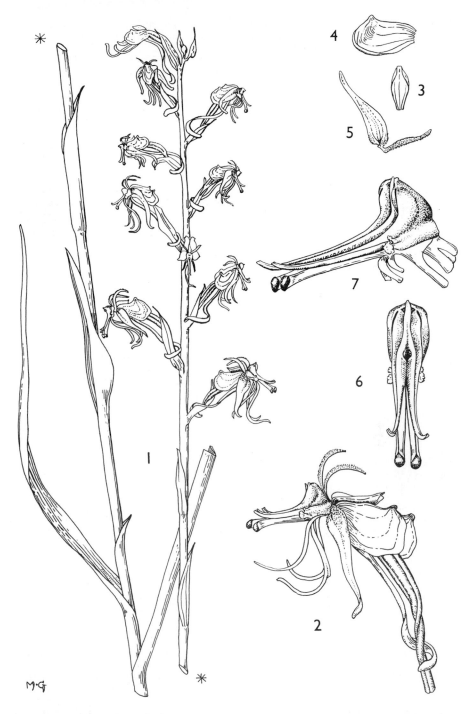

Tab. 34. HABENARIA STRANGULANS. 1, habit (×1); 2, flower, side view (×3); 3, dorsal sepal (×2); 4, lateral sepal (×2); 5, petal (×2); 6, column, front view (×6); 7, column, side view (×6), 1–7 from *Pollitt* in *GHS* 42189. Drawn by Mary Grierson. From Kew Bull.

Tab. 35. HABENARIA SUBAEQUALIS. 1, habit (×1); 2, flower, side view (×3); 3, dorsal sepal (×3); 4, lateral sepal (×3); 5, petal (×3); 6, flower, tepals removed to show column and lip, side view (×2); 7, column, front view (×6); 8, column, side view (×6); 9, staminode (×14), 1–9 from *Wild* 2806. Drawn by Mary Grierson. From Kew Bull.

Zimbabwe. C: Goromonzi Distr., Domboshawa Hill, 1500 m, fl. 2.iii.1961, *Greatrex* 135 (K; SRGH). E: Mutare Distr., Himalaya Range, Engwa, 2100 m, fl. 2.iii.1954, *Wild* 4455 (K; SRGH); Nyanga Distr., Troutbeck, North Downs, 1980 m, fl. 8.iii.1981, *Philcox, Leppard, Duri & Urayai* 8931 (K).

Known only from Zimbabwe, where it is apparently frequent in the Eastern Highlands, but is also found on the watershed plateau near Harare. Damp submontane or plateau grassland, usually in marshy ground, almost always amongst rocks; 1500–1980 m.

This species is unusual in Sect. *Replicatae* in having white flowers - the sepals, petals and lip are all described as white.

63. **Habenaria tortilis** P.J. Cribb in Kew Bull. **34**: 323 (1979). Type from Tanzania.

Terrestrial herb c. 60 cm tall. Leaves 8–16, the lowermost sheathing, white with dark green reticulate veining; intermediate leaves semi-erect, c. 9 cm × 4.5 mm, linear, acute, slightly folded; uppermost leaves to 9 cm long and 1 mm wide, narrowly linear, not appressed to stem. Inflorescence c. 12 × 2 cm, fairly densely many-flowered; bracts 17–20 mm long. Flowers green. Pedicel and ovary 13–20 mm long. Dorsal sepal reflexed, 4 × 2.5 mm, elliptic; lateral sepals deflexed, 7 × 4 mm, oblique with a lateral apiculum. Petals 2-lobed almost to base; upper lobe 4.5 × 0.5–1 mm, linear; lower lobe 8 × 3 mm, oblong. Lip 3-lobed almost to base; all lobes c. 5 × 0.5 mm, linear; spur 8 mm long, swollen at tip. Anther 2 mm high; canals 3 mm long, the last 1 mm upturned. Stigmatic arms c. 2 mm long.

Zambia. N: Mbala Distr., rocks on Sumbawanga Road near Kawimbe, 1650 m, fl. 1.iii.1957, *Richards* 8417 (K). S: Kafue Gorge, in forest, fl. iii.1966, *G. Williamson* 303 (K).

Also in Tanzania. Wet, sandy soil and gritty patches between flat rocks; to c. 1650 m.

The two specimens from Zambia differ from the type in some respects - the flowers are slightly smaller and the inflorescence less dense.

64. **Habenaria tubifolia** la Croix & P.J. Cribb in Kew Bull. **48**, 2: 359 (1993). TAB. **36**. Type: Zambia, Hauser's Farm, *Richards* 5063 (K, holotype).

Erect terrestrial herb 18–20 cm tall; roots not seen. Stem leafy. Leaves numerous, the lower 10–12 clustered near stem base, up to 8 cm long and 5–8 mm wide, tubular with a sheathing base c. 4 mm wide, the bases overlapping; the c. 6 uppermost leaves bract-like, appressed to stem, 7–20 × 4–8 mm, lanceolate or ovate, acuminate. Inflorescence slightly shorter than pedicel and ovary, 5–7 × 4–5 cm, fairly laxly 6–9-flowered; bracts 12–13 × 4 mm, lanceolate, acute, the margins erose. Flowers spreading, pale green. Pedicel and ovary arched, 15 mm long. Sepals reflexed; dorsal sepal 5 × 2 mm, oblong; lateral sepals 8–9 × 3–4 mm, oblique with a lateral apiculum. Petals 2-lobed almost to the base; upper lobe erect, c. 5 × 0.2 mm, almost filiform; lower lobes spreading, c. 12 × 2.5–3 mm, curved lanceolate, long acute, slightly papillose at base. Lip reflexed, lying ± parallel to ovary, 3-lobed to c. 1 mm from base, all lobes fleshy, linear, somewhat papillose; mid-lobe 11 × 0.5 mm, obtuse; side lobes 8 mm long, slightly narrower than mid-lobe, acute; spur 25–28 mm long, ± parallel to pedicel and ovary, slender, the apical 5 mm thickened and slightly bent up. Anther 1.5 mm high; canals and side lobes of rostellum 3.5–4 mm long, porrect, upcurved. Stigmatic arms c. 3 mm long, distinctly shorter than anther canals, rather stout, the tips truncate and slightly upturned.

Zambia. N: Hauser's Farm, 1800 m, 28.iii.1955, *Richards* 5063 (K).

Apparently endemic, known only from this collection. Open bush with *Uapaca* trees; 1800 m.

This is another rather nondescript member of Sect. *Replicatae*. It bears some resemblance to *H. strangulans* Summerh. but the crowded, tubular leaves are distinctive. Also, the inflorescence is fewer-flowered and the spur is longer and not wrapped round the pedicel as in *H. strangulans*.

65. **Habenaria unguilabris** B.R. Adams in Orchid Review **98**: 17 (1990). Type: Zimbabwe, near Doma Safari Area, *Adams* 526 (K; SRGH, holotype).

Terrestrial herb to 30 cm tall; tubers c. 2 cm long, ellipsoid. Leaves 9–11, the lowermost 5–6 erect to spreading, 2.5–16.5 cm long and 4–8 mm wide, linear or linear-lanceolate, acute, conduplicate, keeled; upper leaves 17–55 mm long,

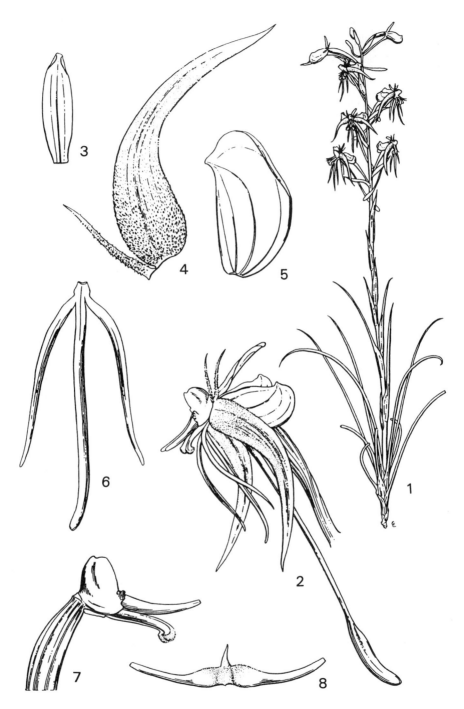

Tab. 36. HABENARIA TUBIFOLIA. 1, habit ($\times\frac{2}{3}$); 2, flower ($\times$3); 3, dorsal sepal ($\times$5); 4, petal ($\times$5); 5, lateral sepal ($\times$5); 6, lip ($\times$5); 7, column ($\times$5); 8, rostellum ($\times$5), 1–8 from *Richards* 5063. Drawn by Eleanor Catherine. From Kew Bull.

appressed to stem. Inflorescence 8–11 × 5 cm, rather laxly 10–14-flowered; bracts 10–18 mm long. Flowers green, white in centre, erect to spreading. Pedicel and ovary 18–25 mm long, ± arched. Sepals reflexed; dorsal sepal 6 × 2.3 mm, oblanceolate, obtuse, convex, somewhat rugose on outside; lateral sepals 9.3 × 6 mm, obliquely obovate, keeled, with a lateral apiculum. Petals 2-lobed almost to base; upper lobe erect, 6 × 0.7 mm, linear; lower lobe hinged at base, 12.5 × 2.8 mm, lanceolate-ligulate, somewhat falcate; both lobes minutely papillose towards base. Lip 3-lobed above a deeply channelled, elongate claw 10 mm long, geniculate at point of origin of spur; mid-lobe c. 10 × 1.2 mm, linear, curving up; side lobes 6–6.5 mm long, 1.5 mm wide at base, lanceolate-ligulate, obtuse, incurved; spur arising 6 mm from base of lip, 12–13 mm long, slightly clavate, slightly twisted, curved back. Column c. 3 mm high; anther canals 4 mm long. Stigmas porrect, 4.5 mm long, the apices convex-capitate. Rostellum mid-lobe triangular, acute, scarcely overtopping the anther; side lobes 5.5 mm long.

Zimbabwe. N: Makonde Distr., near southern boundary of Doma Safari Area, c. 5.7 km south of Chenanga Camp, 1110 m, fl. in cult. 11.iv.1989, *Adams* 526 (K; SRGH).
Endemic. Clearings in *Brachystegia boehmii* woodland on shallow, sandy soil; 1100 m.

66. **Habenaria weberiana** Schltr. in Bot. Jahrb. Syst. **53**: 516 (1915). —Summerhayes in F.T.E.A., Orchidaceae: 94 (1968). —Grosvenor in Excelsa **6**: 83 (1976). —Williamson, Orch. S. Centr. Africa: 52 (1977). —la Croix et al., Orch. Malawi: 80 (1991). Type from Tanzania.
Habenaria huillensis var. *weberiana* (Schltr.) Geerinck in Bull. Jard. Bot. Nat. Belg. **52**: 343 (1982); in Fl. Afr. Centr., Orchidaceae pt. 1: 115 (1984).

Terrestrial herb 30–60 cm tall. Tubers 1–2 × 0.5–1 cm, ± globose or ellipsoid, sparsely hairy. Stem leafy. Leaves 6–13, the lowermost sheathing, the next 2–7 suberect, the largest 8–28 × 0.5–2.5 cm, linear or linear-lanceolate, acute, with 3 fairly prominent longitudinal veins; upper leaves becoming bract-like. Inflorescence to c. 30 × 3–4 cm, rather laxly to fairly densely several- to many-flowered; bracts 8–20 × 2–3.5 mm, shorter than pedicel and ovary. Flowers green, white in centre, with an unpleasant smell. Pedicel and ovary 15–28 mm long, semi-erect, slightly upcurved. Dorsal sepal reflexed, 4.5–6.5 × 2–3 mm, elliptic; lateral sepals deflexed, 6–8.5 × 3.5–7 mm, obliquely obovate, the apiculum lateral. Petals 2-lobed almost to base; upper lobes 4–6.5 × c. 0.5 mm, recurved and sometimes crossing, narrowly linear, white, ciliolate; lower lobes pendent, hinged, 7–13 × 1.5–2 mm, lanceolate, acute. Lip 3-lobed with an undivided base 2–3 mm long; mid-lobe 8.5–12 mm long; side lobes 6–8 mm long; all lobes linear, c. 0.5 mm wide; spur 10–20 mm long, slender with 1 twist in the middle, not much swollen at apex, ± parallel to pedicel and ovary. Anther 2–3 mm high; canals porrect, 4.5–6 mm long, slightly upturned at tips. Stigmas porrect, 3.5–6 mm long, slender, the tips enlarged and truncate. Rostellum mid-lobe c. 1.5 mm long, triangular.

Zambia. N: Mbala Distr., Mbulu River, 1620 m, fl. 2.ii.1962, *Richards* 15978 (K). W: Mwinilunga, fl. 6.xii.1958, *Holmes* 118 (K; SRGH). C: Serenje Distr., Kundalila Falls, fl. ii.1971, *G. Williamson* 2224 (K; SRGH). **Zimbabwe**. C: Harare Distr., west of Epworth Rocks, 1360 m, fl. 19.ii.1950, *Greatrex* in *GHS* 26898 (K; PRE; SRGH). **Malawi**. N: Nkhata Bay Distr., Mzenga Dambo, 580 m, fl. 10.ii.1986, *la Croix* 801 (K; MAL). C: Kasungu National Park, in wet grass at edge of dam, 1150 m, fl. 9.viii.1987, *Field* in *la Croix* 1046 (K).
Also in Zaire, Tanzania and Angola. Dambo, wet grassland; 850–1620 m.

67. **Habenaria welwitschii** Rchb.f. in Flora **48**: 179 (1865). —Rolfe in F.T.A. **7**: 235 (1898). —Summerhayes in F.T.E.A., Orchidaceae: 83 (1968). —Geerinck in Fl. Afr. Centr., Orchidaceae pt. 1: 108 (1984). —la Croix et al., Orch. Malawi: 80 (1991). Type from Angola.
Habenaria leptostigma Schltr. in Bot. Jahrb. Syst. **38**: 146 (1906). Type: Mozambique, Beira, *Schlechter* 6995 (B†, holotype).
Habenaria keiliana Kraenzl. in Bot. Jahrb. Syst. **43**: 393 (1909). Type from Rwanda.

Terrestrial herb 27–72 cm tall; tubers c. 1.5 × 0.5 cm, ellipsoid. Stem leafy. Leaves 6–12, the lowermost usually sheathing, the next 3–6 semi-spreading, distichous, 6–20 × 0.5–2 cm, linear or narrowly lanceolate, somewhat folded; uppermost leaves bract-like. Inflorescence 6–18 × 3–4.5 cm, fairly laxly 10–30-flowered; bracts 15–18 mm long. Flowers greenish-cream or green with pure white or pale yellow petals. Pedicel and ovary suberect or slightly arched, 18–25 mm long. Dorsal sepal usually reflexed but occasionally almost erect, 5–6.5 × 2–2.5 mm; lateral sepals deflexed, 7–9 × 4–6 mm, obliquely obovate with a lateral apiculum. Petals 2-lobed almost to base; upper lobe recurved, 4–6 × 0.5 mm, linear, ciliolate; lower lobes spreading, 4–5.5 × 1.5–3 mm, obliquely obovate. Lip 3-lobed almost to base; mid-lobe 8–13.5 × 0.5 mm, linear; side lobes 4–5.5 × 1–1.5 mm, oblong-lanceolate, obtuse; spur 10–16 mm long, slightly incurved, slender, abruptly swollen for apical 3 mm. Anther c. 2.5 mm high, the loculi reclinate; canals porrect, 5–6 mm long, the tips upcurved. Stigmatic arms 5–7 mm long, porrect, slender, the tips enlarged and truncate. Rostellum mid-lobe 1.5–2 mm long.

Malawi. N: Mzimba Distr., Mzuzu, Marymount, 1360 m, fl. 4.vii.1975, *Pawek* 9807 (K; MO; UC). [This specimen is abnormal with the anterior petal lobes antheriform]. S: Mulanje Mt., Nandalanda Path, from Chinzama to Tuchila, c. 2000 m, fl. 13.iv.1985, *Jenkins* 4 (K). **Mozambique**. MS: grassy flats near Beira, fl. iv.1895, *Schlechter* 6995 (B†).
Also in Zaire, Burundi, Rwanda, Angola, Uganda and Tanzania. Grassland; 0–2000 m.

68. **Habenaria singularis** Summerh. in Kew Bull. **20**: 167 (1966). —Grosvenor in Excelsa **6**: 83 (1976). TAB. **37**. Type: Zimbabwe, Chimanimani Mts., *Ball* 238 (K; SRGH, holotype).
      Habenaria sp. No. 3, Goodier & Phipps in Kirkia **1**: 52 (1961).

Terrestrial herb 25–45 cm tall, glabrous. Leaves 6–11, the lowermost 1–2 sheath-like; stem leaves 3–7, recurved, usually 4–9 × 1–2 cm, lanceolate or oblong-lanceolate, acute; the uppermost grading to the bracts. Inflorescence 9–10 × 3–5 cm, fairly laxly 4–25-flowered; bracts 12–18 mm long. Flowers green, erect or sub-erect. Pedicel c. 10 mm, long, ovary 7–11 mm long. Dorsal sepal erect, 4 × 2.25 mm, elliptic, obtuse, convex; lateral sepals deflexed, 5–6 × 5 mm, obliquely triangular-ovate, the upper edge straight, the lower rounded. Petals 2-lobed almost to base; upper lobe erect, 3.5–4 × 0.5 mm, linear or ligulate, adnate to dorsal sepal; lower lobe much longer, ascending or erect, 10–11 × 1 mm, rather fleshy, ± papillose. Lip 3-lobed to base, deflexed or slightly incurved, all lobes ± equal, 9–11.5 × 0.5 mm, linear, obtuse; spur c. 11 mm long, usually recurved and standing above flower, swollen at apex. Anther erect, 3 mm high; canals c. 2 mm long. Stigmatic arms deflexed or porrect, 3–4 mm long, slender with an enlarged apex, densely papillose-pubescent. Rostellum 3-lobed, mid-lobe erect, c. 1 mm long, side lobes deflexed, 4.5–5 mm long, rather zigzag, 2–3 mm longer than anther canals.

Zimbabwe. E: Chimanimani Mts., c. 1500 m, fl. 21.ii.1954, *Ball* 238 (K; SRGH). C: Harare Distr., Cleveland, fl. iii.1944, *Greatrex* 45 (K; SRGH).
Apparently endemic, known only from the Eastern Highlands of Zimbabwe and the watershed plateau near Harare. Grassland; c. 1500 m.

69. **Habenaria calvilabris** Summerh. in Kew Bull. **16**: 303, fig. 20 (1962). —Grosvenor in Excelsa **6**: 83 (1976). —Williamson, Orch. S. Centr. Africa: 56 (1977). —Geerinck in Fl. Afr. Centr., Orchidaceae pt. 1: 139 (1984). TAB. **38**. Type: Zambia, Mbala Distr., Lake Chila, *Richards* 12449 (K, holotype).

Terrestrial herb 30–90 cm tall, glabrous except for roots. Tubers c. 1.5 × 1 cm, ellipsoid. Stem leafy along its length. Leaves 5–10, the lowermost sheath-like, the 3–6 above these suberect, c. 10–20 × 1–3.5 cm, lanceolate or oblong-lanceolate, acute; uppermost leaves grading into the bracts. Inflorescence 6–30 × 3–5 cm, fairly densely several- to many-flowered; bracts 5–22 mm long. Flowers with a strong scent variously described as attractive or unpleasant or sickly; suberect, greenish-white or green with lip mid-lobe white. (A note on the type specimen says lip is orange and white.) Pedicel 10 mm long; ovary 12–16 mm long, 6-ribbed, densely papillose. Dorsal sepal erect, 9–11 × 4–5.5 mm, elliptic, obtuse, convex, the margins ± incurved; lateral sepals spreading, 10–13.5 × 6–8.5 mm, obliquely ovate or oblong-ovate, apiculate or subacute, densely papillose-rugulose outside. Petals 2-lobed to base, upper lobe adnate to dorsal sepal, 8–10.5 × 0.7–1.3 mm, linear; lower lobe porrect, 12–16.5 × 1.5–2.2 mm, oblong-lanceolate, obtuse. Lip divided to 1.5–2.5 mm from base, the

Tab. 37. HABENARIA SINGULARIS. 1, lower stem (×1); 2, inflorescence (×1); 3, flower, three-quarter view (×3); 4, dorsal sepal (×3); 5, lateral sepal (×3); 6, petal (×3); 7, column, three-quarter view (×3); 8, column, front view (×6), 1–8 from *Greatrex* 45. Drawn by Mary Grierson. From Kew Bull.

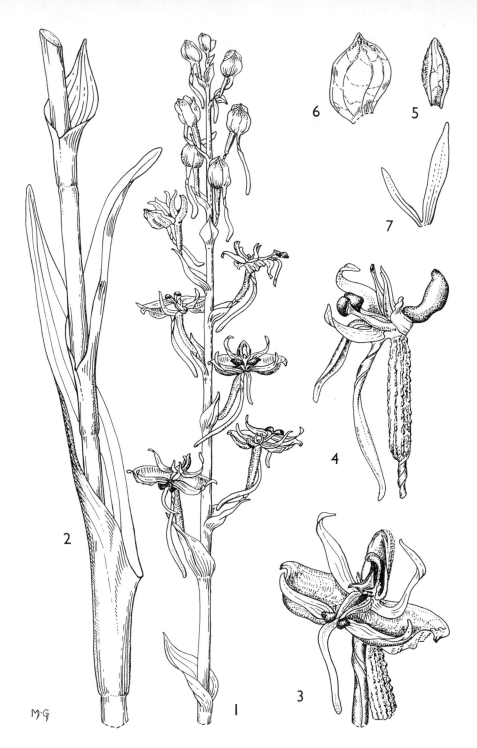

Tab. 38. HABENARIA CALVILABRIS. 1, inflorescence (×1); 2, part of lower stem (×1), 1 & 2 from *Richards* 12449; 3, flower, three-quarter front view (×2), from *Richards* 10854; 4, flower, tepals removed, side view (×2); 5, dorsal sepal (×2); 6, lateral sepal, outer surface (×2); 7, petal (×2), 4–7 from *Richards* 12449. Drawn by Mary Grierson. From Kew Bull.

basal undivided part adnate to stigmatic arms; mid-lobe deflexed, 10–12.5 × 1–2 mm, ligulate or linear-spathulate, rather fleshy; side lobes spreading, 11–13.5 × 2.5–4.5 mm, curved-lanceolate or oblong-lanceolate, acute; spur pendent, slightly recurved, 17–29 mm long, slightly inflated at apex. Anther erect, 5.5–7 mm high, apex obtuse or slightly apiculate; canals 3–3.5 mm long; auricles erect, elliptic, obtuse; stigmatic arms 5.5–7 mm long, porrect, apex enlarged and truncate. Rostellum mid-lobe erect, 2–4 mm long, triangular-linear.

Zambia. N: Mbala Distr., Lake Chila, 1650 m, fl. 27.i.1960, *Richards* 12449 (K). S: Mazabuka Distr., Simasunda, Mapanza, 1060 m, fl. 21.ii.1957, *E.A. Robinson* 2135 (K). **Zimbabwe**. N: Makonde Distr., 1360 m, fl. 7.iii.1947, *Wild* 1863 (K; SRGH). C: Harare Distr., in grasslands or open bush all round Harare (Salisbury), fl. 6.iii.1946, *Greatrex* 44 (K; SRGH).

Also in Zaire and Burundi. In marshy dambo, long grass in swamp, vlei edge, grassland and mixed woodland; 1060–1650 m.

Similar to *H. rautaneniana* Kraenzl. but the leaves are shorter and broader and the inflorescence is usually longer and narrower with more flowers. The anther is not so tall and the lobes of the lip are quite glabrous. The densely papillose outer surface of sepals gives buds a glistening appearance.

70. **Habenaria pasmithii** G. Will. in Kirkia **12**: 199, fig. 1 (1983). Type: Botswana, by Okavanga River, 18°25', 22°01', *P.A. Smith* 2719 (K; PRE; SRGH, holotype).

Erect herb to 90 cm high; tubers c. 12 × 8 mm, ellipsoid. Leaves c. 10, the lowermost 1–2 leaves sheathing, the mid stem leaves erect, to 29 × 1.2 cm, linear, acute; the upper leaves grading into the bracts. Inflorescence 12–16 × 4.5–8 cm, fairly densely 12–16-flowered; bracts 20–24 × 6 mm, lanceolate, acute. Flowers suberect, green with white lip. Ovary and pedicel 25–42 mm long, straight, glabrous. Sepals all minutely verrucose; dorsal sepal erect, 9–10 × 3–5 mm, elliptic, obtuse or apiculate; lateral sepals spreading, 9–12 × 7 mm, obliquely ovate. Petals 2-lobed ± to base; upper lobe erect, 10 × 0.6 mm, linear, not adnate to dorsal sepal; lower lobe c. 20 × 1 mm, linear, curved. Lip adnate to stigmatic arms at base, 3-lobed with basal undivided part c. 1.5 mm long; mid-lobe c. 12 × 1 mm, ligulate or linear, porrect; side lobes spreading, 14–16 × 2–2.5 mm, lanceolate, acute; spur pendent or horizontal, 20–23 mm long, twisted in middle, not inflated at apex. Anther loculi 4 mm high; canals 3–4.5 mm long; stigmatic arms 11 mm long, the apex clavate and truncate. Mid-lobe of rostellum 4 mm high.

Botswana. N: Water meadow by Okavanga River, 18°25', 22°01', fl. 22.ii.1979, *P.A. Smith* 2719 (K; PRE; SRGH); swamp by Khiandiadavhu River, *P.A. Smith* 2759 (SRGH). **Zambia**. W: 132 km south of Mwinilunga on road to Kabompo, fl. 25.i.1975, *Brummitt, Chisumpa & Polhill* 14129 (K).

Not known elsewhere. Water meadows or slow-flowing water to 60 cm deep, growing with grasses, sedges and aquatic herbs.

Differs from *H. rautaneniana* in having minutely papillose sepals, a glabrous lip and a spur which is straight but not inflated at apex; and from *H. calvilabris* in having narrower leaves, non-papillose ovary and narrower petal lobes.

Note. Some specimens of *H. rautaneniana* have slightly papillose ovaries and some of *H. calvilabris* have quite narrow leaves.

71. **Habenaria rautaneniana** Kraenzl. in Bull. Herb. Boissier, sér. 2, **2**: 941, fig. 19 (1902). — Summerhayes in F.T.E.A., Orchidaceae: 103 (1968). —Grosvenor in Excelsa **6**: 83 (1976). —Williamson, Orch. S. Centr. Africa: 56 (1977). —Stewart et al., Wild Orch. South. Africa: 82 (1982). —la Croix et al., Orch. Malawi: 80 (1991). Type from Namibia.
Habenaria trachychila Kraenzl. in Bull. Herb. Boissier, sér. 2, **4**: 1007 (1904). Type from South Africa (Transvaal).

Terrestrial herb to 80 cm tall; tubers to 3.5 × 2 cm, ovoid or ellipsoid, tomentose. Stem leafy. Leaves 5–11, the largest erect, 18–33 × 0.5–1.5 cm, linear-lanceolate, conduplicate, sheathing at base. Inflorescence to 24 × 7 cm, fairly densely many-flowered; bracts 1.5–3.5 cm long. Flowers yellow-green, the side lobes of lip white at base. Ovary and pedicel 5 cm long. Dorsal sepal erect, 9.5–15 × 3.5–5 mm, elliptic, obtuse, convex; lateral sepals spreading, 11–18 × 6–8.5 mm, obliquely ovate, acuminate. Petals 2-lobed almost to base; upper lobe 9–15 mm long, narrowly linear, adnate to dorsal sepal; lower lobe 14–23 × 0.5–1.5 mm, linear, spreading upwards. Lip 3-lobed to 2–3 mm from base; mid-lobe 11.5–20 × 1–1.5 mm, linear, somewhat deflexed; side lobes ± spreading, 11–17 × 1.5–3 mm, lanceolate-ligulate, subacute, the apices slightly upturned, densely pubescent or papillose at base; spur 14–23 mm long, ± recurved with 1 twist in middle, swollen to c. 2 mm diameter in apical half. Anther erect, 6.5–9.5 mm tall, canals c. 2 mm long, partly adnate to column, upcurved at tips. Stigmatic arms 5.5–7.5 mm long, club-shaped, curving down over base of lip. Rostellum mid-lobe 2.5–4 mm long, narrowly triangular, acute.

Zimbabwe. N: Makonde Distr., Rukute Farm, Doma Area, 1200 m, fl. 10.ii.1963, *Jacobsen* 2142 (PRE). C: Harare Distr., University Vlei, fl. 11.ii.1975, *Beasley* 261 (K; SRGH). **Malawi**. N: Mzimba Distr., c. 8 km north of Mpherembe, 1150 m, fl. 2.iii.1987, *la Croix* 993 (K; MAL). S: Zomba Distr., Mbonde Dambo, flood plain grassland, fl. 3.ii.1955, *Jackson* 1455 (K).

Also in Tanzania, South Africa (Transvaal) and Namibia. Marshy ground, dambos or seasonally flooded grassland, often growing in slowly running water; c. 1100–1150 m.

72. **Habenaria cirrhata** (Lindl.) Rchb.f. in Flora **48**: 180 (1865). —Rolfe in F.T.A. **7**: 248 (1898). —H. Perrier in Fl. Madag., Orch. 1: 54 (1939). —Summerhayes in F.W.T.A. ed. 2, **3**: 198 (1969); in F.T.E.A., Orchidaceae: 112 (1968). —Williamson, Orch. S. Centr. Africa: 58 (1977). —Geerinck in Fl. Afr. Centr., Orchidaceae pt. 1: 123 (1984). —la Croix et al., Orch. Malawi: 81 (1991). Type from Madagascar.
    Bonatea cirrhata Lindl., Gen. Sp. Orchid. Pl.: 327 (1835).
    Habenaria schweinfurthii Rchb.f., Otia Bot. Hamburg.: 58 (1878). Type from Sudan.
    Habenaria zenkeriana Kraenzl. in Bot. Jahrb. Syst. **19**: 247 (1894). —Rolfe in F.T.A. **7**: 247 (1898). Type from Cameroon.
    Habenaria longistigma Rolfe in F.T.A. **7**: 248 (1898). Type from Tanzania.
    Habenaria dawei Rolfe in Bull. Misc. Inform., Kew 191**2**: 134 (1912). Type from Uganda.
    Habenaria megistosolen Schltr. in Bot. Jahrb. Syst. **53**: 512 (1915). Type from Tanzania.

Robust terrestrial herb 50–130 cm tall, glabrous except for roots. Tubers to 4.5 × 2.5 cm, ovoid or ellipsoid. Stem leafy. Leaves 9–13, rarely more, the lowermost 1–2 sheathing, the mid-stem leaves ± spreading, 7–22 × 3.5–9 cm, lanceolate, ovate or almost orbicular; the uppermost leaves grading into bracts. Inflorescence 4–37 cm long, laxly 2–12-flowered; bracts 2–6.5 cm long, lanceolate, acute. Flowers green or greenish-white. Pedicel c. 5 cm long, ovary 3–4 cm long. Dorsal sepal erect, 19–25 × 7–10 mm, elliptic, very convex; lateral sepals deflexed, 20–30 × 10–15 mm, rolled lengthwise. Petals 2-lobed almost to base, upper lobe 17–25 mm long, less than 0.5 mm wide, filiform, adnate to dorsal sepal; lower lobe 50–90 mm long, c. 1 mm wide, curving forwards and upwards. Lip projecting forwards, 3-lobed with an undivided base 2–4 mm long; mid-lobe 28–40(65) × 1 mm, side lobes 20–30(45) × c. 1 mm, all ± filiform; spur 13–22 cm long, slender, inflated to 3–4 mm diameter in apical third. Anther erect, 7–10 mm high; canals 16–20 mm long, the tips upturned. Stigmas 15–22 mm long, slender, truncate at apex, projecting forwards.

Zambia. N: Mbala Distr., Mbala–Mpulunga road, near Kasulo turning, 1480 m, fl. 4.ii.1952, *Richards* 756 (K). W: Solwezi Distr., fl. 16.i.1960, *Holmes* 0190 (K; SRGH). **Malawi**. N: Mzimba Distr., c. 11 km north of Mpherembe, 1280 m, fl. 10.iii.1978, *Pawek* 14026A (K; MO). S: Ntcheu Distr., Msasa escarpment, Dedza–Golomoti road, 1250 m, fl. 19.iii.1955, *Exell, Mendonça & Wild* 1033a (BM).

Widespread in tropical Africa and Madagascar. *Brachystegia* woodland, often on termite mounds, and in grassland with scattered shrubs; 1100–1500 m.

73. **Habenaria clavata** (Lindl.) Rchb.f. in Flora **48**: 180 (1865). —Rolfe in Dyer, F. C. **5**, 3: 134 (1912). —Summerhayes in Bot. Not. **1937**: 183 (1937). —Suessenguth & Merxmüller, Contrib. Fl. Marandellas Distr.: 83 (1951). —Summerhayes in F.W.T.A. ed. 2, **3**: 196 (1968); in F.T.E.A., Orchidaceae: 109 (1968). —Grosvenor in Excelsa **6**: 83 (1976). —Williamson, Orch. S. Centr. Africa: 59 (1977). —Stewart et al., Wild Orch. South. Africa: 92 (1982). —Geerinck in Fl. Afr. Centr., Orchidaceae pt. 1: 124 (1984). —la Croix et al., Orch. Malawi: 81 (1991). Type from South Africa.
    Bonatea clavata Lindl. in Hook. Companion Bot. Mag. **2**: 208 (1837).

Terrestrial herb 20–80 cm tall; tubers to 3.5 × 2 cm, ovoid or ellipsoid, woolly. Stem leafy. Leaves 8–13, the lowermost 1–3 sheath-like; the 4–5 mid-stem leaves suberect or spreading, the largest 7–13 × 1.5–4 cm, ovate or lanceolate; the upper ones grading into the bracts. Inflorescence 4–27 × 6–9 cm, several- to many-flowered; bracts to c. 30 mm long. Flowers pale green or yellow-green, white in centre. Pedicel and ovary erect or incurved, 35–45 mm long. Dorsal sepal erect, 11–19 × 6–7 mm, ovate, apiculate, convex; lateral sepals deflexed, 12–19 × 6–9 mm, oblique, rolled up lengthwise. Petals 2-lobed ± to base, upper lobe 9–15 mm long, filiform, adnate to dorsal sepal; lower lobe 25–40 mm long, c. 1 mm wide, horn-like, curving up. Lip 3-lobed with an undivided base c. 2 mm long; mid-lobe 17–26 mm long; side lobes 12–21 mm long; all less than 1 mm wide, narrowly linear; spur 3–5 cm long, parallel to ovary and pedicel, the apical third much swollen. Anther erect, 4–6 mm high, canals 8–14 mm long, porrect, the tips upturned. Stigmas porrect, 8–13 mm long, slender, widened and truncate at apex.

Zambia. N: Mbala Distr., path from Lunzua road to Chisungu Escarpment, 1500 m, fl. 25.i.1962, *Richards* 15948 (K). W: Ndola, fl. 18.i.1970, *Fanshawe* 10730 (K; SRGH). C: Lusaka Distr., c. 30 km east of Lusaka, fl. ii.1967, *G. Williamson* 260 (K; SRGH). **Zimbabwe**. N: Hurungwe National Park; 257.5 km from Harare on road to Chirundu, 1200 m, fl. 20.ii.1981, *Philcox, Leppard & Duri* 8794 (K). C: Harare Distr., fl. i.1946, *Greatrex* 24 (K; PRE; SRGH). E: Nyanga, Inyangani, 1700 m, fl. 5.ii.1931, *Norlindh & Weimarck* 4892 (K). S: Masvingo (Fort Victoria), fl. 1909, *Monro* 956 (BM). **Malawi**. N: North Viphya, Mzuzu–Choma road, 1300 m, fl. 2.iii.1986, *la Croix* 813 (K; MAL).
Also in Nigeria, Cameroon, Zaire, Ethiopia, Tanzania and South Africa. *Brachystegia* woodland, montane grassland, mopani bush; 1000–1700 m.

74. **Habenaria cornuta** Lindl. in Companion Bot. Mag. **2**: 208 (1837). —Rolfe in Dyer, F.C. **5**, 3: 132 (1912). —Summerhayes in Kew Bull. **8**: 576 (1954); in F.W.T.A. ed. 2, **3**: 198 (1968); in F.T.E.A., Orchidaceae: 106, fig. 20 (1968). —Grosvenor in Excelsa **6**: 83 (1976). — Williamson, Orch. S. Centr. Africa: 57 (1977). —Stewart et al., Wild Orch. South. Africa: 92 (1982). —Geerinck in Fl. Afr. Centr., Orchidaceae pt. 1: 119 (1984). —la Croix et al., Orch. Malawi: 82 (1991). TAB. **39**. Type from South Africa.
    Habenaria ceratopetala A. Rich. in Ann. Sci. Nat., sér. 2, **14**: 267, tab. 16,4 (1840); in Tent. Fl. Abyss. **2**: 295 (1852). —Rolfe in F.T.A. **7**: 231 (1898). Type from Ethiopia.
    Habenaria ruwenzoriensis Rendle in J. Bot. **33**: 280 (1895). —Rolfe in F.T.A. **7**: 233 (1898). —Summerhayes in F.W.T.A. **2**: 412 (1936). Type from Uganda.
    Habenaria subcornuta Schltr. in Bot. Jahrb. Syst. **53**: 510 (1915). Types from Tanzania.
    Habenaria orthocaulis Schltr. in Bot. Jahrb. Syst. **53**: 519 (1915). Type from Tanzania.

Terrestrial herb 20–60(80) cm tall; tubers to 3.5 × 2 cm, ellipsoid or ovoid, woolly. Stem leafy. Leaves 9–15, sometimes distichous, the lowermost 1–2 sheath-like, the middle ones suberect or spreading, the largest 2–10 × 0.7–4.5 cm, linear to ovate, the uppermost grading into the bracts. Inflorescence 5–19 cm long, laxly to densely few- to many-flowered; bracts leafy, to 27 mm long. Flowers green or yellow-green, the lip occasionally white. Pedicel and ovary 18–28 mm long. Dorsal sepal erect, 5–16 × 4–8 mm, ovate, convex; lateral sepals deflexed, 6–16 × 5–10 mm, obliquely ovate, rolled up lengthwise. Petals 2-lobed almost to base, upper lobe 6–15 mm long, ± filiform, adnate to dorsal sepal; lower lobe 20–50 mm long, 1–2 mm wide at base, horn-like, usually curling up but occasionally erect and straight. Lip 3-lobed from undivided base 1–2 mm long; mid-lobe 9.5–19 mm long, to 1 mm wide, linear; side lobes diverging, 8–18 mm long, the basal part flattened, 1–2.5 mm wide, usually with 1–6 short teeth, the apical part tapering; spur 14–28 mm long, very slender but suddenly swollen in apical 5 mm. Anther erect, 3–7.5 mm long; canals 4.5–8.5 mm long. Stigmas 3–8 mm long, porrect, slender, widened and truncate at apex. Rostellum mid-lobe triangular, c. 1 mm long.

Zambia. N: Mbala Distr., Lake Chila, 1500 m, fl. 4.i.1952, *Richards* 610 (K). C: c. 16 km east of Lusaka, fl. 2.iii.1975, *Williamson & Gassner* 2484 (K; SRGH). E: Isoka Distr., Nyika Plateau, north bank of Chire River, fl. iii.1966, *G. Williamson* 292 (K; SRGH). S: Choma Distr., Siamambo, 1300 m, fl. 20.iii.1958, *E.A. Robinson* 2810 (K; SRGH). **Zimbabwe**. N: Makonde Distr., Mwami (Miami), fl. iv.1926, *Rand* 42 (BM). C: Harare Distr., Denda, c. 1270 m, fl. 15.iii.1946, *Greatrex* in *GHS* 50266 (K; SRGH). E: Mutare Distr., Vumba, 1060 m, fl. 5.iii.1955, *Greatrex* in *GHS* 50248 (K; SRGH). **Malawi**. N: Nyika Plateau, edge of Juniper Forest, c. 2100 m, fl. 13.iii.1977, *Grosvenor & Renz* 1205 (K; SRGH). S: Zomba Plateau, Chitinji Marsh, 1800 m, fl. 6.iii.1983, *la Croix* 468 (K).
Widespread in tropical Africa, also in South Africa. Woodland and montane grassland, marshy ground and dambos; 1000–2300 m.
A very variable species, usually easily recognised by the toothed base of the lip side lobes; occasionally the teeth are absent, but the base is always flattened.

Tab. 39. HABENARIA CORNUTA. 1, habit (×1); 2, flower, tepals removed (×2); 3, dorsal sepal (×2); 4, lateral sepal, uncurled (×2); 5, petal (×2); 6, column, with part of lip, spur and ovary (×4); 7, column, front view (×4), 1–7 from *Milne-Redhead & Taylor* 8655A. Drawn by Heather Wood. From F.T.E.A.

75. **Habenaria gonatosiphon** Summerh. in Kew Bull. **14**: 134 (1960); in F.T.E.A., Orchidaceae: 113 (1968). —Moriarty, Wild Fls. Malawi: t. 26, 1 (1975). —Williamson, Orch. S. Centr. Africa: 57 (1977). —Geerinck in Fl. Afr. Centr., Orchidaceae pt. 1: 126 (1984). —la Croix et al., Orch. Malawi: 83 (1991). TAB. **40**. Type from Tanzania.

Terrestrial herb 25–75 cm tall; tubers to 3.5 × 1.5 cm, ovoid or ellipsoid. Stem slender, leafy. Leaves 5–10, the lowermost 1–2 sheath-like, the mid-stem leaves semi-erect, the largest 3.5–11 × 1.5–3.5 cm, ovate or lanceolate; uppermost grading into the bracts. Inflorescence 3–14 × 5–8 cm, fairly laxly to fairly densely 2–10-flowered; bracts to 3.5 cm long. Flowers white or cream, sometimes green-tinged. Pedicel and ovary 4–6 cm long. Dorsal sepal erect, 9.5–12 × 6–7 mm, elliptic, obtuse, convex; lateral sepals deflexed but spreading, 10–15 × 6–8.5 mm, obliquely ovate, not usually rolled lengthwise. Petals 2-lobed ± to base, upper lobe erect, 8.5–11 × 1–2 mm, linear, adnate to dorsal sepal; lower lobe 13–22 × 1.5–2 mm, curving upwards and outwards. Lip 3-lobed, with undivided basal part c. 2 mm long; mid-lobe 12–18 × 1–3 mm, ligulate, obtuse; side lobes diverging, 10–15 × 2–3 mm, narrowly lanceolate, acute; spur 4.5–6.5 cm long, projecting forwards at base for about 1 cm, then geniculately recurved, slender but swollen near apex. Anther erect, 4.5–6 mm high; canals 2–3 mm long, porrect. Stigmas c. 6 mm long, truncate at apex. Rostellum mid-lobe triangular, c. 3.5 mm long.

Zambia. N: Mbala Distr., Nkali Dambo, Kawimbe Road, 1760 m, fl. 18.i.1970, *Sanane* 1476A (K). **Malawi**. N: Mzimba Distr., South Viphya, c. 20 km NNW of Chikangawa, c. 1660 m, fl. 21.i.1979, *Phillips* 4672B (K). C: Dedza Distr., Kangoli (Kanjoli) Hill, fl. 17.i.1967, *Salubeni* 493 (K; SRGH).
Also in Tanzania and Zaire. Marshy ground, *Brachystegia/Uapaca* woodland; 1500–2200 m.

76. **Habenaria harmsiana** Schltr. in Bot. Jahrb. Syst. **53**: 511 (1915). —Summerhayes in F.T.E.A., Orchidaceae: 110 (1968). —Grosvenor in Excelsa **6**: 83 (1976). —Williamson, Orch. S. Centr. Africa: 59 (1977). —la Croix et al., Orch. Malawi: 83 (1991). Type from Tanzania.

Terrestrial herb 30–70 cm tall; tubers to 4 × 1.5 cm, ovoid or ellipsoid, woolly. Stem leafy. Leaves 7–12, the lowermost 1–2 sheath-like, the mid-stem leaves suberect, the largest 6.5–12 × 1.5–4 cm, lanceolate; the uppermost grading into the bracts. Inflorescence 5–28 × 8–18 cm, fairly laxly 2–25-flowered; bracts to 5 cm long. Flowers spreading or curving up, mainly white, scented. Pedicel 4–5 cm long, ovary 1.5–2 cm long. Dorsal sepal erect, 12–16 × 6–8 mm, elliptic, convex; lateral sepals spreading, 14–20 × 10–14 mm, obliquely obovate, rolled lengthwise. Petals 2-lobed from a triangular, undivided base 4–6 mm long, with a knob at the junction of the lobes; upper lobe 7–15 × 0.5 mm, flliform, adnate to dorsal sepal at tip; lower lobe curving up, 25–41 mm long, horn-like, tapering. Lip 3-lobed almost to base; mid-lobe pendent, 23–30 × 0.5 mm, filiform; side lobes projecting fowards, 15–19 mm long, also filiform; spur pendent, 4–7 cm long, slender with the apical 1 cm abruptly inflated. Anther erect, 5–6.5 mm high, canals 14–19 mm long, porrect, the tips upturned. Stigmas porrect, 12–18 mm long, slender, the apices enlarged and truncate. Rostellum mid-lobe c. 3 mm long, triangular.

Zambia. E: Chipata, 1200 m, fl. 10.ii.1959, *King* 460 (K; SRGH). **Malawi**. N: Nkhata Bay Distr., Mzenga Dambo, c. 650 m, fl. 10.i.1983, *la Croix* 419 (K). C: Nkhota Kota Distr., Sani Road, fl. 22.ii.1953, *Jackson* 1092 (K). S: Zomba Distr., Ngokwe, fl. 6.ii.1955, *Jackson* 1466 (K). **Mozambique**. N: Marrupa, beyond airport on road to Nungo, c. 800 m, fl. 18.ii.1982, *Jansen & Boane* 7818 (K; PRE). Z: Supiao, fl. 9.ii.1905, *le Testu* 613 (K). MS: Manica, fl. 1920, *Honey* 651 (K).
Also in Tanzania. Boggy grassland; 650–1200 m.

77. **Habenaria holubii** Rolfe in F.T.A. **7**: 249 (1898). —Summerhayes in F.W.T.A. ed. 2, **3**: 196 (1968); in F.T.E.A., Orchidaceae: 109 (1968). —Grosvenor in Excelsa **6**: 83 (1976). —Williamson, Orch. S. Centr. Africa: 57 (1977). —la Croix et al., Orch. Malawi: 84 (1991). Type: Zimbabwe, Leshumu Valley, *Holub* (K, holotype).
Habenaria rhopaloceras Schltr. in Warb., Kunene-Sambesi-Exped. Baum: 207 (1903). Type from Angola.
Habenaria valida Schltr. in Bot. Jahrb. Syst. **38**: 148 (1906). Type from Togo.
Habenaria brevilabris Kraenzl. in Bot. Jahrb. Syst. **51**: 371 (1914). Type from Zaire.
Habenaria hennigiana Schltr. in Bot. Jahrb. Syst. **53**: 511 (1915). Type from Tanzania.
Habenaria clavata sensu Geerinck in Fl. Afr. Centr., Orchidaceae pt. 1: 124 (1984).

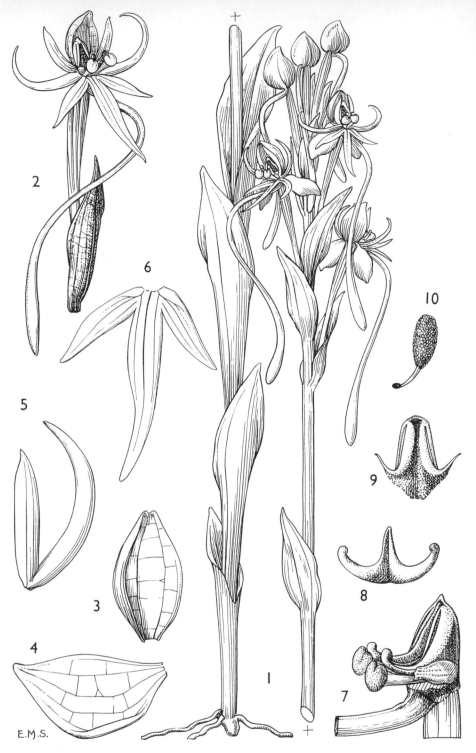

E.M.S.

Tab. 40. HABENARIA GONATOSIPHON. 1, habit (×1); 2, flower (×1); 3, dorsal sepal (×3); 4, lateral sepal (×3); 5, petal (×3); 6, lip (×3); 7, column, side view (×4); 8, rostellum, opened out (×4); 9, anther, pollinia removed (×4); 10, pollinarium (×4), 1–10 from *Milne-Redhead & Taylor* 7982. Drawn by Margaret Stones. From Hooker's Icon. Pl.

Terrestrial herb 25–80 cm tall; tubers to 5 × 2.5 cm, ovoid or ellipsoid. Stem leafy. Leaves 7–11, the lowermost 1–2 sheath-like, the mid-stem leaves suberect, the largest 7–15 × 2–4.5 cm, ovate or lanceolate; the uppermost grading into the bracts. Inflorescence 4–20 × 7–13 cm, laxly or fairly densely few- to several-flowered; bracts to 6 cm long. Flowers greenish or green with white lip, sometimes sweetly scented. [*Williamson & Gassner's* field notes say that the woodland form is heavily scented, the grassland form is not.] Pedicel and ovary 4.5–7.5 cm long, curved up. Dorsal sepal erect, 14–20 × 6–10 mm, elliptic, convex; lateral sepals deflexed, 18–24 × 8–15 mm, rolled lengthwise. Petals 2-lobed almost to base, upper lobe 15–20 × 0.5 mm, filiform, adnate to dorsal sepal; lower lobe curving up, 20–40 × 1–3 mm, horn-like, rather fleshy. Lip 3-lobed to c. 2 mm from base; mid-lobe deflexed, 16–35 × 1 mm, linear; side lobes 6–12 × 1–3 mm, lanceolate, the edge often toothed, clasping column; spur 5.5–7.5 cm long, pendent, swollen at apex. Anther erect, 5–8 mm high; canals 13–19 mm long, porrect, the tips upturned; stigmas porrect, 13–18 mm long, slender but enlarged and truncate at apex. Rostellum mid-lobe 3–5 mm long, triangular.

Zambia. B: Sesheke Distr., c. 96 km west of Katima Mulilo, fl. i.1973, *G. Williamson* 2243 (K; SRGH). N: Kasama, 1300 m, fl. 19.i.1962, *J.M. Wright* 327A (K). W: Ndola Distr., turn-off to Sacred Lake near St. Anthony's Mission, c. 1200 m, fl. 13.ii.1975, *Hooper & Townsend* 25 (K). E: Lundazi Distr., between Lundazi and Mzimba, fl. 13.ii.1957, *Angus* 1524 (K; SRGH). S: Choma, 1300 m, fl. 9.i.1957, *E.A. Robinson* 2115 (K). **Zimbabwe**. W: Leshumu Valley, *Holub* s.n. (K). **Malawi**. C: Dowa Distr., Kongwe Forest Reserve, 1500 m, fl. 7.iii.1982, *Brummitt, Polhill & Banda* 16372 (K; MAL). S: Mangochi Distr., Phirilongwe, 810 m, fl. 17.iii.1985, *Johnston-Stewart* 404 (K). Widespread in tropical Africa. Grassland, woodland, dambo; 810–1500 m.

78. **Habenaria kassneriana** Kraenzl. in Bot. Jahrb. Syst. 48: 388 (1912). —Summerhayes in F.T.E.A., Orchidaceae: 111 (1968). —Williamson, Orch. S. Centr. Africa: 57 (1977). —Geerinck in Fl. Afr. Centr., Orchidaceae pt. 1: 122 (1984). Type from Zaire.

Terrestrial herb 30–45 cm tall; tubers to 4.5 × 3 cm, ovoid to globose, densely tomentose. Stem leafy. Leaves 6–8, the lowermost sheath-like, the mid-stem leaves semi-erect or recurved, the largest 5–8 × 1.5–4.5 cm, ovate or lanceolate. Inflorescence 4–10 × 16–17 cm, laxly 2–6-flowered; bracts to 5.5 cm long. Flowers spreading, white or cream, strongly scented at night. Pedicel and ovary almost straight, 5.5–8.5 cm long. Dorsal sepal erect, 18–25 × 10–13 mm, elliptic, acuminate, very convex; lateral sepals deflexed, 20–30 × 10–15 mm, obliquely ovate, rolled up lengthwise. Petals 2-lobed almost to base, upper lobe 18–27 × 0.5 mm, narrowly linear, adnate to dorsal sepal; lower lobes curving out and up, 40–98 × 1–2 mm, fleshy, horn-like, tapering. Lip 3-lobed, with undivided basal part 2–3 mm long; mid-lobe deflexed, 30–45 × 1–2 mm, linear; side lobes diverging, 40–60 mm long, linear, slightly narrower than mid-lobe; spur pendent, 8.5–14.5 cm long, much swollen in apical quarter. Anther erect, 10–15 mm high; canals porrect, upturned at tips, 20–30 mm long. Stigmas porrect, c. 20 mm long, slender but enlarged and truncate at apices. Rostellum 3–6 mm long, triangular, acute.

Zambia. B: Kaoma (Mankoya), fl. 15.i.1960, *Gilges* 912 (K; NDO). N: Mbala Distr., Lumi River Marsh, 1500 m, fl. 16.i.1964, *Richards* 18779 (K). C: Kabwe Distr., between Chipepo and Lukanga River, 60 km west of Kabwe, 1140 m, fl. 20.i.1973, *Kornaś* 3057 (K). Also in Zaire and Tanzania. Dambo, flooded grassland; 1140–1500 m.

79. **Habenaria laurentii** De Wild., Notices Pl. Utiles Congo: 325 (1904). —Summerhayes in F.W.T.A. ed. 2, **3**: 198 (1968); in F.T.E.A., Orchidaceae: 111 (1968). —Grosvenor in Excelsa **6**: 83 (1976). —Williamson, Orch. S. Centr. Africa: 57 (1977). —Geerinck in Fl. Afr. Centr., Orchidaceae pt. 1: 125, pl. 17 (1984). —la Croix et al., Orch. Malawi: 85 (1991). Type from Zaire.

Terrestrial herb 25–75 cm tall; tubers to 3.5 × 2.5 cm, ellipsoid or ovoid. Stem leafy. Leaves 6–15, the lowermost 1–2 reduced to funnel-shaped sheaths; the mid-stem leaves spreading, the largest 6–22 × 2.5–7 cm, ovate, acute; the uppermost grading into the bracts. Inflorescence 6–30 × 10–15 cm, fairly laxly 3–15-flowered; bracts 3–8 cm long. Flowers greenish, or sepals green, petals and lip white. Pedicel and ovary 6–8.5 cm long, curving up. Dorsal sepal erect, 20–24 × 6–9 mm, elliptic, convex;

lateral sepals deflexed, 22–28 × c. 10 mm, rolled lengthwise. Petals 2-lobed ± to base, upper lobe 19–22 mm long, filiform, adnate to dorsal sepal; lower lobe 38–85 × 1.5–3 mm, horn-like, fleshy, curving upwards and outwards. Lip 3-lobed to base, mid-lobe 24–36 mm long; side lobes 20–30 mm long; all narrowly linear; spur 5.5–8 mm long, slender but swollen to 3.5–5 mm diameter at apex, usually curved backwards, occasionally curved forwards. Anther erect, 5–7 mm long; canals porrect, upturned at tips, 13–20 mm long. Stigmas porrect, 13–18 mm long, slender but widened and truncate at tips. Rostellum 3.5–5.5 mm long, triangular.

Zambia. N: Mbala Distr., Kasulo Farm, 1500 m, fl. 21.xii.1951, *Richards* 496 (K). W: Solwezi Distr., fl. 21.i.1960, *Holmes* 0199 (K; SRGH). C: c. 14 km east of Lusaka, in thicket, fl. 2.iii.1975, *Williamson & Gassner* 2480 (K; SRGH). **Zimbabwe**. N: Makonde Distr., Doma area, fl. 15.ii.1963, *Jacobsen* 2161 (PRE). C: Harare Distr., fl. 3.iii.1946, *Greatrex* 95 (K; PRE; SRGH). **Malawi**. N: Mzimba Distr., Nyika approach, c. 5 km after turn-off to Thazima, 1500 m, fl. 30.i.1987, *la Croix* 947 (K). S: Blantyre Distr., Mt. Soche, c. 1000 m, path to summit, fl. 22.iii.1986, *Jenkins* 12 (K).
Also in Cameroon, Central African Republic, Zaire, Nigeria, Senegal, French Guinea and Kenya. *Brachystegia* woodland, thicket; 1000–1580 m.

80. **Habenaria mira** Summerh. in Kew Bull. **17**: 513, fig. 2 (1964). —Grosvenor in Excelsa **6**: 83 (1976). —Geerinck in Fl. Afr. Centr., Orchidaceae pt. 1: 120 (1984). TAB. **41**. Type: Zimbabwe, Chimanimani Mts., *Ball* 460 (K, holotype; SRGH).

Terrestrial herb 40–65 cm tall. Stem leafy. Leaves 8–12, the lowermost reduced to sheaths, the mid-stem leaves spreading, 16–20 × 5–7 cm, lanceolate or oblanceolate, acute or acuminate. Inflorescence 17–39 × 6–10 cm, fairly densely many-flowered. Flowers suberect or spreading, green and white. Ovary and pedicel straight or slightly curved, 2.5–3 cm long. Dorsal sepal erect or reflexed, c. 11 × 6 mm, elliptic, obtuse; lateral sepals deflexed, rolled lengthwise, 13–14 × 7–10 mm, obliquely semi-orbicular. Petals 2-lobed ± to base; upper lobe erect, 10.5–12.5 × 0.7 mm, narrowly linear; lower lobe 30–50 mm long, c. 3 mm wide at base, horn-like, curving up. Lip 3-lobed ± to base, mid-lobe c. 20 × 1 mm, pendent then curving up; side lobes diverging, 8–12 × 2–4 mm, oblanceolate, acute, lacerate on both sides in apical half; spur pendent, 23–28 mm long, inflated to 1.5–2 mm diameter in apical third. Anther erect, 5 mm high; canals 8–11 mm long; auricles oblong, rugulose. Stigmas porrect or slightly incurved, 6–7 mm long, slender, enlarged at apex. Rostellum mid-lobe c. 3 mm long.

Zambia. N: Mbala Distr., Mukoma, 900 m, fl. 7.iv.1962, *Richards* 16280 (K). **Zimbabwe**. E: Chimanimani Mts., Bridal Veil Falls, 1360 m, fl. 19.ii.1955, *Ball* 460 (K; SRGH).
Also in Zaire. Riverine forest, woodland near stream; 900–1360 m.

81. **Habenaria stenorhynchos** Schltr. in Bot. Jahrb. Syst. **20**, Beibl. 50: 33 (1895). —Rolfe in Dyer, F.C. **5**, 3: 135 (1912). —Summerhayes in F.T.E.A., Orchidaceae: 106 (1968). —Grosvenor in Excelsa **6**: 83 (1976). —Williamson, Orch. S. Centr. Africa: 57, t. 42 (1977). —Stewart et al., Wild Orch. South. Africa: 91 (1982). —la Croix et al., Orch. Malawi: 85 (1991). Type from South Africa.
Habenaria dactylostigma Schltr. in Bot. Jahrb. Syst. **53**: 509 (1915), non Kraenzl. (1901), nom. illegit. Type from Tanzania.

Slender terrestrial herb 15–45 cm tall; tubers to 2.5 × 1.5 cm, ellipsoid or ovoid. Stem leafy. Leaves 4–8, the basal 1–2 usually sheath-like, the mid-stem leaves semi-spreading, the largest 3.5–8.5 × 0.7–2 cm, ovate or lanceolate; the upper ones grading into the bracts. Inflorescence to 11 cm long, laxly 1–7-flowered; bracts usually slightly shorter. Flowers sweetly scented, white with green veined sepals. Pedicel and ovary 15–25 mm long. Dorsal sepal erect, 10–11 × 2.5–4.5 mm, convex, acute; lateral sepals deflexed, c. 12 × 6 cm, obliquely obovate, not rolled lengthwise. Petals 2-lobed ± to base, upper lobe 8–10 mm long, erect; lower lobe 10–15 mm long, curving forwards and up, both lobes narrowly linear, less than 1 mm wide. Lip 3-lobed almost to base; mid-lobe deflexed, 12–14 × 1 mm, linear; side lobes 5.5–8.5 × 0.5 mm, ± filiform, projecting forwards but curving down slightly; spur 8–10 mm long, pendent, swollen at apex. Anther erect, c. 4 mm high; canals 7–8 mm long, porrect, the tips upcurved. Stigmas 5.5–8 mm long, projecting forwards, enlarged and truncate at apex. Rostellum mid-lobe 2.5–4 mm high, triangular-linear, acute.

Tab. 41. HABENARIA MIRA. 1, habit (×⅕); 2, inflorescence (×1); 3, leaf (×1); 4, dorsal sepal (×2); 5, lateral sepal (×2); 6, petal (×2); 7, lip and column (×2), 1–7 from *Ball* 460. Drawn by Mary Grierson. From Kew Bull.

Zambia. N: Mbala Distr., c. 80 km south of Mbala, fl. 15.i.1961, *E.A. Robinson* 4266 (K; SRGH). W: Mwinilunga Distr., NE of Dobeka Bridge, fl. 30.xii.1937, *Milne-Redhead* 3880 (K). C: Chakwenga Headwaters, 100–129 km east of Lusaka, fl. 10.i.1964, *E.A. Robinson* 6167 (K). **Zimbabwe**. C: Harare Distr., in vlei by Ruwa River, 1360 m, fl. 21.i.1946, *Wild* 707 (K; SRGH). S: Masvingo (Fort Victoria), 900 m, fl. 1.ii.1959, *Kleinenberg* 31 (K; SRGH). **Malawi**. N: Nyika National Park, Chelinda Rock dambo, c. 2250 m, fl. ii.1968, *Williamson, Ball & Simon* 813 (K; SRGH); South Viphya, Luchilemu Dambo, 1280 m, fl. 4.iv.1984, *la Croix* 587 (K).

Also in Tanzania, Angola and South Africa. Dambo, boggy grassland; 1280–2300 m.

82. **Habenaria goetzeana** Kraenzl. in Bot. Jahrb. Syst. **28**: 173 (1980). —Summerhayes in F.T.E.A., Orchidaceae: 114 (1968). —Williamson, Orch. S. Centr. Africa: 61 (1977). —la Croix et al., Orch. Malawi: 86 (1991). Type from Tanzania.

Kryptostoma goetzeana (Kraenzl.) Geerinck in Bull. Jard. Bot. Nat. Belg. **52**: 149 (1982); in Fl. Afr. Centr. Orchidaceae pt. 1: 144 (1984).

Slender terrestrial herb 15–35 cm tall. Tubers 1–1.5 × 0.5–1 cm, ovoid, woolly. Leaves 5–9, the lowermost 1–2 sheathing, the next few suberect, 1.5–3.5 × 0.5–1 cm, lanceolate; the upper ones becoming bract-like. Inflorescence to about 12 × 4 cm, laxly 2–9-flowered; bracts usually slightly shorter. Flowers green. Pedicel and ovary 15–25 mm long, arched. Dorsal sepal erect, 11–12 × 6–8 mm, elliptic, obtuse; lateral sepals spreading, partly adnate to column and base of lip, 10–12 × 4–6 mm, obliquely lanceolate. Petals 2-lobed with an undivided basal part 4–6 mm long; posterior lobe 8–12 × 3–4 mm, adnate to dorsal sepal; anterior lobe erect, 16–21 × c. 1 mm, linear, glabrous, fleshy. Lip 3-lobed with a basal undivided part 3–5.5 mm long; mid-lobe deflexed, 7–10 × 2 mm, linear; side lobes deflexed or projecting forwards and diverging, 13–15 × 0.8–0.9 mm, linear; spur 20–30 mm long, curved forwards and up, forming a complete loop, somewhat swollen near apex. Anther loculi 2–3 mm long, reclinate, joined by a U-shaped connective 2 mm high and 11 mm long; canals very short. Stigmatic arms porrect, 3–4 mm long, enlarged and truncate at apex.

Zambia. N: Chinsali Distr., c. 8 km east of Ishiba Ngandu (Shiva Ngandu), 1960 m, fl. 18.i.1972, *Kornaś* 0900 (K). W: Mwinilunga, fl. 3.xii.1958, *Holmes* 0116 (K; SRGH). **Malawi**. N: Nyika National Park, c. 3 km past Thazima, 1650 m, fl. 13.ii.1987, *la Croix* 961 (K). C: Dedza Distr., c. 12 km past Dedza on Lilongwe road, c. 1500 m, fl. 3.ii.1982, *Gassner & Cribb* 207 (K).

Also in Zaire and Tanzania. Rough grassland and woodland, often in clearings and on shallow soil over rocks, edge of woodland; 1500–1650 m.

83. **Habenaria tentaculigera** Rchb.f. in Flora **50**: 101 (1867). —Rolfe in F.T.A. **7**: 230 (1898). —Summerhayes in Bot. Not. **1937**: 187 (1937); in F.T.E.A., Orchidaceae: 114, fig. 21 (1968). —Moriarty, Wild Fls. Malawi: t. 26, 3 (1975). —Grosvenor in Excelsa **6**: 83 (1976). —Williamson, Orch. S. Centr. Africa: 61 (1977). —la Croix et al., Orch. Malawi: 88 (1991). TAB. **42**. Type from Angola.

Habenaria antunesiana Kraenzl. in Bot. Jahrb. Syst. **28**: 173 (1900). Type from Angola.

Habenaria ludens Kraenzl. in Bot. Jahrb. Syst. **51**: 374 (1914). —Schltr. in Bot. Jahrb. Syst. **53**: 510 (1915). Type: Zambia, Lukanda R., *Kassner* 2132 (B†, holotype; K).

Kryptostoma tentaculigera (Rchb.f.) Geerinck in Bull. Jard. Bot. Nat. Belg. **52**: 149 (1982); in Fl. Afr. Centr. Orchidaceae pt. 1: 143 (1984).

Terrestrial herb 20–60 cm tall. Tubers 1.5–5 × 1–2 cm, ovoid or ellipsoid, woolly. Stem leafy. Leaves 5–11, the lowermost 1–2 sheathing; the next few suberect or spreading, 3–8 × 1.5–3 cm, lanceolate, acute; the upper ones grading into the bracts. Inflorescence 5–26 × 5–6 cm, laxly 3–17-flowered; bracts 10–30 mm long. Flowers green or yellow-green. Pedicel and ovary 15–30 mm long, ± straight. Dorsal sepal 16–20 × 15 mm, ovate, very convex, curving forwards, the tip recurved; lateral sepals projecting forwards, 12–15 × 3–6 mm, obliquely lanceolate, adnate to column and base of spur for about half their length. Petals erect, 2-lobed, with an undivided basal part c. 10 mm long; posterior lobe 7–8.5 × c. 3 mm, lanceolate, the margins ciliate, adnate to dorsal sepal; anterior lobe 20–33 × 1 mm, linear, fleshy. Lip 3-lobed, the base adnate to the column for 6–9 mm, the free part deflexed, 3-lobed with an undivided part 3–4 mm long; mid-lobe ± reflexed, 7–9 × 1–2 mm, linear, fleshy; side lobes 14–22 × 1 mm, linear, curving forwards; spur 20–22 mm long, bent forwards with a knee-like bend just below the base, then recurved, the apex swollen. Anther

Tab. 42. HABENARIA TENTACULIGERA. 1, habit (×⅓); 2, lower stem and base of plant (×1); 3, inflorescence (×1); 4, dorsal sepal, side view (×2); 5, lateral sepal (×2); 6, petal (×2); 7, column, with lip, spur and ovary, side view (×2); 8, column, with spur and part of lip and ovary, oblique view (×2), 1–8 from *Milne-Redhead & Taylor* 8680. Drawn by Heather Wood. From F.T.E.A.

loculi c. 3 mm high, reclinate, joined by a U-shaped connective c. 4 mm high and 14–16 mm long; canals c. 3 mm long, porrect, then upcurved. Stigmatic arms 4–5 mm long, clavate, decurved.

Zambia. N: Chinsali Distr., Ishiba Ngandu (Shiwa Ngandu), fl. ii.1969, *G. Williamson* 1420 (K). W: Mwinilunga, fl. 30.xi.1958, *Holmes* 0106 (K; SRGH). C: Lusaka Distr., c. 13 km west of Lusaka, fl. 9.ii.1975, *Williamson & Gassner* 2374 (K). E: Isoka Distr., Nyika Plateau, 2200 m, fl. xii.1966, *G. Williamson* 241 (K; SRGH). **Zimbabwe**. C: Marondera (Marandellas), fl. xii.1931, *Myres* 364 (K). E: Nyanga (Inyanga), 1800 m, fl. 18.xi.1948, *Chase* 4076 (K; SRGH). **Malawi**. N: Nyika National Park, Chelinda Bridge, 2250 m, fl. 8.i.1974, *Pawek* 7892B (K; UC). C: Kasungu Distr., near Chamama, 1000 m, fl. 16.i.1959, *Robson & Jackson* 1229 (K). S: Zomba Mt., 1850 m, 25.i.1959, *Robson & Jackson* 1324 (K).

Also in Angola, Zaire and Tanzania. Marshy grassland and seepage areas amongst rocks, and in *Brachystegia* woodland; 1000–2250 m.

84. **Habenaria argentea** P.J. Cribb in Kew Bull. **32**: 141 (1977). Type: Zambia, Kawambwa Distr., Ntumbachushi Falls, *Williamson & Simon* 620 (K, holotype).
    Habenaria sp. Williamson, Orch. S. Centr. Africa: 51 (1977) [*Williamson & Drummond* 1649].

Slender terrestrial herb 28–60 cm tall; tubers c. 12 × 5 mm, ellipsoid. Stem leafy. Leaves 6–7, the lowermost 2 sheathing, the next 2 suberect, 6–11 × 0.4–1 cm, narrowly lanceolate; upper leaves becoming bract-like. Inflorescence 3–16 × 2–5 cm, racemose or sub-umbellate, 3–15-flowered. Flowers white. Pedicel 12–20 mm long; ovary 10–15 mm long, silvery papillose. Dorsal sepal erect, 5 × 3 mm, ovate; lateral sepals 5.5 × 3 mm, semi-erect. Petals 2-lobed almost to base, upper lobe erect, adnate to dorsal sepal, 5 × 0.8 mm, narrowly linear; lower lobe 3 mm long, narrowly linear, somewhat S-shaped. Lip porrect, 3-lobed to c. 1.5 mm from base; mid-lobe 4 × 1.5 mm, linear; side lobes 2 mm long, curved; spur pendent, very slender, 3.5–8.5 cm long. Anther erect, 3 mm high; canals c. 1 mm long, upcurved. Stigmatic arms porrect, 1.5 mm long, truncate at apex. Rostellum 3 mm high, triangular.

Zambia. N: Kawambwa Distr., Ntumbachushi Falls, fl. xii.1967, *Williamson & Simon* 620 (K). W: Mwinilunga Distr., Sinkabolo Swamp, 1200 m, fl. 20.xi.1962, *Richards* 17428 (K). C: Kabwe Distr., Mpunde, c.53 km NW from Kabwe, 1170 m, fl. 7.xii.1972, *Kornaś* 2778 (K).

As yet known only from Zambia. Swampy grassland; 1170–1300 m.

85. **Habenaria macroplectron** Schltr. in Warb., Kunene-Sambesi-Exped. Baum: 206 (1903). — Williamson, Orch. S. Centr. Africa: 61 (1977). —Geerinck in Fl. Afr. Centr., Orchidaceae pt. 1: 103 (1984). Type from Angola.

Terrestrial herb 25–75 cm tall; tubers c. 2 × 1.7 cm, ± globose. Leaves 8–10, the lowermost 2 forming funnel-shaped sheaths, the next few semi-spreading, the largest 10.5–14 × 2.5–4 cm, lanceolate, pale green, rather fleshy, the base loosely funnel-shaped. Inflorescence c. 12 × 8 cm, fairly densely 5–10-flowered; bracts 5–6 cm long, leafy, with transparent, papillose hairs. Pedicel c. 6 cm long, ovary 1–1.5 cm long. Sepals white, the tips golden-yellow; petals and lip yellow, whitish at base; spur green and white. Dorsal sepal reflexed, 10–12 × 4 mm; lateral sepals spreading, 11–12 × 4–5 mm, obliquely semi-ovate. Petals 2-lobed, upper lobe 8.5–11 × 1.3–2 mm, oblong or oblanceolate, curved; lower lobe 12–14 × 1–1.5 mm, oblong, curved; both lobes erect. Lip 3-lobed to 2 mm from base; mid-lobe 16–18 × 0.5 mm, linear, the last 4 mm bent forward; side lobes 12–13 × 1–1.5 mm, lanceolate; spur 5–7.5 cm long, pendent, slender, the last 1.5 cm swollen. Anther 2–3 mm high; canals 9–10 mm long. Stigmatic arms 9–11 mm long; rostellum mid-lobe c. 2 mm long.

Zambia. N: Mbala Distr., c. 80 km south of Mbala, fl. 15.i.1961, *E.A. Robinson* 4267 (K). W: Mwinilunga Distr., by River Kalalima (Kamwezhi), fl. 3.i.1938, *Milne-Redhead* 3935 (K). C: Serenje Distr., Kundalila Falls, fl. iii.1969, *G. Williamson* 1546 (K; SRGH).

Also in Zaire and Angola. Marshy grassland; c. 1360 m.

86. **Habenaria mirabilis** Rolfe in F.T.A. **7**: 572 (1898). —Summerhayes in Hook., Ic. Pl. 35, t. 3442 (1943); in F.T.E.A. Orchidaceae: 117 (1968). —Williamson, Orch. S. Centr. Africa: 61 (1977). —Geerinck in Fl. Afr. Centr., Orchidaceae pt. 1: 140 (1984). TAB. **43**. Type: Zambia, Mbala Distr., Fwambo, *Carson* 9 (K, holotype).
   Habenaria insignis Rolfe in F.T.A. **7**: 234 (1898), non Schltr. (1895), nom. illegit. Type as for species.

Slender terrestrial herb 20–50 cm tall; tubers 2–3 × 1 cm, ellipsoid. Leaves 3–6, the lowermost 1–2 sheathing, the mid-stem leaves clasping the stem, 3–6.5 × 1–2 cm, lanceolate, acute; the upper ones bract-like. Inflorescence to 12 × 6.5 cm, densely few to several-flowered; bracts leafy, 1.5–3.5 cm long. Flowers sweetly scented, sepals green, petals and lip whitish-green. Pedicel and ovary c. 2.5 cm long, somewhat curved. Dorsal sepal erect, 11–15 × 5.5–8 mm, lanceolate or ovate, acute, convex; lateral sepals spreading and curving up, 12.5–19.5 × 7–9 mm, obliquely lanceolate, acute. Petals divided almost to base, both lobes erect; upper lobe 9.5–13 × 1–2 mm, ligulate, curved; lower lobe 5.5–11 × 1–2 mm, linear or ligulate, very fleshy. Lip velvety-papillose, 3-lobed with an undivided base 1–3.5 mm long; mid-lobe 14–27 × 4.5–7 mm, oblong-lanceolate, acute; side lobes diverging, 10–16 × 2.5–5.5 mm, oblanceolate; spur 12–15 mm long, pendent, swollen to c. 3 mm diameter at apex. Anther erect, 5–6.5 mm high; canals 3.5–5 mm long, porrect. Stigmatic arms 4.5–6.5 mm long, porrect, enlarged and truncate at the tips. Rostellum mid-lobe c. 2 mm long, triangular.

Zambia. N: Mbala Distr., Nkali (Kali) Dambo, c. 1500 m, fl. 26.i.1952, *Richards* 780 (K). W: Mwinilunga Distr., near R. Kalalima (Kamwezhi), fl. 3.i.1938, *Milne-Redhead* 3936 (K).
Also in Zaire and Tanzania. Marshy ground, sometimes in standing water; 1500–1740 m.

87. **Habenaria walleri** Rchb.f., Otia Bot. Hamburg.: 98 (1881). —Rolfe in F.T.A. **7**: 247 (1898). —Summerhayes in F.W.T.A. ed. 2, **3**: 198 (1968); in F.T.E.A., Orchidaceae: 116 (1968). —Moriarty, Wild Fls. Malawi: t. 26, 2 (1975). —Geerinck in Fl. Afr. Centr., Orchidaceae pt. 1: 140 (1984). —la Croix et al., Orch. Malawi: 88 (1991). Type: Malawi, Manganja Hills, 1865, *Waller* (K, holotype).
   Habenaria soyauxii Kraenzl. in Bot. Jahrb. Syst. **16**: 93 (1892). Type from Gabon.

Robust terrestrial herb 40–80 cm tall; tubers 2–4 × 1–2 cm, ovoid or ellipsoid. Stem leafy. Leaves 7–10, the lowest 1–2 sheath-like, the mid-stem leaves erect or semi-spreading, the largest 6–14 × 1–3.5 cm, lanceolate, acute; the upper ones bract-like. Inflorescence 8–28 × 7.5–10 cm, rather laxly 2 to several-flowered; bracts leafy, to 8 cm long, shorter than the pedicel and ovary. Flowers showy, scented, sepals green, petals and lip white. Pedicel c. 6.5 cm long, ovary 2.5–3 cm long, minutely hairy on ribs. Dorsal sepal erect, 11–16 × 10–11 mm, ovate, acute, convex; lateral sepals deflexed, c. 20 × 9 mm, obliquely semi-ovate. Petals 2-lobed almost to base; upper lobe 11–16 × 1.5–3 mm, linear, adnate to dorsal sepal; lower lobe deflexed, 20–32 × 3.5–5.5 mm, lanceolate, acute. Lip 3-lobed, with an undivided base c. 4 mm long; mid-lobe 15–28 × 2.5–6 mm, oblong, obtuse; side lobes 20–32 × 4–7 mm, the inner edge straight, the outer curved; the apex acute; spur 13–17 cm long, pendent, inflated in apical third. Anther erect, 7–9 mm high; canals 3–5 mm long. Stigmatic arms 8–12 mm long, porrect, club-shaped. Rostellum mid-lobe 4 mm high, triangular, acute.

Zambia. N: Mbala Distr., Saisi Valley, 1500 m, fl. i.1933, *Gamwell* 135 (BM). **Malawi**. N: Mzimba Distr., c. 8 km south of Euthini, towards Vuvumbwe Bridge, 1170 m, fl. 20.i.1978, *Pawek* 13648 (K; MAL; MO; SRGH; UC). C: near Kasungu Hill, 1100 m, fl. 14.i.1959, *Robson & Jackson* 1159 (K). S: Thyolo Distr., Bvumbwe, 1100 m, fl. 2.i.1983, *la Croix* 383 (K).
Also in Sudan, Nigeria, Cameroon, Gabon, Congo, Zaire, Burundi, Uganda, Kenya and Tanzania. Grassland, usually wet; 1100–1170 m.

88. **Habenaria aberrans** Schltr. in Bot. Jahrb. Syst. **53**: 505 (1915). —Summerhayes in F.T.E.A., Orchidaceae: 120 (1968). —Grosvenor in Excelsa **6**: 82 (1976). —la Croix et al., Orch. Malawi: 89 (1991). Type from Tanzania.

Terrestrial herb 18–30 cm tall; tubers c. 15 × 10 mm, ovoid or ellipsoid, villous. Basal leaf 1, appressed to ground, 2.5–4 × 2.5–3.5 cm, broadly ovate, apiculate, cordate at base, dark green mottled with white, glabrous or with a few scattered hairs on upper surface, ciliate on edge. Stem light green with 3–4 bract-like leaves c. 10 mm long.

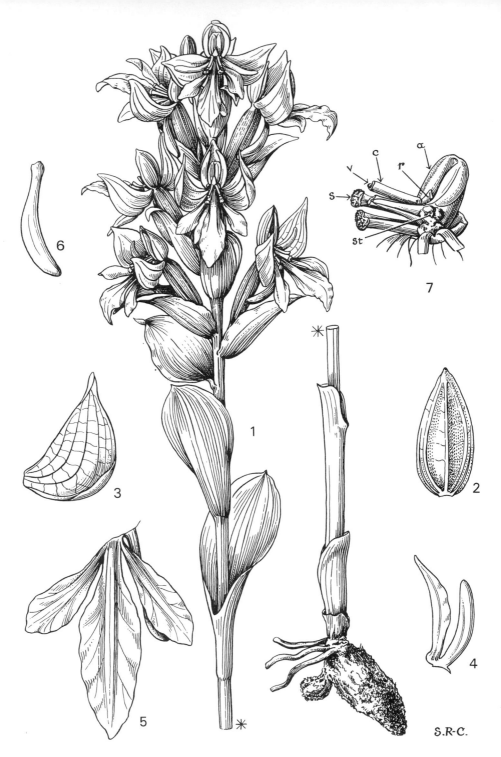

Tab. 43. HABENARIA MIRABILIS. 1, habit (×1); 2, dorsal sepal (×2); 3, lateral sepal (×2); 4, petal (×2); 5, lip, flattened out (×2); 6, spur (×2); 7, column, with perianth members cut off (×3); a, anther loculus; c, anther canal; r, rostellum mid-lobe; s, stigma; st, staminode; v, viscidium. Drawn by Stella Ross-Craig. From Hooker's Icon. Pl.

Stem, sheaths, ovary, bracts and sepals all pubescent. Inflorescence 3.5–7.5 × 2–3 cm, fairly densely 3–11-flowered; bracts 6–7 mm long, lanceolate, acuminate. Flowers non-resupinate, primrose-yellow. Ovary and pedicel 10 mm long. Dorsal sepal 6.5–8 × c. 4 mm, ovate, convex, ± at right angles to ovary; lateral sepals of similar length but 2.5–3.5 mm wide, obliquely lanceolate. Petals 2-lobed in apical third, 7–9 × c. 3.5 mm; posterior lobe (the one nearest intermediate sepal) 3.5 × 2.5–3 mm; anterior lobe 2–2.5 × 1 mm. Lip 8–10 mm long, 3-lobed about halfway, the lobes diverging slightly, the apices rounded; mid-lobe 4–5.5 × 1.5–2 mm; side lobes 3.5–4.5 × 0.6–1 mm; spur 10.5–12 mm long, bent up so that it lies parallel to the flower but above it, slightly swollen towards the end but with an acute apex. Anther c. 2.5 mm high; canals c. 1 mm long. Stigmatic arms c. 1.5 mm long.

Zimbabwe. N: Makonde Distr., c. 1160 m, fl. 4.ii.1965, *Jacobsen* 2690 (PRE). **Malawi**. N: Rumphi Distr., Livingstonia Escarpment, fl. 16.ii.1947, *Benson* 1231 (K). C: Nkhota Kota (Nkhotakota) Escarpment, fl. 28.ii.1953, *Jackson* 1133 (K). **Mozambique**. N: Mandimba, fl. 26.i.1942, *Hornby* 3533 (K). Z: Mucanga, fl. 16.ii.1905, *le Testu* 686 (BM; K).
Also in Tanzania. *Brachystegia* woodland, on escarpments and dry hillsides; 900–1250 m.

89. **Habenaria villosa** Rolfe in F.T.A. **7**: 228 (1898). —Summerhayes in F.T.E.A., Orchidaceae: 119 (1968). —la Croix et al., Orch. Malawi: 100 (1991). Type from Tanzania.

Terrestrial herb c. 35 cm tall. Stem erect, slender with a single basal foliage leaf and c. 7 bract-like leaves 10–20 mm long. Basal leaf appressed to ground, 4–7 × 4–6 cm, heart-shaped, glabrous above but ciliate on edge, silvery with dark green veins. Stem, bract-like leaves, bracts, ovary and sepals all pubescent. Inflorescence 5 × 3 cm, fairly densely 7–12-flowered; bracts c. 7 mm long. Flowers non-resupinate, rather campanulate in shape; sepals yellow-green, petals and lip clear lemon-yellow. Ovary and pedicel suberect, 8–11 mm long. Dorsal sepal 7–7.5 × 3–4 mm, ovate, convex; lateral sepals 7–8 × 3 mm. Petals 8.5–10 × 3.5–4 mm, 2-lobed in apical third; posterior lobe 3.5 × 2 mm; anterior lobe 2 × c. 1 mm, the tips of the lobes slightly reflexed. Lip 9–11 × 4 mm, 3-lobed about halfway along; mid-lobe 5–6 × 1–2.5 mm; side lobes 4–5.5 × 1–2 mm; spur 13–15 mm long, curving downwards, somewhat swollen at apex. Anther c. 2 mm high; canals 1.5 mm long. Stigmas club-shaped, 1.25 mm long.

Malawi. S: Thyolo Distr., Bvumbwe, 1150 m, fl. 21.ii.1984, *la Croix* 554 (K).
Also in Tanzania. In longish grass in dry dambo; 1150 m.

90. **Habenaria odorata** Schltr. in Bot. Jahrb. Syst. **53**: 502 (1915). —Summerhayes in F.T.E.A., Orchidaceae: 121 (1968). —la Croix et al., Orch. Malawi: 96 (1991). Type from Tanzania.

Terrestrial herb 30–40 cm tall; tubers c. 2 cm long, ovoid, villous. Stem erect with a single leaf at the base and 4–5 rather loose sheaths scattered along its length; sheaths 2.5–3.5 cm long. Basal leaf appressed to ground, 3–6 × 3.5–6.5 cm, ± orbicular, cordate at base, fleshy. Stem with 4–5 rather loose sheaths, 2.5–3.5 cm long. Inflorescence 4–13 cm long, laxly 2–6-flowered; bracts 15–25 mm long. Flowers sweetly scented; sepals green, petals and lip white. Pedicel 15–20 mm long, ovary slender, 15–20 mm long. Dorsal sepal erect, 9–14 × 2.5–6 mm, ovate, convex; lateral sepals rolled up lengthways, reflexed, 10–15 × 3 mm (7 mm when flattened). Petals 9–12 × 3–4 mm, entire, falcate, adnate to dorsal sepal. Lip 3-lobed to 1–2 mm from base; mid-lobe 11–13 × 2–5 mm, oblanceolate, acute; side lobes 12–19 × 4–9 mm, obliquely oblanceolate; spur 2.5–3.5(6.5) cm long, somewhat inflated at apex. Anther erect, c. 4 mm high; canals very short. Stigmatic arms 4–5 mm long, porrect, rather stout, truncate. Mid-lobe of rostellum c. 1.5 mm long.

Zambia. W: Mwinilunga Distr., slope east of Kasombo R., fl. 20.xii.1937, *Milne-Redhead* 3784 (K).
Malawi. N: Nyika National Park, Zovochipolo, among rocks, 2200 m, fl. 8.i.1983, *la Croix* 411 (K).
Also in Tanzania. Grassland and montane grassland; up to 2200 m.

91. **Habenaria nicholsonii** Rolfe in F.T.A. **7**: 571 (1898). —Williamson, Orch. S. Centr. Africa: 67 (1977). —Geerinck in Fl. Afr. Centr., Orchidaceae pt. 1: 62 (1984). Type: Zambia, near Fort Young, Sept. 1896, *Nicholson* s.n. (K, holotype).
Habenaria debeerstiana sensu Geerinck in Fl. Afr. Centr., Orchidaceae pt. 1: 62 (1984).

Terrestrial herb c. 40 cm tall. Stem erect, with 2 basal foliage leaves and about 6 sheath-like leaves scattered along its length. Basal leaves appressed to ground, the larger 3–3.5 × 3–4.5 cm, ± orbicular, cordate at base. Inflorescence 10–16 × 5–8 cm, laxly 3–8-flowered; bracts up to 18 mm long. Sepals green, petals and lip white. Ovary and pedicel c. 25 mm long. Dorsal sepal erect, 11 × 5–5.5 mm, elliptic, convex; lateral sepals 12 × 2–2.5 mm, reflexed, rolled up lengthways. Petals 11 × 2.5 mm, entire, falcate, adnate to dorsal sepal. Lip 3-lobed to c. 2 mm from base; mid-lobe 13–16 × 2.3–3 mm; side lobes 18–22 × 5–8.5 mm; spur 20–21 mm long, swollen in apical half. Anther 2–3.5 mm high; stigmatic arms 5 mm long.

Zambia. W: Kabompo-Chilnata, fl. 20.i.1963, *Holmes* 0380 (K). C: Kabwe (Broken Hill), fl. 4.i.1961, *Morze* 38 (K). E: near Fort Young, fl. ix.1896, *Nicholson* s.n. (K). Also in Zaire. Miombo woodland.

92. **Habenaria debeerstiana** Kraenzl. in Bull. Soc. Roy. Bot. Belg. **38**: 67 (1900). —Grosvenor in Excelsa **6**: 83 (1976). —Williamson, Orch. S. Centr. Africa: 67 (1977). —la Croix et al., Orch. Malawi: 91 (1991). Type from Zaire.
 Habenaria tiesleriana Kraenzl. in Bot. Jahrb. Syst. **43**: 394 (1909). Type: Mozambique, Chifumbazi, *Tiesler* 37 (B†, holotype).
 Habenaria nicholsonii var. *debeerstiana* (Kraenzl.) Geerinck in Bull. Jard. Bot. Nat. Belg. **52**: 146 (1982); in Fl. Afr. Centr., Orchidaceae pt. 1: 62 (1894).

Terrestrial herb 30–60 cm tall; tubers 2–5 cm long, ovoid, villous. Stem erect, with 2 basal foliage leaves and 4–7 sheath-like leaves scattered along its length. Basal leaves appressed to ground, the larger 3–6.5 × 3.5–7 cm, heart-shaped. Inflorescence 11–17 × 3–6 cm, laxly up to 20-flowered; bracts 11–14 mm long. Flowers sweetly scented, the sepals green, petals and lip white. Ovary and pedicel 23–30 mm long. Dorsal sepal erect, 9–12 × 4–6 mm, ovate, convex; lateral sepals 8–15 × 2.5–4 mm, reflexed, rolled lengthways. Petals 9–12 × 3–5 mm, entire, falcate, acute, adnate to dorsal sepal. Lip 3-lobed ± to base; mid-lobe 12–15 × 3–5 mm, oblanceolate, acute; side lobes 14–19 × 3–9 mm, obliquely oblanceolate, acute; spur 8–9 mm long, slightly swollen at apex, blunt. Anther 3.5 mm high; canals 1.5 mm long. Stigmatic arms c. 5 mm long.

Zambia. N: Mbala Distr., near middle Lunzua Falls, c. 1300 m, fl. immat. 12.i.1975, *Brummitt & Polhill* 13754 (K). C: c. 11 km east of Lusaka, fl. & fr. 3.ii.1958, *King* 414 (K). **Zimbabwe**. N: Makonde Distr., fl. 20.i.1963, *Jacobsen* 2075 (PRE). C: Harare Distr., fl. 26.i.1946, *Greatrex* 105 (PRE). **Malawi**. N: Rumphi Distr., c. 8 km SE of Thazima, 1350 m, fl. 17.i.1987, *la Croix* 937 (K; MAL). C: Lilongwe Distr., Chirikanda, fl. 29.i.1953, *Jackson* 1051A (K). S: Blantyre Distr., fl. 1887, *Last* s.n. (K). **Mozambique**. T: Chifumbazi, *Tiesler* 37 (B†).
 Also in Zaire, Burundi and Tanzania. *Brachystegia* woodland, often on stony soil, and on edge of dambos; 1050–1350 m.
 H. debeerstiana, *H. nicholsonii* and *H. odorata* are very similar, obviously closely related and may be conspecific. However, the differences, although small, seem to be consistent and it is easy to key them out, and so they are being kept separate here. *H. odorata* has a single, fleshy basal leaf and the stem sheaths are longer and looser than in the other two species. *H. debeerstiana* usually has more flowers in the inflorescence than the others and almost always has a very short spur, less than 1 cm long. *H. nicholsonii* tends to have larger flowers than the others and like *H. debeerstiana* has 2 basal leaves. Colonies of *H. odorata* and *H. debeerstiana* seen in the field in Malawi were consistently 1- and 2-leafed respectively. If these species do turn out to be conspecific *H. nicholsonii* will be the correct name.

93. **Habenaria adolphi** Schltr. in Bot. Jahrb. Syst. **53**: 503 (1915). —Mansfeld in Fedde, Repert. Spec. Nov. Regni Veg., Beih. 68, t. 24/93 (1932). —Summerhayes in F.T.E.A., Orchidaceae: 134 (1968). —Geerinck in Fl. Afr. Centr., Orchidaceae pt. 1: 64, pl. 11 (1984). —la Croix et al., Orch. Malawi: 90 (1991). Type from Tanzania.

Terrestrial herb 25–50 cm tall; tubers globose or ellipsoid, tomentose. Stem erect, with 2 basal leaves and several bract-like leaves scattered along its length. Basal leaves appressed to ground, 4–7 × 5–8 cm, broadly ovate, reniform or orbicular, cordate at base; bract-like leaves up to 4 cm long. Inflorescence 4–12 cm long, densely 3–15-flowered. Flowers white or greenish-white, suberect. Ovary and pedicel c. 2 cm long, bracts of similar length. Sepals all with 1–3 toothed keels on the veins outside; dorsal sepal erect, 13–16 × 6.5–9 mm, ovate, acute, convex; lateral sepals deflexed, 15–18 × 7 mm, obliquely semi-orbicular, acuminate. Petals 2-lobed ± to base; posterior lobe

13–16 mm long, less than 1 mm wide, erect; anterior lobe 20–35 mm long, 1–2 mm wide at base, horn-like, curving upwards. Lip projecting, 3-lobed ± to base, all lobes 1–2 mm wide, linear; mid-lobe 18–27 mm long; side lobes 14–20 mm long; spur 20–25 mm long, very swollen at apex, ± parallel to ovary. Anther 5–7 mm high, erect, apiculate; canals slender, 6–9 mm long. Stigmatic arms 7–10 mm long, slender but suddenly widened and truncate at apex.

Zambia. W: Mwinilunga Distr., *Milne-Redhead* 3820 (PRE). **Malawi**. N: Mzimba Distr., c. 32 km west of Mzuzu, Lunyangwa River Bridge, 1150 m, fl. 2.ii.1974, *Pawek* 8154 (K; MAL; MO; SRGH; UC).
Also in Zaire and Tanzania. Woodland and grassland, usually on rocky hillsides; 1100–1200 m.

94. **Habenaria lindblomii** Schltr. in Notizbl. Bot. Gart. Berl. **8**: 224 (1922). —Summerhayes in F.T.E.A., Orchidaceae: 133 (1968). —Grosvenor in Excelsa **6**: 83 (1976). —Williamson, Orch. S. Centr. Africa: 67 (1977). —la Croix et al., Orch. Malawi: 93 (1991). Type from Kenya.

Terrestrial herb 30–60 cm tall; tubers 2–4 cm long, ovoid, villous. Stem erect, with 2 basal leaves and 5–10 bract-like leaves scattered along its length. Basal leaves appressed to ground, the larger 4–6.5 × 4.5–9 cm, ovate to reniform, apiculate, fleshy; bract-like leaves up to 5 cm long. Inflorescence 7–19 × 3.5–7 cm, laxly or densely 8- to many-flowered. Flowers pale green, white in centre, fragrant. Pedicel 9–12 mm long, ovary 12–20 mm long with 6 toothed keels, suberect or arching. Dorsal sepal erect, 9–15 × 4–8 mm, ovate, convex, with 3 toothed keels on outside; lateral sepals deflexed, 10–16 × 2.5–6.5 mm, obliquely lanceolate, acuminate, sometimes with the apex hooked, with a central toothed keel on the outside. Petals 2-lobed ± to base; posterior lobe erect, 9–16 × 0.5–1.2 mm, linear; anterior lobe 14–25 mm long, c. 1 mm wide at base, curving up. Lip 3-lobed almost to base, all lobes linear, fleshy; mid-lobe 16–18.5 × 1–1.3 mm; side lobes 14.5–20 mm long, slightly narrower than mid-lobe; spur 17–23 mm long, very swollen at apex. Anther 3.5–6.5 mm high, erect, apiculate; canals c. 3 mm long, slightly upturned. Stigmatic arms 4–5 mm long, porrect, rather stout, truncate at apex, papillose below.

Zambia. N: Kasama Distr., c. 45 km east of Kasama, fl. 28.i.1962, *E.A. Robinson* 4916 (K). W: Mwinilunga Distr., Kalenda Dambo, fl. 1.i.1938, *Milne-Redhead* 3906 (K). **Zimbabwe**. N: Makonde Distr., Umvukwe Mts., 1360 m, fl. ii.1954, *Pollitt* in *GHS* 46271 (K; SRGH). C: Harare, fl. 4.ii.1946, *Greatrex* 88 (K; PRE). **Malawi**. N: Mzimba Distr., just beyond Kizutu Bridge, 1180 m, fl. 9.ii.1969, *Pawek* 1695 (K). C: Kasungu Distr., near Chamama, 1000 m, fl. 16.i.1959, *Robson & Jackson* 1226B (K).
Also in Zaire, Kenya and Tanzania. Open grassland usually on dambo margins, or in *Brachystegia* woodland; 1000–1500 m.

95. **Habenaria mechowii** Rchb.f. in Flora **65**: 532 (1882). —Rolfe in F.T.A. **7**: 226 (1898). —Summerhayes in F.T.E.A., Orchidaceae: 134 (1968). —Williamson, Orch. S. Centr. Africa: 67 (1977). —Geerinck in Fl. Afr. Centr., Orchidaceae pt. 1: 68 (1984). —la Croix et al., Orch. Malawi: 94 (1991). Type from Angola.

Robust terrestrial herb 45–60 cm tall; tubers 3–6 cm long, ovoid to ellipsoid. Stem erect, with 2 basal leaves and 5–13 bract-like leaves scattered along its length. Basal leaves appressed to ground, 4.5–6.5 × 6–8 cm, reniform or ovate, cordate at base, apiculate or rounded, rather fleshy; bract-like leaves up to 5 cm long. Inflorescence 10–20 cm long, loosely or fairly densely 2–9-flowered; bracts 35–40 mm long. Flowers pale green. Pedicel 20 mm long; ovary arched, 25–30 mm long with 6 toothed keels. Sepals all with 3 toothed keels on outside; dorsal sepal erect, 20–28 × 9–15 mm, ovate, convex; lateral sepals 22–27 × 7–9 mm, obliquely lanceolate with curved apiculus, deflexed and rolled up lengthways. Petals 2-lobed ± to base, the lobes horn-like, c. 1 mm wide; posterior lobe 20–45 mm long, erect; anterior lobe 33–75 mm long, curved outwards and upwards. Lip 3-lobed almost to base, projecting forwards and curving up, the lobes linear, fleshy, 1–2 mm wide; mid-lobe 35–45 mm long, side lobes 25–40 mm long; spur 3.5–6.5 cm long, very swollen, up to 5 mm diameter near apex. Anther 9–10 mm high, erect, apiculate; canals 7–8 mm long, slender, the last 2 mm upturned. Stigmatic arms 7.5–12 mm long, slender but enlarged and truncate at apex.

Zambia. N: Mbala Distr., Lunzua Agric. Station, 1650 m, fl. 28.ii.1957, *Richards* 8388 (K). E:
Luangwa (Loangwa) R., Bar Missale, fl. ii.1897, *Nicholson* s.n. (K). **Malawi**. N: South Viphya,
Lwafwa Drift, grassy bank at edge of dambo, 1650 m, fl. & fr. 31.i.1987, *la Croix* 949 (K). C:
Kasungu Distr., near Chamama, 1000 m, fl. 16.i.1959, *Robson & Jackson* 1226A (K). S: Zomba
Plains, 750–900 m, *Whyte* s.n. (K).
Also in Zaire, Tanzania and Angola. Wet grassland and dambo margins, and in *Brachystegia*
woodland; 750–1650 m.

96. **Habenaria edgarii** Summerh. in Bot. Mus. Leafl. Harv. Univ. **10**: 278 (1942), as '*edgari*';
in F.T.E.A., Orchidaceae: 135 (1968). —Williamson, Orch. S. Centr. Africa: 67 (1977).
—Geerinck in Fl. Afr. Centr., Orchidaceae pt. 1: 75 (1984). Type: Zambia, Mwinilunga
Distr., slope east of Kewumbo (Kaoomba) River, *Milne-Redhead* 3781 (BR; K, holotype).

Terrestrial herb 30–65 cm tall; tubers up to 5.5 cm long, ovoid or ellipsoid, tomentose.
Stem erect, with 2 basal leaves and many loosely sheathing stem-leaves. Basal leaves
appressed to ground, the larger 3–5 × 4–7 cm, reniform, apiculate or rounded;
sheathing stem-leaves c. 5 cm long. Inflorescence laxly 2–4-flowered; bracts 40–50 mm
long. Flowers erect, sepals green, petals and lip white. Pedicel 35–60 mm long, ovary
25–30 mm long. Dorsal sepal erect, 20–30 × 7–11 mm, elliptic; lateral sepals deflexed,
24–35 × 7–10 mm, obliquely semi-ovate, acuminate; all sepals 3-nerved, the middle
nerve with a scabrous keel. Petals 2-lobed ± to base; posterior lobe erect or spreading,
24–33 × c. 2 mm, linear; anterior lobe spreading, 40–60 mm long, c. 1 mm wide at base,
becoming subulate. Lip 3-lobed to c. 2 mm from base; mid-lobe pendent, 30–40 × 2 mm,
linear; side lobes divergent, 40–55 mm long, 2 mm wide at base, tapering to become
subulate; spur 14–25 cm long, pendent, tucked into bracts; slender but swollen to c. 3
mm in diameter near apex, but tapering at tip. Anther erect, 13–14 mm high; canals 6–7
mm long. Auricles c. 3 mm long. Stigmatic arms 8–11 mm long, stout, coiled at apex.

Zambia. N: Mbala Distr., Ndundu, 1740 m, fl. 24.i.1962, *Richards* 15944 (K). W: Mwinilunga
Distr., slope east of Kewumbo (Kaoomba) R., fl. 22.xii.1937, *Milne-Redhead* 3781 (K).
Also in Zaire and Tanzania. Wet dambo and dambo margins, in long grass and in woodland,
also in dry ground at side of gorge, amongst rocks; 1500–1750 m.

97. **Habenaria armatissima** Rchb.f., Otia Bot. Hamburg.: 98 (1881). —Rolfe in F.T.A. **7**: 227
(1898). —Summerhayes in Bull. Misc. Inform., Kew **1932**: 339 (1932); in F.W.T.A. ed. 2,
3: 198 (1968); in F.T.E.A., Orchidaceae: 128 (1968). —Grosvenor in Excelsa **6**: 82 (1976).
—Williamson, Orch. S. Centr. Africa: 66 (1977). —Stewart et al., Wild Orch. South.
Africa: 94 (1982). —Geerinck in Fl. Afr. Centr., Orchidaceae pt. 1: 76 (1984). —la Croix
et al., Orch. Malawi: 90 (1991). TAB. **44**. Type from Ethiopia.

Terrestrial herb up to 80 cm tall; tubers up to 5 cm long, ovoid or ellipsoid. Stem
erect, with 2 basal leaves and several bract-like stem leaves along its length. Basal leaves
appressed to the ground, the larger 12–18 × 17.5 × 25 cm, ovate to reniform, cordate
at base, apiculate or rounded, rather fleshy; stem-leaves c. 3 cm long. Inflorescence
16–30 × 7–12 cm, densely 10- to many-flowered; bracts 15–30 mm long. Flowers white,
± spreading. Pedicel 15–20 mm long, ovary arched, 30–35 mm long. Dorsal sepal
erect, 11–16 × 5–8 mm, ovate, convex, acuminate, the tip sometimes slightly reflexed;
lateral sepals reflexed, 13–17 × 4–7 mm, obliquely ovate, acuminate. Petals 2-lobed ±
to base, both lobes filiform, less than 1 mm wide at base; posterior lobe 11–14 mm
long, ± adnate to dorsal sepal; anterior lobe 30–40 mm long, deflexed or spreading.
Lip 3-lobed with a basal, undivided part 2–3 mm long; mid-lobe linear, 15–20 mm
long, less than 1 mm wide; side lobes 30–45 mm long, filiform; all lobes pendent; spur
8.5–21.5 cm long, slender or somewhat sigmoid. Anther 6–9 mm high, erect,
apiculate; canals 3.5–5 mm long, upcurved. Stigmatic arms 5.5–10 mm long, porrect,
club-shaped, truncate at tip. Rostellum 4–6 mm high, narrowly triangular, acute.

Zambia. N: Mbala Distr., Chinakila to Loye Flats, 1200 m, fl. 10.i.1965, *Richards* 1946 (K). W:
Ndola Distr., fl. 18.i.1957, *Fanshawe* 2942 (K; NDO). **Zimbabwe**. E: Chimanimani Distr.,
Umvumvumwe Gorge, at old drift, 700 m, fl. 20.i.1957, *Chase* 6295 (K; SRGH). **Malawi**. N:
Nkhata Bay Distr., lower Kandoli Mts. near Lumpheza R., 620 m, fl. 28.ii.1987, *la Croix* 987 (K).
S: Chikwawa (Shibisa), R. Shire, fl. iii.1859, *Kirk* s.n. (K).
Widespread in tropical Africa and South Africa. Grassland and dry deciduous woodland,
often on termite mounds; 600–1200 m.

Tab. 44. HABENARIA ARMATISSIMA. 1, habit (×⅓); 2, inflorescence (×1); 3, dorsal sepal, side view (×2); 4, lateral sepal (×2); 5, petal (×2); 6, column, with part of lip, spur and ovary (×2), 1–6 from *Polhill & Paulo* 1370. Drawn by Heather Wood. From F.T.E.A.

98. **Habenaria subarmata** Rchb.f., Otia Bot. Hamburg.: 98 (1881). —Rolfe in F.T.A. **7**: 227 (1898). —Summerhayes in F.T.E.A., Orchidaceae: 128 (1968). —Grosvenor in Excelsa **6**: 83 (1976). —Williamson, Orch. S. Centr. Africa: 73 (1977). —Geerinck in Fl. Afr. Centr., Orchidaceae pt. 1: 72 (1984). —la Croix et al., Orch. Malawi: 98 (1991). Type: Mozambique, near Tete, *Kirk* s.n. (K, holotype).

Robust terrestrial herb 30–60 cm tall; tubers 1–2 cm long, ± globose, pubescent. Stem erect, with 2 basal leaves and 6–10 bract-like leaves along its length. Basal leaves appressed to the ground, 7.5–16 × 7–22 cm, reniform to suborbicular, apiculate to slightly emarginate, pale green, rather fleshy; bract-like leaves up to 5 cm long. Inflorescence 10–27 × 5–9 cm, densely 12- to many-flowered; bracts 10–20 mm long, lanceolate, acute. Flowers white, semi-erect or spreading. Pedicel 10 mm long, ovary 25–30 cm long with 6 lightly toothed keels. Dorsal sepal erect, 7.5–11 × 4–6 mm, ovate, convex; lateral sepals reflexed, 10–12 × 4–6 mm, very obliquely ovate. Petals 2-lobed ± to base; posterior lobe 7.5–9.5 × c. 1 mm, linear, lying beside dorsal sepal; anterior lobe 25–40 mm long, filiform, spreading or deflexed. Lip 3-lobed to 1–2 mm from base; mid-lobe 12–18 × 1 mm, linear, deflexed; side lobes 25–50 mm long, filiform, ± deflexed; spur 5–7.5 cm long, slender, incurved. Anther 4.5–7.5 mm high, erect, apiculate; canals upcurved, 3–4 mm long. Stigmatic arms 5–7 mm long, porrect, club-shaped, truncate at apex.

Caprivi Strip. Eastern Caprivi, Ipalela Is., 394 m, fl. 15.iii.1976, *Du Preez* 15 (PRE). **Zambia**. B: Shesheke Distr., Masese Forest Reserve, 1050 m, fl. 3.iii.1975, *Brummitt, Chisumpa & Polhill* 14247 (K). N: Mbala Distr., Kalambo Falls, 1200 m, fl. 8.ii.1965, *Richards* 19606 (K). C: Mulungushi, fl. 9.ii.1964, *Fanshawe* 8251 (K). S: banks of Zambezi R. above Victoria Falls, fl. 26.i.1975, *Williamson & Gassner* 2363 (K). **Zimbabwe**. W: Victoria Falls Distr., fl. i.1904, *Allen* 139 (K). S: Masvingo (Fort Victoria), fl. 1909–12, *Monro* (BM). **Malawi**. N: Mzimba Distr., c. 10 km north of Mpherembe, 1180 m, fl. 7.ii.1987, *la Croix* 955 (K). C: Lilongwe Nature Sanctuary, Zone A, fl. 28.i.1985, *Patel & Banda* 1968 (K; MAL). **Mozambique**. T: near Tete, fl. 1.ii.1860, *Kirk* s.n. (K).
Also in Zaire, Kenya and Tanzania. Dry deciduous woodland; 350–1200 m.
This species closely resembles *H. armatissima*, differing only in the slightly smaller flower and shorter spur.

99. **Habenaria trilobulata** Schltr. in Bot. Jahrb. Syst. **26**: 332 (1899). —Summerhayes in Kew Bull. **2**: 125 (1948); in F.T.E.A., Orchidaceae: 131 (1968). —Grosvenor in Excelsa **6**: 83 (1976). —Stewart et al., Wild Orch. South. Africa: 95 (1982). —la Croix et al., Orch. Malawi: 98 (1991). Type: Mozambique, "25 Miles Station", *Schlechter* 12251 (B†, holotype; K; PRE; S). *Habenaria quadrifila* Schltr. in Bot. Jahrb. Syst. **53**: 504 (1915). Type from Tanzania.

Slender terrestrial herb 20–35 cm tall; tubers 1–2.5 × 1–2 cm, ellipsoid or ± globose, tomentose. Stem erect, with 2 basal leaves and 5–8 bract-like stem leaves scattered along its length. Basal leaves appressed to the ground, the larger 4.5–7.5 × 5.5–7 cm, ovate to ± orbicular, acute or apiculate, cordate at base, silvery-green with dark green reticulate veining, occasionally dark green; bract-like stem leaves up to 2 cm long. Inflorescence 6–13 cm long, c. 5 cm in diameter, laxly 3–12-flowered; bracts 8–10 mm long. Flowers white or pale green. Ovary and pedicel arched, 20–25 mm long. Dorsal sepal erect, 5–6.5 × 4–5 mm, ovate, convex; lateral sepals reflexed, 5–7 × 2–3 mm, obliquely semi-elliptic. Petals 2-lobed to 3–4 mm from base; posterior lobe erect, 3–4 × c. 2 mm, ovate, rounded, papillose; anterior lobe 17–27 mm long, filiform, the base c. 0.5 mm wide, deflexed or spreading and curving up. Lip 3-lobed with a basal undivided part 2–3 mm long; mid-lobe decurved, 6–8 × 1 mm, linear; side lobes 20–27 mm long, filiform, similar to anterior petal lobes; spur 2–3.5 cm long, slightly inflated at apex, somewhat incurved. Anther c. 3 mm high, erect, apiculate; canals c. 4 mm long. Stigmas 2–3 mm long, club-shaped, lying along base of lip. Rostellum projecting in front of anther, 3-lobed at apex, the mid-lobe longer than the side lobes.

Zimbabwe. E: Vumba, Witchwood, 800 m, fl. 23.iii.1955, *Ball* 513 (K; SRGH). **Malawi**. N: Nkhata Bay Distr., foot of Kandoli Mts., 650 m, fl. 28.ii.1987, *la Croix* 988 (K; MAL). **Mozambique**. Z: Milange-Quelimane road, fl. 20.v.1949, *Barbosa & Carvalho* 2770 (K). MS: forest near Dondo, 60 m, fl. 16.iv.1961, *Ball* 929 (K; SRGH); Garuso Mt., 800 m, *Jacobsen* 3858 (PRE).
Also in Kenya, Tanzania and South Africa (E Transvaal). *Brachystegia* or mixed deciduous woodland and riverine forest; 60–800 m.

100. **Habenaria dregeana** Lindl. in Ann. Nat. Hist. **4**: 314 (1840). —Rolfe in F.C. **5**, 3: 135 (1912). —Summerhayes in Kew Bull. **17**: 521 (1964); in F.T.E.A., Orchidaceae: 132 (1968). —Grosvenor in Excelsa **6**: 83 (1976). —Stewart et al., Wild Orch. South. Africa: 96 (1982). —Geerinck in Fl. Afr. Centr., Orchidaceae pt. 1: 70 (1984). Type from South Africa.
Habenaria calva (Rchb.f.) Rolfe in F.T.A. **7**: 226 (1898). Type from Angola.
Habenaria friesii Schltr. in R.E. Fries, Wiss. Ergebn. Schwed. Rhod.-Kongo-Exped. 1911–12, **1**: 240 (1916). Type from Zaire.

Terrestrial herb 15–30 cm tall; tubers up to 2 × 1 cm, ellipsoid, tomentose. Stem erect, with 2 basal leaves and several to many bract-like stem leaves along its length. Basal leaves appressed to the ground, the larger 3.5–5 × 4.5–5.5 cm, ovate to reniform, cordate at base, apiculate to rounded; stem leaves c. 25 mm long. Inflorescence 7.5–18 × 1.5–2 cm, narrowly cylindrical, densely many-flowered; bracts 13–20 mm long. Ovary and pedicel 10–11 mm long. Flowers yellow, yellow-green or green. Dorsal sepal erect, 4–7 × 3–4 mm, ovate, acute, convex; lateral sepals spreading, 4.5–7 × c. 3 mm, obliquely ovate, acuminate. Petals 2-lobed ± to base; posterior lobe 3.5–6 × 1–2 mm, adnate to dorsal sepal; anterior lobe 1–3 × c. 0.5 mm, curving upwards. Lip projecting forwards, 3-lobed with a basal undivided part 1–2 mm long, the lobes usually glabrous but rarely papillose or ciliolate; mid-lobe 4–6 mm long, c. 1 mm wide, linear; side lobes shorter and narrower, 1.5–4.5 × c. 0.5 mm; spur 8.5–10 mm long, tapering from a wide base but swollen at the apex. Anther erect, 1.5–2 mm high; canals c. 1 mm long. Stigmas porrect, narrowly club-shaped, 2–3 mm long.

Zimbabwe. C: Goromonzi Distr., fl. 26.iii.1961, *Herd* in *GHS* 122618 (K; SRGH). E: Chimanimani Distr., open woodland, 1500 m, fl. 25.iii.1953, *Crook* 461 (K; SRGH).
Also in Zaire, Uganda, Angola and South Africa. Short grassland and open woodland; c. 1500 m.

101. **Habenaria galactantha** Kraenzl. in Bot. Jahrb. Syst. **48**: 387 (1912). —Summerhayes in F.T.E.A., Orchidaceae: 126 (1968). —Grosvenor in Excelsa **6**: 83 (1976). —Williamson, Orch. S. Centr. Africa: 67 (1977). —Geerinck in Fl. Afr. Centr., Orchidaceae pt. 1: 66 (1984). —la Croix et al., Orch. Malawi: 91 (1991). Type from Tanzania.

Terrestrial herb 45–80 cm tall; tubers up to 2.5 cm long, ± globose to ovoid. Stem erect, with 2 basal leaves and c. 6 sheath-like stem leaves along its length. Basal leaves appressed to the ground, 4–8 × 2.5–6 cm, broadly ovate to orbicular, apiculate, sometimes cordate at base; stem leaves up to 5.5 cm long. Inflorescence 9–18 cm long, rather laxly 5–13-flowered; bracts 12–25 mm long. Ovary and pedicel 20–30 mm long. Sepals greenish-white, papillose or pubescent on outside, petals and lip white. Dorsal sepal erect, 6–9 × 3.5–5 mm, ovate, convex; lateral sepals deflexed, 8–11 × 3.5–6 mm, obliquely semi-ovate, acute. Petals 2-lobed almost to base; posterior lobe 5.5–9.5 × 3–4 mm, curved lanceolate, acute, lying inside dorsal sepal; anterior lobe deflexed, 6.5–12 × 2.5–3.5 mm, oblanceolate, obtuse. Lip 3-lobed to 1–2 mm from base, all lobes oblanceolate, obtuse; mid-lobe 7–13 × 2.5–3 mm; side lobes 5.5–12 × 2.5–4 mm, slightly incurved; spur 3–11.5 cm long, slender but swollen at apex, often caught inside bracts. Anther 4–5.5 mm high, erect, apiculate; canals very short, c. 0.5 mm long. Stigmas 2–3 mm long, club-shaped.

Zambia. N: Hauser's Farm near Mbala, c. 1500 m, fl. 24.i.1955, *Richards* 4259 (K). W: Solwezi Distr., fl. 20.i.1960, *Holmes* 0250 (K; SRGH). **Malawi.** N: North Viphya, near Mphompha on road to Uzumara, 1300 m, fl. 30.i.1986, *la Croix* 790 (K). C: Lilongwe Distr., Bunda Forest, c. 1180 m, fl. 18.i.1974, *Allen* 463 (K). S: Zomba Distr., road to Lake Chilwa, c. 700 m, fl. 29.xii.1970, *Moriarty* 405 (MAL).
Also in Zaire, Burundi and Tanzania. *Brachystegia* woodland; 700–1500 m.

102. **Habenaria macrura** Kraenzl. in Bot. Jahrb. Syst. **16**: 152 (1892). —Rolfe in F.T.A. **7**: 229 (1898). —Summerhayes in F.W.T.A. **2**: 410 (1936); in ed. 2, **3**: 198 (1968); in F.T.E.A., Orchidaceae: 126 (1968). —Williamson, Orch. S. Centr. Africa: 67 (1977). —Geerinck in Fl. Afr. Centr., Orchidaceae pt. 1: 73 (1984). —la Croix et al., Orch. Malawi: 94 (1991). Type from Angola.
Habenaria pentaglossa Kraenzl. in Bot. Jahrb. Syst. **33**: 55 (1902). Type from Tanzania.

Terrestrial herb 15–70 cm tall; tubers up to 5 cm long, ellipsoid or almost globose, tomentose. Stem erect, with 2 basal leaves and 5–6 rather loose sheath-like stem leaves. Basal leaves appressed to the ground, 3–7.5 × 2–5.5 cm, ovate to almost orbicular, slightly cordate at base, acute, rounded or apiculate; stem leaves up to 5 cm long. Inflorescence laxly 1–11-flowered; bracts 25–40 mm long. Flowers white with a spicy, carnation-like scent. Pedicel 20–35 mm long, ovary 15–35 mm long. Dorsal sepal erect, 9–13 × 4.5–8 mm, ovate, acute, papillose on the outside; lateral sepals reflexed, 9–16.5 × 4–8.5 mm, obliquely lanceolate, apiculate, slightly rolled up lengthways. Petals 2-lobed almost to base; posterior lobe erect, 7.5–14 × 2.5–7.5 mm, curved lanceolate, ± adnate to dorsal sepal; anterior lobe deflexed or spreading, 9.5–20 × 2.5–7 mm, obliquely oblanceolate, rounded or subacute. Lip 17–27 mm long, 3-lobed with an undivided base 2.5–6.5 mm long; mid-lobe 11–23 × 3.5–5.5 mm, oblanceolate, acute or obtuse; side lobes 9–23 × 3–8 mm, obliquely oblanceolate, rounded; spur 9–20 cm long, slender, pendent, usually tucked into the bracts and stem sheaths. Anther 5–6 mm high, erect, apiculate; canals c. 1 mm long. Stigmatic arms 3–4 mm long, porrect, stout, the ends truncate.

Zambia. W: Mwinilunga Distr., south of Matonchi, fl. 20.xii.1937, *Milne-Redhead* 3750 (K). C: Kabwe (Broken Hill), fl. 15.i.1961, *Morze* 41 (K). E: Lundazi Distr., Nyika Plateau, 2100 m, fl. 3.i.1959, *Richards* 10431 (K). S: Mazabuka Distr., Mabwingombe Hills, fl. 5.ii.1960, *White* 684 (FHO; K). **Zimbabwe**. N: Makonde Distr., fl., *Greatrex* 82 (PRE). **Malawi**. N: Mzimba Distr., south of Mpherembe, c. 1300 m, fl. 17.i.1987, *la Croix* 940 (K; MAL). S: Blantyre Distr., Nyambadwe Hill, fl. 4.i.1945, *Benson* 1079 (K).
Widespread in tropical Africa. Montane grassland and *Brachystegia* woodland on stony soil; 1200–2400 m.

103. **Habenaria stylites** Rchb.f. & S. Moore in J. Bot. **16**: 136 (1878). —Rolfe in F.T.A. **7**: 226 (1898). —Summerhayes in Kew Bull. **17**: 525 (1964); in F.T.E.A., Orchidaceae: 129 (1968). —Williamson, Orch. S. Centr. Africa: 66 (1977). —la Croix et al., Orch. Malawi: 97 (1991). Type from Kenya.

Terrestrial herb 20–55 cm tall; tubers up to 2.5 × 1.5 cm, ellipsoid or ± globose, tomentose. Stem erect, with 2 basal leaves and 4–9 appressed bract-like stem leaves along its length. Basal leaves appressed to the ground, 3–9 × 2–11 cm, ovate to reniform, cordate at base, apiculate or rounded, fleshy; stem leaves up to 3 cm long. Inflorescence 3–20 × 3.5–6.5 cm, rather densely but sometimes loosely few- to many-flowered; bracts 10–20 mm long. Flowers suberect or curving out, white, scented. Ovary and pedicel 20–30 mm long. Dorsal sepal erect, 7–12 × 3–7 mm, elliptic, convex; lateral sepals spreading, 10–16 × 5–8 mm, obliquely semi-ovate, acuminate. Petals 2-lobed almost to base, posterior lobe erect, 6–12 × 1–2 mm, linear, adnate to dorsal sepal; anterior lobe 8.5–15 mm long, c. 1 mm wide, narrowly linear or filiform, spreading outwards and upwards. Lip 12–21 mm long, 3-lobed with an undivided basal part 2–11 mm long; mid-lobe 6–14 × 1.5–5 mm, linear, ligulate or oblong-lanceolate, acute or subacute; side lobes 5.5–19 × 1–2.5 mm, slightly curved, acute; spur 2.5–6 cm long, slender, parallel to ovary. Anther 4 mm high, erect, rounded at apex; canals 2–2.5 mm long, porrect or upcurved. Stigmas 4.5–7 mm long, porrect, club-shaped with a truncate apex. Rostellum mid-lobe 2–4 mm long, narrowly triangular.

Claw of lip as long as, or longer than the lobes; lobes of lip relatively broad · subsp. *johnsonii*
Claw of lip shorter than lobes; lobes of lip narrow · · · · · · · · · · · · · · · · · · subsp. *rhodesiaca*

Subsp. **rhodesiaca** Summerh. in Kew Bull. **17**: 527 (1964). —Grosvenor in Excelsa **6**: 83 (1976). Type: Zimbabwe, Harare Distr., Rumani, *Wild* 3763 (K, holotype; SRGH).

Lip claw 3–5.6 mm long; mid-lobe 7–11 × 1.5–3 mm; side lobes 5.5–14 × 1–2 mm; spur 2.6–3(5) cm long.

Zambia. E: Chipata Distr., Chamchenga Dam forest, near Chadiza, 1200 m, fl. 3.iii.1973, *Kornaś* 3393 (K). S: c. 60 km south of Lusaka, fl. ii.1967, *G. Williamson* 261 (K). **Zimbabwe**. N: Mazowe Distr., Shamva, 900 m, fl. 10.ii.1957, *Chase* 6309 (K; SRGH). C: Harare Distr., Rumani,

1260 m, fl. 22.ii.1952, *Wild* 3763 (K; SRGH). S: Chibi Distr., near Madzivire Dip, c. 6 km north of Runde (Lundi) R., fl. 4.v.1962, *Drummond* 7927 (K; SRGH). **Malawi**. S: Mangochi Distr., Phirilongwe Mt., 950 m, fl. 17.iii.1985, *Johnston-Stewart* 403 (K). **Mozambique**. N: Mandimba, fl. 24.ii.1942, *Hornby* 3710 (PRE). MS: Chimoio Distr., 16 km north of Vanduzi, fl. 27.iv.1962, *Chase* 7702 (K; SRGH).

Only in the Flora Zambesiaca area. *Brachystegia* woodland, often on rocky slopes and on old termite mounds; 900–1300 m.

Subsp. **johnsonii** (Rolfe) Summerh. in Kew Bull. **17**: 527 (1964). Type: Mozambique, Mts. east of Lake Malawi (Nyasa), *Johnson* s.n. (K, holotype).
  Habenaria johnsonii Rolfe in F.T.A. **7**: 57 (1898).
  Habenaria narcissiflora Kraenzl. in Bot. Jahrb. Syst. **30**: 282 (1901). Type from Tanzania.

Claw of lip 8–11 mm long; mid-lobe 6–8 × 3–5 mm; side lobes 7.5–8 × 2–3 mm; spur 2–4.5 cm long.

Mozambique. N: Mts. east of Lake Malawi (Nyasa), *Johnson* s.n. (K).
Also in Tanzania. Open grassland; 800 m.
Subsp. *stylites*, which has larger flowers with a longer spur, occurs in Kenya and Tanzania.

104. **Habenaria lithophila** Schltr. in Bot. Jahrb. Syst. **53**: 504 (1915). —Summerhayes in Kew Bull. **17**: 525 (1964); in F.T.E.A., Orchidaceae: 132 (1968). —Williamson, Orch. S. Centr. Africa: 67 (1977). —Stewart et al., Wild Orch. South. Africa: 96 (1982). —Geerinck in Fl. Afr. Centr., Orchidaceae pt. 1: 73 (1984). —la Croix et al., Orch. Malawi: 93 (1991). Type from Tanzania.
  Habenaria macowaniana Kraenzl. in Bot. Jahrb. Syst. **16**: 150 (1892), non (Rchb.f.) N.E.Br. (1889) nom. illegit. Type from South Africa.
  Habenaria petri Schltr. in Ann. Transvaal Mus. **10**: 245 (1924). Type as for *H. macowaniana* Kraenzl.

Terrestrial herb 10–30 cm tall; tubers 1.5–2.5 cm long, ellipsoid, tomentose. Stem erect, with 2 basal leaves and several to many stem leaves along its length. Basal leaves 2, appressed to the ground, 1.5–5.5 × 1.5–7 cm, ovate to reniform, apiculate or rounded, cordate at base; stem leaves ± appressed to stem, 15–25 mm long, linear or lanceolate. Inflorescence 4.5–9 × 2–3 cm, fairly densely many-flowered; bracts c. 10 mm long. Flowers green or yellow-green, suberect or spreading. Ovary and pedicel 8–10 mm long. Dorsal sepal erect, 5–7.5 × 2.5–4 mm, ovate, convex, papillose on the margins and keels; lateral sepals deflexed or spreading, 5.5–8.5 × 2–3.5 mm, obliquely lanceolate, acute or acuminate. Petals 2-lobed ± to base, glabrous, puberulous or ciliolate; posterior lobe erect, 4–6.5 × 1–2 mm, linear-ligulate; anterior lobe 4.5–10 × 0.5–1 mm, narrowly linear, curving upwards. Lip projecting forwards, glabrous or puberulous, 3-lobed almost to the base, all lobes linear; mid-lobe 6–9.5 × 1–1.5 mm; side lobes 5.5–8 mm long, c. 0.5 mm wide; spur 8–17 mm long, slender, slightly swollen towards apex, parallel to ovary. Anther erect, c. 2 mm high; canals c. 1 mm long, slightly upcurved. Stigmas 1.5–3 mm long, clavate, porrect or lying along base of lip.

Malawi. N: Nyika Plateau, c. 2400 m, fl. & fr. 14.ii.1960, *Holmes* 0234 (K; SRGH).
Also in Zaire, Tanzania and South Africa. Stony montane grassland; c. 2400 m.

105. **Habenaria livingstoniana** la Croix & P.J. Cribb in Kew Bull. **48**, 2: 359 (1993). TAB. **45**. Type: Malawi, Rumphi Distr., Livingstonia Escarpment, *I. & E. la Croix* 1015 (K, holotype; MAL).
  Habenaria sp., la Croix et al., Orch. Malawi: 101 (1991).

Terrestrial herb 15–25 cm tall; tubers c. 2 × 1 cm, ellipsoid, smooth. Stem ribbed, with 2 basal leaves and 5–9 bract-like stem leaves. Basal leaves appressed to the ground, the larger up to 6 × 6 cm, dark green, glabrous, ± orbicular, apiculate; the lowermost stem leaves loosely sheathing, the rest up to 2 cm long, lanceolate, acute, pale green, appressed to stem. Inflorescence compact, c. 3.5 × 3.5 cm, more or less pyramidal, densely c. 20-flowered; bracts 10 × 5 mm, lanceolate, acute. Flowers glistening white, spur greenish at tip. Ovary and pedicel arched, 15 mm long. Dorsal sepal erect, 8 × 6 mm, ovate, convex, acute; lateral sepals spreading, 8 × 4 mm, obliquely ovate, acuminate. Petals 2-lobed to 1 mm from base; posterior lobe 8 × 1 mm, linear, acute, adnate to dorsal sepal; anterior lobe 13 mm long, filiform, curving

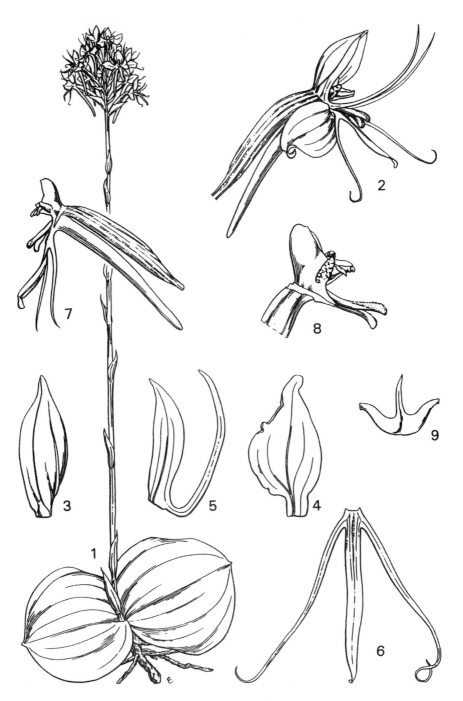

Tab. 45. HABENARIA LIVINGSTONIANA. 1, habit ($\times\frac{2}{3}$); 2, flower ($\times$3); 3, dorsal sepal ($\times$5); 4, lateral sepal ($\times$5); 5, petal ($\times$5); 6, lip ($\times$5); 7, column and lip, side view ($\times$3); 8, column ($\times$5); 9, rostellum ($\times$5), 1–9 from *la Croix* 1015. Drawn by Eleanor Catherine. From Kew Bull.

up. Lip projecting forwards, 3-lobed to 1 mm from base; mid-lobe 10–12 × 1.5 mm, linear-ligulate, acute; side lobes about the same length but becoming filiform from a base c. 0.5 mm wide; spur 15 mm long, tapering from a wide mouth, parallel to ovary. Anther erect, c. 2 mm high; canals 2 mm long, upcurved. Stigmas 4 mm long, club-shaped, lying along lip. Mid-lobe of rostellum narrowly triangular, 3 mm long.

Malawi. N: Rumphi Distr., Livingstonia Escarpment, 950 m, fl. 6.iii.1987, *I. & E. la Croix* 1015 (K; MAL).
Apparently endemic, known only from the eastern escarpment of the Nyika Plateau. Stony soil on slope in *Brachystegia* woodland; 950–1000 m.

106. **Habenaria mosambicensis** Schltr. in Bot. Jahrb. Syst. **26**: 331 (1899). Type: Mozambique, c. 16 km from Beira, *Schlechter* s.n. (B†, holotype).

Terrestrial herb c. 30 cm tall. Stem erect, with 2 basal leaves and several lax, acuminate sheathing stem leaves. Basal leaves appressed to the ground, 3–3.5 cm long, suborbicular, obtuse, glabrous. Inflorescence 10–13 × 2–3 cm, cylindrical, 10–20-flowered. Sepals green, petals and lip white. Lateral sepals c. 12 × 4 mm, obliquely ovate-lanceolate, subacute; dorsal sepal similar, with keeled mid-vein. Petals 2-lobed, both lobes erect, narrowly linear, acute, glabrous; posterior lobe subfalcate, adnate to dorsal sepal; anterior lobe 13–15 mm long. Lip 3-lobed, the lobes glabrous, filiform-linear, the side lobes slightly longer than the mid-lobe, similar to the anterior petal lobe; spur c. 10 mm long, pendent, swollen towards apex, obtuse. Anther apiculate, canals upturned. Stigmas 5–6 mm long. Rostellum mid-lobe subulate, acuminate.

Mozambique. MS: c. 16 km from Beira, fl. v.1895, *Schlechter* s.n. (B†).
Not known elsewhere.
From Schlechter's description, this species sounds very similar to *H. lindblomii*, differing only in the mid-lobe of the lip being slightly shorter, rather than slightly longer than the side lobes, and in the spur being only 1 cm long.

107. **Habenaria tysonii** Bolus in J. Linn. Soc., Bot. **25**: 166, 167 (1890). —Rolfe in F.C. **5**, 3: 136 (1912). —Grosvenor in Excelsa **6**: 83 (1976). —Stewart et al., Wild Orch. South. Africa: 95 (1982). Type from South Africa.

Terrestrial herb 20–30 cm tall. Stem erect, with 2 basal leaves and many bract-like stem leaves. Basal leaves appressed to the ground, 2.5–3 × 2.5–5 cm, suborbicular or reniform, apiculate, fleshy; stem leaves c. 2 cm long, lanceolate, acuminate. Inflorescence c. 9 cm long, fairly laxly many-flowered; bracts 6–10 mm long. Flowers green, spreading. Ovary and pedicel 12–14 mm long. Dorsal sepal erect, 5–8 mm long, ovate, convex; lateral sepals 8–9 mm long, semi-ovate, acuminate. Petals 2-lobed, posterior lobe slightly shorter than dorsal sepal, falcate-linear, acute, slightly ciliate; anterior lobe c. 10 mm long, filiform, curved, hispid. Lip 3-lobed almost to base, somewhat hispid; mid-lobe 10–12 mm long, linear, acute, curved; side lobes slightly shorter and narrower, diverging; spur 12–14 mm long, inflated towards apex. Stigmas 2 mm long, clavate, oblong.

Zimbabwe. E: Chimanimani (Melsetter), Sibu, Charter Forest Estate, fl. 10.iv.1968, *Ball* 1187 (SRGH).
Also in South Africa. In grass on rocky hillsides.
This species resembles *H. dregeana* but has larger flowers.

108. **Habenaria decurvirostris** Summerh. in Bot. Mus. Leafl. Harv. Univ. **10**: 275 (1942); in Kew Bull. **17**: 523 (1964). —Williamson, Orch. S. Centr. Africa: 64 (1977). Type: Zambia, Mwinilunga Distr., Dobeka Bridge, 14.xii.1937, *Milne-Redhead* 3664 (K, holotype).

Terrestrial herb 20–30 cm tall; tubers c. 2 cm diameter, globose. Stem erect, with 1 basal leaf and several to many bract-like stem leaves. Basal leaf appressed to the ground, 2–3 × 2.5–3.5 cm, heart-shaped; stem leaves c. 15 mm long, linear, acute. Inflorescence 10–11 × 3 cm, densely many-flowered; bracts c. 15 mm long. Flowers yellow-green. Pedicel 3–6 mm long, ovary arched, 8–11 mm long. Dorsal sepal erect, 3–3.5 × 2.5 mm, ovate, obtuse, convex; lateral sepals deflexed, 5.5 × 3–3.5 mm, obliquely semi-ovate, acute. Petals 2-lobed ± to base; posterior lobe erect, 3.5 × 0.7

mm, lanceolate-linear, adnate to dorsal sepal; anterior lobe 5.5–6.5 × 0.5–0.8 mm, curving downwards, linear, obtuse, rather fleshy. Lip rather fleshy, 3-lobed almost to base; mid-lobe 4.5–5.5 × 1–3 mm, spathulate-ligulate, the lower edge slightly recurved; side lobes 4.5–5.5 × 0.8–1.2 mm, slightly diverging, ligulate or oblanceolate; spur 9–10 mm long, swollen at apex, pendent or ± incurved. Column 3 mm high; anther loculi parallel, 2 mm long; canals 1.5 mm long. Stigmatic arms 2–2.5 mm long, curving down.

Zambia. W: Mwinilunga Distr., c. 1 km SW of Dobeka Bridge, fl. 14.xii.1937, *Milne-Redhead* 3664 (K).
Apparently endemic to the Mwinilunga area, so far known from only one collection. Sandy ground in open, near edge of *Cryptosepalum* woodland.

109. **Habenaria hirsutissima** Summerh. in Kew Bull. **17**: 521 (1964). TAB. **46**. Type: Mozambique, 10 km from Mutuáli, *Gomes e Sousa* 4189 (K, holotype; PRE).

Terrestrial herb 20–30 cm tall, ± whole plant hairy; tubers 8–17 × 7–10 mm, ellipsoid or globose. Stem erect, with 1 basal leaf and several bract-like stem leaves along its length. Basal leaf appressed to the ground, c. 3 × 3 cm, orbicular, cordate, ± glabrous above and below, but densely hairy on margins. Inflorescence 5 × 3 cm, fairly laxly c. 7-flowered; bracts c. 6 mm long. Flowers greenish-white. Ovary and pedicel 12–15 mm long. Dorsal sepal erect, 8 × 6 mm, ovate, subacute, convex; lateral sepals patent, 8.5 × 4 mm, obliquely semi-orbicular, convex, subacute or apiculate; all sepals very hairy. Petals 2-lobed ± to base, long ciliate; posterior lobe 7 × 1.5 mm, erect, curved; anterior lobe 8–8.5 × 0.5 mm, patent or ascending. Lip hairy, 3-lobed almost to base; mid-lobe deflexed, 10 × 1.5 mm, linear; side lobes 12–13 × 0.5 mm, curving up like anterior petal lobe; spur 2 cm long, slightly inflated at apex, parallel to ovary, then slightly recurved. Anther 2.5–3 mm high; canals 1.75 mm long. Stigmatic arms 2 mm long, thick.

Mozambique. N: Mutuali–Malema road, 10 km from Mutuali, fl. ii.1954, *Gomes e Sousa* 4189 (K).
In the type description, Summerhayes said that this species, which is known from only one collection in Mozambique, is similar to *H. leucotricha* but has hairy petals and lip and the anterior petal lobe is patent or ascending, not deflexed. The anther is shorter and the canals relatively longer than in *H. leucotricha.*

110. **Habenaria leucotricha** Schltr. in Bot. Jahrb. Syst. **53**: 506 (1915). —Mansfeld in Fedde, Repert. Spec. Nov. Regni Veg., Beih. 68, t. 29/113 (1932). —Summerhayes in F.T.E.A., Orchidaceae: 123 (1968). —Williamson, Orch. S. Centr. Africa: 61 (1977). —la Croix et al., Orch. Malawi: 92 (1991). Type from Tanzania.

Terrestrial herb 10–20 cm tall; tubers 1–2 cm long, ellipsoid, tomentose. Stem hairy with 1 basal leaf and 2–3 bract-like stem leaves. Basal leaf appressed to the ground, 2.5–4 × 6 cm, heart-shaped, silvery-green with dark green veining, the upper surface covered with scattered white hairs c. 4 mm long. Inflorescence 3–10 cm long and c. 2.5 cm wide, densely 3–12-flowered; bracts 10–12 mm long, whitish-green, densely hairy. Flowers suberect, white with the tips of the lower petal lobes and lip side lobes green. Ovary and pedicel c. 13 mm long. Dorsal sepal 8–12 × 5 mm, elliptic, obtuse, convex; lateral sepals spreading, 9–11 × 3.5–4.5 mm, obliquely elliptic-lanceolate, acute; all sepals hairy on outside. Petals 2-lobed ± to base, glabrous; posterior lobe erect, 8–10 × 1–1.8 mm, curved linear-lanceolate, acute, ± adnate to dorsal sepal; anterior lobe deflexed, 8–12 × c. 0.5 mm, linear. Lip 3-lobed to c. 1 mm from base; mid-lobe 9.5–12 × 1.5–2.5 mm, deflexed, linear, obtuse; side lobes 9–14 × 1 mm, upcurved; spur 12–16.5 mm long, the basal 5 mm straight, then sharply incurved, the terminal 5 mm abruptly inflated but tapering to an acute apex, or else the whole spur ± straight, the apex inflated but obtuse. Anther 3–4.5 mm high, erect, apiculate; canals 1–1.5 mm long, slightly upcurved. Stigmatic arms 1–2 mm long, the apices enlarged and truncate. Rostellum 1.5–2 mm long, narrowly triangular.

Spur geniculate, apex acute; side lobes of lip up to 12 mm long · · · · · · · · · · · ·var. *leucotricha*
Spur straight or only slightly bent, apex obtuse; side lobes of lip up to 14 mm long · · · · · · · ·
· var. *recticalcar*

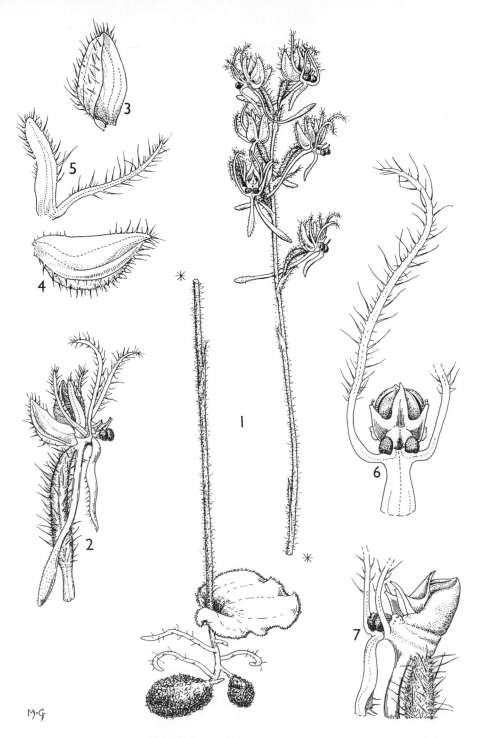

Tab. 46. HABENARIA HIRSUTISSIMA. 1, habit (×1); 2, flower (×2); 3, dorsal sepal (×4); 4, lateral sepal (×4); 5, petal (×4); 6 & 7, base of lip and column, in front and lateral views respectively (×6), 1–7 from *Gomes e Sousa* 4189. Drawn by Mary Grierson. From Kew Bull.

Var. **leucotricha**

Spur sharply geniculate, swollen for last 5 mm but tapering to an acute apex; side lobes of lip up to 12 mm long.

Zambia. W: Solwezi, fl. 26.xii.1959, *Holmes* 0183 (K; SRGH). C: Serenje, fl. 23.i.1951, *Fanshawe* 1852 (K; NDO). **Malawi**. N: Rumphi Distr., Livingstonia Escarpment, 900 m, fl. 25.i.1986, *la Croix* 786 (K).
Also in Tanzania. Woodland, often on rocky soil; 900–1200 m.

Var. **recticalcar** la Croix & P.J. Cribb in Kew Bull. **48**, 2: 361 (1993). Type: Zambia, Mbala Distr., Kalambo Falls, *Richards* 19617 (K, holotype).

Spur straight or only slightly bent, swollen towards end but with apex obtuse; side lobes of lip up to 14 mm long.

Zambia. N: Mbala Distr., Kalambo Falls, 1200 m, fl. 9.ii.1965, *Richards* 1961 (K). W: Kitwe, fl. 9.i.1968, *Mutimushi* 2419 (K; NDO).
Endemic. Woodland on stony ground; 1200 m.
There are several collections of this variety in the Kew Herbarium from the Kalambo Falls area in northern Zambia.

111. **Habenaria kabompoensis** G. Will. in J. S. African Bot. **46**, 4: 330, fig. 2 (1980). Type: Zambia, east bank of Kabompo R., *Williamson & Simon* 1766 (SRGH, holotype; K).
  Habenaria sp. Williamson, Orch. S. Centr. Africa: 63 (1977).
  Habenaria pilosa var. *kabompoensis* (G. Will.) Geerinck in Bull. Jard. Bot. Nat. Belg. **52**: 146 (1982); in Fl. Afr. Centr., Orchidaceae pt. 1: 78 (1984).

Terrestrial herb 13–20 cm tall; tubers c. 1 cm long, ovoid, villous. Stem erect, with 1 basal leaf and 1 bract-like stem leaf; stem clothed in white hairs 4–5 mm long. Basal leaf appressed to the ground, c. 3 × 3 cm, ovate-orbicular, cordate at base, acute, the upper surface fairly densely covered with white hairs c. 5 mm long, margins densely ciliate. Inflorescence c. 8 × 3 cm, laxly 6-flowered. Flowers suberect; sepals green, petals and lip white. Ovary and pedicel 10 mm long, slightly arched. Dorsal sepal erect, c. 7 × 3 mm, ovate, convex, acute; lateral sepals 8 × 2.8 mm, obliquely semi-ovate; all sepals hairy on outside. Petals 2-lobed almost to base; posterior lobe erect, 6–8 × 0.5–0.7 mm, linear, adnate to dorsal sepal; anterior lobe 18–20 × 0.3–0.7 mm, filiform, curving outwards and upwards. Lip 3-lobed ± to base; mid-lobe pendent, 12–14 × 1 mm, linear-ligulate, obtuse; side lobes 18–20 × 0.5–0.9 mm, curving up and out like lower petal lobes; spur 8–11 mm long, pendent, swollen at apex. Anther erect, c. 2 mm tall; canals 2 mm long, porrect; stigmatic arms decurved, c. 1 mm long, fairly stout. Rostellum mid-lobe 2 mm long.

Zambia. W: East bank of Kabompo R., on quartzite ridge, fl. xii.1969, *Williamson & Simon* 1766 (SRGH; K); Solwezi Distr., between Kafue and Muchindamu R. (Mushindamu R.), fl. i.1962, *Holmes* 0322 (SRGH).
Also in Zaire. Woodland, often on quartzite ridges or laterite.
This species is closely related to *H. leucotricha* Schltr., differing from it in the longer petal lobes and lip side lobes, and in the spur, which is shorter than the mid-lobe of the lip.

112. **Habenaria pilosa** Schltr. in Bot. Jahrb. Syst. **53**: 506 (1915). —Mansfeld in Fedde, Repert. Spec. Nov. Regni Veg., Beih. 68, t. 32/126 (1932). —Summerhayes in F.T.E.A., Orchidaceae: 122 (1968). —Grosvenor in Excelsa **6**: 83 (1976). —Williamson, Orch. S. Centr. Africa: 62 (1977). —Geerinck in Fl. Afr. Centr., Orchidaceae pt. 1: 77 (1984). —la Croix et al., Orch. Malawi: 96 (1991). Type from Tanzania.

Terrestrial herb 9–25 cm tall; tubers up to 2.5 × 1 cm, ovoid or globose, villous. Stem hairy, with 1 basal leaf and usually with 1–2 bract-like stem leaves. Basal leaf appressed to the ground, 1.5–3.5 × 2–4.5 cm, heart-shaped, green with purplish margin, sparsely hairy or glabrous on upper surface, the margin ciliate. Inflorescence 8–11 × 3.5 cm, laxly 6–14(30)-flowered; bracts 5 mm long. Flowers spreading, pale green. Ovary and pedicel 12–15 mm long. Dorsal sepal erect, 4.5–6.5 × 2.5–4 mm, ovate, convex; lateral sepals spreading, 5.5–7 × 2–3 mm, obliquely semi-ovate. Petals 2-lobed ± to base; posterior lobe erect, 4–6 × 1–1.5 mm, curved linear, sparsely hairy; anterior lobe 6–14 mm long, c. 0.5 mm wide, upcurved, narrowly linear, glabrous. Lip glabrous, 3-lobed almost to base; mid-lobe 6–12 × c. 1.5 mm; side lobes 7.5–14 mm long, upcurved, about half as wide as mid-lobe; spur pendent or slightly incurved, usually swollen towards apex, 9–27 mm long. Anther 2–3 mm high, rounded; canals 0.8–1.5 mm long, upcurved. Stigmatic arms 1–2 mm long, decurved, club-shaped.

Zambia. N: Kasama Distr., fl. 28.i.1962, *E.A. Robinson* 4914 (K). W: Mwinilunga Distr., NW of junction of Kamweshi and Dobeka Rivers, fl. 17.ii.1937, *Milne-Redhead* 3716 (K). C: c. 17 km east of Lusaka, fl. 14.i.1975, *Williamson & Gassner* 2347 (K). **Zimbabwe**. E: Mutare Distr., Nyamakari R., Burma Valley, c. 750 m, fl. 28.ii.1960, *Chase* 7307 (K; SRGH). **Malawi**. S: Blantyre Distr., Michiru Mt., 900 m, fl. 24.i.1989, *Jenkins* 68 (K).
Also in Zaire and Tanzania. *Brachystegia* woodland; c. 900 m.

113. **Habenaria rhopalostigma** Kraenzl., Orchid. Gen. Sp. 1: 342 (1898). —Rolfe in F.T.A. 7: 248 (1898). —Summerhayes in F.T.E.A., Orchidaceae: 125 (1968). —Williamson, Orch. S. Centr. Africa: 63 (1977). —Geerinck in Fl. Afr. Centr., Orchidaceae pt. 1: 75 (1984). —la Croix et al., Orch. Malawi: 96 (1991). Type: Zambia, Mbala (Abercorn) Distr., Fwambo, *Carson* 31 (K, holotype).
Habenaria nephrophylla Schltr. in Bot. Jahrb. Syst. **53**: 507 (1915). Type from Tanzania.

Terrestrial herb up to 40 cm tall; tubers 1–2 cm long, ovoid or ellipsoid, densely tomentose. Stem erect, with 1 basal leaf and 3–6 loose, more or less sheathing stem leaves. Basal leaf appressed to the ground, 3–4.5 × 3.5–5 cm, orbicular or reniform, rounded or apiculate, cordate at base, rather fleshy; sheathing leaves 20–35 mm long. Inflorescence short, 2–4 cm long, with 3–6 flowers closely arranged; bracts loose, c. 25 mm long. Dorsal sepal and spur green, rest of flower white. Pedicel 35 mm long, ovary 45 mm long. Dorsal sepal erect, 13–17 × 8–10 mm, ovate, convex, acute; lateral sepals deflexed, 17–25 × 10 mm, obliquely oblanceolate with a hooked lateral apiculus c. 3 mm long. Petals 2-lobed ± to base; posterior lobe erect, 13–17 × 1 mm, linear, ± adnate to dorsal sepal; anterior lobe 5–9.5 cm long, 1.5 mm wide at the base, becoming filiform, pendent. Lip 3-lobed with an undivided base 4–6 mm long, all lobes pendent, filiform; mid-lobe 2–4 cm long; side lobes 5–10 cm long; spur 10–17 cm long, pendent, slender but slightly swollen towards apex. Anther 8–10 mm high, erect; canals 13–15 mm long, curving up and making an angle of c. 60° with the stigmatic arms. Auricles 2–3 mm long. Stigmatic arms 13–17 mm long, projecting forwards, club-shaped. Rostellum 6–10 mm high.

Zambia. N: Kasama Distr., Mungwi, fl. 7.i.1962, *E.A. Robinson* 4242 (K). W: Mwinilunga Distr., just north of Mwinilunga, fl. 26.xi.1937, *Milne-Redhead* 3407 (K). C: Mkushi R., xii.1966–i.1967, *Odgers* 242 (K). **Malawi**. N: Rumphi Distr., 40 km north of Rumphi on road to Nyika, 1200 m, fl. 10.i.1974, *Pawek* 7962 (K; MAL; MO).
Also in Zaire, Burundi and Tanzania. *Brachystegia* woodland; 1200–1350 m.

114. **Habenaria holothrix** Schltr. in Warb., Kunene-Samb.-Exped. Baum: 204 (1903). —Mansfeld in Fedde, Repert. Spec. Nov. Regni Veg., Beih. 68, t. 27/106 (1932). —Summerhayes in F.T.E.A., Orchidaceae: 122 (1968). —Grosvenor in Excelsa **6**: 83 (1976). —Williamson, Orch. S. Centr. Africa: 61 (1977). Type from Angola.

Terrestrial herb 10–30 cm tall; tubers 1–1.5 cm long, ellipsoid, sparsely tomentose. Stem erect, with spreading hairs, with 1 basal leaf and 2–3 bract-like leaves. Basal leaf appressed to the ground, 1–3 × 1.5–3 cm, broadly ovate to orbicular, cordate at base, apiculate, glabrous or with scattered hairs on upper surface, ciliate on margin; bract-like stem leaves c. 1 cm long, hairy. Inflorescence 1–11 × 1.5 cm, laxly 2 to c. 20-flowered; bracts 3–7 mm long; all softly hairy. Flowers green or yellow-green, sometimes fragrant. Ovary and pedicel 7–11 mm long, softly hairy. Dorsal sepal erect, 3–4 × 2–3 mm, ovate, obtuse, convex; lateral sepals deflexed, 3.5–4 × 1.5 mm, obliquely lanceolate, obtuse or subacute; all sepals with spreading hairs on outside. Petals glabrous, 2-lobed to 1–1.5 mm from base; posterior lobe erect, 2–3 × c. 1 mm, curved-lanceolate, subacute; anterior lobe erect, 0.5–5 × c. 0.25 mm, filiform. Lip glabrous, curving down, 3-lobed to 0.5–1 mm from base; mid-lobe 2.8–6 mm long, c.

0.8 mm wide; side lobes filiform, 1.5–5.5 mm long; spur 7–10 mm long, slightly
inflated in apical half, parallel to ovary. Anther c. 1.5 mm tall, slightly incurved,
rounded; canals 0.5–1 mm long, upcurved. Stigmatic arms c. 1 mm long, stout,
truncate at apex. Rostellum mid-lobe triangular, less than 1 mm long.

Zambia. C: c. 5 km east of Lusaka, fl. ii.1968, *G. Williamson* 859 (K). **Zimbabwe**. C: Harare
Distr., Chakoma, 1360 m, fl. 27.i.1950, *Greatrex* in *GHS* 26745 (K; SRGH).
Also in Tanzania and Angola. In seasonally wet grassy depressions (vleis) ; c. 1360 m.

115. **Habenaria nyikensis** G. Will. in Pl. Syst. Evol. **134**: 58 (1980). —la Croix et al., Orch. Malawi:
      95 (1991). Type: Malawi, Nyika Plateau, *Williamson, Ball & Simon* 375 (K, holotype; SRGH).
      Habenaria sp. (*Williamson, Ball & Simon* 375) Williamson, Orch. S. Centr. Africa: 63 (1977).

Terrestrial herb 20–30 cm tall; tubers c. 2.5 cm long, ellipsoid. Stem erect, with 1
basal leaf and 20–30 bract-like stem leaves. Basal leaf appressed to the ground, c. 2.5
× 2.5 cm, ± orbicular, cordate at base, rounded or apiculate, fleshy with a hyaline
margin; bract-like stem leaves 10–12 mm long, linear-lanceolate. Inflorescence 10–15
× 3 cm, densely 20–30-flowered; bracts c. 17 mm long. Flowers green. Ovary and
pedicel arched, c. 10 mm long. Dorsal sepal erect, 5 × 2 mm, ovate, convex; lateral
sepals deflexed, c. 6 × 3 mm, obliquely ovate. Petals 2-lobed almost to base; posterior
lobe 4 × 3 mm, curved lanceolate, adnate to dorsal sepal; anterior lobe 7 × 1.5 mm,
linear, acute, curving up. Lip decurved, 3-lobed to base; mid-lobe 8 × 1 mm, linear,
acute; side lobes of similar length but less than half as wide, curving back; spur 9–10
mm long, twisted in the middle, horizontal. Anther c. 2 mm high, erect; canals c. 1
mm long. Stigmatic arms c. 2 mm long, decurved and forming an angle of c. 45° with
anther canals. Rostellum triangular, 0.8 mm high.

Malawi. N: Nyika Plateau, close to main road c. 8 km south of Lake Kaulime, fl. ii.1968,
Williamson, Ball & Simon 375 (K; SRGH); South Viphya, just outside southern edge of Forest
Reserve, c. 1660 m, fl. 3.i.1983, (flowers abnormal) *la Croix* 386 (K).
Apparently endemic to the plateaux of northern Malawi. In short open montane grassland
and *Brachystegia* woodland, usually on stony soil; 1660–2250 m.
This species is unusual in that it forms large colonies, but very rarely flowers. The authors
and others have found large numbers of leaves matching those of this species in many localities
on the Nyika Plateau in Malawi, but have never seen flowers or any signs of flowering. Plants
with peloric flowers, apparently referable to *H. nyikensis*, have been found on the South Viphya
Plateau, also in northern Malawi.

116. **Habenaria unifoliata** Summerh. in Bot. Mus. Leafl. Harv. Univ. **10**, 9: 276 (1942). —
      Williamson, Orch. S. Centr. Africa: 63 (1977). —la Croix et al., Orch. Malawi: 99 (1991).
      Type: Zambia, Mwinilunga Distr., NE of Dobeka Bridge, *Milne-Redhead* 3886 (K, holotype).

Terrestrial herb 35–45 cm tall; tubers 20–25 × 15–20 mm, ellipsoid, villous. Stem
erect, with 1 basal leaf and several bract-like stem leaves. Basal leaf appressed to the
ground, 4–7 × 5–8 cm, reniform, cordate at base; bract-like stem leaves 10–15 mm
long. Inflorescence 15–22 × 2.5–4 cm, rather laxly 20 to many-flowered; bracts c. 12
mm long. Flowers green or yellow-green. Ovary and pedicel 10–18 mm long, semi-
erect and slightly arched. Dorsal sepal erect, 4.8–6.3 × 3–4 mm, ovate, convex; lateral
sepals reflexed, 6–7 × 2–3.5 mm, obliquely ovate, apiculate. Petals 2-lobed almost to
base; posterior lobe erect, 4–8 × 0.5–1 mm, linear; anterior lobe 9–11 mm long, less
than 0.5 mm wide, filiform, curving up. Lip 3-lobed to c. 2 mm from base; mid-lobe
deflexed, 6.5–9 × 1–1.5 mm, linear; side lobes c. 11 mm long, upswept or reflexed,
similar to lower petal lobes; spur 12–13.5 mm, inflated at apex with slight spiral twist
in middle, parallel to ovary. Anther c. 3 mm high, erect; canals c. 1.5 mm long,
porrect. Stigmatic arms 1–2 mm long.

Zambia. N: Mbala Distr., Itimbwe Gorge, c. 1500 m, fl. 3.i.1960, *Richards* 12060 (K). W:
Mwinilunga Distr., NE of Dobeka Bridge, in *Cryptosepalum* woodland, fl. 30.xii.1937, *Milne-
Redhead* 386 (K). C: c. 65 km north of Serenje, fl. xii.1968, *G. Williamson* 402 (K). E: Chipata
Distr., Kapatamoyo, 1200 m, fl. 5.i.1959, *Robson* 1039 (K). **Malawi**. N: Mzimba Distr., c. 43 km
west of Mzuzu, beyond Mtwalo, 1300 m, fl. 25.i.1964, *Pawek* 1662 (K).
Also in Angola. *Brachystegia* and *Cryptosepalum* woodlands, in pockets in rocks and on cliffs;
1250–1600 m.

117. **Habenaria velutina** Summerh. in Kew Bull. **17**: 528 (1964). —Williamson, Orch. S. Centr. Africa: 63 (1977). TAB. **47**. Type: Zambia, Kasama Distr., *E.A. Robinson* 4325 (K, holotype).

Terrestrial herb 25–35 cm tall, almost the whole plant velvety hairy. Tubers 7–25 × 5–12 mm, ellipsoid or globose. Stem erect with 1 basal leaf and 1–3 appressed bract-like leaves. Basal leaf appressed to the ground, 2.5–3.5 × 2.5–3.5 cm, ovate or orbicular, cordate at base, rounded at apex, densely pubescent above, glabrous below; stem leaves very pubescent. Inflorescence 10–15 × 3–3.5 cm, fairly laxly 10–20-flowered; bracts 6–11 mm long, all densely pubescent. Flowers arching or sub-patent, pale green. Ovary and pedicel 12–17 mm long, arched, densely pubescent. Dorsal sepal erect, 6–7 × 4–4.5 mm, ovate, convex; lateral sepals 7.5–8 × 2.5–3.5 mm, obliquely lanceolate; all sepals hairy outside. Petals 2-lobed with an undivided basal part 3–3.5 mm long; posterior lobe c. 5 × 1.3 mm, linear, erect; anterior lobe patent, c. 11 mm long, filiform. Lip glabrous, 3-lobed with a claw 1.4–2 mm long; mid-lobe 7.5 × 1 mm, linear, convex; side lobes 10–12 × 0.5 mm, filiform; spur c. 2.5 cm long, inflated in apical third, pendent, slightly incurved. Anther 3–4 mm high, erect; canals 2–3 mm long. Stigmas 1.5–2 mm long, stout, clavate, curved, deflexed. Rostellum standing in front of anther, 3.5–4 mm long, 3-lobed in upper half only; mid-lobe c. 2.3 mm long; side lobes 1.2 mm long.

Zambia. N: Kasama Distr., Mungwi, fl. 31.i.1961, *E.A. Robinson* 4325 (K). C: c. 10 km east of Lusaka, 1270 m, fl. 17.i.1956, *King* 277 (K).
Known only from Zambia. In grassland, often near streams; c. 1270 m.
This species is closely related to *H. trilobulata* Schltr., which has a similar rostellum structure. It differs mainly in the velvety pubescence and the single basal leaf.

118. **Habenaria verdickii** (De Wild.) Schltr. in Bot. Jahrb. Syst. **53**: 508 (1915). —Williamson, Orch. S. Centr. Africa: 63 (1977). —Geerinck in Fl. Afr. Centr., Orchidaceae pt. 1: 63 (1984). —la Croix et al., Orch. Malawi: 99 (1991). TAB. **48**. Type from Zaire.
   Bonatea verdickii De Wild. in Ann. Mus. Congo, Bot., ser. 4, 1: 24 (1902).
   Habenaria monophylla Schltr. in Warb., Kunene-Samb.-Exped. Baum: 206 (1903). Type from Angola.
   Habenaria cordifolia Summerh. in Kew Bull. **17**: 517, fig. 4 (1964); in F.T.E.A., Orchidaceae: 124 (1968). Type from Tanzania.

Terrestrial herb 20–30 cm tall; tubers c. 2.5 cm long, ellipsoid, tomentose. Stem erect, with 1 basal leaf and 3–5 bract-like leaves. Basal leaf appressed to the ground, 4.5–5 × 4–7.5 cm, orbicular to reniform, cordate at base, apiculate or obtuse, glabrous; stem leaves up to 3 cm long. Inflorescence 6–10 × 4 cm, laxly 2–11-flowered; bracts 13–20 mm long, lanceolate, acuminate. Flowers white or yellow-green. Ovary and pedicel 15–20 mm long, slightly arched. Dorsal sepal erect, 9–16 × 4.5–7 mm, ovate, convex; lateral sepals deflexed, 11–16.5 × 3.5–5 mm, obliquely lanceolate, acute; all sepals with 3 toothed keels on outside. Petals 2-lobed to base; posterior lobe erect, 7.5–18 × 0.5–1 mm, linear; anterior lobe 12–28 × c. 1 mm, curving up. Lip porrect, 3-lobed almost to base; mid-lobe 8–17.5 × 0.5–1.5 mm, linear; side lobes 8–16.5 × c. 1 mm; spur 10–18 mm long, inflated at apex, parallel to ovary. Anther 4–6 mm high, erect, apiculate; canals 2.5–5.5 mm long. Stigmatic arms 4–6.5 mm long, thickened and truncate at apex, porrect. Mid-lobe of rostellum 2.25–4 mm long, thickened and truncate at apex, porrect.

Zambia. N: Mbala Distr., Kanyika Flats, Chinakila, 1200 m, fl. 12.i.1965, *Richards* 19497 (K). W: Solwezi Distr., Kafue-Musaka F.R., fl. 10.i.1962, *Holmes* 0315 (K; SRGH). **Zimbabwe**. N: Makonde Distr., Doma area, 1200 m, fl. 20.i.1963, *Jacobsen* 2070 (PRE). **Malawi**. N: Nyika National Park, Zovochipolo grassland, 2210 m, fl. 25.ii.1985, *Dowsett-Lemaire* 273 (K). C: Dowa Distr., Kongwe Mt., fl. 7.iii.1982, *Brummitt, Polhill & Banda* 16372 (K). S: Blantyre Distr., Salisbury Road, c. 4 km from Nyambadwe, fl. 30.i.1945, *Benson* 1155 (K).
Also in Zaire and Angola. In grassland, dry dambo edges and woodland, sometimes on rocky outcrops; 1100–2200 m.
Summerhayes described *H. cordifolia* as similar to *H. verdickii* but with smaller flowers and shorter anther canals and stigmas. However, plants from the Nyika Plateau in Malawi have flowers intermediate in size between these two species and so it would seem reasonable to conclude that *H. cordifolia*, which is represented by very few specimens, is a small flowered form of *H. verdickii*.
   H. verdickii also resembles *H. lindblomii* Schltr. but has a single basal leaf while the latter has two. Also *H. lindblomii* tends to have larger flowers with more slender stigmatic arms, larger leaves and more flowers in the inflorescence. Should they prove to be conspecific, *H. verdickii* is the correct name.

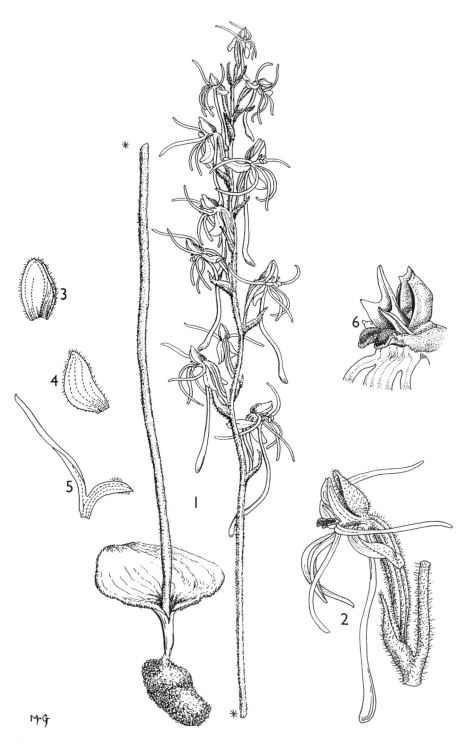

Tab. 47. HABENARIA VELUTINA. 1, habit (×1); 2, flower and bract (×2); 3, dorsal sepal (×2);
4, lateral sepal (×2); 5, petal (×2); 6, column and base of lip, three-quarter profile (×5), 1–6
from *Robinson* 4325. Drawn by Mary Grierson. From Kew Bull.

Tab. 48. HABENARIA VERDICKII. 1, habit (×1); 2, flower (×2), 1 & 2 from original drawing by *McLoughlin*; 3, dorsal sepal (×4); 4, lateral sepal (×4); 5, petal (×4); 6, column and base of lip, lateral view (×4), 3–6 from *McLoughlin* 47A. Drawn by Mary Grierson. From Kew Bull.

119. **Habenaria perpulchra** Kraenzl. in Bot. Jahrb. Syst. **51**: 373 (1914). —Summerhayes in Kew Bull. **12**: 109 (1957); in F.T.E.A., Orchidaceae: 125 (1968). —Williamson, Orch. S. Centr. Africa: 63 (1977). —Geerinck in Fl. Afr. Centr., Orchidaceae pt. 1: 67 (1984). Type from Zaire.
Habenaria platymera Schltr. in Bot. Jahrb. Syst. **53**: 502 (1915). Type from Tanzania.

Terrestrial herb 15–35 cm tall; tubers up to 2.5 × 1.5 cm, ovoid or ellipsoid, tomentose. Stem erect, with 1 basal leaf and 2–7 loosely sheathing stem leaves. Basal leaf appressed to ground, up to 4.5 cm long and broad, ovate, orbicular or reniform, cordate at base, fleshy; stem leaves 2–3 cm long. Inflorescence up to 7 cm long, fairly densely 1–5-flowered. Flowers white, erect. Pedicel and ovary 2.5–4 cm long, the ovary with 6 toothed wings. Sepals all with 1–3 keels on outside; dorsal sepal erect, 11–16 × c. 6 mm, elliptic, obtuse, convex; lateral sepals deflexed, 17–20 × 6–8 mm, obliquely lanceolate, acute. Petals 2-lobed almost to base; posterior lobe erect, 12.5–14 × 2–3 mm; anterior lobe 14–20 × 3.5–5 mm, both lobes oblanceolate from a narrow base, acute. Lip projecting forwards, 3-lobed to c. 2 mm from base, all lobes oblanceolate from a narrow base; mid-lobe 15–21 × 4.5–9 mm; side lobes 15–21 × 5–7.5 mm, somewhat incurved; spur 24–30 mm long, pendent, the apical third very swollen. Anther 5–6 mm high, curved back, apiculate; canals 2–3.5 mm long, slightly incurved. Stigmatic arms 5.5–7 mm long, porrect, stout, enlarged and truncate at apex. Rostellum narrowly triangular, acute, c. 4 mm long.

Zambia. N: Kasama Distr., road to Chief Mwamba's village, 1320 m, fl. 25.ii.1962, *Richards* 16166 (K). **Malawi**. N: c. 40 km west of Karonga, fl. 22.iii.1953, *J. Williamson* 244 (BM). Also in Zaire and Tanzania. Upland grassland; 1300–1400 m.

17. PLATYCORYNE Rchb.f.

Platycoryne Rchb.f. in Bonplandia **3**: 212 (1855). —Summerhayes in Kew Bull. **13**: 58–73 (1958).

Terrestrial herb with fleshy or tuberous roots and leafy, unbranched stems. Flowers resupinate, usually yellow, orange or greenish, rarely white, often in short, dense heads. Sepals free, the dorsal sepal forming a hood with the petals. Lip entire or 3-lobed, spurred. Column erect, anther erect with parallel loculi, canals adnate to side lobes of rostellum. Rostellum large, usually placed in front of anther which is often over-topped by the mid-lobe; lateral lobes porrect, usually projecting beyond the hood but sometimes short and shoulder-like. Stigmatic processes thickened with rounded ends.

A genus of 20–25 species in tropical Africa and Madagascar. Most species form colonies where they occur, which is usually in wet areas.

1 Lip entire · 2
– Lip 3-lobed or toothed · 7
2. Dorsal sepal 9 mm long or less · 3
– Dorsal sepal at least 9.5 mm long · 5
3. Flowers in a dense head; side lobes of rostellum slender, porrect · · · · · · 2. *buchananiana*
– Flowers spread out on rhachis; side lobes of rostellum shoulder-like · · · · · · · · · · · · · 4
4. Dorsal sepal 6–8 mm long; lip 5–7 mm long; petals adnate to dorsal sepal · · 1. *protearum*
– Dorsal sepal 5–6 mm long; lip less than 5 mm long; petals free · · · · · · · · · · · · 6. *affinis*
5. Spur more than 5 cm long · 11. *macroceras*
– Spur less than 2 cm long · 6
6. Petals and lip c. 2 mm wide · 3. *pervillei*
– Petals and lip more than 4.5 mm wide · 10. *latipetala*
7. Lip lobed near base · 8
– Lip lobed about halfway along its length · 12
8. Petals 2-lobed · 8. *guingangae*
– Petals entire · 9
9. Dorsal sepal less than 3 mm long · 9. *isoetifolia*
– Dorsal sepal more than 4 mm long · 10

10. 3–9 leaves forming tuft at base of stem, remainder spaced along stem · · · · · · · · 5. *crocea*
– Leaves spaced out along stem, not forming basal tuft · 11
11. Dorsal sepal 7–11 mm long; side lobes of lip triangular or tooth-like · · · · · · 4. *mediocris*
– Dorsal sepal less than 5.5 mm long; side lobes of lip linear-ligulate · · · · · · 12. *micrantha*
12. Dorsal sepal 4.5–5 mm long; spur less than 2 cm long · · · · · · · · · · · · · · · 7. *brevirostris*
– Dorsal sepal c. 8 mm long; spur 3 cm long · 13. *trilobata*

1. **Platycoryne protearum** (Rchb.f.) Rolfe in F.T.A. **7**: 257 (1898). —Summerhayes in Kew Bull.
13: 60 (1958); in F.T.E.A., Orchidaceae: 142 (1968). —Williamson, Orch. S. Centr. Africa:
69 (1977). —Geerinck in Fl. Afr. Centr., Orchidaceae pt. 1: 160 (1984). —la Croix et al.,
Orch. Malawi: 105 (1991). Type from Angola.

Terrestrial herb 18–50 cm tall; tubers 10–20 × 5–10 mm, globose or ellipsoid.
Leaves 4–8, the lowermost sheath-like, the next 2–3 above these 2–6 cm long and 3–8
mm wide, lanceolate, acute; the remaining leaves spaced out along the stem, erect,
bract-like. Inflorescence laxly 1–5-flowered, up to 12 cm long. Flowers greenish-
orange, erect. Pedicel 12–20 mm long, ovary arched, 12–15 mm long; bracts 17–22
mm long. Dorsal sepal erect, 6–8 × 3.5–6 mm, ovate, convex; lateral sepals deflexed,
6.5–8 × 2–3 mm, obliquely lanceolate. Petals 5–7 × 1.5–3.5 mm, obliquely triangular-
lanceolate, adnate to dorsal sepal. Lip deflexed, 5–7 × 2.5–3 mm, entire, fleshy,
convex, narrowly oblong; spur 2–3.5 cm long, slender but swollen towards apex,
usually enclosed within the bracts. Anther 2.5–4.5 mm high, erect with a long
apiculus; canals 2–3 mm long. Stigmatic arms projecting forwards, 2–2.5 mm long,
stout. Rostellum 4–4.5 mm high, placed in front of anther and overtopping it; mid-
lobe acute, side lobes very short and shoulder-like.

Var. **protearum**

Zambia. N: Mbala Distr., Lake Chila, fl. & fr. 12.xii.1949, *Bullock* 2090 (K). W: Mwinilunga
Distr., peaty dambo, fl. & fr. 10.xi.1958, *Holmes* 085 (K; SRGH). C: Mkushi Dambo, fl. x.1967, *G.
Williamson* 315 (K). E: Lundazi Distr., Nyika Plateau, c. 2200 m, fl. 11.ii.1960, *Holmes* 0212 (K).
Zimbabwe. W: Matobo, 1400 m, fl. i.1954, *Miller* 2037 (PRE; SRGH). **Malawi**. N: Nyika National
Park, Chowo Rocks, 2200 m, fl. 11.i.1983, *la Croix & Dowsett-Lemaire* 431 (K); Mzimba Distr.,
Katoto Dambo near Mzuzu, 1250 m, fl. 17.xii.1986, *la Croix* 904 (K; MAL).
Also in Zaire, Tanzania and Angola. Wet grassland and marshy or seepage areas amongst
rocks; 1050–2200 m.
Plants from the Nyika Plateau in Malawi tend to be small, the inflorescence usually 1-flowered
and the flowers predominantly green.

Var. **recurvirostra** G. Will. in Pl. Syst. Evol. **134**: 60 (1980). Type: Zambia, Ntumbachushi,
Williamson & Simon 340 (SRGH, holotype).

Differs from typical variety in having the mid-lobe of the rostellum curving
backwards and being shorter than the anther thecae. The whole plant is much
smaller.

Zambia. N: Ntumbachushi, c.17.6 km west of Kawambwa, fl. xii.1967, *Williamson & Simon* 340
(SRGH); Mbala Distr., Ndundu Dambo, fl. xii.1967, *Williamson & Simon* 674 (SRGH).

2. **Platycoryne buchananiana** (Kraenzl.) Rolfe in F.T.A. **7**: 257 (1898) pro parte. —Summerhayes
in F.T.E.A., Orchidaceae: 143 (1968). —Grosvenor in Excelsa **6**: 84 (1976). —Williamson,
Orch. S. Centr. Africa: 70 (1977). —Geerinck in Fl. Afr. Centr., Orchidaceae pt. 1: 158
(1984). —la Croix et al., Orch. Malawi: 102 (1991). Type: Malawi, without precise locality,
Buchanan 1155 (B†, holotype; K).
Habenaria buchananiana Kraenzl. in Bot. Jahrb. Syst. **19**: 247 (1894).
Habenaria stolzii Kraenzl. in Bot. Jahrb. Syst. **48**: 386 (1912), non Schltr. Type from Tanzania.
Habenaria ipyanae Schltr. in Bot. Jahrb. Syst. **53**: 493 (1915). Type from Tanzania.

Slender terrestrial herb 20–60 cm tall; tubers c. 1.5 × 1 cm, ovoid, villous. Leaves 5–9,
spaced along stem, the lower ones ± spreading, 3–7 × 0.5–1 cm, lanceolate, acute; the

upper ones erect and bract-like. Inflorescence short, up to 5.5 cm long, densely (1)4–12(20)-flowered. Flowers, including the ovary, bright orange or yellow, sometimes scented, erect or spreading. Ovary and pedicel 10–20 mm long; bracts of similar length, yellow or orange towards tip. Dorsal sepal erect, 5–9 × 4–7 mm, broadly ovate, convex; lateral sepals reflexed, 5–10 × 2–3 mm, obliquely elliptic, apiculate. Petals erect, 3.5–8 × 1–3 mm, adnate to dorsal sepal, curved lanceolate. Lip entire, 5–10 × 1–2 mm, ligulate, fleshy, pendent or curving forwards; spur 10–20 mm long slightly swollen towards apex. Anther c. 4 mm high, erect, apiculate; canals 2–3 mm long, porrect. Stigmatic arms 2–3 mm long, stout with enlarged, truncate ends, horizontal or slightly decurved. Rostellum variable in length, standing in front of anther.

Zambia. N: Mbala Distr., Katula Gorge, 1450 m, fl. 8.i.1952, *Richards* 409 (K). W: c. 2 km east of Ndola, 1200 m, fl. 2.i.1953, *Draper* 18 (K). C: Kabwe Distr., Mpunde Mission, Kelongwe R., 1130 m, fl. 20.i.1973, *Kornaś* 3386 (K). S: Mapanza, Simasunda, 1060 m, fl. 21.ii.1957, *E.A. Robinson* 2137 (K). **Zimbabwe**. N: Makonde Distr., Guruve (Sipolilo), fl. 29.i.1948, *Whellan* 292 (K; SRGH). W: Matobo Distr., Besna Kobila, 1450 m, fl. i.1957, *Miller* 4059 (K; SRGH). C: Makoni Distr., hills c. 8 km west of Rusape, fl. 20.i.1949, *Chase* 4074 (K; SRGH). E: Nyanga Distr., Odzani, fl. 31.i.1958, *Beasley* 44 (K; SRGH). **Malawi**. N: Nkhata Bay Distr., Mzenga Dambo, 600 m, fl. 20.xii.1981, *la Croix* 241 (K). C: lower slopes of Dedza Mt., fl. 16.ii.1972, *Westwood* 656 (K; SRGH). S: Zomba Distr., Old Naisi Road, fl. 17.i.1980, *Blackmore* 1093 (MAL).

Also in Zaire, Burundi, Tanzania and Cameroon. Dambo, wet grassland and on seepage slopes; 500–1850 m.

Plants of this species vary considerably in size, and also in flower colour. Some colonies are entirely orange-flowered while others are all yellow-flowered, more rarely do colonies contain flowers of mixed colours.

3. **Platycoryne pervillei** Rchb.f. in Bonplandia **3**: 212 (1855). —H. Perrier in Fl. Madag., Orch. 1: 66 (1939). —Summerhayes in Kew Bull. **13**: 62 (1958); in F.T.E.A., Orchidaceae: 144 (1968). —Grosvenor in Excelsa **6**: 84 (1976). —la Croix et al., Orch. Malawi: 104 (1991). Type from Madagascar.
    Habenaria tenuicaulis Rendle in J. Linn. Soc., Bot. **30**: 396 (1895). Type from Tanzania.
    Habenaria buchwaldiana Kraenzl. in Bot. Jahrb. Syst. **24**: 503 (1898). Type from Tanzania.
    Platycoryne tenuicaulis (Rendle) Rolfe in F.T.A. **7**: 257 (1898).
    Platycoryne buchananiana sensu Rolfe in F.T.A. **7**: 257 (1898), partim., non (Kraenzl.) Rolfe.
    Habenaria flammea Kraenzl. in Bot. Jahrb. Syst. **28**: 173 (1900). Type from Tanzania.

Terrestrial herb 30–70 cm tall; tubers up to 2.5 × 1 cm, ellipsoid. Leaves 5–8, spaced along stem, the lowermost ± spreading, the largest up to 8 × 1 cm, narrowly lanceolate. Inflorescence compact, up to 5.5 cm long, densely 2–8-flowered. Flowers bright orange, suberect. Ovary and pedicel arching, up to 25 mm long; bracts up to 20 mm long. Dorsal sepal reflexed, 10–14 × 3–4 mm, obliquely lanceolate, acute. Petals erect, 8–12 × c. 2 mm, curved lanceolate-linear, adnate to dorsal sepal. Lip entire, 7–11 × 2 mm, ligulate, curving forwards, fleshy; spur 12–19 mm long, swollen at apex, usually tucked into bracts. Anther 4.5–6.5 mm high, erect, apiculate; canals 3.5–5 mm long, porrect. Stigmatic arms c. 3 mm long, horizontal, clavate; rostellum standing in front of anther, c. 4 mm high, 3-lobed in upper half, all lobes triangular.

Zambia. N: Mbala Distr., Kalambo Falls, fl. xii.1967, *Williamson & Simon* 346 (K; SRGH). **Zimbabwe**. N: Mutoko Distr., Mudzi Dam, 1200 m, fl. 16.ii.1962, *Wild* 5667 (K; SRGH). C: Gweru Distr., c. 12 km south of Gweru, 1400 m, fl. 8.ii.1967, *Biegel* 1908 (K; SRGH). E: Mutare Distr., Zimunya's Reserve, c. 1000 m, fl. 13.iii.1960, *Chase* 7295 (K; SRGH). **Malawi**. S: Blantyre Distr., Matanje Road, 1 km north of Limbe, 1175 m, fl. 5.ii.1970, *Brummitt & Banda* 8411 (K; MAL). **Mozambique**. N: Marrupa, on road to Nungo, c. 800 m, fl. 18.ii.1982, *Jansen & Boane* 7815 (K; PRE). Z: Lugela, road to Moebede, fl. 25.i.1948, *Faulkner* 179 (K). MS: c. 16 km north of Mavita, fl. 1.ii.1962, *Wild* 5632 (K; SRGH).

Also in Kenya, Tanzania and Madagascar. Woodland and damp grassland; 800–1400 m.

4. **Platycoryne mediocris** Summerh. in Kew Bull. **13**: 72 (1958); in F.T.E.A., Orchidaceae: 145 (1968). —Moriarty, Wild Fls. Malawi: t. 29, 2 (1975). —Grosvenor in Excelsa **6**: 84 (1976). —Williamson, Orch. S. Centr. Africa: 73 (1977). —Geerinck in Fl. Afr. Centr., Orchidaceae pt. 1: 157 (1984). —la Croix et al., Orch. Malawi: 104 (1991). Type from Tanzania.

Terrestrial herb 20–50 cm tall; tubers 5–20 × 5–12 mm, ellipsoid or ± globose, sparsely tomentose. Leaves 6–10, spaced along stem, the lowermost 1–2 sheath-like, the rest ± spreading, the largest 2.5–6 × 0.4–0.9 cm, narrowly lanceolate, the upper ones becoming bract-like. Inflorescence compact, less than 5 cm long, densely 2–5(7)-flowered. Flowers bright orange, including ovary. Ovary and pedicel arched, 15–25 mm long, the bracts shorter. Dorsal sepal erect, 7–11 × 5–6 mm, ovate, acute, convex; lateral sepals deflexed, 8–10 × 2–3.5 mm, obliquely lanceolate. Petals erect, 7–10 × 2–3 mm, curved-lanceolate, adnate to dorsal sepal. Lip deflexed, 6.5–10 × 1–2 mm, with 2 tooth-like lobes c. 0.5 mm long at the base; spur 9–17 mm long, swollen at apex, usually tucked into bract. Anther 2.8–5 mm high, erect, apiculate; canals 1–2 mm long, porrect. Stigmatic arms 2–3 mm long, horizontal, clavate; rostellum almost as long as anther, mid-lobe c. 2.5 mm long, side lobes c. 1.2 mm long, truncate.

Zambia. C: 8 km west of Lusaka, on old Mumbwa road, fl. 15.ii.1970, Drummond & Williamson 9604 (K; SRGH). E: by Katete R., at crossing with Great East Road, 1060 m, fl. 16.i.1957, Wright 123 (K). Zimbabwe. N: Chicomba Vlei, Umwindsidale Road, c. 24 km NE of Harare, fl. 28.i.1948, Greatrex in GHS 228731 (K; SRGH). E: Nyamkwarara (Nyumquarara) Valley, 1060–1200 m, fl. & fr. ii.1935, Gilliland 1542 (K). Malawi. N: Mzimba Distr., c.9 km north of Mzambazi, 1260 m, fl. 10.iii.1978, Pawek 13978 (K; MAL; MO). C: Kasungu Distr., in dambo north of town, 1100 m, fl. 14.i.1959, Robson & Jackson 1165 (K). S: Mulanje Distr., Likhubula Valley, c. 900 m, fl. 7.i.1982, la Croix 254 (K).
Also in Zaire, Burundi and Tanzania. Poorly drained grassland; 700–1750 m.

5. **Platycoryne crocea** (Schweinf. ex Rchb.f.) Rolfe in F.T.A. **7**: 257 (1898). —Summerhayes in Kew Bull. **13**: 71 (1958); Kew Bull. **17**: 532 (1964); in F.T.E.A., Orchidaceae: 146 (1968). — Moriarty, Wild Fls. Malawi: t. 29, 1 (1975). —Williamson, Orch. S. Centr. Africa: 71 (1977). —Geerinck in Fl. Afr. Centr., Orchidaceae pt. 1: 162 (1984). —la Croix et al., Orch. Malawi: 102 (1991). Type from Sudan.
 Habenaria crocea Rchb.f., Otia Bot. Hamburg.: 57 (1878).
 Habenaria elegantula Kraenzl. in Bot. Jahrb. Syst. **51**: 376 (1914). Type: Zambia, Ndola, Kassner 2169 (K, holotype; BR).
 Platycoryne elegantula (Kraenzl.) Summerh. in F.W.T.A. **2**: 414 (1936); in Kew Bull. **13**: 70 (1958).
 Platycoryne crocea subsp. elegantula (Kraenzl.) Summerh. in Kew Bull. **17**: 532 (1964).
 Habenaria ochrantha Schltr. in Bot. Jahrb. Syst. **53**: 494 (1915). Type from Tanzania.
 Platycoryne ochrantha (Schltr.) Summerh. in Kew Bull. **2**: 125 (1948); in Kew Bull. **13**: 70 (1958).
 Platycoryne crocea subsp. ochrantha (Schltr.) Summerh. in Kew Bull. **17**: 533 (1964).
 Platycoryne heterophylla Summerh. in Kew Bull. **13**: 69 (1958). —Williamson, Orch. S. Centr. Africa: 74 (1977). Type: Zambia, Mwinilunga Distr., Milne-Redhead 4382 (K, holotype; PRE).

Slender terrestrial herb 12–45 cm tall; tubers 8–18 × 5–10 mm, ellipsoid to globose, densely tomentose. Leaves 5–16, the lowermost 3–9 forming a basal tuft, spreading; the rest spaced along the stem, becoming bract-like. Largest leaf 1.5–6 cm long, 2–8 mm wide, linear or linear-lanceolate. Inflorescence compact, up to c. 3 cm long, densely 1–6(9)-flowered. Flowers yellow-orange to deep orange, suberect. Ovary and pedicel 14–20 mm long, arching; bracts 7–18 mm long. Dorsal sepal erect, 6.5–12 × 3.5–9.5 mm, broadly ovate, very acute, the tip often slightly reflexed, convex; lateral sepals deflexed, 4.5–11.5 × 2–3.5 mm, obliquely lanceolate, acute. Petals erect, 4.5–11 × 1.5–3 mm, curved lanceolate, adnate to dorsal sepal. Lip 5–12 mm long with tooth-like lobes 1–2 mm long at base; mid-lobe 2 mm wide, ligulate, obtuse, fleshy; spur 9–15 mm long, pendent or slightly incurved. Anther 2.5–4.5 mm high, erect, obtuse or apiculate; canals 1–2.5 mm long, porrect. Stigmatic arms 1.5–3 mm long, clavate, horizontal or decurved. Rostellum erect, mid-lobe 1–2.5 mm long, triangular, acute, the side lobes shorter and truncate.

Zambia. N: Mbala Distr., Kalambo Falls, 1050 m, fl. 29.xii.1962, Richards 17082 (K). W: Mwinilunga Distr., by Kalenda Village, 1200 m, fl. 19.xi.1962, Richards 17293 (K). C: Serenje Distr., Kundalila Falls, fl. i.1972, G. Williamson 2157 (K; SRGH). Zimbabwe. N: Makonde Distr., 1200 m, Jacobsen 2079 (PRE). Malawi. N: Rumphi Distr., c. 8 km SE of turning to Thazima, c. 1350 m, fl. 17.i.1985, la Croix 936 (K).
Widespread in tropical Africa. Brachystegia woodland on stony soil, pebbly grassland; 900–1350 m.

Other workers have recognised 2 subspecies in the Flora area; subsp. *elegantula* and subsp. *ochrantha*, but differences between these are not consistent. In Malawi any one colony will contain plants which key out to either subspecies.

Summerhayes in describing *P. heterophylla* differentiated it from *P. crocea* by its larger rostellum mid-lobe and by its relatively broad basal leaves and much narrower cauline leaves. However, all the measurements given fall within the range recognised for *P. crocea*.

6. **Platycoryne affinis** Summerh. in Kew Bull. **13**: 61 (1958). —Grosvenor in Excelsa **6**: 84 (1976). Type: Zimbabwe, Harare, *Wild* 2262 (K, holotype; SRGH).

Slender terrestrial herb 20–40 cm tall; tubers c. 15 × 7 mm, ellipsoid, pubescent. Leaves 3–5, ± evenly spaced along the stem, the largest up to 5 × 0.5 cm, linear-lanceolate. Inflorescence 1.5–4 cm long, fairly laxly 2–4-flowered. Flowers yellow or yellow-green, erect or suberect. Pedicel and ovary 2.5–3 cm long; bracts 7–21 mm long. Dorsal sepal erect, 5–6 × 3–5 mm, broadly ovate, acute, convex; lateral sepals deflexed, 5–6.5 × 2–2.5 mm, obliquely lanceolate. Petals 4–5.5 × 2–3 mm, obliquely triangular or oblong-lanceolate, not adnate to dorsal sepal but attached to column at base. Lip 3.7–4.7 × 2–2.7 mm, ovate or elliptic-ovate, entire but sometimes with obscure lobes at base, fleshy; spur 15–25 mm long, slightly swollen at apex. Anther 2–3 mm high, apiculate, slightly incurved; canals 2 mm long. Stigmatic arms 2 mm long; rostellum c. 3 mm high, the side lobes shoulder-like, much shorter than mid-lobe.

Zimbabwe. W: Matobo Distr., Besna Kobila, 1450 m, fl. xii.1961, *Miller* 8070 (K; SRGH). C: Harare Distr., Ruwa R., 1360 m, fl. 20.xii.1947, *Wild* 2262 (K; SRGH).
Endemic. Wet vlei; 1360–1450 m.
This species resembles *P. protearum* but may be distinguished from it by the usually smaller flowers with free petals, and by the broader lip sometimes with obscure lobes. The species seems prone to producing deformed flowers.

7. **Platycoryne brevirostris** Summerh. in Kew Bull. **13**: 68 (1958). —Williamson, Orch. S. Centr. Africa: 71 (1977). TAB. **49**. Type: Zambia, Mwinilunga Distr., Kalenda Dambo, *Milne-Redhead* 3811 (K, holotype; PRE).

Terrestrial herb 15–30 cm tall; tubers 10–15 × 6–10 mm, ovoid or ellipsoid, sparsely tomentose. Leaves 5–8, the lowermost sheathing with 2–4 suberect leaves clustered above these, up to 5 × 0.4 cm, linear or linear-lanceolate, acute; the remaining 2–3 spaced along the stem. Inflorescence 1.5–6 cm long, laxly 2–7-flowered. Flowers orange or yellow-green, suberect. Ovary and pedicel 15–20 mm long; bracts 7–17 mm long. Dorsal sepal erect, 4.5–5 × 3 mm, elliptic, slightly incurved; lateral sepals 5–6 × 2–2.5 mm, obliquely elliptic-oblong. Petals erect, c. 4.5 × 1.5 mm, curved lanceolate-ligulate, adnate to dorsal sepal. Lip deflexed or slightly incurved, 3-lobed at about halfway, 3.7–5 mm long, 2 mm wide across lobes; mid-lobe 2–4 × 1 mm, side lobes 0.2–1.2 × 0.3 mm; spur 15–18 mm long, slightly swollen at apex. Anther 3–4 mm high; canals 2.5 mm long. Stigmatic arms c. 1.5 mm long. Rostellum 2.5 mm long; mid-lobe c. 0.4 mm high; side lobes 1 mm long, relatively long and narrow.

Zambia. W: Mwinilunga Distr., Kalenda Dambo, fl. 24.xii.1937, *Milne-Redhead* 3811 (K; PRE); Mwinilunga Distr., c. 6 km north of Salujinga turn-off, on granite outcrop, fl. 24.xii.1969, *Williamson & Simon* 1817 (K; SRGH).
Also in Angola. Dambo and granite outcrops, in marshy ground over laterite or rock.
This species resembles *P. trilobata* but has smaller flowers.

Platycoryne guingangae × P. brevirostris.
In the Mwinilunga District of Zambia, hybrids occur between *P. brevirostris* and *P. guingangae*, these are intermediate between the two species and grow alongside them.
Zambia. W: Mwinilunga Distr., Kalenda Dambo, fl. 24.xii.1937, *Milne-Redhead* 3813 (K).

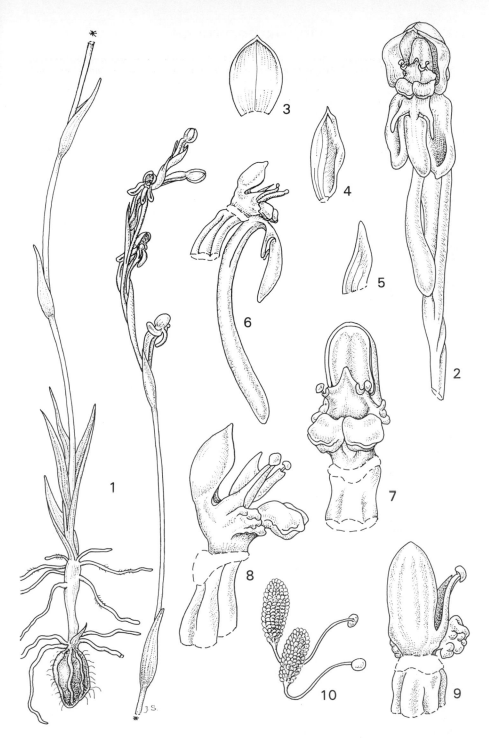

Tab. 49. PLATYCORYNE BREVIROSTRIS. 1, habit (×1), from *Milne-Redhead* 3811; 2, flower (×4); 3, dorsal sepal (×4); 4, lateral sepal (×4); 5, petal (×4); 6, column and lip, side view (×4); 7, column, front view (×6); 8, column, side view (×6); 9, column, back view (×6); 10, pollinia (×6), 2–10 from *Milne-Redhead* 3814 (in K spirit coll. 18509). Drawn by Judi Stone.

8. **Platycoryne guingangae** (Rchb.f.) Rolfe in F.T.A. **7**: 258 (1898). —Summerhayes in Kew Bull.
13: 73 (1958). —Williamson, Orch. S. Centr. Africa: 74 (1977). —Geerinck in Fl. Afr.
Centr., Orchidaceae pt. 1: 156 (1984). Type from Angola.
Habenaria guingangae Rchb.f., Flora **48**: 179 (1865).
Habenaria poggeana Kraenzl. in Bot. Jahrb. Syst. **16**: 207 (1892). Type from Zaire.
Platycoryne poggeana (Kraenzl.) Rolfe in F.T.Á. **7**: 258 (1898).

Terrestrial herb 30–50 cm tall; tubers c. 10 × 8 mm, ovoid. Leaves up to 8, 2–2.5 ×
0.5 mm, lanceolate, usually evenly spread along the stem but occasionally forming a
tuft at the base. Inflorescence short, closely 1–6-flowered. Flowers orange, erect.
Ovary and pedicel c. 15 mm long; bracts 10–15 mm long. Dorsal sepal erect, 6.5–7 ×
c. 5.5 mm, ovate, convex; lateral sepals 6.3–8.7 × 2.5 mm, obliquely lanceolate. Petals
adnate to dorsal sepal, c. 7 mm long, with an anterior lobe near the base; posterior
lobe c. 6.3 × 2 mm, anterior lobe 0.3–2 × 0.2 mm. Lip 5.5–6.5 mm long, 3-lobed in
the basal half; mid-lobe 4.5–5 × 1–1.5 mm; side lobes 1.8–2 × 0.2 mm; spur 10.5–13.5
mm long, pendent, swollen at apex. Anther c. 4 mm tall; canals 2.5 mm long.
Stigmatic arms 2–2.5 mm long. Rostellum 2–3 mm long; mid-lobe 1.3–2 mm, side
lobes c. 1.5 mm.

Zambia. N: Chinsali Distr., Ishiba Ngandu (Shiwa Ngandu), fl. 21.xii.1964, *E.A. Robinson* 6328
(K). W: Mwinilunga Distr., just south of R. Kalalima (Kamwezhi), fl. 23.xii.1937, Milne-Redhead
3800 (K; PRE).
Also in Gabon, Zaire and Angola. Permanently wet dambo, in sandy grassland; 1350–1500 m.
For hybrids between *P. guingangae* and *P. brevirostris*, see under latter species.

9. **Platycoryne isoetifolia** P.J. Cribb in Kew Bull. **32**, 1: 141 (1977). Type: Zambia, Chinsali
Distr., Ishiba Ngandu (Shiwa Ngandu), *Richards* 10672 (K, holotype).
Habenaria sp., Williamson, Orch. S. Centr. Africa: 41 (1977).

Slender terrestrial herb 20–30 cm tall; tubers c. 10 × 7 mm, ovoid, tomentose, up
to 5 cm below ground on the end of a stalk. Leaves 6–10, with up to 9 in a basal tuft
and 1–2 cauline leaves. Basal leaves 12–18 mm long and 1–2 mm wide, narrowly
lanceolate; cauline leaves c. 10 mm long. Inflorescence up to 10 cm long, laxly 1–7-
flowered. Flowers green or yellow-green, suberect. Pedicel c. 10 mm long, ovary c. 10
mm long, making an angle of almost 90° with the pedicel. Bracts c. 7 mm long.
Dorsal sepal erect, 1.5 × 2 mm, ovate, convex; lateral sepals 4 × 1.5 mm, lanceolate,
falcate, adnate to dorsal sepal. Lip 6 mm long, 3-lobed at base; mid-lobe 6 × 2 mm,
ligulate or slightly spathulate, obtuse; side lobes 3 × 1 mm, ligulate, obtuse; spur 15
mm long, pendent, slender but swollen at apex. Anther 1 mm high; canals very short,
incurved. Stigmatic arms 1 mm long, truncate. Rostellum mid-lobe acute, concave,
lying in front of anther loculi; side lobes incurved.

Zambia. N: Chinsali Distr., Ishiba Ngandu (Shiwa Ngandu), 1350 m, fl. 15.i.1959, *Richards* 10672
(K). E: Lundazi Distr., Danger Hill, c. 30 km west of Nyika, fl. ii.1969, *G. Williamson* 1406 (K).
Endemic. Wet or dry dambo; c. 1350 m.

10. **Platycoryne latipetala** Summerh. in Kew Bull. **13**: 64 (1958). —Williamson, Orch. S. Centr.
Africa: 71 (1977). —Geerinck in Fl. Afr. Centr., Orchidaceae pt. 1: 159 (1984). TAB. **50**.
Type: Zambia, Sinkabolo, *Milne-Redhead* 3566 (K, holotype; BR).

Terrestrial herb 30–50 cm tall. Leaves 7–10 spaced along the stem, the 2 lowermost
sheath-like, the largest up to 8 × 0.8 cm, lanceolate. Inflorescence subumbellate,
densely 2–6-flowered. Flowers bright yellow, suberect or slightly spreading. Ovary and
pedicel 13–20 mm long; bracts c. 15 mm long. Dorsal sepal erect, 9.5–10.5 × 6–8 mm,
broadly ovate, convex; lateral sepals spreading and curving up, 9.5–10.5 × 5–5.5 mm,
obliquely ovate-elliptic. Petals 9–10.5 × 4.5–5 mm, broadly curved-lanceolate, adnate
to dorsal sepal. Lip porrect, entire, 7.5–9.5 × 4.5–6 mm, broadly lanceolate or
oblanceolate; spur 13–14.5 mm long, slender, swollen at apex; incurved. Anther 3.5
mm high. Stigmatic arms 2–4 mm long. Rostellum 3.5–4 mm long, the mid-lobe
taller than the side lobes.

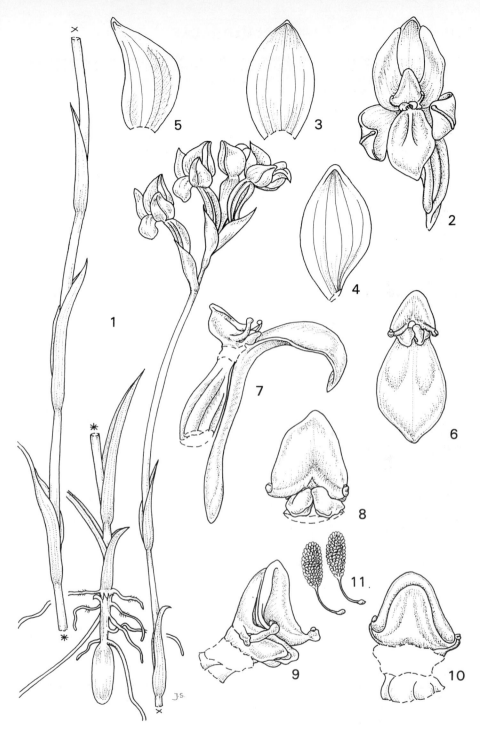

Tab. 50. PLATYCORYNE LATIPETALA. 1, habit (×1), from *Brummitt, Chisumpa & Polhill* 14091; 2, flower (×2); 3, dorsal sepal (×3); 4, lateral sepal (×3); 5, petal (×3); 6, column and lip, front view, spur removed (×3); 7, column and lip, side view (×3); 8, column, front view (×4); 9, column, side view (×4); 10, column, back view (×4); 11, pollinia (×4), 2–11 from *Holmes* 098 (in K spirit coll. 9489). Drawn by Judi Stone.

Zambia. B: Zambezi (Balovale), *Holmes* s.n. (K). W: Mwinilunga Distr., Sinkabolo, fl. 9.xii.1937, *Milne-Redhead* 3566 (K; BR); c. 37 km west of Mwinilunga on Matonchi Road, 1390 m, fl. 24.i.1975, *Brummitt, Chisumpa & Polhill* 14091 (K).
Also in Zaire. Wet, peaty swamp; 1200–1400 m.

11. **Platycoryne macroceras** Summerh. in Kew Bull. **13**: 66 (1958). —Williamson, Orch. S. Centr. Africa: 71 (1977). —Geerinck in Fl. Afr. Centr., Orchidaceae pt. 1: 159 (1984). Type: Zambia, Kalenda, *Milne-Redhead* 3300 (K, holotype; BR; PRE).

Terrestrial herb 20–40 cm tall; tubers up to 17 × 10 mm, ovoid, villous. Leaves 3–6, spaced along the stem, up to 7 × 0.4 cm, linear or linear-lanceolate. Inflorescence 1-flowered, rarely 2-flowered. Flowers cream, creamy-white or lime-yellow (type description says rarely red). Pedicel 4–8 cm long; ovary 1.5 cm long; bracts 2–2.5 cm long. Dorsal sepal erect, 11.5–12 × 8–9 mm, ovate, convex, rounded at the apex or apiculate; lateral sepals spreading, 13–14 × 4–5 mm, curved-lanceolate, oblique. Petals 11.5 × 7–8.5 mm, very curved, obliquely triangular-ovate, adnate to dorsal sepal. Lip entire, deflexed or incurved, 14.5–18 × 3.5–4.5 mm, ligulate with a narrow base, longitudinally convex; spur pendent, 5.5–7 cm long, swollen to a diameter of 2.5–3.5 mm at apex. Anther 5.5 mm high; canals very short. Stigmatic arms 3 mm long. Rostellum 5 mm long; mid-lobe 3 mm high, side lobes almost absent.

Zambia. N: Mbala Distr., top of Kambole Escarpment, 1650 m, fl. 1.ii.1959, *Richards* 10832 (K). W: Mwinilunga Distr., Matonchi Farm, Kalenda Plain, fl. 18.xi.1937, *Milne-Redhead* 3300 (K; BR; PRE).
Also in Zaire. Dry grassland and open dambo, in damp sand amongst grass; 1500–1650 m.

12. **Platycoryne micrantha** Summerh. in Kew Bull. **13**: 71 (1958). —Williamson, Orch. S. Centr. Africa: 72 (1977). Type: Zambia, Mwinilunga Distr., west of Dobeka Bridge, *Milne-Redhead* 2891 (K, holotype).

Terrestrial herb 20–35 cm tall; tubers shortly stalked, c. 1 cm long, ovoid. Leaves 5–7, spaced along the stem, the 2 lowermost ± spreading, 15–25 × 4–8 mm; the rest appressed. Inflorescence c. 15 mm long, densely 1–7-flowered. Flowers bright orange, erect. Pedicel 4 mm long, ovary 10 mm long; bracts c. 7 mm long. Dorsal sepal erect, 4.3–5.3 × 4–4.5 mm, broadly ovate, convex; lateral sepals reflexed, 4.5–5.5 × 2–2.5 mm, obliquely oblong-lanceolate. Petals 4–4.5 × 1.7–2 mm, falcate, free from dorsal sepal. Lip 4.5–5 mm long, 3-lobed in basal half; mid-lobe 3.5–4 × 0.75–1 mm, ligulate, obtuse; side lobes 1–1.5 × 0.4 mm, linear-ligulate, slightly diverging; spur 10–12.5 mm long, swollen and slightly 2-lobed at apex, pendent. Anther 2.5–3 mm high; canals c. 1.5 mm long. Stigmatic arms 1–2 mm long. Rostellum mid-lobe 1–1.5 mm long; side lobes 1–1.3 mm long.

Zambia. N: Mbala Distr., Chinakila, Mpukutu Forest, 1200 m, fl. 14.i.1965, *Richards* 19529 (K). W: Mwinilunga Distr., west of Dobeka Bridge, fl. 27.x.1937, *Milne-Redhead* 2981 (K).
Also in Angola. Marshy grassland; 1200–1300 m.
Summerhayes remarks that this species resembles *P. crocea*, but has smaller flowers with the side lobes of the lip longer and narrower (like those of *P. guingangae*), and leaves not in a basal tuft. It differs from *P. mediocris* in the smaller flowers, narrower lip side-lobes and relatively longer side-lobes of the rostellum.

13. **Platycoryne trilobata** Summerh. in Kew Bull. **13**: 67 (1958). —Williamson, Orch. S. Centr. Africa: 71 (1977). Type from Angola.

Terrestrial herb 30–40 cm tall; tubers c.10 × 5 mm, ovoid. Leaves 5–6, spaced along the stem, the 1–2 lowermost sheathing, the upper ones appressed to stem, the intermediate ones ± spreading, up to 4 × 0.7 cm lanceolate, acute, sheathing at base. Inflorescence up to 8 cm long, laxly 1–4-flowered. Flowers green or green and yellow with an orange lip, erect or suberect. Pedicel 25–30 mm long, ovary 10 mm long; bracts 10–25 mm long. Dorsal sepal erect, c. 8 × 5 mm, ovate, convex; lateral sepals deflexed, c. 7.5 × 2.5 mm, obliquely lanceolate, obtuse. Petals erect, adnate to dorsal sepal, 7 × 2.8 mm, obliquely curved oblong-lanceolate, obtuse, sometimes with a minute lobe on front margin. Lip 5 × 4 mm, deflexed, 3-lobed below the middle. Mid-lobe 4 × 2 mm, oblong, convex; side lobes 1.5 × 0.75 mm, reflexed, narrowly ligulate, obtuse; spur c. 3 cm long, pendent, inflated at apex. Anther 3 mm high; canals 2.5 mm long. Stigmatic arms 3 mm long. Rostellum 5 mm long, taller than the anther; mid-lobe c. 2.5 mm high, side lobes very short.

Zambia. C: Chakwenga Headwaters, fl. 27.x.1963, *E.A. Robinson* 5777 (K).
Also in Angola. Permanently wet dambo.
This species resembles *P. protearum*, having a very similar rostellum, but differs in the 3-lobed lip.

18. ROEPEROCHARIS Rchb.f.

Roeperocharis Rchb.f., Otia Bot. Hamburg.: 104 (1881).

Terrestrial herb with tuberous roots, leafy stem and terminal inflorescence. Flowers green; sepals and petals free. Lip free, 3-lobed (rarely entire), with cylindrical spur. Column erect, the anther loculi divergent, separated by a broad connective; canals hardly developed; pollinaria 2, each with a sectile pollinium, long caudicle and small viscidium. Stigmatic processes each 2-lobed with one lobe projecting down in front of the lip-base, the other lobe upright in front of the anther connective. Rostellum 3-lobed, mid-lobe low, rounded or emarginate, adnate to the connective; side lobes spreading, narrowed towards apex.

A genus of 5 species in eastern Africa, from Ethiopia to Malawi. Two species are found in the Flora Zambesiaca area.

Upper part of petal narrower than lower part, tapering from middle to an acute apex; side lobes
    of lip 9 mm long; spur 5–15 mm long ·1. *bennettiana*
Upper part of petal as broad as or broader than lower part, margin wavy or toothed, apex short
    and ill-defined; side lobes of lip 1.5–4.5 mm long; spur 4–7.5 mm long · · · ·2. *wentzeliana*

1. **Roeperocharis bennettiana** Rchb.f., Otia Bot. Hamburg.: 104 (1881). —Oliv. in Hooker's. Ic.
    Pl. **15**, t. 1500 (1885). —Rolfe in F.T.A. **7**: 250 (1898). —Summerhayes in Kew Bull. **17**: 533
    (1964); in F.T.E.A., Orchidaceae: 149 (1968). —Williamson, Orch. S. Centr. Africa: 74
    (1977). —la Croix et al., Orch. Malawi: 106 (1991). TAB. **51**. Type from Ethiopia.
    Roeperocharis elata Schltr. in Bot. Jahrb. Syst. **53**: 521 (1915). Type from Tanzania.

Robust terrestrial herb 50–100 cm tall; tubers 2–2.5 × 1–2 cm, globose or ellipsoid, glabrous. Stem leafy; leaves up to 10, the 2 lowermost black and sheathing, the rest erect, up to 22 × 1.5 cm, narrowly lanceolate, acute, ribbed, decreasing in size towards the stem apex. Inflorescence 11–24 × 3–4 cm, densely many-flowered. Flowers green. Ovary and pedicel 11–25 mm; bracts leafy, 25–30 mm long, longer than the flowers. Dorsal sepal erect, 6.2–11.5 × 4–6.5 mm, ovate; lateral sepals 6–12.5 × 4.2–6.5 mm, obliquely lanceolate to ovate, acuminate, spreading or reflexed, curving upwards. Petals erect, (5.5)7.5–15 × (2.5)3.5–7 mm, obliquely lanceolate, twisted and folded in the middle, with a distinct acute apex; rather fleshy. Lip 3-lobed, with undivided basal part 2–4.5 mm long; mid-lobe 7.2–16.5 × 1–1.5; side lobes 4–9 × 0.5–1 mm, curving upwards; spur 5–15 mm long, pendent, cylindrical, slightly bifid at apex. Column c. 2.5 mm high; anther connective 3–3.5 mm wide, papillose; loculi diverging, 2.5–3.5 mm long, canals 1.5–2 mm long. Stigmatic arms papillose, 2-lobed from base; upper lobe 2–4 mm long, lower lobe 3–5 mm long.

Zambia. E: Nyika National Park, fl. iii.1967, *Williamson & Odgers* 304 (K). **Malawi**. N: Nyika National Park, edge of thicket on road to Chowo Forest, 2200 m, fl. 21.iv.1982, *la Croix* 322 (K). Also in Ethiopia, Kenya and Tanzania. Montane grassland, usually in drier areas; 2200–2440 m.

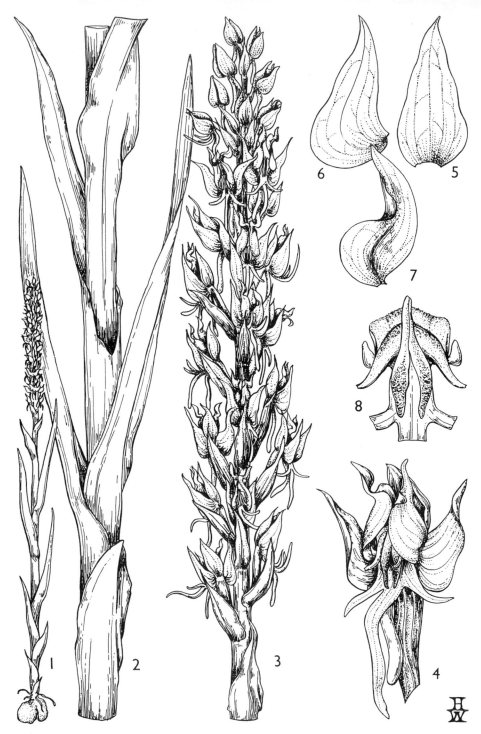

Tab. 51. ROEPEROCHARIS BENNETTIANA. 1, habit ($\times\frac{1}{6}$); 2, lower stem ($\times$1); 3, inflorescence ($\times$1), 1–3 from *Stolz* 2492; 4, flower ($\times$2); 5, dorsal sepal ($\times$3); 6, lateral sepal ($\times$3); 7, petal ($\times$3); 8, column, front view ($\times$3), 4–8 from *Holmes* 0227. Drawn by Heather Wood. From F.T.E.A.

2. **Roeperocharis wentzeliana** Kraenzl. in Bot. Jahrb. Syst. **30**: 283 (1901). —Summerhayes in F.T.E.A., Orchidaceae: 149 (1968). —Williamson, Orch. S. Centr. Africa: 74 (1977). —la Croix et al., Orch. Malawi: 106 (1991). Type from Tanzania.

Terrestrial herb 30–85 cm tall; tubers up to 2 cm long, ellipsoid. Stem leafy; leaves up to 10, the 2 lowermost sheathing, the rest erect, up to 14.5 × 4 cm, lanceolate or linear-lanceolate, acute, ribbed. Inflorescence 5–17 × 2–3 cm, densely many-flowered. Flowers green. Ovary and pedicel 8–10 mm long; bracts leafy, c. 20 mm long, longer than the flower. Dorsal sepal erect, 6.5–10 × 3.5–5 mm, ovate, acute; lateral sepals spreading or reflexed, 8–10 × 4–5 mm, curved upwards, obliquely lanceolate, acuminate. Petals erect, 7.5–11 × 3–4 mm, fleshy, obliquely oblong, twisted and folded in the middle, the upper part serrated, the apex indistinct. Lip 3-lobed; basal undivided part c. 3 mm long; mid-lobe 9.5–12 × 1–1.5 mm, linear, obtuse; side lobes 1.5–4.5 × 0.5 mm, narrowly linear; spur pendent, 4–7 mm, slightly bifid at apex. Column 2–2.5 mm long; anther connective 3–4 mm wide, papillose; loculi diverging, 2–3 mm long, canals 1.5–2 mm long. Stigmatic arms papillose, 2-lobed from the base; upper lobe 1.5–3.5 mm long; lower lobe 3–3.5 mm long.

Zambia. E: Nyika Plateau, fl. iii.1967, *G. Williamson* 296 (K; SRGH). **Malawi**. N: Nyika National Park, c. 2 km north of Chosi Peak, fl. 19.iv.1986, *Philcox, Pope & Chisumpa* 10030 (K). S: Mulanje Mt., Madzeka Basin, 1750–2100 m, fl. 9.iv.1974, *Allen* 533 (K).

Also in Tanzania. Montane grassland, usually in damper areas, montane bog; 1700–2440 m.

19. DISA
by H.P. Linder

Disa P.J. Bergius, Descr. Pl. Cap.: 348 (1767).

Terrestrial herb with perennating tubers. Tubers testicular, rarely absent. Sterile shoots produced occasionally, to 15 cm long with several basal sheaths and several erect conduplicate acute leaves to 40 cm long. Fertile shoots to 100 cm tall. Leaves radical, basal or cauline, linear to ovate, soft to rigid, lax to imbricate, monomorphic to dimorphic, green at flowering time, occasionally hysteranthous, conduplicate to flat, grading into the floral bracts or sharply distinct from the floral bracts. Inflorescence racemose with one to numerous flowers. Flowers not resupinate, resupinate or doubly resupinate, 4–80 mm in diameter, usually red or white, rarely blue or yellow or green. Bracts usually green, ovate, acute or acuminate, rarely dry and brown. Dorsal sepal erect, galeate, rarely cucullate or dish-shaped, usually ovate, rarely oblong or spathulate, with a spur which is obsolete to 8 cm long, slender or rarely massive, rarely flattened or with a dorsal or ventral groove, horizontal, descending or ascending. Lateral sepals lanceolate to broadly ovate, rounded to acuminate, often keeled, patent, sometimes ascending or descending. Petals lanceolate to square, erect or reflexed, acute or truncate, the apex curving over the anther or over the rostellum, or curved out of the galea, often with a complex plane shape. Lip usually lanceolate, rarely spathulate or broadly ovate, occasionally linear. Rostellum 3-lobed; lateral lobes erect, square or horn-like, flat or canaliculate, free or fused to the petals by a long keel; central lobe usually small, fleshy, rarely tall. Anther 2-celled, horizontal, ascending or pendent; viscidia usually separate, borne on the lateral rostellum arms. Stigma pedicellate or sessile, the surface flat, concave or convex, tripulvinate with the rear cushion often smaller than the lateral cushions.

A genus comprising c.123 species, widespread in Africa and Madagascar, reaching the Yemen and Réunion Island.

Disa danieliae Geerinck has recently been described from Kolwezi, in the Shaba Province of Zaire, and from Mexico in Angola, but has not yet been recorded from Zambia. It may be distinguished from *Disa welwitschii* by a subclavate spur which reaches to the base of the dorsal sepal, by a subacute posterior petal lobe, and by a sparse inflorescence.

1. Petals 2-lobed · 2
– Petals not 2-lobed · 17
2. Spur reaching to below the base of the dorsal sepal · 3
– Spur much shorter than the dorsal sepal, or reaching to the base of the dorsal sepal · · 10
3. Spur ± inflated; flowers in a long dense tapering spike, yellow or orange; dorsal sepal more
   than 6 mm long · 4
– Flowers not with the above combination of characters · 5
4. Anterior petal lobe as tall as the posterior petal lobe and curved in front of the anther;
   lateral sepals 8–10 mm long · 4. *satyriopsis*
– Anterior petal lobe smaller than the posterior petal lobe, flanking the anther; lateral sepals
   6–9 mm long · 3. *ochrostachya*
5. Petal lobes sub-equal; posterior petal lobe truncate · · · · · · · · · · · · · · · · · 9. *aequiloba*
– Petal lobes unequal, if equal in length, then the posterior lobe not truncate · · · · · · · · 6
6. Posterior petal lobes shorter than anterior petal lobes; bracts prominent · · · · · · · · · · 7
– Posterior petal lobes as long as or longer than the anterior petal lobes · · · · · · · · · · · 8
7. Spur clavate, reaching to the base of the dorsal sepal · · · · · · · · · · · · · · 11. *miniata*
– Spur slender, reaching to below the base of the dorsal sepal · · · · · · · · · · · 10. *cryptantha*
8. Bracts prominent, partially obscuring the flowers · 8. *celata*
– Bracts not prominent and not partially obscuring the flowers · · · · · · · · · · · · · · · · · 9
9. Spur 8–13 mm long; dorsal sepal 7.5–11.5 mm long; posterior petal lobe truncate, taller
   than the anterior lobe by one third; flowers red · · · · · · · · · · · · · · · · 7. *roeperocharoides*
– Spur 4–8 mm long; dorsal sepal 4–13 mm long; posterior lobe variable; flowers red or pink
   · 6. *welwitschii*
10. Dorsal sepal less than 8 mm long · 11. *miniata*
– Dorsal sepal (12)14–45 mm long · 11
11. Posterior petal lobe straight · 12
– Posterior petal lobe twisted through an S-curve · 15
12. Claw of the dorsal sepal as long as the blade · 13
– Claw of the dorsal sepal about half as long as, or shorter, than the blade · · · · · · · · · · 14
13. Anterior lobe of the petal longer than the claw of the dorsal sepal; dorsal sepal 20 mm long
   or longer · 15. *verdickii*
– Anterior lobe of petal much shorter than the claw of the dorsal sepal; dorsal sepal 20 mm
   long or shorter · 14. *zombica*
14. Petal less than 8 mm long; inflorescence dense; plant from Nyika Plateau (and Njombe,
   Tanzania) · 13. *ukingengsis*
– Petal more than 8 mm long; inflorescence lax; plant not from the above localities · · · · · ·
   · 12. *engleriana*
15. Spur ascending; bracts reaching to middle of dorsal sepal, or shorter · · · · · 18. *erubescens*
– Spur pendent, parallel to the dorsal sepal; bracts usually overtopping the flowers · · · · 16
16. Anterior lobe of the petal shorter than the claw of the dorsal sepal; dorsal sepal more than
   3 cm long, blade curved upwards · 17. *katangensis*
– Anterior lobe of petal as long as the dorsal sepal; dorsal sepal less than 3 cm long, erect ·
   · 16. *ornithantha*
17. Spur, at least at the base, ascending · 18
– Spur, at least at the base, horizontal or pendent · 24
18. Spur more than 15 mm long · 19
– Spur less than 15 mm long · 20
19. Lateral sepals more than 10 mm long · 31. *robusta*
– Lateral sepals less than 10 mm long · 26. *rhodantha*
20. Lateral sepals more than 8 mm long · 29. *caffra*
– Lateral sepals less than 8 mm long · 21
21. Spur curved downwards in the apical half; lateral sepal more than 4 mm long · · · · · · 22
– Spur curved upwards in the apical half; lateral sepal less than 4 mm long · · · · · · · · · 23
22. Lip narrowly oblong, 1.3 mm or less wide; flowers purple · · · · · · · · · · · · 27. *hircicornis*
– Lip obovate, 1.3 mm or more wide; flowers red, white and green · · · · · · · · · 28. *perplexa*
23. Leaves more than 6 mm wide, the margins flat; lateral sepal more than 2.5 mm long; from
   south of the Zambezi River · 37. *zimbabweensis*
– Leaves less than 6 mm wide, the margins undulate or crisped; lateral sepal less than 3 mm
   long; from north of the Zambezi River · 36. *rungweensis*
24. Spur pendent from the base; anther erect · 25
– Spur horizontal at the base; anther deflexed · 27

25. Spur as long as the dorsal sepal, 4–10 mm long; flowers white-mottled $\cdots\cdots$ 5. *fragrans*
– Spur shorter than dorsal sepal, 0.8–4.5 mm long; flowers red or yellow, not mottled $\cdots$ 26
26. Flowers red; spur 1.5–4.5 mm long, usually reaching below base of galea $\cdot$ $\cdot$2. *polygonoides*
– Flowers yellow; spur 0.8–1.5(2.5) mm long, usually only reaching to the base of the galea
$\cdots\cdots\cdots\cdots\cdots\cdots\cdots\cdots\cdots\cdots\cdots\cdots\cdots\cdots\cdots\cdots\cdots\cdots$1. *woodii*
27. Spur straight, or the apex curved upwards, usually massive (one third as thick as the height of the galea) or reduced to a small cone $\cdots\cdots\cdots\cdots\cdots\cdots\cdots\cdots\cdots\cdots\cdots$ 28
– Spur at length curving downwards, slender, usually as long as or longer than sepals $\cdot\cdot$ 32
28. Lateral sepals longer than the dorsal sepal; flowers with some green colouring $\cdot$21. *aperta*
– Lateral sepals shorter than the dorsal sepal; flowers never with green colouring $\cdots\cdots$ 29
29. Spur reduced to a small cone, only half as long as the dorsal sepal or shorter 20. *similis*
– Spur about as long as the dorsal sepal $\cdots\cdots\cdots\cdots\cdots\cdots\cdots\cdots\cdots\cdots\cdots$ 30
30. Spur base not in line with the base of the dorsal sepal, cylindrical and often subclavate; petals obtuse; flowers purple $\cdots\cdots\cdots\cdots\cdots\cdots\cdots\cdots\cdots$ 25. *equestris*
– Spur base in line with the base of the dorsal sepal; petals acute; flowers white with purple-blue mottling $\cdots\cdots\cdots\cdots\cdots\cdots\cdots\cdots\cdots\cdots\cdots\cdots\cdots\cdots\cdots$ 31
31. Plants with inflorescence drying much paler than the leaves; upper and lower margins of the spur parallel; inflorescence lax $\cdots\cdots\cdots\cdots\cdots\cdots\cdots\cdots\cdots$ 23. *dichroa*
– Plants drying an even colour; spur usually tapering; inflorescence dense $\cdot$ $\cdot$22. *aconitoides*
32. Spur more than twice as long as the sepals $\cdots\cdots\cdots\cdots\cdots\cdots\cdots\cdots\cdots$ 33. *eminii*
– Spur less than twice as long as the sepals $\cdots\cdots\cdots\cdots\cdots\cdots\cdots\cdots\cdots\cdots$ 33
33. Lip linear-spathulate; petals erect inside the galea, but exserted at the base and contributing to forming the galea $\cdots\cdots\cdots\cdots\cdots\cdots\cdots\cdots\cdots\cdots\cdots$ 34
– Lip not spathulate; petals generally not visible from outside the galea $\cdots\cdots\cdots\cdots$ 35
34. Petals less than 5 mm long; galea constricted in the middle; flowers white $\cdots$ 34. *saxicola*
– Petals 5 mm long or more; galea not constricted in the middle; flowers pink $\cdot\cdot$ 35. *patula*
35. Lateral sepals 3.5–6(7.5) mm long, if more than 6.5 mm long the flowers are purple and petals oblanceolate $\cdots\cdots\cdots\cdots\cdots\cdots\cdots\cdots\cdots\cdots\cdots\cdots\cdots\cdots$ 36
– Lateral sepals usually more than 7 mm long, if 7 mm long or less, flowers rose to pink coloured and petals lanceolate $\cdots\cdots\cdots\cdots\cdots\cdots\cdots\cdots\cdots\cdots\cdots\cdots$ 39
36. Spur massive, the diameter about one third the height of the galea; petals fused to rostellum by a long keel $\cdots\cdots\cdots\cdots\cdots\cdots\cdots\cdots\cdots\cdots\cdots$ 24. *nyikensis*
– Spur slender, diameter less than a quarter the height of galea; petals not fused to rostellum by a keel $\cdots\cdots\cdots\cdots\cdots\cdots\cdots\cdots\cdots\cdots\cdots\cdots\cdots\cdots$ 37
37. Petals curved over the rostellum, forming two passages to the spur, obliquely ovate in outline; spur often abruptly decurved at the base $\cdots\cdots\cdots\cdots\cdots\cdots$ 32. *versicolor*
– Petals not as above, obovate-oblong in outline; spur only decurved about half-way down its length $\cdots\cdots\cdots\cdots\cdots\cdots\cdots\cdots\cdots\cdots\cdots\cdots\cdots\cdots\cdots$ 38
38. Lip narrowly oblong, 1.3 mm or less wide; flowers purple $\cdots\cdots\cdots\cdots$ 27. *hircicornis*
– Lip obovate, 1.3 mm or more wide; flowers red, white and green $\cdots\cdots\cdots$ 28. *perplexa*
39. Lip lorate to obovate, fleshy, upper surface convex; petal a with large basal anticous lobe flanking the stigma $\cdots\cdots\cdots\cdots\cdots\cdots\cdots\cdots\cdots\cdots\cdots\cdots\cdots$ 19. *cornuta*
– Lip linear to narrowly elliptic, flat; petal narrowly lanceolate without a large basal anticous lobe $\cdots\cdots\cdots\cdots\cdots\cdots\cdots\cdots\cdots\cdots\cdots\cdots\cdots\cdots\cdots$ 40
40. Lateral sepals less than 13 mm long; plants less than 60 cm tall $\cdots\cdots\cdots\cdots$ 29. *caffra*
– Lateral sepals more than 13 mm long; plants more than 50 cm tall $\cdots\cdots\cdots$ 30. *walleri*

1. **Disa woodii** Schltr. in Ann. Transvaal Mus. **10**: 247 (1924). —Linder in Bull. Jard. Bot. Belg. **51**: 273 (1981). —Stewart et al., Wild Orch. S. Africa: 141 (1982). Type from South Africa. *Disa polygonoides* sensu Grosvenor in Excelsa **6**: 80 (1976) quoad specim. *Ball 572.*

Terrestrial herb with separate fertile and sterile shoots, perennating by testicular tubers. Sterile shoot not always developed, c. 5 cm long when present, with 3–4 semi-erect, acute, linear to lorate leaves up to 25 cm long. Fertile shoots 15–70 cm long, cauline leaves 6–25, sheathing at the base, free blade more or less conduplicate, linear to narrowly lanceolate and acute. Inflorescence dense, 3–17 cm long, cylindrical, 80–100-flowered. Floral bracts narrowly ovate, acute to subacuminate, usually as tall as the flowers with the apices reflexed. Flowers bright yellow, often with apices of the parts orange-tinted. Dorsal sepal forming an erect galea, 5–7 mm tall, 3–4 mm wide and c. 1 mm deep, elliptic to subobovate, obtuse. Spur pendent from just above the base of the galea, 0.8–1.5(2.5) mm long. Lateral sepals 6–7 mm long, narrowly oblong, acute. Petals 4–6 mm long, narrowly obovate, acute. Lip c. 6 mm long, lorate to linear. Rostellum simple. Anther erect, c. 3.5 mm long. Stigma with well developed lateral lobes. Ovary 5–10 mm long.

Zimbabwe. E: Chipinge (Chipinga), fl. 27.iv.1956, *Ball* 572 (K; SRGH). **Mozambique**. M: Maputo, fl. x.1979, *Schäfer* 7001 (K; UPS).
Also from South Africa. Grows in well-drained to wet grassland, occasionally as a weed at roadsides.
This species is very closely related to *Disa polygonoides*, and herbarium specimens may be difficult to determine. However, fresh material can be readily distinguished on flower colour.

2. **Disa polygonoides** Lindl., Gen. Sp. Orchid. Pl.: 349 (1838). —Hook.f. in Curtis, Bot. Mag.:
    t. 6532 (1880). —Kraenzlin, Orchid. Gen. Sp. **1**: 747 (1900). —Schlechter in Bot. Jahrb.
    Syst. **31**: 222 (1901). —Bolus, Icon. Orchid. Austro-Afric **2**: t. 84 (1911). —Rolfe in F.C. **5**,
    3: 226 (1913). —Compton, Fl. Swaziland: 157 (1976). —Linder in Bull. Jard. Bot. Belg. **51**:
    269 (1981). —Stewart et al., Wild Orch. S. Africa: 141 (1982). Type from South Africa.

Terrestrial herb with separate fertile and sterile shoots, perennating by testicular tubers. Sterile shoots rare, produce 3–4 semi-erect leaves c. 25 cm long, linear to lorate, acute. Fertile shoots 15–70 cm long; cauline leaves imbricate, few if sterile shoots are present, numerous if not, largest at the base of the shoot, to 25 cm long, linear to narrowly lanceolate, acute. Inflorescence dense, 3–17 cm long, cylindrical, 80–100-flowered. Floral bracts narrowly ovate, acute to subacuminate, as tall as the flowers with the apices reflexed. Flowers brick-red, rarely orange. Dorsal sepal forming an erect galea, 5–7 mm long, 3–4 mm wide and c. 1 mm deep, elliptic to subobovate, obtuse. Spur pendent from just above the base of the galea, reaching to below the base of the galea, 1.5–3.3–4.5 mm long, cylindrical. Lateral sepals 6–7 mm long, narrowly oblong, acute. Petals 4–6 mm long, narrowly obovate, acute. Lip c. 6 mm long, lorate to linear, rounded. Rostellum simple. Anther erect, c. 3.5 mm long. Stigma with well developed lateral lobes. Ovary 5–10 mm long.

Mozambique. GI: Zandamela, *Stephens* s.n. (PRE).
Also in Swaziland and South Africa. Commonly in damp or wet grassland; sea level to 1000 m. Rare in the Flora Zambesiaca area.

3. **Disa ochrostachya** Rchb.f. in Flora **48**: 181 (1865). —N.E. Br. in F.T.A. **7**: 279 (1898). —
    Kraenzlin, Orchid. Gen. Sp. **1**: 749 (1900). —Schlechter in Bot. Jahrb. Syst. **31**: 217
    (1901). —Summerhayes in F.W.T.A. **2**: 414 (1936); in Bot. Not. **1937**: 188 (1937). —Piers,
    Orchids E. Africa: 61 (1968). —Summerhayes in F.T.E.A., Orchidaceae: 164 (1968). —
    Geerinck in Bull. Soc. Roy. Bot. Belg. **107**: 68 (1974). —Stewart in Agnew, Upland Kenya
    Wild Flowers: 746 (1974). —Grosvenor in Excelsa **6**: 80 (1976). —Williamson, Orch. S.
    Centr. Africa: 83 (1977). —Linder in Bull. Jard. Bot. Belg. **51**: 280 (1981). —Geerinck in
    Fl. Afr. Centr., Orchidaceae pt. 1: 186 (1984). —la Croix et al., Orch. Malawi: 112 (1991).
    Type from Angola.
    Disa aurantiaca Rchb.f. in Flora **50**: 98 (1867). Type from Angola.
    Disa adolphi-friderici Kraenzl. in Bot. Jahrb. Syst. **43**: 131 (1901). Type from Zaire.
    Disa ochrostachya var. *major* Rendle in J. Linn. Soc., Bot. **37**: 222 (1905). Type from Uganda.

Sterile shoots to 15 cm tall, with 2–3 erect, acute narrowly elliptic conduplicate leaves 15–20 × 4 cm. Fertile shoots slender to robust, 40–90 cm tall; cauline leaves imbricate, the basal leaf completely sheathing, the remainder with small free blades 5–9 × 1–3(4) cm, acute or apiculate. Inflorescence dense, 15–30 cm long, cylindrical, tapering, 50–200-flowered. Bracts 1.5–2.5 cm long, acute to acuminate, with the apices reflexed. Flowers c. 1 cm in diameter, bright yellow or yellow with orange mottling. Dorsal sepal obovate, obtuse, galea 6–9 mm long, 4–7 mm wide and 2–3 mm deep, curved to face down. Spur pendent from the middle of the galea, 7–12 mm long, ± inflated in the lower half, subacute. Lateral sepals 6–9 × 3–6 mm, narrowly oblong, obtuse to subacute, spreading sideways. Petals bifid; posterior lobe 5–7.5 × 1–2 mm, obtuse or subacute, erect behind the anther; anterior lobe 3–6 mm long, narrow and reaching two thirds up the length of the main petal and flanking the anther, rarely taller. Lip 5–9 mm long, linear, rarely irregularly toothed, pendent. Stigma 0.5–1 mm tall, lateral lobes well developed. Ovary 1–1.5 cm long.

Platylepis glandulosa

Zeuxine ballii

Epipactis africana

Brachycorythis buchananii

Schwartzkopffia lastii

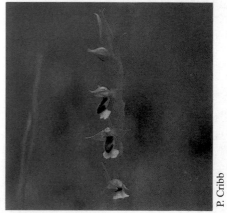

Neobolusia stolzii

Plate 1

Schizochilus sulphureus

E. la Croix

Holothrix longiflora

E. la Croix

Holothrix papillosa

E. la Croix

Stenoglottis zambesiaca

A. Gassner

Bonatea steudneri

J. Tanner

Centrostigma clavatum

A. Gassner

Plate 2

Cynorkis kirkii   E. la Croix

Habenaria hologlossa   E. la Croix

Habenaria insolita   E. la Croix

Habenaria splendens   E. la Croix

Habenaria trachypetala   P. Leedal

Habenaria amoena   C. Grey-Wilson

Plate 3

Habenaria leucoceras

Habenaria schimperiana

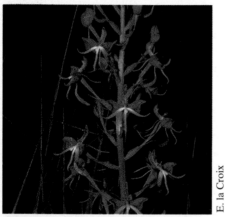

Habenaria rautaneniana

Habenaria clavata

Habenaria holubii

Habenaria walleri

Plate 4

Habenaria lindblomii   P. Cribb

Habenaria subarmata   E. la Croix

Platycoryne buchananiana   P. Cribb

Roeperocharis bennettiana   E. la Croix

Disa engleriana   E. la Croix

Disa perplexa   P. Crilbb

Plate 5

P. Cribb

Disa walleri

E. la Croix

Disa robusta

E. la Croix

Brownleea maculata

A. Gassner

Herschelianthe baurii

I. la Croix

Monadenia brevicornis

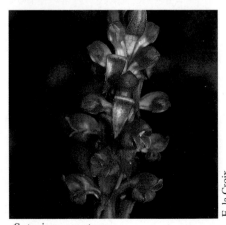

E. la Croix

Satyrium sceptrum

Plate 6

Satyrium anomalum

E. la Croix

Satyrium trinerve

P. Cribb

Disperis dicerochila

C. Grey-Wilson

Vanilla polylepis

F. Piers

Nervilia kotschyi

E. la Croix

Nervilia crociformis & N. kotschyi

P. Cribb

Plate 7

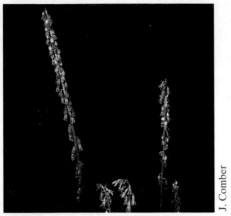

Epipogium roseum

J. Comber

Calanthe sylvatica

E. la Croix

Liparis nervosa

C. Grey-Wilson

Malaxis weberbaueriana

E. la Croix

Stolzia repens

D. Menzies.

Bulbophyllum maximum

J. Tanner

Plate 8

Zambia. N: Kasama, Mungwi, fl. 18.xii.1960, *E.A. Robinson* 4198 (K; M; SRGH). E: Nyika Plateau, fl. 1.i.1964, *Benson* 396 (BR; K; LISC). **Zimbabwe**. E: Nyanga (Inyanga) Village, fl. 30.i.1931, *Norlindh & Weimarck* 4766 (BM; BR; K). **Malawi**. N: Nyika Plateau, fl. ix.1968, *G. Williamson* 1040 (K; SRGH).
Also in Cameroon, Kenya, Uganda, Tanzania, Rwanda, Zaire and Angola.
Widespread in the montane grasslands of subsaharan Africa; mostly 1500–2600 m.

4. **Disa satyriopsis** Kraenzl. in Bot. Jahrb. Syst. **28**: 177 (1901). —Linder in Bull. Jard. Bot. Belg. **51**: 284 (1981). —la Croix et al., Orch. Malawi: 113 (1991). Holotype from Tanzania, *Goetze* 577 (B†). Neotype: Zambia, Nyika Plateau, Feb. 1968, *Williamson, Simon & Ball* 357 (SRGH, neotype chosen by Linder loc. cit. 1981).
    Disa ochrostachya var. *latipetala* G. Will. in Plants. Syst. Evol. **134**: 60 (1980). Type as above.

Terrestrial herb with separate sterile and fertile shoots. Sterile shoots slender to robust, up to 15 cm tall, with 2–3 narrowly elliptic, acute, erect, conduplicate leaves, up to 20 cm long and 4 cm wide. Fertile shoots 40–70 cm tall; leaves imbricate, the basal 1–3 leaves small, completely sheathing, rounded to obtuse; the remainder 4–8 cm long, ovate to narrowly ovate, acute. Inflorescence dense, 10–30 cm long, c. 150-flowered. Bracts 1.5–2.5 cm long, lanceolate, acuminate, the apex reflexed, dry. Flowers c. 1 cm in diameter, orange-red with darker mottling. Dorsal sepal galeate, 9–10 mm long and 3–4 mm wide, obovate, obtuse, curved to face ± downwards. Spur pendent, clavate, subacute, 7.5–10 mm long. Lateral sepals 8–10 mm long, narrowly ovate, subacute, spreading or reflexed. Petals 2-lobed; the posterior lobe erect, 7–8.5 mm long, lorate, acute; the anterior lobe originating about halfway up the posterior lobe, equalling the posterior lobe in length, the front margin crenulate or undulate, curved in front of the anther. Lip ± pendent, 7–10 mm long, lorate, acute. Rostellum c. 1.5 mm tall. Anther erect, 4 mm long. Stigma 3 mm in diameter. Ovary 1 cm long.

Zambia. N: Nyika Plateau, road to Zambian Rest House, fl. 5.ii.1968, *Ball* 1071 (SRGH). **Malawi**. N: Nyika Plateau, fl. 9.ii.1960, *Holmes* 0201 (K; SRGH).
Also recorded from the Southern Highlands and Western Province of Tanzania, and from Burundi. Montane grasslands; 1800–2500 m, sometimes very common locally.
Although this species is closely allied to the preceding species, they can be readily separated on petal and spur shape.

5. **Disa fragrans** Schltr. in Bot. Jahrb. Syst. **20**, Beibl. Nr. 50: 40 (1895). —Kraenzlin, Orchid. Gen. Sp. **1**: 748 (1900). —Schlechter in Bot. Jahrb. Syst. **31**: 223 (1901). —Rolfe in F.C. **5**, **3**: 225 (1913). —Grosvenor in Excelsa **6**: 80 (1976). —Linder in Bull. Jard. Bot. Belg. **51**: 286 (1981). —Stewart et al., Wild Orch. S. Africa: 142 (1982). Type from South Africa.
    Disa leucostachys Kraenzl. in Bot. Jahrb. Syst. **30**: 285 (1901). —Schlechter in Bot. Jahrb. Syst. **53**: 540 (1915). —Goodier & Phipps in Kirkia **1**: 53 (1961). Type from Tanzania.
    Monadenia junodiana Kraenzl. in Vierteljahrsschr. Naturf. Ges. Zürich **74**: 108 (1929). Type from South Africa.

Subsp. **fragrans** —la Croix et al., Orch. Malawi: 111 (1991).

Terrestrial herb with separate sterile and fertile shoots. Sterile shoots to 7 cm tall, usually with 2 semi-erect, narrowly elliptic, acute, conduplicate leaves, 7–20 cm long and 1–3 cm wide. Fertile shoots robust, 7–50 cm tall; leaves imbricate, the lower 2–3 reduced to sheaths; the remainder with free blades, 3–10 cm long, narrowly lanceolate to ovate, acute. Inflorescence dense, 3–10 cm long, cylindrical, 30–150-flowered. Floral bracts 1.5 cm long, narrowly ovate, acute, to the apices frequently reflexed. Flowers strongly scented, white, pale lilac or rarely crimson, usually with mottling darker than the base colour, c. 8 mm in diameter. Dorsal sepal galeate, 4.5–7 mm long and 2.5–4 mm wide and 1–2 mm deep, oblong to elliptic-oblong, obtuse, erect, the upper margin incurved. Spur arising one third of the galea length above the galea base, (4)7.5–10 mm long, always reaching to below the galea base, slender to apically clavate. Lateral sepals 5–6.5 mm long, oblong or obtuse to subacute, spreading sideways. Petals entire, 3.5–6.5 mm long, oblanceolate, obtuse, erect with the apices curved over the anther. Lip 3–6 mm long, narrowly oblanceolate. Rosetellum c. 1 mm tall, anther semi-erect, 2.5–3 mm long. Stigma 0.5–1 mm tall. Ovary c. 1 cm long.

Zimbabwe. E: Nyanga (Inyanga), Rhodes Estate, Hill Fort, fl. 19.iv.1953, *Chase* 4982 (BM; BOL; K; SRGH). **Malawi**. S: Mulanje Mt., Chilemba Peak, fl. 7.vi.1962, *Richards* 16562 (K). **Mozambique**. MS: Gorongosa Mt., fl. 4.vii.1955, *Schelpe* 5492 (BM; BOL).
Also known from the Southern Highlands of Tanzania and South Africa. Short montane-grassland, where it can form dense stands.
The flowers are somewhat variable in colour and generally very strongly scented.
Subsp. *deckenii* (Rchb.f.) H.P. Linder extends from northern Tanzania through Kenya to Ethiopia. It is distinguished from the typical subspecies by the acute petals, the rose-coloured to crimson flowers, and the shorter, 4–6 mm long, spur.

6. **Disa welwitschii** Rchb.f. in Flora **48**: 181 (1865). —N.E. Br. in F.T.A. **7**: 280 (1898). — Kraenzlin, Orchid. Gen. Sp. **1**: 752 (1900). —Schlechter in Bot. Jahrb. Syst. **31**: 219 (1901). —Summerhayes in Kew Bull. **17**: 543 (1964). —Piers, Orchids E. Africa: 66 (1968). —Summerhayes in F.T.E.A., Orchidaceae: 160 (1968). —Geerinck in Bull. Soc. Roy. Bot. Belg. **107**: 70 (1974). —Grosvenor in Excelsa **6**: 80 (1976). —Williamson, Orch. S. Centr. Africa: 80 (1977). —Linder in Bull. Jard. Bot. Belg. **51**: 301 (1981). —Stewart et al., Wild Orch. S. Africa: 143 (1982). —Geerinck in Fl. Afr. Centr., Orchidaceae pt. 1: 183 (1984). —la Croix et al., Orch. Malawi: 114 (1991). Type from Angola.

Terrestrial herb with separate sterile and fertile shoots. Sterile shoots to 15 cm tall, with 2–4 suberect conduplicate leaves, 10–30 × 1–4 cm, linear-elliptic, acute or apiculate; basal sheath, often blotched or red-striped. Fertile shoots 20–100 cm tall; leaves imbricate to subimbricate, the basal 2 reduced to short sheaths blotched with red; the remainder 3–10 cm long, lanceolate, subacuminate, mostly sheathing. Inflorescence dense, 3–14 cm long, 20–100-flowered. Floral bracts inconspicuous, as long as the ovaries, taller than ovaries at base of inflorescence, narrowly ovate, acuminate. Flowers bright red to carmine, or pink. Dorsal sepal subrhomboid; galea 4.5–11 mm long, narrowly oblong to lanceolate, subacute. Spur pendent from below the middle of the galea, reaching to below the base of the sepal, slender, 3.5–9 mm long, cylindrical to rarely subclavate. Lateral sepals suboblique, 5–12 × 3–6(8) mm, oblong-ovate, obtuse to rounded. Petals 2-lobed; anterior lobe 3–7(8.5) × 2.5–6 mm, ovate, spreading, the upper margin incurved; posterior lobe equalling anterior lobe in length, or exceeding it by one third, truncate or ovate lanceolate. Lip 4–12 mm long, linear. Anther 3–5 mm long. Stigma equally 3-lobed. Ovary 10–20 mm long.

Flowers red or carmine; lateral sepals 5–9 mm long; posterior petal lobe generally one third taller than anterior lobe; from south and south-central Africa · · · · · · · subsp. *welwitschii*
Flowers pink; lateral sepals 7–12 mm long; posterior petal lobe as tall as or slightly taller than anterior lobe; from East, Central and West Africa · · · · · · · · · · · · · · · subsp. *occultans*

Subsp. **welwitschii**
  Disa welwitschii var. *buchneri* Schltr. in Bot. Jahrb. Syst. **31**: 219 (1901). Syntypes from Angola.
  Disa calophylla Kraenzl. in Bot. Jahrb. Syst. **33**: 58 (1901). Type from Tanzania.
  Disa ignea Kraenzl. in Bot. Jahrb. Syst. **33**: 57 (1902). Type from Tanzania.
  Disa hyacinthina Kraenzl. in Bot. Jahrb. Syst. **48**: 390 (1912). Type: Malawi, Msambia, *Fromm-Muenzner* 199 (B†, holotype).
  Disa breyeri Schltr. in Ann. Transvaal Mus. **10**: 247 (1924). Type from South Africa.

Zambia. B: Kaoma (Mankoya), fl. 15.i.1960, *Gilges* 908 (K). N: Mporokoso Distr., Lumangwe Falls, fl. 8.i.1960, *Richards* 12098 (K; SRGH). W: Mwinilunga Distr., by River Kasompa, fl. 28.x.1937, *Milne-Redhead* 2991 (BR; K; PRE). C: Mkushi, fl. 9.i.1958, *E.A. Robinson* 2304 (K). E: Chipata-Katete, fl. 9.i.1959, *Robson* 1116 (K). S: Mazabuka, Siamambo Forest Reserve, fl. 30.i.1960, *White* 6597 (K; SRGH). **Zimbabwe**. C: Makoni, fl. 19.i.1976, *Best* 1255 (BOL; K; SRGH). E: Stapleford, fl. 22.ii.1946, *Wild* 940 (K; SRGH). **Malawi**. N: Nyika Plateau, fl. xii.1966, *G. Williamson* 226 (K; SRGH). C: near Kasungu Hill, fl. 14.i.1959, *Robson* 1163 (K; LISC; SRGH). S: Shire Highlands, fl. ?, *Buchanan* 313 (K). **Mozambique**. N: Imala, fl. 14.i.1964, *Torre & Paiva* 9969 (LISC).
Also in Tanzania, Burundi, Zaire, Angola and South Africa. Mostly recorded from damp grasslands, usually dambos, between 900 and 2400 m, common.

Subsp. **occultans** (Schltr.) H.P. Linder in Bull. Jard. Bot. Belg. **51**: 306 (1981). Type from Kenya.
  Disa occultans Schltr. in Notizbl. Bot. Gart. Berlin **8**: 225 (1922). Type as above.
  Disa welwitschii var. *occultans* (Schltr.) Geerinck, Bull. Jard. Bot. Belg. **52**: 342 (1982). Type as above.

Disa subaequalis Summerh. in Bull. Misc. Inform., Kew **1936**: 221 (1936); in F.W.T.A. **2**: 414 (1936). Type from Uganda.
Disa tanganyikensis Summerh. in Kew Bull. **17**: 541 (1964); in F.T.E.A., Orchidaceae: 161 (1968). Type from Tanzania.

Plants generally taller than in subsp. *welwitschii*; dorsal sepal 5–11 mm long; lateral sepals 7–11.5(13) × 4–6(8) mm; petal posterior lobe generally as tall as or slightly taller than the anterior lobe, ± ovate lanceolate, acute; flowers pink, and variations on pink, spur 5–9 mm long.

Zambia. N: Mporokoso Distr., Mporokoso-Senga Hill Rd., fl. 19.i.1960, *Richards* 12421 (BR; K; SRGH). W: Mwinilunga Distr., by River Kewumbo (R. Kaoomba), fl. 12.xii.1937, *Milne-Redhead* 3674 (K; PRE). **Malawi**. N: Nyika Plateau, near Chilinda Bridge, c. 2300 m, fl. 7.i.1983, *la Croix* 407 (K).
Also recorded from Guinea, Liberia, Nigeria, Cameroon, the Central African Republic, Uganda, Kenya, Tanzania and Zaire. Wet dambo grasslands, but occupying a range of habitats over its whole distribution range.
The subspecies of the *D. welwitschii* complex intergrade in southern Tanzania. In the Flora Zambesiaca area, however, most collections can be determined with relative ease.

7. **Disa roeperocharoides** Kraenzl. in Bot. Jahrb. Syst. **51**: 379 (1914). —Williamson, Orch. S. Centr. Africa: 81 (1977). —Linder in Bull. Jard. Bot. Belg. **51**: 310 (1981). —Geerinck in Fl. Afr. Centr., Orchidaceae pt. 1: 182 (1984). Type from Zaire.

Terrestrial herb with separate fertile and sterile shoots. Sterile shoots to 8 cm tall; leaves conduplicate, 20 × 1–2.5 cm, linear to narrowly oblanceolate, acuminate. Fertile shoots 30–70 cm tall; leaves imbricate, 5–12 cm long, narrowly lanceolate, acuminate. Inflorescence dense, 7–17 cm long, c. 30-flowered. Floral bracts lanceolate, acuminate, somewhat longer than the ovaries. Flowers c. 15 mm in diameter, bright red, the posterior petal lobes and the inside of the galea usually yellow with bright red blotches, the anther purple to yellow. Dorsal sepal erect, 7.5–11.5 × 5–8 mm, rhomboid, subacute; galea usually narrowly oblong, subacute, c. 2 mm deep. Spur pendent, 8–13 mm long, slender, cylindrical. Lateral sepals suboblique, 9–10(12) mm long, ovate-oblong, obtuse to rarely apiculate. Petals 2-lobed; anterior lobe 4–7 × c. 3 mm, ± narrowly ovate rounded, spreading with the upper margin curved over the anther; posterior lobe erect inside the galea, 7–9 mm tall, oblanceolate, truncate, one third taller than the posterior lobe. Lip pendent, 9–10 mm long, linear. Anther erect, c. 4 mm long. Stigma sessile, 3-lobed. Ovary 1.5–2 cm long.

Zambia. W: Solwezi, fl. 10.i.1962, *Holmes* 316 (K; SRGH). C: Lusaka, fl. 1.ii.1957, *Noak* 79 (K; SRGH).
Also recorded from the Shaba Province of Zaire. Mostly recorded from dambo grasslands, rarely from grassy areas in *Brachystegia* woodlands; 1200–1500 m.
This species is closely allied to *D. welwitschii*, but is readily distinguished by the larger flowers and distinctive petal shape.

8. **Disa celata** Summerh. in Kew Bull. **17**: 535 (1964). —Williamson, Orch. S. Centr. Africa: 82 (1977). —Linder in Bull. Jard. Bot. Belg. **51**: 312 (1981). —la Croix et al., Orch. Malawi: 113 (1991). Type from Angola.

Terrestrial herb with separate sterile and fertile shoots. Sterile shoots to 7 cm long; leaves at least 1, conduplicate, 10–20 × 0.5–1 cm, linear-elliptic, acute to subacuminate. Fertile shoots 30–50 cm tall; cauline leaves imbricate, 5–7 × 8–12 mm, acute, closely sheathing with the apices free; basal leaves smaller, obtuse, hyaline or red. Inflorescence dense, 6–8 cm long. Floral bracts c. 2 cm long and 5–10 mm wide, narrowly ovate, subacuminate, overtopping and partially obscuring the flowers. Flowers orange to yellow. Dorsal sepal 5.5–7 mm long, obovate, obtuse; galea erect, 3–5.5 mm deep. Spur pendent from the middle of the galea, slender cylindrical, 5–6.5 mm long. Lateral sepals 7–8 mm long, oblong, subobtuse or shortly apiculate. Petals 2-lobed; anterior lobe, 3.5–5 mm in diameter, rotund, spreading; posterior lobe 4.5–6 mm long, linear, acute, included in the galea. Lip pendent, 5–6.5 mm long, linear. Rostellum c. 1.5 mm tall. Anther erect, c. 3 mm long. Stigma subsessile. Ovary c. 1 cm long.

Zambia. W: Mwinilunga Distr., Sinkabolo Dambo, fl. 21.xii.1937, *Milne-Redhead* 3774 (K). C: Serenje, fl. xii.1968, *G. Williamson* 404 (K; SRGH). E: Nyika Plateau, Zambian Border, fl. xii.1966, *G. Williamson* 232 (K; SRGH). **Malawi**. N: Nyika Plateau, 3 km north of Lake Kaulime, fl. 22.xii.1969, *Ball* 1214 (SRGH).

Also recorded from the Southern Highlands of Tanzania and Angola. Wet marshy grassland in dambos; 1400–2200 m.

This species can easily be confused with *D. welwitschii* or *D. cryptantha*. It may be distinguished from the former by its prominent bracts, and from the latter by its longer posterior petal lobes which equal or exceed the anterior petal lobes.

9. **Disa aequiloba** Summerh. in Bull. Misc. Inform., Kew **1927**: 419 (1927); in F.T.E.A., Orchidaceae: 163 (1968). —Geerinck in Bull. Soc. Roy. Bot. Belg. **107**: 64 (1974). —Williamson, Orch. S. Centr. Africa: 82 (1977). —Linder in Bull. Jard. Bot. Belg. **51**: 313 (1981). —Geerinck in Fl. Afr. Centr., Orchidaceae pt. 1: 188 (1984). Type from Zaire.

Terrestrial herb with separate sterile and fertile shoots. Sterile shoots to 4 cm long, leaves erect, c. 8 cm long and 8 mm wide, linear-lanceolate, apiculate. Fertile shoots 20–50 cm tall; cauline leaves imbricate, 3–6 cm long, subacuminate, the apices free; basal leaves 1–2, reduced to sheaths. Inflorescence dense, 3–8 cm long, cylindrical, 10–30-flowered. Floral bracts c. 1.5 cm long and 5 mm wide, narrowly ovate, acuminate, conspicuous, overtopping the flowers. Flowers white to pale mauve with maroon and purple mottling. Dorsal sepal 4–5 mm tall, obovate-elliptic, obtuse, forming a galea 2.4–4 mm wide and 1.5–2 mm deep. Spur pendent from shortly below the middle of the galea, 3.5–5 mm long, slender cylindrical. Lateral sepals 4–5 mm long, oblong-ovate, obtuse, spreading upwards and curved forwards. Petals 2-lobed, 3.5–5 mm tall and 3–4.5 mm wide, with subequal lobes; anterior lobe c. 0.2–0.5 mm taller and up to twice as wide as the posterior lobe, often longer than the dorsal sepal, the upper margin incurved, hooding the anther; posterior lobes overlapping inside the galea. Lip pendent, 3.5–4.5 mm long, linear. Rostellum erect, c. 1.5 mm tall. Anther erect, 2–3 mm long. Stigma tripulvinate. Ovary c. 1 cm long.

Zambia. N: Kawambwa Distr., Mbereshi Dambo, fl. 11.i.1961, *Holmes* 0289 (K; SRGH). W: Zambezi (Balovale) Distr., Kucheka Valley, fl. 1.xi.1961, *Holmes* 0311 (K; SRGH). C: Serenje Dambo, fl. xii.1967, *Williamson & Simon* 641 (K; SRGH).

Also known from Tanzania, Zaire and Angola. Usually in marshy montane grassland; 1300–2200 m. Some collections were made at lower altitudes in *Brachystegia* woodlands.

10. **Disa cryptantha** Summerh. in Kew Bull. **17**: 537 (1964); in F.T.E.A., Orchidaceae: 163 (1968). —Williamson, Orch. S. Centr. Africa: 82 (1977). —Linder in Bull. Jard. Bot. Belg. **51**: 316 (1981). —Geerinck in Fl. Afr. Centr., Orchidaceae pt. 1: 189 (1984). Type from Tanzania.

Terrestrial herb with separate sterile and fertile shoots. Sterile shoots to 8 cm long; leaves erect, c. 14 cm long and 6 mm wide, linear, acute. Fertile shoots 30–60 cm tall; cauline leaves imbricate, 6.5–7.5 cm long and c. 1 cm wide, acute to subacuminate, sheathing with a free apex; basal leaves entirely sheathing, smaller, reddish. Inflorescence dense, 7–11 mm long, 15–50-flowered. Floral bracts prominent, 2–3 cm long, narrowly ovate, overtopping and partially obscuring the flowers. Flowers brightly flesh-coloured with darker spots. Dorsal sepal broadly elliptic, forming a galea 3.5–5.5 mm tall, 2.5–3.7 mm wide and c. 1.5 mm deep, obtuse, erect. Spur 3.5–5 mm long, slender cylindrical, pendent from near the base of the galea, reaching below the base of the galea. Lateral sepals 3.8–6.5 mm long, obliquely oblong-elliptic, spreading upwards. Petals 2-lobed; anterior lobe 4–5.5 mm long and 1.8–3 mm wide, ovate, taller than the dorsal sepal, the upper margin incurved; posterior lobe 0.5–1 mm shorter than the anterior lobe, acute, 0.5 mm wide. Lip 4–6 mm long, linear, acute, pendent. Rostellum c. 1 mm tall. Anther erect, 2–3 mm long, the connective massive. Stigma 0.5 mm high. Ovary 1–1.5 cm long.

Zambia. N: Kundalila Falls, fl. xii.1967, *G. Williamson* 725 (K; SRGH). W: Mwinilunga Distr., Sinkabolo Dambo, fl. 9.xii.1937, *Milne-Redhead* 3585 (K).

Also recorded from Tanzania and Zaire. Marshy grassland or dambos; 1000–1800 m.

This species is relatively rare.

11. **Disa miniata** Summerh. in Kew Bull. **17**: 539 (1964); in F.T.E.A., Orchidaceae: 164 (1968). —Geerinck in Bull. Soc. Roy. Bot. Belg. **107**: 66 (1974). —Grosvenor in Excelsa **6**: 80 (1976). —Williamson, Orch. S. Centr. Africa: 82 (1977). —Linder in Bull. Jard. Bot. Belg. **51**: 317 (1981). —Geerinck in Fl. Afr. Centr., Orchidaceae pt. 1: 190 (1984). —la Croix et al., Orch. Malawi: 113 (1991). TAB. **52**. Type: Zambia, Mbala (Abercorn), upper end of Lake Chila, 4.i.1952, *Richards* 256 (K, holotype).

Terrestrial herb with separate sterile and fertile shoots. Sterile shoot up to 70 cm tall; leaves 2–3, conduplicate, to 30 × 1 cm, oblanceolate-linear, acute. Fertile shoots 30–60 cm tall; cauline leaves imbricate or subimbricate, 4–8 cm long, lanceolate, acute, closely sheathing; basal leaves somewhat smaller. Inflorescence subdense to dense, 4–16 cm long, 5–40-flowered. Floral bracts 2–3 cm long, narrowly ovate, acute to acuminate, overtopping the flowers. Flowers more or less orange. Dorsal sepal 4.5–6.6 × 2.5–3.6 mm, ovate-elliptic, subacute, erect or slightly curved forwards, galea c. 1.5 mm deep. Spur pendent from above the middle of the galea, reaching to the base of the dorsal sepal 3.5–4.2 mm long, strongly clavate, to 2 mm in diameter at apex. Lateral sepals 5.5–6.6 mm long, ovate, subacute, spreading sideways or upwards. Petals 2-lobed; anterior lobe 3.5–5 mm tall and 3–4 mm wide, the upper margins incurved; posterior lobe 3–4.5 mm tall and 0.5–0.7 mm wide. Lip pendent, 4–6 mm long, linear. Rostellum less than 1 mm tall. Anther erect, with a massive connective. Stigma massive, to 3 mm in diameter. Ovary 1–2 cm long.

Zambia. N: Mbala, Lake Chila, fl. xii.1967, *Williamson & Simon* 678 (BOL; K; SRGH). C: Mpika Distr., 56 km S of Mpika, fl. xii.1967, *Williamson & Simon* 652 (K; LISC; SRGH). **Zimbabwe**. N: Umvukwes, Luwali, fl. 18.xii.1952, *Wild* 3993 (K; SRGH). C: Harare Distr., Rumani Farm, c. 16 km NE of Harare, fl. 14.xii.1946, *Greatrex* in *GHS* 228718 (SRGH). **Malawi**. N: Nyika Plateau, fl. 12.ii.1960, *Holmes* 0225 (K; SRGH). **Mozambique**. Z: Morrumbala, fl. 6.vii.1942, *Torre* 4532 (LISC).
Also recorded from Tanzania and the Shaba Province of Zaire. Wet grassland, dambos or river valleys, usually submontane; above 1000 m.

12. **Disa engleriana** Kraenzl. in Bot. Jahrb. Syst. **33**: 58 (1902). —Summerhayes in F.T.E.A., Orchidaceae: 159 (1968). —Geerinck in Bull. Soc. Roy. Bot. Belg. **107**: 65 (1974). — Williamson, Orch. S. Centr. Africa: 80 (1977). —Linder in Bull. Jard. Bot. Belg. **51**: 321 (1981). —Geerinck in Fl. Afr. Centr., Orchidaceae pt. 1: 180 (1984). —la Croix et al., Orch. Malawi: 115 (1991). Neotype from Tanzania.
Disa subscutellifera Kraenzl. in Bot. Jahrb. Syst. **48**: 389 (1912). Type from Tanzania.

Terrestrial herb with separate fertile and sterile shoots. Sterile shoot up to 60 cm tall; leaves 2–4, to 22 × 2 cm, linear-lanceolate, acute. Fertile shoot up to 60 cm tall; cauline leaves imbricate, 4–8 cm long, closely sheathing the scape, free blades produced occasionally; basal leaves 2–3 cm long. Inflorescence lax to subdense, 6–22 cm long, 3–20-flowered. Floral bracts 3–5 cm long, ovate, acuminate, usually overtopping the flowers. Flowers facing down, pink. Dorsal sepal 14–22 mm long, spathulate; claw 3–6 × 3–4 mm, canaliculate, not always distinct; blade c. 14 mm in diameter, round, obtuse, bowl-shaped and usually curved. Spur pendent from shortly below the middle of the dorsal sepal, reaching to below its base, 13–20 mm long, clavate, obtuse, spreading sideways and upwards. Lateral sepals 13–20 mm long, obliquely oblong, obtuse, spreading sideways and upwards. Petals 2-lobed, 10–15 mm long; anterior lobe 7–11 × 5–8 mm, margin entire or rarely serrate, spreading; posterior lobes 10–15 × 3–5 mm, ovate-spathulate, acute to serrate apically, included in the galea and partially overlapping. Lip 14–18 mm long, linear, pendent. Anther erect, 5–7 mm long, connective longer than the anther. Stigma flat, c. 3 mm in diameter. Ovary c. 2 cm long.

Zambia. N: Kasama Distr., Misamfu, fl. 22.i.1961, *E.A. Robinson* 4290 (K; SRGH). W: 4 miles SW of Ndola, fl. 7.i.1953, *Draper* 21 (K). E: Danger Hill, 18 miles NE of Mpika, fl. ii.1969, *G. Williamson* 1396 (K). **Malawi**. N: Nyika Plateau, fl. 10.ii.1960, *Holmes* 0211 (K; SRGH).
Also from Tanzania and Zaire. Usually in *Brachystegia* and *Uapaca* woodlands, and associated dambos and grassland, on lateritic and sandy soil, not montane; 1000–2000 m.

Tab. 52. DISA MINIATA. 1, habit (×4/9); 2, inflorescence (×1); 3, lower part of stem (×1); 4, flower, front view (×4); 5, flower, back view (×3); 6, dorsal sepal, front view (×3); 7, dorsal sepal, back view (×4); 8, lateral sepal (×4); 9, petal (×4); 10, column with one petal, front view (×4), 1–10 from *Richards* 256. Drawn by Mary Grierson. From Kew Bull.

13. **Disa ukingensis** Schltr. in Bot. Jahrb. Syst. **53**: 539 (1915). —Linder in Bull. Jard. Bot. Belg. **51**: 324 (1981). —la Croix et al., Orch. Malawi: 115 (1991). TAB. **53**. Type from Tanzania.

Terrestrial herb with separate fertile and sterile shoots. Sterile shoot to c. 4 cm tall; leaves up to 5, conduplicate, 7–12 cm long and up to 12 mm wide, linear-lanceolate, erect, acute. Fertile shoot robust, 15–30 cm tall; cauline leaves imbricate, 4.5–5.5 cm long, ovate-lanceolate, acute, sheathing with the apical third free; basal leaves reduced to sheaths 2–3 cm long. Inflorescence dense, 6–10 cm long, 7–15-flowered. Floral bracts c. 3 cm long, ovate, acute to subacuminate, overtopping the flowers. Flowers magenta, the galea paler with darker spots. Dorsal sepal spathulate; claw 3 mm long and 3 mm wide, sometimes difficult to distinguish from the blade; blade 10–13 mm in diameter, round, obtuse, shallowly galeate, margin incurved. Spur pendent from just below the middle of the blade, reaching to the base of the dorsal sepal, 4–6 mm long, clavate, obtuse. Lateral sepals 9–13 mm long, oblong, obtuse to subacute, spreading sideways and upwards. Petals 2-lobed; anterior lobe 6–7 × 4–5 mm, flanking the anther and spreading, the margins incurved and sometimes serrate; posterior lobe 7–8 × 2–3 mm, ± ovate, acute to obtuse, occasionally serrate, erect inside the galea and somewhat overlapping. Lip pendent, 10–11 mm long, linear. Anther erect, 4–5 mm long. Stigma vertical, sessile, tripulvinate. Ovary c. 1.5 cm long.

Zambia. E: Nyika Plateau, fl. iii.1967, *Williamson & Odgers* 281 (K; PRE; SRGH). **Malawi**. N: Nyika Plateau, fl. 11.iii.1977, *Grosvenor & Renz* 1116 (BOL; K; SRGH).
Also recorded from the Southern Highlands of Tanzania. Montane short, dry, grassland; 2100–2800 m.
It is close to *D. engleriana*, with which it was long confused. However, its denser inflorescence and shorter petals (less than 8 mm long) distinguished it from that species.

14. **Disa zombica** N.E. Br. in F.T.A. **7**: 278 (1898). —Kraenzlin, Orchid. Gen. Sp. **1**: 783 (1900). —Summerhayes in Kew Bull. **4**: 432 (1949); in F.T.E.A., Orchidaceae: 158 (1968). —Geerinck in Bull. Soc. Roy. Bot. Belg. **107**: 71 (1974). —Grosvenor in Excelsa **6**: 80 (1976). —Williamson, Orch. S. Centr. Africa: 80 (1977). —Linder in Bull. Jard. Bot. Belg. **51**: 326 (1981). —Geerinck in Fl. Afr. Centr., Orchidaceae pt. 1: 200 (1984). —la Croix et al., Orch. Malawi: 117 (1991). Type: Malawi, summit of Mt. Zomba, *Buchanan* 305 (K, holotype).
Disa nyassana Schltr. in Bot. Jahrb. Syst. **53**: 538 (1915). Type from Tanzania.

Terrestrial herb with separate fertile and sterile shoots. Sterile shoot up to 6 cm long; leaves 2–3, conduplicate, erect, 6–20 × 8–20 cm, linear-elliptic, acute or apiculate. Fertile shoot 20–60 cm tall; cauline leaves imbricate, 4–6(13) cm long, ovate-lanceolate, acute, sheathing with the apical half often free; basal leaves shorter, red-spotted or barred. Inflorescence usually dense, 9–16 cm long, 10–25-flowered. Floral bracts 3.5–4.5 cm long, lanceolate-ovate, acute to acuminate, much overtopping the flowers. Flowers with grey-green lateral sepals and magenta to purple petals and dorsal sepal. Dorsal sepal 14–21 mm long, spathulate; claw 7–11 mm long and c. 3 mm wide, erect, canaliculate; blade 7–11 × 5–10 mm, ovate, obtuse, galeate, ± facing down. Spur pendent from a small hump near the base of the blade, not reaching the base of the dorsal sepal, 4–7 mm long, subclavate to clavate. Lateral sepals 14–18 mm long, obliquely oblong, subacute. Petals 2-lobed; anterior lobe 4–7 × 3–5 mm, ovate, flanking the anther, spreading; posterior lobes 12–18 × 2–3 mm, oblanceolate, acute to subacute, erect inside the galea, the lobes partially overlapping. Lip pendent, 13–17 mm long, linear. Rostellum c. 3 mm tall. Anther erect, 4–5 mm long. Ovary 1–1.5 cm long.

Zambia. N: Nyika Plateau, fl. iii.1967, *Williamson & Odgers* 295 (K; SRGH). **Zimbabwe**. E: Chimanimani Distr., Cashel, fl. 3.iii.1957, *Ball* 629 (K; SRGH). **Malawi**. N: Nyika Plateau, fl. 5.iii.1977, *Pawek* 12481 (K). S: Zomba Mt., fl. 23.ii.1919, *McLoughlin* 125/6 (K; PRE). **Mozambique**. MS: Mt. Tsetserra (Tsetsera), fl. 20.iii.1955, *Ball* 534 (K; SRGH).
Also from Tanzania and Burundi. Montane grassland; 1200–2200 m.
A very distinctive species.

15. **Disa verdickii** De Wild. in Ann. Mus. Congo, Belge, Bot., sér. 4, **1**: 26 (1902). —Geerinck in Bull. Soc. Roy. Bot. Belg. **107**: 69 (1974). —Williamson, Orch. S. Centr. Africa: 80 (1977). —Linder in Bull. Jard. Bot. Belg. **51**: 328 (1981). —Geerinck in Fl. Afr. Centr., Orchidaceae pt. 1: 181 (1984). Type from Zaire.

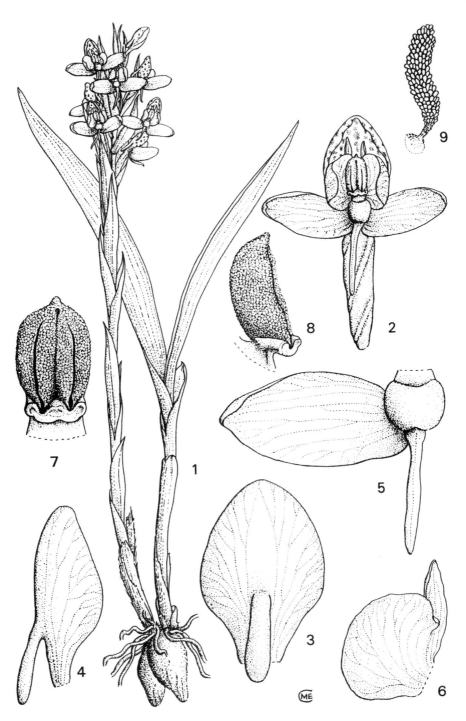

Tab. 53. DISA UKINGENSIS. 1, habit (×⅔); 2, flower (×2); 3, dorsal sepal, dorsal view (×4); 4, dorsal sepal, side view (×4); 5, lateral sepal, stigma and lip (×4); 6, petal (×4); 7, anther and column, front view (×6); 8, anther and column, side view (×6); 9, pollinium (×10), 1–9 from *Cribb & Grey-Wilson* 10767. Drawn by M.E. Church. From Kew Bull.

Terrestrial herb with separate fertile and sterile shoots. Sterile shoots up to 5 cm tall; leaves up to 5, erect, conduplicate, up to 11 × 1.5 cm, linear to elliptic, acute. Fertile shoots 25–50 cm tall; cauline leaves imbricate, 4–7 cm long, ovate-lanceolate, acute to subacuminate, sheathing, apices free; basal leaves reduced to several subacute sheaths c. 3 cm long. Inflorescence subimbricate, 8–20 cm long, 8–25-flowered. Floral bracts ovate-lanceolate, subacuminate, 3–4 cm long, reaching to the middle or overtopping the flowers. Flowers rosy-mauve to magenta-pink. Dorsal sepal 20–25 mm tall, spathulate; claw 11–12 × c. 4 mm, erect, deeply canaliculate towards the apex; blade c. 12 mm in diameter, orbicular, obtuse, galeate, the groove in the claw continued into the blade. Spur pendent from below the middle of the blade, following the curve of the dorsal sepal, 6–7 mm long, clavate. Lateral sepals 20–25 mm long, suboblique, oblong, obtuse to subacute. Petals 2-lobed; anterior lobe 13–15 × 6–8 mm, oblong, flanking the anther, spreading, the margins incurved; posterior lobe 20–22 × 2–3 mm, narrowly oblanceolate to linear, lobes partially overlapping. Lip pendent, 20–23 mm long, linear. Rostellum c. 3 mm tall. Anther erect, 6–9 mm tall; connective longer than the anther sacs. Stigma flat, sessile. Ovary c. 2 cm long.

Zambia. W: Mwinilunga Distr., Dobeka Plain, fl. 14.xii.1937, *Milne-Redhead* 3661 (BR; K; PRE). Also recorded from Zaire and Angola. In wet sandy grassland, and in *Brachystegia* or *Uapaca* woodlands; in submontane grasslands in Angola.

16. **Disa ornithantha** Schltr. in Bot. Jahrb. Syst. **53**: 538 (1915). —Summerhayes in Bot. Not. **1937**: 188 (1937). —Goodier & Phipps in Kirkia **1**: 53 (1961). —Summerhayes in F.T.E.A., Orchidaceae: 158 (1968). —Geerinck in Bull. Soc. Roy. Bot. Belg. **107**: 68 (1974). — Grosvenor in Excelsa **6**: 80 (1976). —Williamson, Orch. S. Centr. Africa: 79 (1977). —Linder in Bull. Jard. Bot. Belg. **51**: 330 (1981). —Geerinck in Fl. Afr. Centr., Orchidaceae pt. 1: 179 (1984). —la Croix et al., Orch. Malawi: 119 (1991). Type from Tanzania.

Terrestrial herb with separate sterile and fertile shoots. Sterile shoots 3–4-leaved; leaves erect, conduplicate, up to 20 × 1 cm, linear, acute. Fertile shoots 19–50 cm tall; cauline leaves imbricate, 4–7 cm long, ovate to lanceolate, acute, sheathing, apices free; lower leaves grading into several short sheaths. Inflorescence 4–9 cm long, 1–5(6)-flowered. Floral bracts acuminate, as tall as or overtopping the flowers, often keeled towards the apex. Flowers red with white posterior petals lobes. Dorsal sepal spathulate; claw 7–12 mm long, narrowly oblong, erect; blade somewhat reflexed, 10–14 mm long, ovate, acute, galea 4 mm deep with the margins incurved. Spur pendent from above the base of the blade, reaching the base of the dorsal sepal, 9–11 mm long, subclavate and subacute. Lateral sepals 17–23 mm long, obliquely oblong, obtuse. Petals 2-lobed; anterior lobe 9–11 mm long, ovate, flanking the anther, spreading, margins incurved; posterior lobes c. 20 mm long, linear, subconduplicate, parallel at the base, then diverging by an S-shaped curve. Lip pendent, 9–14 mm long, linear. Rostellum 2–3 mm tall. Anthers erect, 6–8 mm long. Stigma flat, 2–4 mm in diameter.

Zambia. N: Mbala Distr., Ndundu to Kawimbe road, fl. 19.i.1968, *Richards* 22926 (K). W: Mwinilunga Distr., Dobeka Plain, fl. 14.xii.1937, *Milne-Redhead* 3663 (BR; K; PRE). E: Danger Hill, fl. ii.1969, *G. Williamson* 1397 (BOL; K; SRGH). **Zimbabwe**. E: Heathfield Farm, c. 9 miles SW of Chimanimani (Melsetter), fl. i.1954, *A.O. Crook* 502 (K; SRGH). **Malawi**. N: Nyika Plateau, near Lake Kaulime, fl. 13.iii.1961, *E.A. Robinson* 4487 (K; M; SRGH). S: Mt. Zomba, fl. 25.i.1959, *Robson* 1318 (K; SRGH). **Mozambique**. MS: Manica, Chimanimani Mts., fl. 1973, *Dutton* 34 (LMA). Also from Tanzania, Zaire and Angola. *Brachystegia* woodland and in montane grassland, usually in wet grassland and dambos; 1000–2000 m.

17. **Disa katangensis** De Wild. in Ann. Mus. Congo Belg., Bot., sér. 4, **1**: 25 (1902). — Williamson, Orch. S. Centr. Africa: 80 (1977). —Linder in Bull. Jard. Bot. Belg. **51**: 333 (1981). Type from Zaire.
    Disa erubescens var. *katangensis* (De Wild.) Geerinck in Bull. Soc. Roy. Bot. Belg. **107**: 66 (1974). —Geerinck in Fl. Afr. Centr., Orchidaceae pt. 1: 179 (1984). Type as above.

Terrestrial herb with separate fertile and sterile shoots. Sterile shoots 2–3-leaved; leaves up to 22 × 1 cm, linear, acute, erect. Fertile shoots 40–60 cm tall; cauline leaves imbricate, 4–8 cm long and 12–18 mm wide, sheathing, apices free; basal leaves reduced to sheaths. Inflorescence lax, 8–16 cm long, 2–8-flowered. Floral bracts 5–7 cm long, lanceolate, long-acuminate, the apical part rolled, overtopping the flowers. Flowers red, pink or scarlet, the galea with darker spots. Dorsal sepal spathulate; claw 16–17 × 3–4 mm, erect, canaliculate; blade 13–20 mm long, ovate, subcordate, obtuse, galea c. 8 mm deep, curved forwards. Spur pendent from a ridge below the middle of the dorsal sepal, 13–16 mm long, filiform to rarely subclavate. Lateral sepals 21–36 mm long, obliquely oblong, subobtuse to acute. Petals 2-lobed; anterior lobe 9–15 mm tall, ovate, flanking the anther, spreading, the margins incurved; posterior lobes 22–30 mm long, linear, parallel at the base, then diverging through an S-curve. Lip pendent, 14–22 mm long, linear. Anther erect, 6–9 mm tall; connective longer than the cells. Stigma 3–4 mm in diameter. Ovary c. 2 cm long.

Zambia. N: Ishiba Ngandu (Shiwa Ngandu), fl. ii.1969, *G. Williamson* 413 (BOL; K; SRGH). W: Mwinilunga Distr., near River Musangila, fl. 15.i.1938, *Milne-Redhead* 4321 (BR; K).

Also from Zaire and Angola. In sandy *Brachystegia* and *Uapaca-Monotes* woodlands, and in sandy grassland; c. 1000 m. A rather rare and local species.

18. **Disa erubescens** Rendle in J. Linn. Soc., Bot. **33**: 297 (1895). —N.E. Br. in F.T.A. **7**: 277 (1898). —Kraenzlin, Orchid. Gen. Sp. **1**: 738 (1900). —Schlechter in Bot. Jahrb. Syst. **31**: 216 (1901); in Bot. Jahrb. Syst. **53**: 537 (1915). —Summerhayes in F.W.T.A. **2**: 414 (1936). —Piers, Orchids E. Africa: 57 (1968). —Summerhayes in F.T.E.A., Orchidaceae: 157 (1968). —Stewart in Agnew, Upland Kenya Wild Flowers: 746 (1974). —Williamson, Orch. S. Centr. Africa: 78 (1977). —Linder in Bull. Jard. Bot. Belg. **51**: 335 (1981). —Geerinck in Fl. Afr. Centr., Orchidaceae pt. 1: 175 (1984). —la Croix et al., Orch. Malawi: 117 (1991). Type from Uganda.

Terrestrial herb with separate fertile and sterile shoots. Sterile shoots to 12 cm long; leaves 2–3, conduplicate, to 30 × 2 cm, linear, acute, erect. Fertile shoots 40–90 cm tall; leaves imbricate, 3–8 cm long, lanceolate, acute, sheathing, only the apices free. Inflorescence lax, 5–13 cm long, 1–6(12)-flowered. Floral bracts c. 3 cm long, lanceolate, acute, reaching up to the flowers. Flowers bright red. Dorsal sepal spathulate; claw erect, 11–26 × 3–7 mm, canaliculate; blade 12–20 mm long, ovate-elliptic, acute, galea 3–5 mm deep, facing downwards. Spur ascending from shortly below the middle of the blade, slender, 8–16 mm long. Petals 2-lobed; anterior lobe 7–15 mm long, ovate, flanking the anther, spreading; posterior lobe 20–35 mm long, linear, ascending inside the galea, parallel in the lower half, then diverging by an S-shaped curve. Lip pendent, 9–20 mm long, linear. Rostellum 2 mm long. Anther erect, 5–8 mm long. Stigma sessile, tripulvinate, 3–4 mm in diameter. Ovary c. 2 cm long.

Dorsal sepal less than 3 cm long; plants usually 3–6-flowered · · · · · · · · · · · · subsp. *erubescens*
Dorsal sepal more than 3 cm long; plants with 1–3(12)-flowered · · · · · · · · · · subsp. *carsonii*

Subsp. **erubescens** TAB. 54.
    Disa erubescens var. *leucantha* Schltr. in Bot. Jahrb. Syst. **53**: 537 (1915). Type from Tanzania.

Zambia. N: Kundalila Falls, 15 km SE of Kanona, fl. i.1972, *G. Williamson* 2149 (SRGH). W: Samahina, 12 km east of Kalene Hill, fl. 16.xii.1963, *E.A. Robinson* 6087 (SRGH). **Zimbabwe**. E: Chimanimani Mts., fl. 30.i.1958, *Hall* 223 (BOL). **Malawi**. N: Viphya, Kawendama area, fl. 16.i.1962, *Chapman* 1554 (K; SRGH). C: Mt. Dedza, between radio relay and summit, fl. 3.iii.1977, *Grosvenor & Renz* 1025 (SRGH). S: Mt. Zomba, fl. 27.ii.1977, *Grosvenor & Renz* 985 (K; PRE; SRGH). **Mozambique**. N: Lichinga, *Pettersson* 90 (UPS). Z: Milange, fl. 19.i.1966, *Correia* 507 (LISC).

Widespread in subsaharan Africa, from the Sudan, Kenya, Uganda, Cameroon, Nigeria, Tanzania, Rwanda, Burundi and Zaire. In wet grassland or seepages in the montane zone; mostly 1800–2400 m, but also at lower altitudes.

Subsp. **carsonii** (N.E. Br.) H.P. Linder in Bull. Jard. Bot. Belg. **51**: 340 (1981). —la Croix et al., Orch. Malawi: 118 (1991). Type: Zambia, Fwambo, 1894, *Carson* 22 (K, holotype).
    Disa carsonii N.E. Br. in F.T.A. **7**: 277 (1898). —Kraenzlin, Orchid. Gen. Sp. **1**: 737 (1900). —Schlechter in Bot. Jahrb. Syst. **53**: 537 (1915). —Williamson, Orch. S. Centr. Africa: 78 (1977). Type as above.
    Disa stolzii Schltr. in Bot. Jahrb. Syst. **53**: 537 (1915). —Summerhayes in F.T.E.A., Orchidaceae: 156 (1968). —Williamson, Orch. S. Centr. Africa: 78 (1977). Type from Tanzania.
    Disa erubescens var. *carsonii* (N.E. Br.) Geerinck in Fl. Afr. Centr., Orchidaceae pt. 1: 178 (1984). Type as above.

Tab. 54. DISA ERUBESCENS subsp. ERUBESCENS. 1, inflorescence (×1); 2, lower part of scape, and leaves (×1); 3, dorsal sepal (×2); 4, lateral sepal (×1½); 5, petal (×4); 6, lip (×4), 1, 3–6 from *Whyte* 348. Drawn by Stella Ross-Craig. From F.W.T.A.

Plants similar to subsp. *erubescens*, but the flowers are larger, and the dorsal sepal more than 3 cm long. Inflorescences are variable, usually 1–3-flowered, but are sometimes very robust and up to 12-flowered.

Zambia. N: Mbala Distr., near Saisi Bridge, fl. 1.i.1963, *Richards* 17094 (K). C: Mkushi Distr., Moffat Farm, fl. 29.xii.1959, *Marr-Levin* G10, in *GHS* 224272 (SRGH). E: Nyika Plateau, fl. xii.1966, *G. Williamson* 222 (K). **Malawi**. N: Nyika Plateau, Lake Kaulime, fl. 5.ii.1968, *Ball* 1067 (K; SRGH). S: Mulanje Mt., Chambe, eastern edge of plateau, fl. 12.ii.1979, *Blackmore, Brummitt & Banda* 362 (K). **Mozambique**. N: Lichinga (Vila Cabral), fl. 28.ii.1964, *Torre & Correia* 10876 (LISC).

Also from Tanzania, Burundi and Zaire. In damp or wet montane grassland, and in dambos at lower altitudes in *Brachystegia* woodland.

This subspecies is surprisingly variable, and variation between populations is without apparent pattern.

19. **Disa cornuta** (L.) Sw. in Kongl. Vetensk. Acad. Nya Handl. **21**: 210 (1800). —Thunberg, Fl. Cap., ed. 2 (Schultes): 7 (1823). —Lindley, Gen. Sp. Orchid. Pl.: 349 (1838). —Kraenzlin, Orchid. Gen. Sp. **1**: 767 (1900). —Schlechter in Bot. Jahrb. Syst. **31**: 256 (1901). —Rolfe in F.C. **5**, 3: 231 (1913). —Linder in Contrib. Bolus Herb., No. 9: 94 (1981). —Stewart et al., Wild Orch. South Africa: 129 (1982). Type from South Africa.
   Orchis cornuta L., Pl. Rar. Afr.: 109 (1763). Type as above.
   Satyrium cornutum (L.) Thunb., Prodr. Pl. Cap.: 5 (1794). Type as above.
   Disa macrantha Sw. in Kongl. Vetensk. Acad. Nya Handl. **21**: 210 (1800). —Thunberg, Fl. Cap., ed. 2 (Schultes): 8 (1823). —Lindley, Gen. Sp. Orchid. Pl.: 349 (1823). —Rolfe in F.C. **5**, 3: 232 (1913). —Grosvenor in Excelsa **6**: 80 (1976). Type not found.
   Disa aemula Bolus in J. Linn. Soc., Bot. **22**: 59 (1885). —Schlechter in Bot. Jahrb. Syst. **31**: 257 (1901). Type from South Africa.
   Disa cornuta var. *aemula* (Bolus) Kraenzl., Orchid. Gen. Sp. **1**: 768 (1900). Type as above.

Terrestrial herb up to 1 m tall, perennating by tubers. Leaves densely imbricate, lanceolate to narrowly ovate, acute to acuminate, shortly sheathing at the base. Inflorescence imbricate, robust, 15–40 cm long; flowers numerous. Floral bracts lanceolate, acuminate, as tall as or overtopping the flowers. Flowers facing down, galea more or less purple on the outside, lateral sepals and inside of galea greenish silvery-white, lip and petals greenish, lip often with a large purple base, colours generally variable. Dorsal sepal deeply galeate, apiculate; galea 12–18 mm tall, 8–12 mm deep. Spur from a horizontal conical base, gently or rarely sharply decurved, 1–2 cm long, cylindric, emarginate. Lateral sepals patent, 12–16 mm long, narrowly to broadly oblong, obtuse to acute. Petals 8–10 mm long, lorate, acute to acuminate, often with anther apical lobe which may be obscurely or distinctly 2-lobed, petals falcate with a 5 mm long basal anticous lobe flanking the stigma. Lip patent, 5–10 mm long, lorate to obovate, rounded, fleshy. Rostellum lateral lobes erect, canaliculate, fused to the petals, central lobe a small fleshy body between the lateral lobe. Anther borne horizontally. Stigma horizontal, equally tripulvinate. Ovary 15–20 mm long, twisted.

Zimbabwe. E: Nyanga Distr., plateau SW of Nyangani (Inyangani), Pungwe R., fl. 18.i.1951, *Chase* 3592 (K; LISC; SRGH).
Also from South Africa. In short montane grassland.

20. **Disa similis** Summerh. in Kew Bull. **17**: 537 (1964). —Williamson, Orch. S. Centr. Africa: 82 (1977). —Linder in Contrib. Bolus Herb., No. 9: 94 (1981). —Stewart et al., Wild Orch. South Africa: 129 (1982). Type: Zambia, Mwinilunga, R. Kalalima (Kamwezhi), 14.x.1937, *Milne-Redhead* 2780 (K, holotype).

Erect terrestrial herb, 25–50 cm tall, slender, perennating by tubers. Leaves lax to imbricate, 3–8, lower leaves to 12 cm long, upper leaves to 1.5 cm long, lanceolate, acute. Inflorescence lax, 7–18 cm long, 10–30-flowered. Floral bracts narrowly ovate, acuminate, about as long as ovary. Flowers borne horizontally, pale violet, blue or mauve, the lip with a yellow apex. Dorsal sepal erect, rotund, rounded, forming a galea 7–9 mm tall, 5.5–9 mm wide and c. 3 mm deep. Spur c. 3 mm long, conical, subacute, laterally flattened, straight. Lateral sepals 7–9 mm long, oblong to obovate, acute to rounded, patent. Petals erect, 3.5–4.5 mm long, obliquely oblong, truncate or rounded, with a 1–2 mm high basal anticuous lobe flanking the anther. Lip 3.5–4.5 mm long, oblong, rounded, with two side lobes at the base curved up over

163. ORCHIDACEAE 175

the stigma. Rostellum erect, fused to the petals by keels almost as tall as the lateral lobes, c. 2 mm tall, central lobe obsolete. Anther to 2 mm long. Stigma flat, tripulvinate, almost square, shortly stipitate. Ovary twisted, 10–13 mm long.

Zambia. N: Kasama Distr., Mungwi, fl. 7.xii.1960, *E.A. Robinson* 4161 (K; M; SRGH). W: Mwinilunga, fl. viii.1960, *Holmes* 02A (K; SRGH). C: Mkushi Hotel Dambo, fl. x.1967, *Williamson & Simon* 537 (K; SRGH).
Also from Angola and South Africa. In wet dambos; c. 1400 m.

21. **Disa aperta** N.E. Br. in F.T.A. **7**: 286 (1898). —Kraenzlin, Orchid. Gen. Sp. **1**: 785 (1900). — Summerhayes in F.T.E.A., Orchidaceae: 175 (1968). —Williamson, Orch. S. Centr. Africa: 88 (1977). —Linder in Contrib. Bolus Herb., No. 9: 84 (1981). —la Croix et al., Orch. Malawi: 120 (1991). Type: Zambia, Fwambo, S of Lake Tanganyika, *Carson* 51 (K, holotype).
Disa equestris var. *concinna* (N.E. Br.) Kraenzl. in Bot. Jahrb. Syst. **31**: 265 (1901), partly, excluding *D. concinna* N.E. Br. and *D. goetzeana* Kraenzl.

Erect, terrestrial herb, 25–45 cm tall, slender, perennating by tubers. Leaves usually 3, mostly sheathing, the lowermost foliate, 4–8 cm long, lanceolate to narrowly lanceolate, acute to subacuminate, lax. Inflorescence lax, 10–18 cm long, 10–30-flowered. Floral bracts narrowly ovate, acute to acuminate, mauve and possibly dry, about as long as the ovary. Flowers facing somewhat down, sepals mauve-blue to greenish-brown speckled purple, petals white to greenish-brown, lip white with faint spots, spur white. Dorsal sepal elliptic, acute, forming a galea 6–9(11) mm long, 3.5–5(6) mm wide and 3–4 mm deep. Spur 3–4 mm long, conical to sub-cylindrical, obtuse to rounded, usually horizontal, originating above the base of the galea. Lateral sepals patent, 7.5–11.5 mm long, oblique, oblong-elliptic, subacute, apiculate. Petals 5.5–6(7) mm long, falcate, acute, triangular in outline, erect inside the galea. Lip patent, 5.5–8.5 mm long, linear-elliptic, rounded. Rostellum 2 mm tall, fused to petals by long keels. Anther 2 mm long. Stigma subsessile, tripulvinate, almost square. Ovary 10–15 mm long.

Zambia. N: Mbala (Abercorn), Nkali Dambo, fl. xii.1967, *Williamson & Simon* 349 (K; SRGH). W: Mwinilunga Distr., Kalenda Plain, fl. 18.xi.1937, *Milne-Redhead* 3306 (BR; K; PRE). **Malawi**. N: Mzimba Distr., 7 km. SW of Chikangawa, fl. 24.xii.1978, *Phillips* 4470 (K). S: Zomba Distr., Mpalanganga Estate, c. 1100 m, fl. 16.xii.1984, *Hutson* in *la Croix* 678 (K).
Also from Tanzania. In dambos in *Brachystegia* woodland; 1000–2300 m.

22. **Disa aconitoides** Sond. in Linnaea **19**: 91 (1847). —Kraenzlin, Orchid. Gen. Sp. **1**: 780 (1900). —Schlechter in Bot. Jahrb. Syst. **31**: 255 (1901). —Rolfe in F.C. **5**, 3: 223 (1913). —Linder in Contrib. Bolus Herb.No. 9: 80 (1981). —Stewart et al., Wild Orch. S. Afr.: 130, photo (1983). —Geerinck in Fl. Afr. Centr., Orchidaceae pt. 1: 202 (1984). —la Croix et al., Orch. Malawi: 119 (1991). Type from South Africa.

Subsp. **concinna** (N.E. Br.) H.P. Linder in Contrib. Bolus Herb., No. 9: 91 (1981). —la Croix et al., Orch. Malawi: 119 (1991). Type: Malawi, Mt. Zomba, Malosa, Dec. 1896, *Whyte* s.n. (K, lectotype).
Disa concinna N.E. Br. in F.T.A. **7**: 284 (1898). —Schlechter in Bot. Jahrb. Syst. **53**: 543 (1915). —Summerhayes in Bot. Not. **1937**: 187 (1937). —Goodier & Phipps in Kirkia **1**: 53 (1961). —Summerhayes in F.T.E.A., Orchidaceae: 173 (1968). —Geerinck in Bull. Soc. Roy. Bot. Belg. **107**: 64 (1974). —Grosvenor in Excelsa **6**: 80 (1976). —Williamson, Orch. S. Centr. Africa: 82 (1977). Type as above.
Disa bisetosa Kraenzl. in Bot. Jahrb. Syst. **51**: 379 (1914). —Williamson, Orch. S. Centr. Africa: 88 (1977). Type: Zambia, Mulungushi (Malangashi) River, Dec. 1907, *Kassner* 2067 (B†, holotype; K).

Terrestrial herb perennating by tubers, rarely with separate sterile and fertile shoots. Sterile shoots to 7 cm long, usually 2-leaved. Fertile shoots erect, 25–60 cm tall; cauline leave 5–10, imbricate to lax, the lower half to third sheathing, blades 4.5–8(14) cm long, lanceolate to narrowly ovate, obtuse or acute. Inflorescence slender, lax to subimbricate, 8–30 cm long, 15–70-flowered. Floral bracts about as long as the ovary, narrowly ovate to narrowly lanceolate, apex occasionally reflexed, dry, papery. Flowers facing down at an angle of 45°, pale mauve and spotted darker mauve. Dorsal sepal spathulate; galea 5–8 mm tall, 1.5–3.5 mm wide and 2–3 mm deep, ovate to elliptic, obtuse to acuminate. Spur massive, 4–6 mm long, conical to rarely straight, laterally flattened; apex usually rounded, rarely acute. Lateral sepals oblique, 5–7 mm long, narrowly oblong, acute with apiculi 2–4(6) mm long. Petals erect, 3.5–5 mm long, oblique, falcate, approximately oblong in outline, subacute. Lip 4.5–5.5 mm long, narrowly elliptic, subacute, upper surface convex and papillate. Rostellum erect, c. 2 mm tall, lateral lobes fused to petals by long keels. Anther horizontal, c. 2 mm long. Stigma flat, almost square.

Zambia. N: Mbala (Abercorn), Chilongowelo, fl. 30.xii.1954, *Richards* 3796A (K). W: Mwinilunga Distr., Matonchi Farm, fl. 18.xi.1962, *Richards* 17282 (K; SRGH). C: Lusaka Distr., 27 miles E of the town, fl. xii.1967, *G. Williamson* 766 (K; SRGH). E: Nyika Plateau, fl. xii.1966, *G. Williamson* 221 (K; SRGH). **Zimbabwe**. C: Harare, Enterprise, fl. 30.xii.1945, *Greatrex* in *GHS* 14222 (K; SRGH). E: Chimanimani Distr., Glencoe Forest Reserve, fl. 27.xi.1967, *Simon & Ngoni* 1333 (K; SRGH). **Malawi**. N: Nyika Plateau, 2.i.1976, *Phillips* 823 (K). C: Mchinji (Ft. Manning) Distr., near Tamanda Mission, fl. 8.i.1959, *Robson* 1106 (BM; K; LISC; PRE; SRGH). S: Mt. Zomba, fl. 24.xii.1944, *Benson* 1045 (K; PRE). **Mozambique**. MS: Rotanda, fl. 26.xii.1965, *Torre & Correia* 13304 (LISC).

Also from Zaire. *Brachystegia* woodland, montane grassland, and dambos; 1000–2000 m.

This subspecies is variable in the length of the apiculi on the lateral sepals. The apiculi are quite short in plants occurring in the south of the range of the subspecies, and reach their greatest length in plants from the north. Subsp. *aconitoides* is restricted to Africa south of the Limpopo, while subsp. *goetzeana* occurs in Tanzania, Kenya and Ethiopia. Subsp. *concinna* differs from the other 2 subspecies in having apiculi more than 1 mm in length. Subsp. *aconitoides* may be distinguished from subsp. *goetzeana* by its floral bracts which are at least as tall as the flowers.

23. **Disa dichroa** Summerh. in Kew Bull. **17**: 549 (1964). —Linder in Contrib. Bolus Herb., No. 9: 94 (1981). TAB. **55**. Type: Zambia, Sunzu, Kalambo Farm, 9.i.1955, *Richards* 3978 (K, holotype).

    Disa concinna var. *dichroa* (Summerh.) Geerinck in Bull. Soc. Roy. Bot. Belg. **107**: 64 (1974); in Fl. Afr. Centr., Orchidaceae pt. 1: 203 (1984). Type as above.

Terrestrial herb perennating by testicular tubers. Plants erect, slender, 35–60 cm tall. Cauline leaves 3–6, sheathing at the base with a free narrowly elliptic to lanceolate blade 4–7 cm long; all leaves about the same size, drying grey-black while the inflorescence dries pale creamy-brown. Inflorescence lax, spotted mauve, 20–30 cm long, 25–45-flowered. Floral bracts about as long as the ovaries, lanceolate to narrowly lanceolate, acuminate, twisted. Flowers pale purple to mauve. Dorsal sepal spathulate; galea 6.5–9.5 mm tall, 3–4.5 mm wide and c. 2 mm deep, elliptic-oblong, acute. Spur from base of galea, straight, 6–8 mm long, laterally flattened and with the upper and lower margins parallel, (2)3–3.5 mm apart. Lateral sepals 6–9 mm long, oblique, narrowly oblong, acute; apiculi 0.5–1 mm long. Petals 5.5–6 × 1.5–2.5 mm, falcate, acute, attached to the rostellum by a long keel. Lip 5–7 mm long, narrowly elliptic, rounded to subacute, upper surface rounded and papillate. Rostellum c. 2 mm tall, anther c. 3 mm long. Stigma horizontal and almost square.

Zambia. N: Mbala (Abercorn), road to Kasulu Farm, fl. 4.i.1969, *Sanane* 390 (K). W: Solwezi, source of Kifubwe (Chifubwa) R., fl. 5.i.1962, *Holmes* 0317 (K; SRGH). C: Mkushi Distr., Fiwila, fl. 9.i.1958, *E.A. Robinson* 2700 (K; SRGH).

Also from Zaire. *Brachystegia* woodland.

This species is superficially similar to *D. aconitoides*.

24. **Disa nyikensis** H.P. Linder in Contrib. Bolus Herb., No. 9: 96 (1981). —la Croix et al., Orch. Malawi: 122 (1991). Type: Zambia, Nyika Plateau, Dec. 1966, *G. Williamson* 223 (K, holotype).

Terrestrial herb perennating by testicular tubers. Plants 20–45 cm tall. Cauline leaves imbricate, all subequal in size, sheathing at the base; blades free, semi-erect, (2.5)6–8 cm long, narrowly elliptic to lanceolate, acute. Inflorescence subimbricate, 6–12(18) cm long, 10–20(40)-flowered. Floral bracts green, as long as or longer than the ovaries, acute to acuminate. Flowers pale mauve with darker spots. Dorsal sepal elliptic, acute; galea 5.5–6.5 mm long, 3 mm wide and c. 2 mm deep. Spur 4–6 mm long, cylindric-conical, horizontal at the point of origin and usually sharply decurved about halfway down its length. Lateral sepals 4–5.5 mm long, oblong, oblique, subacute to rounded, shortly apiculate. Petals 2.8–4 mm long, obovate, falcate. Lip

Tab. 55. DISA DICHROA. 1, inflorescence (×1); 2, lower part of stem and tuber (×1); 3, flower, lateral view (×3); 4, dorsal sepal, lateral view (×4); 5, lateral sepal (×4); 6, petal (×4); 7, column and two petals, three-quarter profile (×6), 1–7 from *Richards* 3987. Drawn by Mary Grierson. From Kew Bull.

3.5–4.5 mm long, elliptic, upper surface convex. Rostellum erect. Anther 1.5 mm long. Stigma horizontal, almost square. Ovary c. 1 cm long.

Zambia. E: Nyika Plateau, fl. 24.xii.1962, *Fanshawe* 7221 (K). **Malawi**. N: Nyika Plateau, Kasaramba View, fl. 22.xii.1969, *Pawek* 3307 (K).
Endemic to the high mountains about the northern end of Lake Malawi. Montane grassland; 2500 m.
It is probably the high-altitude form of *Disa equestris* Rchb.f.

25. **Disa equestris** Rchb.f. in Flora **48**: 181 (1865). —N.E. Br. in F.T.A. **7**: 284 (1898). —Eyles in Trans. Roy. Soc. S. Africa **5**: 335 (1916). —Summerhayes in Bot. Not. **1937**: 187 (1937). — Suessenguth & Merxmuller, Contrib. Fl. Marandellas Distr.: 156 (1951). —Kraenzlin, Orchid. Gen. Sp. **1**: 783 (1900). —Schlechter in Bot. Jahrb. Syst. **31**: 254 (1901) excl. var. *concinna*. —Goodier & Phipps in Kirkia **1**: 53 (1961). —Summerhayes in F.T.E.A., Orchidaceae: 172 (1968), excl. syn. —Geerinck in Bull. Soc. Roy. Bot. Belg. **107**: 65 (1974). —Grosvenor in Excelsa **6**: 80 (1976). —Linder in Contrib. Bolus Herb., No. 9: 99 (1981). —Geerinck in Fl. Afr. Centr., Orchidaceae pt. 1: 201 (1984). —la Croix et al., Orch. Malawi: 121 (1991). TAB. **56**. Type from Angola.
Disa huillensis Fritsch in Bull. Herb. Boissier, sér 2, **1**: 116 (1901). Type from Angola.

Terrestrial herb perennating by testicular tubers. Plants erect, 35–60 cm tall. Cauline leaves imbricate, grading into the floral bracts, sheathing at the base; free blades up to 15 cm long, lanceolate to narrowly lanceolate, acute to acuminate. Inflorescence lax, 8–20 cm long, 10–40-flowered. Floral bracts green to somewhat papery, ovate, acute to subacuminate, variable in length but usually shorter than the ovaries. Flowers violet to purple, rarely the lip or the flowers whitish. Dorsal sepal rotund, apiculate; galea 6–8 mm long, 5–8(10) mm wide and 3–4(6) mm deep. Spur straight with the apex often upcurved, 6–8 mm long, subconic or cylindric to usually somewhat laterally compressed, apex subclavate. Lateral sepals 6–8 mm long, oblong, rarely ovate to obovate, rounded, apiculi blunt. Petals 4.5 × c. 1.5 mm, falcate, truncate. Lip 4.5–6 mm long, oblong, rounded, upper surface convex and papillate. Rostellum attached to the petal by a keel. Anther c. 3 mm long. Stigma almost horizontal and square. Ovary (8)10–13(15) mm long.

Zambia. N: Luapula Province, Mbereshi-Kawambwa road, fl. 18.i.1960, *Richards* 12397 (BR; K; SRGH). W: Mwinilunga Distr., Sinkabolo Dambo, fl. 9.xii.1937, *Milne-Redhead* 3569 (BM; BR; K; PRE). C: Serenje Dambo, fl. xii.1967, *G. Williamson* 637 (K; SRGH). **Zimbabwe**. N: Umvukwes, Dawsons, Luwali, Mazowe (Mazoe), fl. 18.xii.1952, *Wild* 3995 (BOL; K; SRGH). C: Makoni Distr., Maidstone, 10 km from Rusape, fl. 5.i.1931, *Norlindh & Weimarck* 4111 (BM; K; S). E: Chimanimani Mts., fl. i.1955, *Ball* 450 (K; SRGH). **Malawi**. N: Nyika, fl. xii.1966, *G. Williamson* 151 (K; SRGH). S: Shire Highlands, *Last* s.n. (K). **Mozambique**. N: Lichinga (Vila Cabral), Massangulo, fl. i.1933, *Gomes e Sousa* 1213 (K; LMA).
Also from Nigeria, Cameroon, Central African Republic, Zaire, Tanzania and Angola. Usually in wet dambo grassland, but also in montane grassland in Zimbabwe and Nigeria.

26. **Disa rhodantha** Schltr. in Bot. Jahrb. Syst. **20**, Beibl. 50: 17 (1895); in Bot. Jahrb. Syst. **31**: 254 (1901). —Rolfe in F.C. **5**, 3: 229 (1913). —Grosvenor in Excelsa **6**: 80 (1976). —Linder in Contrib. Bolus Herb., No. 9: 94 (1981). —Stewart et al., Wild Orch. S. Africa: 121 (1982). Type from South Africa.

Terrestrial herb with separate fertile and sterile shoots; perennating by testicular tubers. Sterile shoots 3–4-leaved; leaves erect, conduplicate, often petiolate, 20–30 cm long, narrowly elliptic to linear, acute. Fertile shoots 8–16-leaved; leaves cauline, subimbricate, grading into the floral bracts, mostly sheathing; blades 3–10 cm long, narrowly lanceolate, acuminate. Inflorescence dense, 10–15(20) cm long, 15–30-flowered. Flowers facing down, pink to red, lip yellow to pink. Dorsal sepal galeate; galea erect, 7–10 mm long, 4–5 mm wide and 3.5–4.5 mm deep, oblong, acute. Spur slender, 7–17 mm long, cylindrical from a conical base, ascending, apex somewhat swollen, obtuse. Lateral sepals 7–10 mm long, oblong, obtuse to acute, apiculi up to 1.5 mm long. Petals 5.5–8 mm long, obliquely narrowly oblong, somewhat variable in shape, apex curved over the rostellum, occasionally serrate. Lip pendent, 6–7 mm long, narrowly oblanceolate to oblanceolate, rarely narrowly elliptic, apical margins sometimes faintly serrate. Rostellum 2–2.5 mm tall, lateral lobes parallel. Anther 2–3 mm long, cells ovate. Stigma shortly stipitate. Ovary erect, 1–1.5 cm long.

Tab. 56. DISA EQUESTRIS. 1, habit (×1), from *Milne-Redhead & Taylor* 7961; 2, sterile stem (×1), from *Brunt* 260a; 3, flower (×3); 4, flower, dorsal sepal and one petal removed (×3); 5, dorsal sepal, side view (×4); 6, lateral sepal (×4); 7, petal (×4); 8, lip (×4); 9, column with one petal, front view (×8), 3–9 from *Milne-Redhead & Taylor* 7961. Drawn by Heather Wood. From F.T.E.A.

Zimbabwe. E: Nyanga Distr., Mt. Nyangani (Inyangani), fl. 4.ii.1961, *Beasley* 38 (K; SRGH).
Also from South Africa. Montane grassland.
The Zimbabwean population is distinct from the South African populations in the marginally
longer spur, which is not apically deflexed. Insufficient material is known at present to consider
describing it as a distinct subspecies.

27. **Disa hircicornis** Rchb.f., Otia Bot. Hamburg. **2**: 105 (1881). —N.E. Br. in F.T.A. **7**: 283
    (1898). - •Kraenzlin, Orchid. Gen. Sp. **1**: 758 (1900). —Schlechter in Bot. Jahrb. Syst. **31**:
    242 (1901); **53**: 538 (1915). —Summerhayes in F.T.W.A. **2**: 414 (1936). —Goodier &
    Phipps in Kirkia **1**: 53 (9160). —Summerhayes in F.T.E.A., Orchidaceae: 171 (1968); in
    Bot. Not. **1937**: 187 (1937). —Grosvenor in Excelsa **6**: 80 (1976). —Linder in Contrib.
    Bolus Herb., No. 9: 124 (1981). —Stewart et al., Wild Orch. S. Africa: 122 (1982). —
    Geerinck in Fl. Afr. Centr., Orchidaceae pt. 1: 193 (1984). —la Croix et al., Orch. Malawi:
    123 (1991). Type: Malawi, Manganja Hills, near Mount Soche, *Kirk* s.n. (K, holotype).
       Disa laeta Rchb.f., Otia Bot. Hamburg. **2**: 106 (1881). —Kraenzlin, Orchid. Gen. Sp. **1**:
    753 (1900). —Rolfe in F.C. **5**, 3: 229 (1913). Type from South Africa.
       Disa culveri Schltr. in Bot. Jahrb. Syst. **20**, Beibl. 50: 17 (1895). Type from South Africa.
       Disa amblyopetala Schltr. in Bot. Jahrb. Syst. **53**: 542 (1915). —Grosvenor in Excelsa **6**: 80
    (1976). Type from Tanzania.

Terrestrial herb with separate fertile and sterile shoots; perennating by testicular
tubers. Sterile shoots usually 2-leaved; leaves shortly petioled, semi-erect,
conduplicate, to 30 cm long, narrowly elliptic to linear, acute. Fertile shoots 30–60
cm tall; cauline leaves imbricate, mostly sheathing and mottled or barred red at the
bases, 5–20 cm long, lanceolate, acute. Inflorescence dense, 6–20 cm long, 20–70-
flowered. Floral bracts as or taller than the flowers, lanceolate, acuminate, the
apices usually reflexed. Flowers facing down at an angle of 45°, deep dusty-purple or
other shades of purple-red. Dorsal sepal galeate; galea 5–7.5 mm long, 4–6 mm wide
and 1.5–2 mm deep, elliptic, rarely ovate to obovate, acute to obtuse. Spur slender,
8–13 mm long, cylindric from a conical base, horizontal at the base but soon sharply
deflexed (apparently ascending at base because galea faces down). Lateral sepals 7.5
mm long, oblong, rarely ovate, acute. Petals erect, 4–5.5 mm long, narrowly obovate-
oblong, truncate, oblique, rarely obovate, acute. Lip 4–5.5 mm long, linear to
narrowly elliptic, pendent. Rostellum 1.5–2 mm tall, lateral lobes parallel. Anther
1.5–2 mm long. Stigma stipitate. Ovary c. 1 cm long.

Zambia. N: Kasama Distr., 90 km E of Kasama, fl. 26.xi.1960, *E.A. Robinson* 4112 (K; M;
SRGH). W: c. 30 km E of Mwinilunga, fl. 26.x.1966, *Leach & Williamson* 13482 (K; SRGH). C:
Lusaka, fl. 16.xii.1972, *Strid* 2698 (GRA; K). E: Nyika Plateau, fl. xii.1966, *G. Williamson* 184 (K).
Zimbabwe. C: Harare, Chacoma, fl. 27.i.1950, *Greatrex* in *GHS* 26742 (K; SRGH). E: Stapleford,
fl. 22.ii.1946, *Wild* 943 (K; SRGH). **Malawi**. N: South Viphya, Luwawa Dam, 25.ii.1982, *Brummitt,
Polhill & Banda* 16118 (K). C: Chionjeza (Chiungiza), fl. 2.ii.1959, *Robson* 1533 (BM; K; LISC;
PRE; SRGH). S: Zomba Mt., fl. 25.i.1959, *Robson* 1321 (K). **Mozambique**. Z: Gurué, fl. 4.ii.1968,
Torre & Correia 16886 (LISC).
Also from Nigeria, Cameroon, Kenya, Uganda, Tanzania, Zaire, Angola and South Africa.
Swampy and wet grassland; 1500–1800(2700) m.
The flower colour of this species is extremely variable (see Williamson, Orchids of S. Central
Africa for photographic documentation of this variation).

28. **Disa perplexa** H.P. Linder in Contrib. Bolus Herb., No. 9: 128 (1981). —la Croix et al.,
    Orch. Malawi: 123 (1991). Type: Zambia, Kasama, Mungwi, 30.xi.1960, *E.A. Robinson* 4126
    (K, holotype).

Terrestrial herb perennating by tubers. Fertile shoots 30–90 cm tall; cauline leaves
imbricate, mostly sheathing, 4–8 cm long, lanceolate, acute. Inflorescence dense,
(5)10–25 cm long, cylindric, flowers numerous. Floral bracts 1.5–2 cm long, narrowly
ovate to lanceolate, acuminate, overtopping the flowers or rarely the apex deflexed.
Flowers facing down, white; sepals with green and purple spots. Dorsal sepal galeate;
galea 5.5–6.5 mm long, 4–4.5 mm wide and c. 2 mm deep, elliptic-ovate, subacute. Spur
9–11 mm long, cylindric, horizontal from a small conical base, soon decurved,
(apparently ascending at the base because the galea faces downwards). Lateral sepals
5–5.5 mm long, oblong, obtuse. Petals 4.5–5 mm long, oblong, obliquely acute, mid-vein
falcately curved. Lip 3–5 × 1.3–2 mm, elliptic-spathulate, subacute. Rostellum 1.5–2 mm
tall, lateral lobes parallel. Anther 1.5–2 mm long. Stigma stipitate. Ovary c. 1 cm long.

Zambia. N: Kasama Distr., Mungwi, fl. 30.xi.1960, *E.A. Robinson* 4126 (K; SRGH). **Zimbabwe.** E: Chimanimani Distr., 1 mile W of Chipinge (Chipinga), fl. 10.xii.1949, *Chase* 1864 (BOL; K; SRGH). **Malawi.** S: Mt. Zomba, Chitinje, fl. 7.ix.1976, *Welsh* 368 (K).

Also from Nigeria, Kenya and Tanzania. Wet grassland.

This species is very close to *D. hircicornis*, and may prove to be better treated at varietal rank under it. It is distinguished only by flower colour, flowering time and habitat. While fresh material is quite distinct, herbarium material can be difficult to determine with certainty.

29. **Disa caffra** Bolus in J. Linn. Soc., Bot. **25**: 171 (1889). —Kraenzlin, Orchid. Gen. Sp. **1**: 820 (1900). —Schlechter in Bot. Jahrb. Syst. **31**: 258 (1901). —Rolfe in F.C. **5**, 3: 230 (1913). —Linder in Contrib. Bolus Herb., No. 9: 130 (1981). —Stewart et al., Wild Orch. S. Africa: 122 (1982). —Geerinck in Fl. Afr. Centr., Orchidaceae pt. 1: 194 (1984). —la Croix et al., Orch. Malawi: 122 (1991). TAB. **57**. Type from South Africa.

Disa perrieri Schltr. in Fedde, Repert. Spec. Nov. Regni Veg., Beih. 33: 100 (1925). —H. Perrier in Fl. Madag., Orch. 1: 174 (1939). Type from Madagascar.

Disa compta Summerh. in Kew Bull. **17**: 545 (1964). —Williamson, Orch. S. Centr. Africa: 86 (1979). Type: Zambia, Mwinilunga Distr., Sinkabolo Dambo, 20.x.1937, *Milne-Redhead* 2870 (K, holotype).

Terrestrial herb perennating by testicular tubers, rarely with separate sterile shoots. Sterile shoots c. 2-leaved; leaves erect, up to 13 cm long, linear-lanceolate, acute. Fertile stems 25–50 cm tall; cauline leaves mostly sheathing and 4–6 cm long, 2 leaves with free semi-erect narrowly lanceolate to linear blades up to 12 cm long. Inflorescence lax to sub-dense, 6–13 cm long, 8–18-flowered. Floral bracts as tall as the flowers, narrowly ovate, acuminate, the apices often reflexed. Flowers borne horizontally, pink, sometimes reddish or purplish, or with purple spots. Dorsal sepal galeate; galea erect, 8–12 mm long, 6–10 mm wide and 4–6 mm deep, ovate, acute to obtuse. Spur horizontal at the base, at length gradually decurved, slender, 8–13–15 mm long, cylindric from a conical base. Lateral sepals, 8–13 mm long, elliptic, rotund, acute. Petals 6–9 mm long, narrowly oblong, obliquely acute, erect. Lip 6–8 mm long, narrowly lanceolate to narrowly elliptic, acute. Rostellum lateral lobes parallel, c. 3 mm tall. Anther 3–4 mm long. Stigma shortly stipitate. Ovary c. 1.5 cm long.

Zambia. N: Kasama Distr., Mungwi, fl. 26.xii.1960, *E.A. Robinson* 4010 (K; SRGH). W: Mwinilunga, fl. 3.xii.1958, *Holmes* 0115 (BR; K; SRGH). **Malawi.** N: Nyika Plateau, fl. xii.1966, *G. Williamson* 167 (K; SRGH).

Also from S Tanzania, Angola, South Africa and Madagascar. Wet grassland, usually dambos; 1400–1700 m.

30. **Disa walleri** Rchb.f., Otia Bot. Hamburg. **2**: 105 (1881). —N.E. Br. in F.T.A. **7**: 282 (1892). —Kraenzlin, Orchid. Gen. Sp. **1**: 751 (1900). —Schlechter in Bot. Jahrb. Syst. **31**: 238 (1901). —Summerhayes in Bot. Not. **1937**: 189 (1937); in F.T.E.A., Orchidaceae: 169 (1968). —Grosvenor in Excelsa **6**: 80 (1976). —Williamson, Orch. S. Centr. Africa: 86 (1977). —Linder in Contrib. Bolus Herb., No. 9: 132 (1981). —Geerinck in Fl. Afr. Centr., Orchidaceae pt. 1: 194 (1984). —la Croix et al., Orch. Malawi: 124 (1991). TAB. **58**. Type: Malawi, Manganja Hills, *Waller* s.n. (K, holotype).

Disa zombaensis Rendle in Trans. Linn. Soc., Bot. ser. 2, **4**: 47 (1894). Type: Malawi, Mt. Zomba, *Whyte* 3 (BM, holotype).

Disa leopoldii Kraenzl. in Bull. Soc. Roy. Bot. Belg. **38**: 218 (1899). Type from Zaire.

Disa princeae Kraenzl. in Bot. Jahrb. Syst. **28**: 370 (1900). Type from Tanzania.

Terrestrial herb with separate fertile and sterile shoots; perennating by testicular tubers. Sterile shoots to 14 cm tall, 1–4-leaved; leaves erect, conduplicate, 15–30 cm long, elliptic to oblanceolate, acute. Fertile shoots 60–100 cm tall; cauline leaves imbricate, grading into the floral bracts, mostly sheathing with small free blades; blades 7–8 cm long, lanceolate to narrowly ovate, acute to subacuminate. Inflorescence lax to sub-dense, 15–25 cm long, 15–30-flowered. Floral bracts overtopping the flowers, 2.5–4.5 cm long, lanceolate, acute to acuminate. Flowers borne horizontally, mauve to purple, the galea spotted. Dorsal sepal galeate; galea, rounded, erect 12–18 mm long, 8–15 mm wide and 12–16 mm deep, ovate. Spur borne horizontally from a broad conical base, soon decurved, 12–17 mm long, slender in the middle, the apex clavate or subclavate. Lateral sepals 13–20 mm long, sub-oblique, oblong, sub-acute to rounded. Petals 15–18 mm long, narrowly lanceolate, acute, erect. Lip 11–15 mm long,

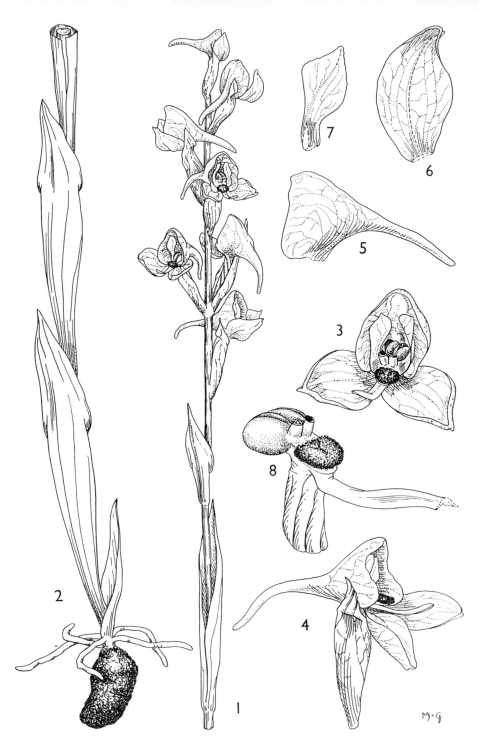

Tab. 57. DISA CAFFRA. 1, inflorescence (×1); 2, lower part of stem and tuber (×1); 3, flower, front view (×2); 4, flower, lateral view (×2); 5, dorsal sepal, lateral view (×2); 6, lateral sepal (×3); 7, petal (×3); 8, column and part of lip, three-quarter profile (×5), 1–8 from *Milne-Redhead* 2870. Drawn by Mary Grierson. From Kew Bull.

Tab. 58. DISA WALLERI. 1, habit (×$\frac{1}{6}$); 2, lower part of stem (×1); 3, lower part of inflorescence (×1); 4, dorsal sepal, side view (×1); 5, lateral sepal (×1); 6, petal (×1); 7, lip (×1); 8, column, side view (×4); 9, column, front view (×4); 10, pollinia (×3), 1–10 from *Polhill & Paulo* 1371. Drawn by Heather Wood. From F.T.E.A.

linear, rounded or truncate. Rostellum lateral lobes parallel, c. 2 mm tall. Anther 5–6 mm long. Stigma subsessile. Ovary 1.5–2.5 cm long.

Zambia. N: Mbala (Abercorn), fl. 5.ii.1957, *Richards* 8059 (K). W: Mwinilunga, fl. 12.xii.1958, *Holmes* 0126 (BOL; K; SRGH). E: Chipata (Fort Jameson), fl. iii.1952, *Benson* 22 (LISC). S: Mazabuka Distr., Mabwingombe Hills, fl. 5.ii.1960, *White* 6839 (FHO; K). **Zimbabwe**. N: Hurungwe Reserve, fl. 13.i.1950, *Collins* in *GHS* 26667 (K; SRGH). C: Harare, Chakoma, fl. 27.i.1950, *Greatrex* in *GHS* 26741 (K; PRE; SRGH). E: Nyanga (Inyanga) Downs, fl. 29.i.1931, *Norlindh & Weimarck* 4655 (BM; K; S). **Malawi**. N: north Mzimba, fl. ii.1968, *G. Williamson* 356 (K). S: Zomba Plains, *Whyte* s.n. (K).

Also from Zaire and Tanzania. *Brachystegia* woodland or secondary montane grassland, in semi-shade in damp places; 1000–2300 m.

The larger-flowered forms of this species are very beautiful and distinct, but the smaller-flowered forms are easily confused with *Disa caffra*.

31. **Disa robusta** N.E. Br. in F.T.A. **7**: 282 (1898). —Kraenzlin, Orchid. Gen. Sp. **1**: 779 (1900). —Schlechter in Bot. Jahrb. Syst. **53**: 538 (1915). —Summerhayes in Kew Bull. **17**: 547 (1964); in F.T.E.A., Orchidaceae: 168 (1968). —Williamson, Orch. S. Centr. Africa: 84 (1977). —Linder in Contrib. Bolus Herb., No. 9: 136 (1981). —Geerinck in Fl. Afr. Centr., Orchidaceae pt. 1: 204 (1984). —la Croix et al., Orch. Malawi: 123 (1991). Syntypes: Malawi, Blantyre, Wilandi (?=Ndirandi), 6.vii.1879, *Buchanan* 23; Mt. Zomba, *Whyte* s.n., Mt. Malosa, *Whyte* s.n. (K).

Disa praestans Kraenzl. in Bot. Jahrb. Syst. **33**: 59 (1902). Type from Tanzania.
Disa coccinea Kraenzl. in Bot. Jahrb. Syst. **33**: 59 (1902). Type from Tanzania.

Terrestrial herb with separate fertile and sterile shoots; perennating by testicular tubers. Sterile shoots to 10 cm tall, 2–4-leaved; leaves semi-erect, conduplicate, up to 30 cm long, lanceolate to narrowly elliptic, acute. Fertile shoots 60–80 cm tall; cauline leaves imbricate, almost completely sheathing, grading into the floral bracts, 6–9 cm long, obtuse to acute. Inflorescence sub-dense, 7–15(30) cm long, 10–20(40)-flowered. Floral bracts ovate to narrowly ovate, acute to acuminate, reaching to the middle of the flowers. Flowers facing down at an angle of 45°, orange-red. Dorsal sepal galeate; galea rounded, 10–16 mm long, 9–12 mm wide and 7–8 mm deep, broadly ovate. Spur horizontal from a broadly conical base, 25–35 mm long, constricted in the middle, the apex sub-clavate, usually upcurved, ascending due to the angle of the flowers. Lateral sepals 12–18 mm long, broadly ovate to broadly oblong, subacute. Petals 11–18 mm long, oblanceolate, acute, erect, falcate. Lip 7–12 mm long, narrowly oblong, acute. Rostellum lateral lobe parallel, 2–3 mm tall. Anther 4–5 mm long. Stigma stipitate. Ovary 2–3 cm long.

Zambia. E: Nyika Plateau, road to Rest House, fl. 2.i.1959, *Richards* 10397 (K; SRGH). **Malawi**. N: Viphya, Chikangawa, fl. 31.i.1956, *Chapman* 362 (K). C: Dedza Mt., fl. 3.iii.1974, *Westwood* 679 (K). S: Zomba Plateau, Mulungusi Dam, fl. xii.1976, *Welsh* 77 (K).

Also from Tanzania and Zaire. Locally common in montane grassland; (1000)2000–3200 m.
Readily distinguished by the ascending spur.

32. **Disa versicolor** Rchb.f. in Flora **48**: 181 (1865). —N.E. Br. in F.T.A. **7**: 283 (1898). —Kraenzlin, Orchid. Gen. Sp. **1**: 754 (1900). —Schlechter in Bot. Jahrb. Syst. **31**: 240 (1901). —Summerhayes in Bot. Not. **1937**: 189 (1937). —Goodier & Phipps in Kirkia **1**: 53 (1961). —Compton, Fl. Swaziland: 158 (1976). —Grosvenor in Excelsa **6**: 80 (1976). —Linder in Contrib. Bolus Herb., No. 9: 139 (1981). —Stewart et al., Wild Orch. S. Africa: 122 (1982). Type from Angola.

Disa macowanii Rchb.f., Otia Bot. Hamburg. **2**: 106 (1881). —Kraenzlin, Orchid. Gen. Sp. **1**: 754 (1900). —Rolfe in F.C. **5**, 3: 288 (1913). Type from South Africa.
Disa hemisphaerophora Rchb.f., Otia Bot. Hamburg. **2**: 106 (1881). Type from South Africa.

Terrestrial herb with separate fertile and sterile shoots; perennating by testicular tubers. Sterile shoots 4–7 cm tall, (3)5-leaved; leaves semi-erect, conduplicate, rarely petiolate, 15–30(40) cm long, very narrowly elliptic, acute. Fertile shoots (20)30–50(70) cm tall; cauline leaves usually imbricate, mostly sheathing, grading apically into the floral bracts, to 12 cm long, lanceolate to narrowly lanceolate, acute. Inflorescence dense, 6–20(30) cm long, many-flowered. Floral bracts c. 1 cm long, lanceolate, acuminate, often dry. Flowers usually facing down; colours variable, pink to scarlet or reddish, often mottled, sepals often turning brown at anthesis, lip and petals green from a purple base. Dorsal sepal angled downwards, galeate; galea 4.5–6.5 mm long, 3.5–4.5 mm wide and 3–4.5 mm deep, ovate to broadly elliptic, obtusely or bluntly apiculate. Spur from the middle of the galea, straight or more usually sharply deflexed at the base, (4)5–7 mm long, cylindrical. Lateral sepals 4.5–6.5 mm long, narrowly oblong to lanceolate, acute. Petals 4–5 mm long, obliquely ovate, acute, erect, partially exserted from the galea, deflexed inwards in about the middle. Lip 4.5–5.5 mm long, lorate to narrowly oblanceolate, obtuse. Rostellum lateral lobes angled forwards. Anther 2.2 mm long. Stigma pedicellate. Ovary c. 8 mm long.

Zimbabwe. E: Stapleford Forest Reserve, fl. 21.i.1945, *Hopkins* in *GHS* 13226 (K; SRGH). **Mozambique**. MS: Tsetserra, fl. 7.ii.1955, *Exell, Mendonça & Wild* 243 (LISC).
Also in Angola and South Africa. Montane grassland; c. 2000m.
Zimbabwean material differs from the common South African form in flower colour and length and orientation of the spur, and may approach *D. maculomarronina* McMurtry more closely than *D. versicolor*. More fieldwork needs to be done to clarify the relationships of this material.

33. **Disa eminii** Kraenzl. in Bot. Jahrb. Syst. **19**: 248 (1894). —N.E. Br. in F.T.A. **7**: 282 (1898). —Kraenzlin, Orchid. Gen. Sp. **1**: 759 (1900). —Schlechter in Bot. Jahrb. Syst. **31**: 298 (1901). —Summerhayes in F.T.E.A., Orchidaceae: 168 (1968). —Williamson, Orch. S. Centr. Africa: 84 (1977). —Linder in Contrib. Bolus Herb., No. 9: 154 (1981). —Geerinck in Fl. Afr. Centr., Orchidaceae pt. 1: 196 (1984). Type from Tanzania.
Disa stolonifera Rendle in J. Linn. Soc., Bot. **37**: 222 (1905). Type from Uganda.

Terrestrial or epiphytic, stoloniferous herb. Plants erect, slender, 30–60 cm tall. Leaves imbricate at base of the plant; blades semi-erect, conduplicate, up to 40 cm long, linear, acute, becoming narrowly ovate to lanceolate and grading into the floral bracts towards the stem apex. Inflorescence lax, 8–15 cm long, 15–30-flowered. Floral bracts acute, shorter than the ovaries. Flowers bright pink to cherry-red. Dorsal sepal galeate; galea erect, 9–11 mm long and c. 3 mm deep, ovate to elliptic, acute. Spur 20–25 mm long, cylindrical, horizontal or gently down-curved. Lateral sepals 9–11 mm long, oblong-elliptic, obtuse. Petals erect, 7–7.5 mm long, oblong to obovate, obtuse. Lip 9–11 mm long, obovate to elliptic. Rostellum erect, 1–1.5 mm tall. Anther horizontal, 2 mm long. Ovary 20–25 mm long.

Zambia. W: Mwinilunga Distr., Matonchi Farm, Sinkabolo Dambo, fl. xii.1969, *G. Williamson* 441 (K; SRGH).
Also from Uganda, Tanzania and Burundi. Wet swamp forest, usually on decaying vegetation; 1100–2100 m.
Known from only one locality in the Flora Zambesiaca area.

34. **Disa saxicola** Schltr. in Bot. Jahrb. Syst. **20**, Beibl. Nr. 50: 41 (1895). —N.E. Br. in F.T.A. **7**: 281 (1898). —Kraenzlin, Orchid. Gen. Sp. **1**: 781 (1900). —Schlechter in Bot. Jahrb. Syst. **31**: 271 (1901). —Rolfe in F.C. **5**, 3: 244 (1913). —Fl. Pl. S. Afr. 13 t. 495 (1933). —Summerhayes in Bot. Not. **1937**: 188 (1937). —Goodier & Phipps in Kirkia 1: 53 (1961). —Summerhayes in Kew Bull. **17**: 551 (1964); in F.T.E.A., Orchidaceae: 171 (1968). —Compton, Fl. Swaziland: 158 (1976). —Grosvenor in Excelsa **6**: 80 (1976). —Williamson, Orch. S. Centr. Africa: 86 (1977). —Linder in Contrib. Bolus Herb., No. 9: 213 (1981). —Stewart et al., Wild Orch. S. Africa: 108 (1982). —la Croix et al., Orch. Malawi: 124 (1991). Type from South Africa.
Disa uliginosa Kraenzl. in Bot. Jahrb. Syst. **30**: 285 (1901). Type from Tanzania.

Terrestrial herb, rarely epilithic or epiphytic; perennating by tubers. Plants flexuose, 10–40 cm tall, base of stem often with a sheath of old leaf fibres. Leaves erect or falcate, conduplicate, 7–20 cm long, linear lanceolate, acute to acuminate, grading into the floral bracts. Inflorescence lax, 3–15 cm long. Floral bracts variable in length, lanceolate to rarely narrowly ovate, acuminate. Flowers white with purple markings on the hood. Dorsal sepal galeate; galea 4–8 mm long, c. 3 mm wide and c. 3 mm deep, rounded to apiculate, constricted in the middle. Spur from a conical base, usually gently down-curved, 4–10 mm long, terete, acute. Lateral sepals erect, 3.5–4.5 mm long, narrowly oblong to oblong, truncate. Lip 3–5 mm long, linear–spathulate. Rostellum square, central lobe small. Anther semi-pendent, c. 1 mm long. Stigma flat. Ovary 5–15 mm long, mottled with purple.

Zambia. E: Nyika Plateau, fl. 18.ii.1961, *Richards* 14393A (K). **Zimbabwe**. E: Chimanimani Mts., fl. iii.1957, *Goodier* 174 (K; SRGH). **Malawi**. N: Nyika Plateau, Vipiri Rocks, fl. 23.ii.1976, *Phillips* 1253 (K). C: Dedza Mt., fl. 3.iii.1977, *Grosvenor & Renz* 1022 (BOL; K; SRGH). S: Sangano Hill, Kirk Range, fl. 31.i.1959, *Robson* 1394 (K; LISC; SRGH). **Mozambique**. N: Lichinga (Vila Cabral), near Massangulo, fl. 25.ii.1964, *Torre & Paiva* 10827 (LISC).

Also from Tanzania, South Africa, Swaziland and Lesotho. Usually associated with rock outcrops and cliffs, often on moss covered rock and on tree trunks in scrub forests; 1000–2800 m.

35. **Disa patula** Sond. in Linnaea **19**: 94 (1847). —Kraenzlin, Orchid. Gen. Sp. **1**: 773 (1900). —Rolfe in F.C. **5**, 3: 245 (1913). —Compton, Fl. Swaziland: 157 (1976). —Linder in Contrib. Bolus Herb., No. 9: 132 (1981). Type from South Africa.

Var. **transvaalensis** Summerh. in Bull. Misc. Inform., Kew **1938**: 148 (1938). —Goodier & Phipps in Kirkia **1**: 53 (1961). —Grosvenor in Excelsa **6**: 80 (1976). —Linder in Contrib. Bolus Herb., No. 9: 258 (1981). —Stewart et al., Wild Orch. S. Africa: 113 (1982). Type from South Africa.
    Disa gerrardii Rolfe in F.C. **5**, 3: 243 (1913). Type from South Africa.

Terrestrial herb perennating by testicular tubers. Fertile shoots 25–60 cm tall, the bases sometimes sheathed in fibrous leaf-remains. Leaves erect, conduplicate, longest near the shoot base, up to 20 cm long, linear to narrowly lanceolate, acute. Inflorescence slender, 5–25 cm long, cylindrical. Floral bracts usually slightly longer than the ovaries, lanceolate to narrowly lanceolate, acuminate. Flowers pink, petals with purplish spots near the apex. Dorsal sepal galeate; galea shallowly emarginate, 6–10 mm long, c. 2.5 mm wide and c. 2 mm deep, apiculate or rounded. Spur horizontal, at length rarely down-curved, 8–10 mm long, terete, acute. Lateral sepals 6–10 mm long, narrowly oblong, rounded to obtuse with fleshy acute apiculi. Petals erect, 5–10 mm long, obliquely narrowly ovate to lanceolate, emarginate to apiculate, rarely acute. Lip c. 9 mm long, linear with the apex somewhat widened. Rostellum lateral lobes square, central lobe finger-like, taller than the lateral lobes. Stigma with the margins taller than the receptive surface. Ovary 1–2 cm long.

Zimbabwe. E: Chimanimani Mts., Stonehenge, fl. 31.i.1958, *Hall* 240 (BOL; K; SRGH).
Also in South Africa and Swaziland. Frequent in montane grassland; 600–2400 m.
Disa patula var. *patula*, restricted to South Africa, differs from var. *transvaalensis* in possessing a longer dorsal sepal (more than 10 mm long) and a shorter spur.

36. **Disa rungweensis** Schltr. in Bot. Jahrb. Syst. **53**: 543 (1915). —Summerhayes in F.T.E.A., Orchidaceae: 175 (1968). —Williamson, Orch. S. Centr. Africa: 88 (1977). —Linder in Contrib. Bolus Herb. No. 9: 263 (1981). —la Croix et al., Orch. Malawi: 125 (1991). TAB. **59**. Type from Tanzania.

Terrestrial herb perennating by testicular tubers. Plants 10–25 cm tall. Basal leaves 4–6, semi-erect, 1–3.5 cm long, narrowly elliptic, acute, undulate or crisped on the margins. Cauline leaves mostly sheathing, lanceolate, acute, grading into the floral bracts. Inflorescence lax, 2.5–4.5 cm long, 2–10-flowered. Floral bracts usually as tall as the flowers, ovate to lanceolate, acute to acuminate. Flowers variable in colour from greenish-white to pinky-mauve to brown, facing downwards. Dorsal sepal galeate; galea 2.5–3 mm long and 1.5–3 mm wide, rounded to obtuse, almost hemisphaerical. Spur 2.5–3.2 mm long, cylindric, horizontal or ascending, up-curved at the apex and irregularly clavate. Lateral sepals c. 2.5 mm long, oblong to ovate, subacute. Petals erect, somewhat curved over the rostellum and partially included within the galea, 1.8–2 mm long, obliquely quadrate-obovate. Lip 1.3–1.5 mm long, narrowly oblong, rounded. Rostellum low, flat; lateral lobes square, the central lobe almost obsolete. Anther horizontal with a massive connective. Stigma massive, pedicellate. Ovary 5–8 mm long.

Malawi. N: Nyika Plateau, Chilinda Rock, fl. ii.1968, *G. Williamson* 363 (K; SRGH).
Also from Tanzania. Restricted to the high mountains at the northern end of Lake Malawi. On rocky outcrops, often in wet grassland in rock crevices.
Very close to *D. zimbabweensis* H.P. Linder.

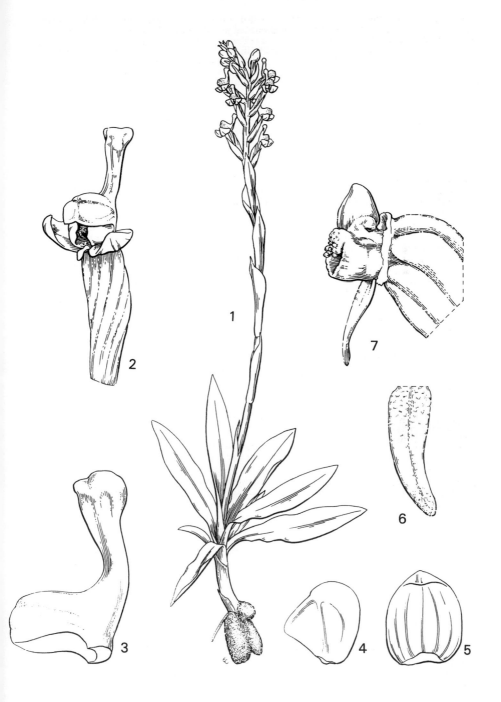

Tab. 59. DISA RUNGWEENSIS. 1, habit (×1); 2, flower (×6); 3, dorsal sepal, side view (×10); 4, petal (×10); 5, lateral sepal (×10); 6, lip (×20); 7, column and lip (×15), 1–7 from *la Croix* 326 and *Pawek* 9113. Drawn by Eleanor Catherine.

37. **Disa zimbabweensis** H.P. Linder in Contrib. Bolus Herb., No. 9: 267 (1981). —Stewart et al., Wild Orch. S. Africa: 105 (1982). Type: Zimbabwe, Nyanga (Inyanga) Downs, *Norlindh & Weimarck* 4650 (K, holotype).

   Disa rungweensis var. *rhodesiaca* Summerh. in Bot. Not. **1937**: 188 (1937). Type as above.
   Disa rungweensis sensu Goodier & Phipps in Kirkia **1**: 53 (1961).
   Disa rungweensis subsp. *rhodesiaca* (Summerh.) Summerh. in Kew Bull. **17**: 551 (1964).
   —Grosvenor in Excelsa **6**: 80 (1976). Type as above.

Terrestrial herb perennating by testicular tubers. Plants (10)15–30 cm tall. Lower cauline leaves semi-erect, often clustered at stem base, 4–8 cm long, narrowly elliptic to narrowly ovate; margins plane or very rarely crisped. Upper cauline leaves sheathing, lax or imbricate, lanceolate, acute. Inflorescence dense, 3–12 cm long, more than 10-flowered. Floral bracts about as tall as the flowers, lanceolate to ovate, usually acuminate. Flowers pale pink to purple. Dorsal sepal galeate; galea 2.5–4.5 mm long and c. 2 mm wide, almost hemisphaerical. Spur horizontal or ascending from the base, 2.7–4 mm long, cylindric; apex irregularly clavate and usually up-curved. Lateral sepal 2.5–3.5 mm long, oblong to ovate, subacute. Petals erect, partially included within the galea, 2–2.8 mm long, obliquely quadrate-obovate. Lip 1.7–2.4 mm long, narrowly oblong, rounded. Rostellum lateral lobes square and the central lobe much reduced. Anther almost horizontal. Stigma pedicellate, almost as tall as the rostellum. Ovary 5–10 mm long.

   Zimbabwe. E: Nyanga (Inyanga), fl. 1.ii.1958, *Beasley* 51 (K; SRGH). **Mozambique**. MS: Tsetserra, fl. 7.ii.1955, *Exell, Mendonça & Wild* 231 (LISC).
   Also from South Africa. Montane grassland, in rocky outcrops and rock crevices; above 1800 m.

20. **BROWNLEEA** Lindl.
by H.P. Linder

Brownleea Lindl. in Lond. J. Bot. **1**: 16 (1842).

Terrestrial, rarely epiphytic or epilithic herb with perennating tubers. Tubers more or less testicular, pubescent to woolly, up to 5 cm long. Plants mostly erect, slender, 5–60 cm tall. Scape usually nitid, the base rarely sheathed in old leaf fibres. Basal sheath(s) pale brown, up to 15 cm long, obtuse to acute, often ± mucronate; leaves 1–3(5), scattered on the scape, erect to spreading, the blade up to 22 cm long, linear to ovate, usually nitid, more or less plicate with veins somewhat prominent. Inflorescence capitate or laxly spicate; flowers 1–numerous, c. 5–30 mm in diameter, white to mauve or pale with darker mauve spots; bracts erect, shorter than to much longer than the ovary, acuminate, green. Dorsal sepal, c. 3–13 mm long, lanceolate, acuminate, falcate in side view, the apex often reflexed; a spur produced from dorsal sepal base, horizontal, then straight or down-curved, 0.5–5 cm long. Lateral sepals flat, oblique, narrowly elliptic or rarely oblong, acute. Petals fused by their backs to the dorsal sepal, oblanceolate, oblong or almost pandurate, occasionally with a basal anticous lobe, the upper margin flat or crenulate, the apex acute or rounded. Lip minute, erect from a broad base in front of the stigma, 0.1–2 mm long. Rostellum tall, erect; lateral lobes more or less lorate, concave, each with an apical viscidium; central lobe a small fleshy body at the base of the lateral lobes (except in *B. coerulea*, where it is finger-like); staminodes as large as or smaller than the lateral lobes. Anther horizontal, the apex usually up-curved; caudicles almost as long as the 2 pollen masses; viscidia 2, globular, separate. Stigma sessile at the base of the rostellum.

A small genus of about 16 species in South and tropical Africa and Madagascar.

1. Flowers in a terminal head, or at least strongly congested; petals patent, apically expanded, crenulate on the margins · **4.** *galpinii*
 – Flowers in a lax to ± dense spike; petals not patent nor expanded apically and not crenulate or only slightly crenulate on the margins · 2
2. Spur 6–8 mm long; inflorescence a lax, several-flowered spike · · · · · · · · · · **1.** *maculata*
 – Spur up to 5 mm long; inflorescence a dense, many-flowered spike, rarely lax · · · · · · · 3
3. Spur straight or gently down-curved; dorsal sepal 5–6 mm long · · · · · · · · **3.** *mulanjiensis*
 – Spur sharply down-curved; dorsal sepal 3–5 mm long · · · · · · · · · · · · · · · · **2.** *parviflora*

Tab. 60. BROWNLEEA MACULATA. 1, habit (×1), from *la Croix* 183; 2, flower, front view (×3); 3, dorsal sepal (forming hood), side view (×3); 4, lateral sepal (×3); 5, petal (×3); 6, lip (×8); 7, column and lip, side view (×8); 8, column, from above, with anther removed (×8); 9, anther loculi, with pollinia (×8) 2–9 from *Wild* 2913 (in K spirit coll. 31175). Drawn by Judi Stone from type collection.

Tab. 61. BROWNLEEA PARVIFLORA. 1, habit (×1); 2, flower, side view (×5); 3, dorsal sepal, side view (×10); 4, lateral sepals (×10); 5, petal (×10); 6, lip (×20); 7, column and lip, side view (×20); 8, column and lip, front view (×20), 1–8 from *Milne-Redhead & Taylor* 10326. Drawn by Heather Wood. From F.T.E.A.

1. **Brownleea maculata** P.J. Cribb in Kew Bull. **32**: 147, fig.7 (1977). —Linder in J. S. African Bot. **47**: 30 (1981). —la Croix et al., Orch. Malawi: 109 (1991). TAB. **60**. Type: Mozambique, Chimanimani Mountains, *Wild* 2913 (K, holotype; SRGH).
   Brownleea sp. no.1 Grosvenor in Excelsa **6**: 79 (1976).

Terrestrial herb 15–35 cm tall, plants often somewhat flexuose; tubers densely villous. Basal sheath fibrous, up to 10 cm long, acute, often absent. Leaves (2)3, shortly sheathing at the base; blades somewhat ascending, 11x 4 cm, lanceolate to narrowly ovate, acute to acuminate, the veins emergent. Inflorescence ± lax, up to 10 cm long, several to many-flowered. Bracts, lanceolate to narrowly lanceolate, acuminate; the lowest bract up to 4.5 cm long, usually leaf-like, the remainder mostly less than 2 cm long. Flowers c. 8 mm in diameter, horizontal; sepals purplish mauve, petals white, lateral sepals and petals purple-spotted and striped. Dorsal sepal galeate; galea 6–7 mm tall, 2.5 mm wide and 1.5–2 mm deep, narrowly ovate, acute. Spur 6–8 mm long, arising from the base of the galea, cylindrical with a short conical base, soon sharply down-curved to recurved, apex subclavate. Lateral sepals spreading forwards, flat, 6–7 × 2–3 mm, oblique, narrowly elliptical, acute. Petals erect, c. 6 × 1.5–2 mm, obovate, subacute, with a basal anticous lobe flanking the stigma. Lip minute, erect, c. 2.8 × 0.5 mm. Rostellum c. 2.5 mm tall; lateral lobes tall, concave; staminodes large; anther c. 2 mm long, sharply up-curved from a horizontal base; stigma vertical. Ovary slender, 1–1.5 cm long.

Zimbabwe. E: Chimanimani Mountains, between Bundi Plain and Upper Valley, fl. 9.iv.1967, *Drummond* 9130 (K; SRGH). **Malawi**. S: Mt. Mulanje (Mlanje), Sombani Plateau, fl. 12.v.1963, *Wild* 6230 (BOL; K; SRGH). **Mozambique**. MS: Chimanimani Mts., fl. 6.vi.1949, *Munch* 176 (LMA).
Known as yet only from the Chimanimani and Mulanje Mountains. In forest shade, often on old rotting tree-trunks, also on rocks and live trees; above 1500 m.

2. **Brownleea parviflora** Harv. ex Lindl. in Lond. J. Bot. **1**: 16 (1842). —Bolus, Icon. Orchid. Austro-Afric. **1**: t. 43 (1893). —Kraenzlin, Orchid. Gen. Sp. **1**: 733 (1900). —Schlechter in Bot. Jahrb. Syst. **31**: 312 (1901). —Rolfe in F.C. **5**, 3: 260 (1913). —Summerhayes in Kew Bull. **20**: 169 (1966); in F.W.T.A. ed. 2, **3**: 201 (1968); in F.T.E.A., Orchidaceae: 179 (1968). — Grosvenor in Excelsa **6**: 79 (1976). —Williamson, Orch. S. Centr. Africa: 90 (1977). —Linder in J. S. African Bot. **47**: 31 (1981). —la Croix et al., Orch. Malawi: 110 (1991). TAB. **61**. Type from South Africa.
   Disa alpina Hook.f. in J. Linn. Soc., Bot. **7**: 220 (1864). Type from Cameroon.
   Disa parviflora (Harv. ex Lindl.) Rchb.f., Otia Bot. Hamburg. **2**: 119 (1881). Type as for *B. parviflora*.
   Disa presussi Kraenzl. in Bot. Jahrb. Syst. **17**: 64 (1893). Type from Cameroon.
   Disa apetala Kraenzl. in Engl., Pflanzenw. Ost-Afr. **C**: 153 (1895). Type from Tanzania.
   Brownleea alpina (Hook.f.) N.E. Br. in F.T.A. **7**: 287 (1898). Type as for *Disa alpina*.
   Brownleea apetala (Kraenzl.) N.E. Br. in F.T.A. **7**: 287 (1898). Type as for *Disa apetala*.
   Brownleea gracilis Schltr. in Bot. Jahrb. Syst. **53**: 545 (1915). Type from Tanzania.
   Brownleea perrieri Schltr. in Fedde, Repert. Spec. Nov. Regni Veg., Beih. 33: 102 (1924). Type from Madagascar.
   Brownleea transvaalensis Schltr. in Ann. Transvaal Mus. **10**: 250 (1924). Type from South Africa.

Terrestrial herb perennating by testicular tubers. Plants erect, 20–60 cm tall, occasionally with fibrous leaf remains around the stem base. Basal sheath 4–12 cm long, muricate or scabrid. Leaves (2)3(5), grading into the floral bracts, bases sheathing; blades semi-erect, 8–20 cm long, narrowly lanceolate, very acute to acuminate, midrib and often the side-veins prominent beneath. Inflorescence 4–12 cm long, densely 20–60-flowered. Floral bracts longer than the flowers, narrowly lanceolate to lanceolate, acuminate. Flowers white with a slight green or brownish tint. Dorsal sepal galeate; galea (2)3–5 mm tall, 1.5–2 mm wide and 0.8–1 mm deep, ovate, acuminate, falcate in side view. Spur horizontal at the base, soon sharply down-curved, 3–5 mm long, cylindric, frequently clavate. Lateral sepals oblique, (2)3–5 mm long, oblong, acute, the bases frequently fused. Petals (1.5)2.5–3.5 mm long, oblong to almost square, obliquely acute. Lip minute, 0.5–1 mm long. Rostellum c. 1 mm tall; lateral lobes caniculate; staminodes almost as tall as the lateral lobes. Anther reflexed, stigma almost horizontal. Ovary 5–10 mm long.

Zambia. E: Nyika Plateau, fl. iii.1967, *Williamson & Odgers* 294 (K; SRGH). **Zimbabwe**. E: Chimanimani Distr., Mt. Peni, fl. 16.iii.1981, *Philcox, Leppard, Duri & Urayai* 8990 (K). **Malawi**. N: Nyika Plateau, fl. 13.iii.1961, *E.A. Robinson* 4477 (K). S: Zomba Mt., fl. 30.iii.1981, *la Croix* 127 (K). **Mozambique**. MS: Mt. Tsetsera, fl. 3.iii.1954, *Wild* 4472 (K; SRGH).

Also from Cameroon, Kenya, Tanzania, Swaziland, South Africa and Madagascar. Montane grassland.

A rather distinctive species.

3. **Brownleea mulanjiensis** H.P. Linder in Kew Bull. **40**: 125, fig.1 (1985). —la Croix et al., Orch. Malawi: 109 (1991). TAB. **62**. Type: Malawi, Mt. Mulanje, *Adamson* 434 (K, holotype; BM).

Terrestrial herb perennating by testicular tubers. Plants erect, to 50 cm tall, slender, somewhat flexuose. Basal sheaths 2, the outer to c. 3 cm long, hyaline; the inner to 10 cm long, scabrid to shortly pilose, dark-brown or black. Leaves 1–2, shortly sheathing; blade erect, to 12 × 2 cm, narrowly lanceolate, acute, veins prominent. Inflorescence a dense spike to 5 cm long, many-flowered. Floral bracts as long as or longer than the ovaries, narrowly lanceolate, acuminate. Flowers resupinate, lilac or pale mauve. Dorsal sepal galeate; galea 5–6 mm long, falcate in side view. Spur horizontal at its base, straight or gently down-curved, 3.5–4.5 mm long, cylindrical. Lateral sepals patent, oblique, 3–5 mm long, narrowly oblong, acute. Petals 4.5–5.5 × 3–4 mm, oblong, somewhat falcate, fused to the dorsal sepal, the upper margin somewhat crenate. Lip minute, 2–3 mm long. Rostellum c. 1.5 mm tall; lateral lobes erect, adjacent, canaliculate, flanked by two large staminodes. Anther reflexed with an up-curved apex; stigma sessile at the base of the rostellum. Ovary c. 1 cm long.

Malawi. S: Mt. Mulanje, Chinzama, fl. 28.iii.1983, *Johnson-Stewart* 478 (K).

Known only from Mt. Mulanje, where it occurs in montane grassland and rocky heathland; 2100–2400 m.

This species is closely related to *B. galpinii*, both species having crenate petals and papillate lateral sepals. They are, however, distinguished by the inflorescence shape and flower colour.

4. **Brownleea galpinii** Bolus, Icon. Orchid. Austro-Afric. **1**: t. 42 (1893). —Kraenzlin, Orchid. Gen. Sp. **1**: 733 (1900). —Schlechter in Bot. Jahrb. Syst. **31**: 310 (1901). —Rolfe in F.C. **5**, 3: 261 (1913). —Summerhayes in Bot. Not. **1937**: 189 (1937). —Grosvenor in Excelsa **6**: 79 (1976). —Linder in J. S. African Bot. **47**: 38 (1981). Type from South Africa.

Brownleea flavescens Schltr. in Ann. Transvaal Mus. **10**: 249 (1924). Type from South Africa.

Subsp. **galpinii**

Terrestrial herb perennating by testicular tubers. Basal sheaths 2, the inner 5–10(14) cm long, ribbed, mottled with small mucronate patches. Leaves 2(4), erect to semi-erect, ± conduplicate, bases sheathing; blades to 6–20 cm long, linear, ribs prominent beneath, upper leaf much smaller than the lower leaf. Inflorescence capitate, often secund, 15–25 × 10–30 mm, 5–25-flowered. Floral bracts somewhat longer than the ovaries, lanceolate, acuminate. Flowers white to cream, often with small purple spots on the petals. Dorsal sepal galeate; galea erect, 5–8 mm tall and c. 2 mm deep, narrowly ovate to lanceolate, acuminate. Spur straight, parallel to the ovary, slender, 3.5–6 mm long, cylindrical. Lateral sepals suboblique, 6–7 mm long, narrowly elliptic, acute. Petals almost as tall as the galea, narrowly oblong-pandurate, erect. Lip minute, erect, 0.8–1.3 mm long. Rostellum erect, c. 3 mm tall; lateral lobes canaliculate; staminodes c. 1 mm in diameter. Anther deflexed with the apex up-curved. Stigma sessile at the base of the rostellum. Ovary c. 1 cm long.

Zimbabwe. E: Nyanga (Inyanga), Mt Nyangani (Inyangani) fl. 7.iii.1981, *Philcox, Leppard, Duri & Urayai*. 8913 (K).

Also from South Africa. Occasional in montane grassland.

Subsp. *major* (Bolus) H.P. Linder is restricted to South Africa and Lesotho, where it occurs at higher altitudes. It is distinguished from the typical subspecies by the larger flowers (lateral sepals (6.5)8–10 mm) and the fan-like, crenulate upper petal lobe.

Tab. 62. BROWNLEEA MULANJIENSIS. 1, plant (×1); 2, flower (×6); 3, dorsal sepal, side view (×6); 4, lateral sepal (×6); 5, petal (×6); 6, column with lip still in position, from front (×12), 1–6 from *Morris* 744. Drawn by C.E. Smith. From Kew Bull.

21. HERSCHELIANTHE Rauschert
by H.P. Linder

Herschelianthe Rauschert in Feddes Repert. **94**: 434 (1983).
Herschelia Lindl., Gen. Sp. Orchid. Pl.: 362 (1838) nom. illegit.

Terrestrial herb perennating by tubers. Tubers testicular, 1–6 cm long. Stem erect, grasslike, 10–100 cm tall, base usually sheathed in old leaf fibres. Basal leaves 5–20, linear, flat or rolled, subsclerophyllous, produced during, before or after flowering; cauline leaves brown, scattered or imbricate on the stem, acuminate. Inflorescence lax, 1–30-flowered. Bracts dry, usually not as long as the flowers, ± broadly ovate, acuminate to setaceous. Flowers resupinate, usually blue or shades of blue, to white with pale blue veins, rarely purplish-red with green parts. Dorsal sepal erect to angled forwards, generally galeate, rounded to acuminate, usually ovate in front view. Spur borne horizontally from the base of the galea, at length straight, decurved or curved upwards, rarely longer than the sepals, cylindrical or conical. Lateral sepals usually patent, 6–30 mm long, lanceolate to ovate, obtuse to acute. Petals with a basal anticous lobe; lobe oblong-ovate, flanking the stigma; petal limb lower part linear or lorate, reflexed and flanking the anther; petal limb upper part lanceolate to expanded fan-like or more or less bifid, gently to abruptly curving upwards behind or near the apex of the anther. Lip ± dissected, rarely entire, usually ovate in outline, rarely spatahulate, always specialized in some way. Anther horizontal or ± pendent, 2-celled with cells parallel; viscidia 2, separate or fused. Rostellum erect, lanceolate, acute, equally 3-lobed, rarely with the lateral lobes canaliculate and the central lobe absent. Stigma horizontal, shortly pedicellate, the odd lobe smaller than the lateral lobes. Ovary patent to ± spreading, 1–1.3 cm long.

A genus of c. 16 species, mainly southern African in distribution but extending into southern Tanzania. The majority of the species are endemic to the Cape Province of South Africa.
    The distinction between *Herschelianthe* and *Disa* is unclear, and perhaps only one genus should be recognized, namely *Disa*. The relationship between them is being investigated, and in the meantime they are treated here as separate genera

1. Petals equalling the dorsal sepal in length; lip almost entire · · · · · · · · · · · · · · 1. *praecox*
– Petals about half as long as the dorsal sepal; lip deeply lacerate · · · · · · · · · · · · · · · · 2
2. Flowers large, lateral sepals 10–18 mm long; inflorescences usually with fewer than 5 flowers
· 2. *baurii*
– Flowers small, lateral sepals less than 10 mm long · 3
3. Inflorescence very lax, with 2–8 widely separated flowers; plants from the Chimanimani
Mts. · 3. *chimanimaniensis*
– Inflorescence densely 4–16-flowered; plants from the Nyika Plateau · · · · · · · 4. *longilabris*

1. **Herschelianthe praecox** (H.P. Linder) H.P. Linder in Kew Bull. **40**: 127 (1985). —la Croix et al., Orch. Malawi: 127 (1991). Type: Zambia, Nyika Plateau, Aug.-Sept. 1967, *G. Williamson* 312 (K, holotype).
    Herschelia praecox H.P. Linder in Bothalia **13**: 382 (1981). Type as above.

Terrestrial herb 20–40 cm tall. Basal leaves c. 6, produced after flowering, c. 30 cm long and 1–2 mm wide, semi-erect, subsclerophyllous. Cauline leaves few, completely sheathing, 2–3 cm long, acuminate. Inflorescence laxly 2–10-flowered, 4–13 cm long. Bracts dry, c. 1 cm long, ovate, acuminate. Flowers white to blue or dark-mauve, occasionally the apices of the petals green. Dorsal sepal galeate; galea erect, 10–12 mm tall, c. 8 mm wide and 4–6 mm deep, acuminate, with an erect to reflexed apex. Spur 3–5 mm long and c. 2 mm in diameter, terete, cylindrical or laterally flattened, horizontal or gradually ascending. Lateral sepals patent, 10–12 mm long, narrowly ovate to lanceolate, subacuminate. Petals with an ovate basal anticous lobe c. 2 mm wide; petal limb c. 12 × 1.5–2 mm, lorate in the lower part and curving up beside the anther within the galea, widened and unequally acutely 2-lobed at the apex with the anterior lobe longer than the posterior lobe. Lip 11–13 mm long, narrowly ovate to lanceolate, curved upwards, margins almost entire to shallowly fimbriate. Anther horizontal, c. 3 mm long, the connective longer than the anther. Rostellum with erect canaliculate lateral lobes 1.5 mm tall. Stigma subsessile, somewhat angled forwards. Ovary 1–1.5 cm long, usually curved.

Zimbabwe. N: Nyika Plateau, fl. 2.ix.1964, *E.A. Robinson* 6259 (K; SRGH). **Malawi**. N: Nyika Plateau, fl. ix.1962, *Tyrer* 966 (BM; SRGH).
Endemic to the Nyika Plateau. Short montane grassland; above 2000 m.
This species overlaps geographically with *H. baurii* (Bolus) Rauschert, but has an earlier flowering time.

2. **Herschelianthe baurii** (Bolus) Rauschert in Feddes Repert. **94**: 434 (1983). —la Croix et al., Orch. Malawi: 126 (1991). TAB. **63**. Type from South Africa.
    Disa baurii Bolus in J. Linn. Soc., Bot. **25**: 174 (1889). —Schlechter in Bot. Jahrb. Syst. **31**: 289 (1901). —L. Bolus in Ann. Bolus Herb. 4 Pl. 11 (1926). Type as above.
    Disa hamatopetala Rendle in Trans. Linn. Soc. ser 2, **4**: 47 (1894). —N.E. Br. in F.T.A. **7**: 286 (1898). —Summerhayes in F.T.E.A., Orchidaceae: 177 (1968). —Williamson, Orch. S. Centr. Africa Pl.: 71 (1977). Type: Malawi, Mt. Mulanje (Mlanje), *Whyte* s.n. (K, lectotype, selected by Linder in Bothalia, 1981).
    Herschelia baurii (Bolus) Kraenzl., Orchid. Gen. Sp. **1**: 804 (1900). —Rolfe in F.C. **5**, 3: 204 (1913). —Linder in Bothalia **13**: 383 (1981). —Stewart et al. in Wild Orch. Southern Africa: 148 (1982). Type as for *Disa baurii*.
    Herschelia hamatopetala (Rendle) Kraenzl., Orchid. Gen. Sp. **1**: 803 (1900). Type as for *Disa hamatopetala*.
    Herschelia bachmanniana Kraenzl., Orchid. Gen. Sp. **1**: 805 (1900). Type from South Africa.
    Herschelianthe hamatopetala (Rendle) Rauschert in Feddes Repert. **94**: 435 (1983). Type as for *Disa hamatopetala*.
    Herschelianthe bachmanniana (Kraenzl.) Rauschert in Feddes Repert. **94**: 434 (1983). Type as *Herschelia bachmanniana*.

Plants 20–40 cm tall; bases often sheathed in old leaf fibres. Basal leaves 5–10, produced after flowering, frequently overtopping the spike, up to 30 cm long and less than 2 mm wide, semi-rigid and subsclerophyllous. Cauline leaves few, completely sheathing, 1.5–2.5 cm long, acuminate, larger towards the base of the stem. Inflorescence lax, rarely subsecund, 2–14-flowered. Bracts rarely as long as the flowers, ovate, acuminate, dry. Flowers varying from pale sky-blue to deep purple-blue, lip frequently a darker blue than the sepals. Dorsal sepal galeate; galea erect, (8)10–20 mm tall, 6–12 mm wide and 5–10 mm deep, ovate, acute. Spur horizontal from the base of the galea, often somewhat ascending, 4–6 mm long, cylindric. Lateral sepals patent, (8)10–18 mm long, oblong to rarely lanceolate, acute to rounded. Petals with a basal, oblong to ovate anticous lobe 1–2.5 mm in wide; petal limb 8–13 mm long, linear, the apex deeply bifid, lacerate or acute. Lip horizontal, at least at base, 10–25 mm long, broadly to narrowly ovate, more or less deeply dissected. Anther horizontal, 2–5 mm long; viscidia separate, ovate. Rostellum erect, equally 3-lobed, 2–3 mm tall. Stigma horizontal, 1 mm tall. Ovary straight or curved, 1–1.5 cm long.

Zambia. E: Nyika Plateau, by road 4 miles SW of Rest House, fl. 25.x.1958, *Robson & Angus* 358 (BM; K; LISC; SRGH). **Zimbabwe**. E: Chimanimani Distr., Mt. Musapa, fl. 10.x.1950, *Wild* 3556 (K; LISC; SRGH). **Malawi**. N: Nyika Plateau, towards Nganda, fl. 23.x.1958, *Robson & Angus* 297 (K; LISC). S: Mt. Mulanje (Mlanje), fl. 24.ix.1929, *Burtt Davy* 22070 (K).
Mozambique. MS: between Skeleton Pass and Martin's Falls, fl 28.ix.1966, *Drummond* 8981 (K; SRGH). M: Ponta do Ouro, fl. 7.vii.1980, *Jansen & de Koning* 7303 (K; LMA).
Also from Zaire, Tanzania (Southern Highlands) and South Africa. Short montane grassland; above 1000 m.
This species is very variable over its range of distribution, and it may be possible to recognize geographical subspecies.

3. **Herschelianthe chimanimaniensis** (H.P. Linder) H.P. Linder in Kew Bull. **40**: 129 (1985). Type: Zimbabwe, Chimanimani, *Ball* 577 (K, holotype; SRGH).
    Herschelia chimanimaniensis H.P. Linder in Bothalia **13**: 384 (1981). Type as above.

Plants 20–40 cm tall. Basal leaves 3–6, produced after flowering, up to 20 cm long and 1–2 mm wide, subsclerophyllous. Cauline leaves 4–8, closely sheathing, up to 2 cm long, acuminate. Inflorescence laxly 2–8-flowered, 3–8 cm long. Bracts dry, not as long

Tab. 63. HERSCHELIANTHE BAURII. 1, habit (×1), from *Newbould & Jefford* 1849; 2, flower (×1); 3, dorsal sepal, side view (×2); 4, lateral sepal (×2); 5, petal (×2); 6, lip (×2); 7, column with one petal, side view (×3); 8, column with one petal, front view (×4), 2–8 from *Robson* 358. Drawn by Heather Wood. From F.T.E.A.

as the flowers, ovate, acuminate. Flowers pinkish-mauve to white, rarely blue. Dorsal sepal galeate; galea erect, 5–7 mm tall, 4 mm wide and 4 mm deep, ovate, acute. Spur horizontal at the base, often gradually ascending, 3–4 mm long, cylindrical or somewhat laterally flattened. Lateral sepals patent, 6–8 mm long, narrowly oblong-ovate, acute. Petals with a basal, ovate anticous lobe 1–1.5 mm wide; petal limb c. 5.5 mm long, lorate, falcately curved beside the anther, included within the galea, the apex acutely bifid. Lip 8–10 mm long, ovate, deeply dissected, the margins curved upwards. Anther horizontal, c. 1.5 mm long. Rostellum lateral lobes canaliculate, erect, 1–1.5 mm tall. Stigma horizontal, 0.6 mm tall. Ovary 5–15 mm long.

Zimbabwe. E: Chimanimani Mt. Range, North End, fl. 8.x.1950, *Munch* 327 (K; SRGH).
Mozambique. MS: Chimanimani Mts., fl. viii.1964, *Whellan* 2145 (SRGH).
Endemic to the montane grassland of the Chimanimani Mts.; 1500–1800 m.
This species is closely related to *H. baurii*, and clearly a segregate from it. It is, however, not very well known, and more data are essential to determine the exact relationships between the two taxa.

4. **Herschelianthe longilabris** (Schltr.) Rauschert in Feddes Repert. **94**: 435 (1983). —la Croix et al., Orch. Malawi: 127 (1991). Type from Tanzania.
    Disa longilabris Schltr. in Bot. Jahrb. Syst. **38**: 150 (1906). —Summerhayes in F.T.E.A., Orchidaceae: 177 (1968). Type as above.
    Herschelia longilabris (Schltr.) Rolfe in Orch. Rev. **27**: 9 (1919). Type as above.
    Herschelia baurii sensu Linder in Bothalia **13**: 383 (1981), pro parte.

Terrestrial herb 12–25 cm tall. Basal leaves numerous, produced after flowering, up to 30 cm long and 1–1.2 mm wide, subsclerophyllous. Cauline leaves 5–7, closely sheathing, imbricate to lax, acuminate. Inflorescence subdensely 4–18-flowered, 5–10 cm long. Bracts dry, not as long as the flowers, ovate, acuminate, dry. Flowers white to pale blue. Dorsal sepal galeate, galea erect, 5–7 mm tall, 4 mm wide and 4 mm deep, ovate, acute to acuminate. Spur horizontal at the base, often somewhat downturned, 3–4 mm long, cylindrical. Lateral sepals patent, 6–9 mm long, narrowly oblong-ovate, acute. Petals with a small basal anticous lobe c. 1 mm in diameter; petal limb c. 5 × 1 mm, lorate in the lower part and curving up beside the anther within the galea, widened and acutely 2-lobed at the apex. Lip 8–10 mm long, lanceolate, deeply dissected. Anther horizontal, 1.5 mm long; viscidia separate. Rostellum erect, 1–1.5 mm tall. Ovary straight or curved, 1–1.5 cm long.

Malawi. N: Nyika Plateau, 4 km from Dembo Bridge to Nganda, fl. 19.ix.1986, *Linder* 3907 (BOL; SRGH).
Also from southern Tanzania. Short montane grassland, 2400-2500 m.

22. MONADENIA Lindl.
by H.P. Linder

Monadenia Lindl., Gen. Sp. Orchid. Pl.: 356 (1838).

Terrestrial herb, perennating by tubers. Tubers testicular. Basal sheaths hyaline, obtuse. Leaves cauline, usually imbricate; leaf bases sheathing; blades usually erect, elliptic to linear-lanceolate, the lowermost the largest, the upper grading into the floral bracts. Inflorescence dense to lax, cylindric to secund. Floral bracts often overtopping the flowers, ovate to lanceolate, acute to acuminate. Flowers resupinate, purplish, brown or green. Dorsal sepal shallowly galeate, 2.5–15 mm long, usually oblong, obtuse, usually with a spur pendent from the base of the galea; spur slender or clavate, obsolete to longer than the galea. Lateral sepals patent or reflexed, often shorter than the galea, usually oblong. Petals obliquely narrowly ovate-oblong, acute to bifid; the broad base enclosing the anther; the apex erect within the galea, or partially exserted from the galea. Lip patent to pendent, linear to elliptic, subfleshy. Anther horizontal to semi-pendent with a single large concrete viscidium and 2 cells. Rostellum simple with a deep notch containing the viscidium and with 2 often well developed lateral flanges flanking the anterior part of the anther. Stigma equally or unequally tripulvinate, shortly stipitate to as tall as the rostellum. Ovary usually twisted, 5–25 mm long.

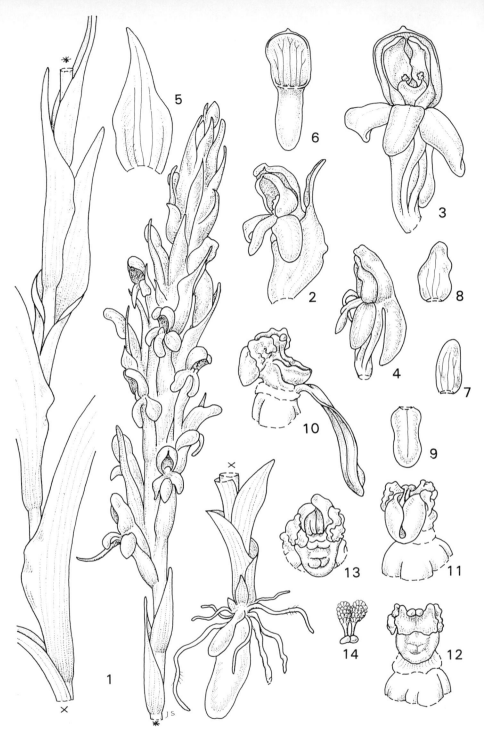

Tab. 64. MONADENIA BREVICORNIS. 1, habit (×1), from *Brummitt, Polhill & Patel* 15943; 2, flower and bract (×2); 3, flower, front view (×3); 4, flower, side view (×2); 5, bract (×2); 6, dorsal sepal (×2); 7, lateral sepal (×2); 8, petal (×2); 9, lip (×2); 10, column and lip, side view (×4); 11, column, back view (×4); 12, column, front view (×4); 13, column, from above, pollinia removed (×4); 14, pollinia (×4), 2–14 from *Hall* 737 (in K spirit coll. 27546). Drawn by Judi Stone.

A genus of 16 species, 15 of which are restricted to the Cape Province of South Africa, one extending into Malawi and Madagascar.

Monadenia brevicornis Lindl., Gen. Sp. Orchid. Pl.: 357 (1838). —Kraenzlin, Orchid. Gen. Sp. **1**: 816 (1900). —Rolfe in F.C. **5**, 3: 192 (1913). —Linder in Bothalia **13**: 360 (1981). — Stewart et al., Wild Orch. Southern Africa: 154 (1982). —la Croix et al., Orch. Malawi: 128 (1991). TAB. **64**. Type from South Africa.
  Disa brevicornis (Lindl.) Bolus in J. Linn. Soc., Bot. **25**: 196 (1889). —Schlechter in Bot. Jahrb. Syst. **31**: 211 (1901). —Bolus, Icones Orch. Austro- Afr. **3**: t.40b (1911). Type as above.

Plants 20–50 cm tall. Leaves numerous, imbricate, up to 15 cm long, narrowly lanceolate to rarely narrowly ovate, acute; lower leaves largest; upper leaves grading into the floral bracts. Inflorescence a lax or cylindrical spike, 4–30 cm long. Bracts reaching or overtopping the flowers, narrowly ovate to ovate, acuminate. Flowers with lime-green petals and lip, lip base maroon, lateral sepals green, dorsal sepal rust-coloured to maroon. Dorsal sepal erect, shallowly galeate, 7–10 mm tall and c. 1 mm deep, narrowly obovate to oblong, obtuse to rounded, apiculate. Spur pendent with the apex curved towards the ovary, 7–11 × 2–3 mm, cylindrical. Lateral sepals reflexed, 5–9 mm long, oblong, obtuse, apiculate. Petals erect, 5–9 mm tall, obliquely narrowly ovate to oblong, obliquely retuse. Lip pendent, 6–8 mm long, narrowly oblong, obtuse. Anther semi-pendent, 1.5–2 mm long. Rostellum 1–2 mm tall, partially flanking the anther, with a deep notch in the front which contains the viscidium. Stigma rear lobe smaller than the lateral lobes, shortly pedicellate, horizontal. Ovary 1–1.5 cm long.

Zimbabwe. E: Nyanga Distr., Bideford Estate, fl. 16.iii.1958, *Beasley* 64 (K). **Malawi**. S: Mulanje (Mlanje) Mt., Chilemba Peak, fl. 17.ii.1982, *Brummitt et al.* 15943 (K).
In South Africa, with a single record from Madagascar. Montane grassland; above 2000 m.

23. SATYRIUM Sw.

Satyrium Sw. in Kongl. Vetensk. Acad. Nya Handl. **21**: 214 (1800), nom. conserv.

Terrestrial herb with globose, ovoid or ellipsoid tubers. Foliage leaves basal or ± basal on flowering stem, or if foliage leaves on separate sterile shoots then flowering stem with sheathing leaves. Inflorescence terminal, few- to many-flowered; bracts often large, frequently reflexed. Flowers non-resupinate. Sepals partly united with petals and lip; petals similar to sepals but usually narrower. Lip thin-textured or fleshy, forming a hood with a wide or narrow mouth, almost always 2-spurred at the base, occasionally 2 vestigial spurs also present; spurs long and slender or short and blunt, rarely absent. Column inside lip, erect, ± incurved; anther dependent from front of column with parallel loculi; pollinaria 2, each with sectile pollinium, caudicle and viscidium; rarely the 2 viscidia are united. Stigma flat or hooded; rostellum 3-lobed, projecting between stigma and anther loculi.

A genus of over 100 species, mostly in mainland Africa, with 5 in Madagascar and 2 in Asia.

1. Leaves 1 or 2 appressed to the ground, stem with a few sheathing leaves only · · · · · · · 2
– Leaves along the stem, if basal then not appressed to ground, or foliage leaves on separate sterile shoot · 6
2 Spurs 24–35 mm long · 2. *kitimboense*
– Spurs less than 20 mm long · 3
3. Flowers white · 4
– Flowers orange-red or rose-pink · 5
4. Petals lacerate; spurs more than 11 mm long · 3. *mirum*
– Petals not lacerate, spurs 8–10 mm long · 1. *carsonii*
5. Basal leaf 1; flowers rose-pink to carmine · 5. *princeae*
– Basal leaves 2; flowers orange-red · 4. *orbiculare*

6. Spurs more than 2 cm long · 7
– Spurs less than 2 cm long, or absent · 11
7. Lip 8–15 mm long; flowers white · 8
– Lip 4–7.5 mm long; flowers green or yellow-green · 10
8. Flowering stem with 2–4 large ± spreading leaves near the base · · · · · · · · · · 8. *hallackii*
– Flowering stem with sheathing leaves only, foliage leaves on a separate sterile shoot · · · 9
9. Spurs 4.5–7.5 cm long; lip 10–15 mm long · 14. *buchananii*
– Spurs 3–4(5) cm long; lip 9–12 mm long · · · · · · · · · · · · · · · · · · · 15. *longicauda*
10. Ovary densely papillose; lip 5.5–7 mm long; spurs of young flowers ascending or straggling
· 19. *riparium*
– Ovary glabrous or sparsely papillate; lip 4.5–5.5 mm long; spurs pendent · · · 21. *volkensii*
11. Lip equalling or longer than spurs · 12
– Lip shorter than spurs · 23
12. Flowers pink or red · 13
– Flowers white or yellowish · 17
13. Spurs absent · 35. *ecalcaratum*
– Spurs present · 14
14. Spurs slender, arching over ovary · 10. *rhynchantoides*
– Spurs parallel to ovary · 15
15. Flowering stem with sheathing leaves, foliage leaves on separate sterile shoot; spurs 2–5 mm
long · 29. *breve*
– Flowering stem with 1 or 2 relatively large leaves near base, no sterile shoot · · · · · · · · 16
16. Lip 8–9 mm long; spurs 4–8 mm long · 31. *oliganthum*
– Lip 13–14 mm long; spurs 8–15 mm long · · · · · · · · · · · · · · · · · · 9. *sphaerocarpum*
17. Ovary and rhachis pubescent · 18
– Ovary and rhachis glabrous · 19
18. Lip 6–8 mm long, spurs 4–7 mm long · 28. *trinerve*
– Lip 4–5.5 mm long, spurs 1–2 mm long · 30. *compactum*
19. Spurs 8–15 mm long, lip 13–14 mm long · · · · · · · · · · · · · · · · · · 9. *sphaerocarpum*
– Spurs up to 8 mm long · 20
20. Flowers yellowish or yellow-green; spurs slender, bulbous at apex · · · · · · · · 23. *microcorys*
– Flowers white; spurs slender or stout but not bulbous at apex · · · · · · · · · · · · · · · 21
21. Lip c. 4 mm long; leaves grouped towards base of stem · · · · · · · · · · · · · · 33. *aberrans*
– Lip 6–14 mm long; leaves spaced along the stem · 22
22. Lip 6–9 mm long; inflorescence elongate; column hairy · · · · · · · · · · · · 27. *amblyosaccos*
– Lip 8.5–14 mm long; inflorescence short and compact; column glabrous · · 32. *paludosum*
23. Foliage leaves on lower part of flowering stem · 24
– Foliage leaves on separate sterile shoot · 32
24. Sepals and petals sharply deflexed, often curled under; lip fleshy with a narrow mouth 25
– Sepals and petals projecting forwards or slightly deflexed, longer than lip; lip thin-textured
with wide mouth · 28
25. Spurs arched in at least some flowers of an inflorescence; bracts not reflexed; ovary c. 12
mm long · 26. *flavum*
– Spurs parallel to ovary; bracts reflexed; ovary 5–9 mm long · · · · · · · · · · · · · · · 26
26. Flowers green or yellow-green; spurs 8–20 mm long · · · · · · · · · · · · · · 22. *chlorocorys*
– Flowers white or clear yellow; spurs 8–12 mm long · 27
27. Flowers white; lip 4–5 mm long; inflorescence 3–15 cm long; largest leaf 5–14.5 × 2–6 cm
· 24. *shirense*
– Flowers white or yellow; lip 5–7 mm long; inflorescence 15–30 cm long; largest leaf 8–30 ×
5–13 cm · 25. *sphaeranthum*
28. Lip 4 mm long; bracts tinged with pink or purple · · · · · · · · · · · · · · · · · 33. *aberrans*
– Lip more than 5 mm long · 29
29. Inflorescence 1–2-flowered; flowers pure white · · · · · · · · · · · · · · · · 34. *afromontanum*
– Inflorescence several- to many-flowered · 30
30. Spurs arching over the ovary · 10. *rhynchantoides*
– Spurs parallel to the ovary · 31
31. Flowering stem with 2 large leaves near the base; other leaves much smaller; lip 8–10.5 mm
long; spurs 12–20 mm long · 6. *macrophyllum*
– Flowering stem not with 2 leaves much larger than others; lip 5–6.5 mm long; spurs 8–12
mm long · 7. *crassicaule*
32. Lip fleshy with a narrow mouth; column short and thick · · · · · · · · · · · · · · · · · · 33
– Lip thin-textured with a wide mouth; column long and slender · · · · · · · · · · · · · 38

33. Ovary densely papillose; spurs 18–25 mm long, diverging from ovary in young flowers · 19. *riparium*
  – Ovary glabrous or very slightly papillate; spurs mostly less than 18 mm long · · · · · · · 34
34. Spurs curved up, 8–11 mm long · 16. *anomalum*
  – Spurs parallel to ovary · 35
35. Flowers green, sometimes tinged with yellow or brown · 36
  – Flowers orange, red or white · 37
36. Spurs 14–22 mm long · 21. *volkensii*
  – Spurs 5–8 mm long · 18. *elongatum*
37. Spurs 6–8 mm long · 17. *confusum*
  – Spurs 8.5–17 mm long · 20. *coriophoroides*
38. Flowers carmine-red; viscidium single; mid-lobe of rostellum shorter than side lobes · 11. *monadenum*
  – Flowers pink, yellowish-white or orange-red; viscidia 2; mid-lobe of rostellum larger than side lobes · 39
39. Flowers pink (rarely yellow-white); lateral sepals lying beside lip so that flower seems wider than high · 13. *neglectum*
  – Flowers apricot, orange or vermilion; lateral sepals deflexed so that flower seems higher than wide · 12. *sceptrum*

1. **Satyrium carsonii** Rolfe in F.T.A. **7**: 265 (1898). —Summerhayes in Kew. Bull. **20**: 170 (1966); in F.W.T.A. ed. 2, **3**: 201 (1968); in F.T.E.A., Orchidaceae: 186 (1968). — Williamson, Orch. S. Centr. Africa: 92 (1977). —Geerinck in Fl. Afr. Centr., Orchidaceae pt. 1: 214 (1984). —la Croix et al., Orch. Malawi: 130, fig. 76 (1991). TAB. **65**. Type: Zambia, Mbala Distr., Fwambo, *Carson* 3 (K, lectotype, Summerhayes 1966).
  Satyrium leucanthum Schltr. in Bot. Jahrb. Syst. **53**: 525 (1915). —Summerhayes in F.T.W.A. ed. 1, **2**: 417 (1936). Type from Tanzania.
  Satyrium nigericum Hutch. in Bull. Misc. Inform., Kew **1921**: 402 (1921). Type from Nigeria.

Terrestrial herb 25–65 cm tall; tubers 1–3 cm long, ovoid or ellipsoid, tomentose. Basal leaves 2, appressed to ground, the larger up to 6 × 10 cm, ovate, orbicular or reniform, glabrous, rather fleshy; stem with c. 4 sheathing leaves, 2–6.5 cm long. Inflorescence 4–16 × 3–6 cm, fairly densely few- to c. 20-flowered. Flowers scented, white, the tips of the spurs green. Ovary c. 10 mm long, with 6 wings; bracts up to 30 × 17 mm, reflexed. Sepals and petals projecting or deflexed, joined shortly at base to each other and to the lip. Median sepal 7–17 × 3–4 mm, oblong, obtuse; lateral sepals 9–20 × 4.5–7 mm, obliquely oblanceolate, obtuse. Petals 7–14 × 2.5–5 mm, oblanceolate, rounded. Lip 9–16 mm long with 2–3 mm reflexed at apex, 11–14 mm wide when spread out, very convex, forming a hood with a prominent keel. Spurs 8–10 mm long, tapering from a wide mouth to an acute apex, parallel to ovary. Column 4–7 mm high, curved; stigma 2–3 × 3–4 mm; rostellum mid-lobe 1.5 × 1.5 mm.

Zambia. N: Mbala Distr., Fwambo, fl. ii.1893, *Carson* 3 (K). W: Solwezi Distr., 8 km E of Kabompo R., fl. 19.xii.1969, *Williamson & Simon* 1772 (K; SRGH). C: Lusaka Distr., 11 km SE of Lusaka, 1200 m, fl. 14.i.1952, *Best* 13 (K). E: Lundazi Distr., foothills of Nyika, fl. ii.1968, *G. Williamson* 828 (K). **Malawi.** N: Nkhata Bay Distr., North Viphya, Mzuzu-Usisya road, 1350 m, fl. 19.i.1986, *la Croix* 780 (K).
  Also in Nigeria, Cameroon, Zaire, Uganda, Kenya and Tanzania. Mixed woodland, dambo edge, roadside grass; 1200–1500 m.

2. **Satyrium kitimboense** Kraenzl. in Bot. Jahrb. Syst. **51**: 380 (1914) [as "*ketumbense*"]. — Summerhayes in Kew Bull. **4**: 432 (1949); in F.T.E.A., Orchidaceae: 186 (1968). — Williamson, Orch. S. Centr. Africa: 92 (1977). —Geerinck in Fl. Afr. Centr., Orchidaceae pt. 1: 211 (1984). —la Croix et al., Orch. Malawi: 131 (1991). Type from Zaire.

Robust plants 35–75 cm high; tubers 2.5–4 × 1–2.5 cm, ovoid or globose, tomentose. Basal leaves 2, appressed to the ground, 4–8 × 6.5–11 cm, broadly ovate to reniform, rather fleshy, glabrous; stem with 3–6 sheathing leaves up to 7 cm long. Inflorescence c. 10 × 7 cm, up to 12-flowered. Flowers white, the spurs green at the tips. Sepals and petals projecting or decurved, joined at the base to each other and to the lip. Median sepal 18–23 × 5–10 mm, oblanceolate; lateral sepals 17–25 × 6.5–12 mm, obliquely

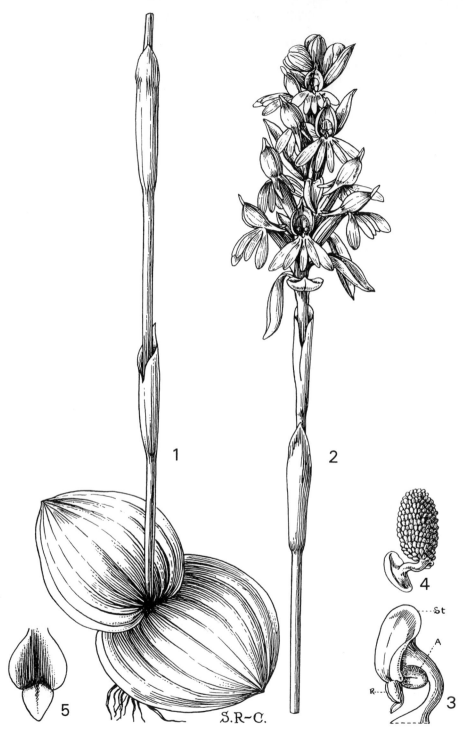

Tab. 65. SATYRIUM CARSONII. 1, base of scape and leaves; 2, inflorescence; 3, column (a,
anther; r, rostellum; st, stigma); 4, pollinium; 5, rostellum. Drawn by Stella Ross-Craig.
From F.W.T.A.

oblong-lanceolate, obtuse. Petals similar to median sepal. Lip 16–23 mm long, very convex and wide-mouthed, the apex reflexed; spurs 24–35 mm long, less than 2 mm wide at base, tapering to an acute apex. Column c. 10 mm long; stigma c. 4.5 × 7.5 mm.

Zambia. N: Mporokoso Distr., road from Senga Hill to Mporokoso, 1380 m, fl. & fr. 16.i.1960, *Richards* 12072 (K). W: Mwinilunga Distr., source of Zambezi R., fl. 13.xii.1963, *E.A. Robinson* 5987 (K). C: Serenje Distr., Kundalila Falls, 1500 m, fl. 12.ii.1973, *Strid* 2915 (K). S: Kalomo, fl. 1.i.1962, *Mitchell* 12/98 (K; NDO). **Malawi**. N: Rumphi Distr., Chiweta Escarpment, 875 m, fl. 24.i.1987, *la Croix* 944 (K). C: Dedza Distr., c. 17 km W of town, near Mozambique border, c. 1500 m, fl. 20.i.1971, *Westwood* 587 (K; MAL). **Mozambique**. N: Marrupa, on road to Nungo, c. 800 m, fl. 18.ii.1982, *Jansen & Boane* 7821 (K).
Also in Zaire, Burundi, Tanzania and Angola. Open woodland; 800–1800 m.

3. **Satyrium mirum** Summerh. in Kew Bull. **20**, 2: 170, fig. 3 (1966). —Grosvenor in Excelsa **6**: 85 (1976). TAB. **66**. Type: Zimbabwe, Chimanimani Distr., between Cashel and Chimanimani (Melsetter) *Ball* 561 (K, holotype; SRGH).

Terrestrial herb 40–50 cm high. Basal leaves 2, appressed to the ground, 5–9 × 5–7 cm, ovate or ± orbicular, apiculate, rather fleshy; stem with 8–10 sheathing, imbricate leaves c. 7–8 cm long. Inflorescence 5–9 × 6 cm, fairly laxly 3–7-flowered. Flowers white. Ovary and pedicel 17–24 mm long, slightly curved; bracts erect, 20–40 mm long, elliptic, acute. Sepals ± deflexed, joined in the basal quarter to the petals and lip, all obscurely toothed at apex. Median sepal 15.5–17.5 × 3.5–5 mm, oblong, rounded; lateral sepals similar but 5–6.5 mm wide. Petals 16–17 × 6–8 mm, elliptic-lanceolate, lacerate on apical margin. Lip erect, ± orbicular, the apex recurved, fimbriate, in all 18–19 mm long, 14–16 mm wide when spread out. Spurs 11.5–14.5 cm long, pendent, slender, tapering from a base 2.5–3 mm wide. Column 10.5 mm high, ± straight; stipe c. 6 mm long; stigma 3 × 6 mm; rostellum mid-lobe 1.6 × 3 mm, side lobes much shorter.

Zimbabwe. E: Chimanimani Distr., "Tank Nek", between Cashel and Chimanimani (Melsetter), 1800 m, fl. 4.iii.1956, *Ball* 561 (K; SRGH).
Endemic. Moorland grassland along ridge; 1800–1880 m.

4. **Satyrium orbiculare** Rolfe in F.T.A. **7**: 266 (1898). —Kraenzlin, Orchid. Gen. Sp. **1**: 676 (1899). —Schlechter in Bot. Jahrb. Syst. **31**: 196 (1901); **53**: 525 (1915). —Summerhayes in F.T.E.A., Orchidaceae: 190 (1968). —Williamson, Orch. S. Centr. Africa: 92 (1977). —Geerinck in Fl. Afr. Centr., Orchidaceae pt. 1: 212 (1984). —la Croix et al., Orch. Malawi: 132, fig. 78 (1991). Type: Zambia, Mbala Distr., Kambole, *Nutt* s.n. (K, holotype).

Terrestrial herb 15–35 cm high; tubers 6–22 × 5–8 mm, almost globose to ellipsoid-fusiform, densely tomentose. Basal leaves 2, appressed to the ground, 3–4 × 4–6 cm, ± orbicular to reniform; stem with 4–7 loosely sheathing leaves up to 3 cm long. Inflorescence 5–7 × 1.3–2.3 cm, fairly densely several- to many-flowered. Flowers bright orange-red, the spurs paler at tips. Pedicel and ovary 6–8 mm long; bracts reflexed, 15 mm long at base of inflorescence. Sepals and petals spreading or slightly decurved, joined to each other and lip for c. 1 mm at base. Median sepal 7–8.5 × 2–3 mm, oblanceolate; lateral sepals 6–8 × 4 mm, obliquely ovate, obtuse. Petals 5–7.5 × 2.5 mm, oblanceolate. Lip 6–7 mm long, very convex, wide-mouthed with a dorsal keel. Spurs 9–10 mm long, slender, parallel to ovary. Column c. 5 mm high; stigma 1 × 2 mm; rostellum mid-lobe about twice as long as side lobes.

Zambia. N: Mbala Distr., Msipazi Road, 1500 m, fl. 31.i.1971, *Sanane* 1514 (K). E: Lundazi Distr., Nyika Plateau, fl. ii.1968, *G. Williamson* 795 (K). **Malawi**. N: Nyika Plateau, Mbuzinandi-Olera road, c. 1700 m, fl. 6.iii.1977, *Pawek* 12493 (K; MAL; MO; SRGH; UC); South Viphya, near Chikangawa, 1650 m, fl. 24.ii.1984, *la Croix* 556 (K).
Also in Zaire, Rwanda and Tanzania. Montane grassland, woodland; 1650–2200 m.

5. **Satyrium princeae** Kraenzl., in Bot. Jahrb. Syst. **33**: 56 (1902). —Summerhayes in Kew Bull. **20**: 172 (1966); in F.T.E.A., Orchidaceae: 189 (1968). —Williamson, Orch. S. Centr. Africa: 92, t. 72 (1977). —la Croix et al., Orch. Malawi: 133 (1991). Type from Tanzania.

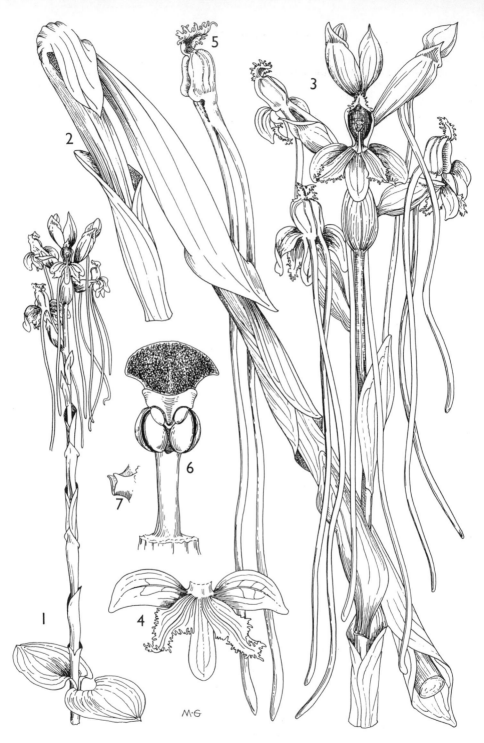

Tab. 66. SATYRIUM MIRUM. 1, habit ($\times\frac{1}{3}$); 2, middle section of stem with sheathing leaves ($\times 1$); 3, inflorescence ($\times 1$); 4, sepals and petals ($\times 1\frac{1}{3}$); 5, lip and spurs, side view ($\times 1\frac{1}{3}$); 6, column ($\times 4$); 7, rostellum, side view ($\times 4$), 1–7 from *Ball* 561. Drawn by Mary Grierson. From Kew Bull.

Slender terrestrial herb 20–40 cm high. Basal leaf 1, appressed to the ground, 3.5–5 × 4.5 cm, heart-shaped; stem with 3–6 sheathing leaves up to 3 cm long. Inflorescence c. 6 × 2 cm, fairly densely up to 20-flowered. Flowers bright rose-pink, bracts purplish. Pedicel and ovary 7–8 mm long; bracts reflexed, up to 15 × 5 mm, lanceolate, acute. Sepals and petals projecting forwards, joined to each other and lip for 2–3 mm. Median sepal 7–9 × 1–1.3 mm, ligulate to oblanceolate, rounded at apex; lateral sepals oblique, of similar length and slightly wider. Petals similar to median sepal but slightly shorter. Lip 6–7 mm long, 6–7 mm wide when spread out, very convex, the apex reflexed. Spurs 11–18 mm long, slender, tapering, parallel to ovary. Column slender, erect, 4.5–6.5 mm high; stigma 1–1.5 × 2 mm; rostellum shortly and almost equally 3-toothed, the middle tooth bifid.

Zambia. E: Lundazi Distr., Nyika Plateau, c. 1900 m, fl. 9.iii.1977, *Grosvenor & Renz* 1101 (K; SRGH). **Malawi**. N: Nyika National Park, Kasaramba Road near Juniper Forest junction, 2300 m, fl. 28.iii.1970, *Pawek* 3391 (K).

Also in Tanzania. Montane grassland, usually in wetter areas; 1900–2400 m.

Tanzanian plants are much larger, with basal leaf up to 11 × 9 cm, and inflorescence to 18 cm long.

6. **Satyrium macrophyllum** Lindl., Gen. Sp. Orchid. Pl.: 338 (1838). —Schlechter in Bot. Jahrb. Syst. **31**: 168 (1901). —Rolfe in F.C. **5**, 3: 166 (1912). —Eyles in Trans. Roy. Soc. S. Africa **5**: 334 (1916). —Grosvenor in Excelsa **6**: 85 (1976). —Hall in Contrib. Bolus Herb., No.10: 96 (1982). —Stewart et al., Wild Orch. South. Africa: 172 (1982). —la Croix et al., Orch. Malawi: 134 (1991). Type from South Africa.

Satyrium cheirophorum Rolfe in F.T.A. **7**: 265 (1898). —Kraenzlin, Orchid. Gen. Sp. **1**: 717 (1900). —Schlechter in Bot. Jahrb. Syst. **31**: 195 (1901). —Summerhayes in Kew Bull. **20**: 172 (1966); in F.T.E.A., Orchidaceae: 191 (1968). Type: Malawi, near Blantyre, *Last* s.n. (K, holotype).

Satyrium buchananii Rolfe in F.T.A. **7**: 270 (1898) non Schltr. (1897). —Kraenzlin, Orchid. Gen. Sp. **1**: 689 (1899), pro parte. Type: Malawi, 5 gatherings by *Buchanan* and *Kirk* (K, syntypes).

Satyrium speciosum Rolfe in F.T.A. **7**: 574 (1898). —Schlechter in Bot. Jahrb. Syst. **31**: 195 (1901). —Summerhayes in Bot. Not. **1937**: 189 (1937); in Mem. N.Y. Bot. Gard. **9**, 1: 80 (1954). —Eyles in Trans. Roy. Soc. S. Africa **5**: 334 (1916). Types as for *S. buchananii* Rolfe.

Satyrium morrumbalaense De Wild. in Pl. Nov. Horti Then., **1**: t. 14 (1904). Type: Mozambique, Morrumbala, *Luja* 424 (BR, holotype).

Terrestrial herb 15–80 cm high; tubers up to 2.5 × 1 cm, ± globose or ovoid, almost glabrous. Leaves 6–10, the lowest 1–3 sheath-like, the next 2 large and spreading, near the base of the stem but not appressed to the ground, 10–24 × 4.5–8 cm, ovate, glossy light green; the remainder spaced along the stem, much smaller and sheath-like. Inflorescence 7–22 × 2.5–4 cm, fairly densely 10- to many-flowered. Flowers pale to deep pink or carmine-red. Ovary and pedicel 10 mm long, tinged with pink; bracts reflexed, up to 3 cm long at base of inflorescence, lanceolate, acute. Sepals and petals projecting forwards or slightly decurved, joined to each other and lip for about a third of their length. Median sepal 10–15 × 1.5–2 mm, ligulate, obtuse; lateral sepals oblique, slightly shorter and 2–3 mm wide. Petals slightly shorter and broader than median sepal. Lip 8–10.5 mm long, 6.5–8.5 mm wide when spread out, very convex and hooded, the apex shortly reflexed, wide-mouthed and thin-textured. Spurs 12–20 mm long, slender, tapering, parallel to ovary. Column up to 8 mm high; stigma 1.5 × 2.5 mm; rostellum equally 3-lobed, the mid-lobe shortly bifid.

Zimbabwe. N: near Mazowe (Mazoe), 1200 m, fl. 21.iv.1946, *Wild* 1037 (K; SRGH). W: Matopos Hills, fl. iii. *Eyles* 1036 (K; PRE). C: Harare (Salisbury), 1360 m, fl. 10.iv.1950, *Greatrex* in *GHS* 27429 (K; SRGH). E: Nyanga (Inyanga), Erin Forest Reserve, 1800 m, fl. 18.iii.1972, *Plowes* 4411 (K; SRGH). **Malawi**. N: Nkhata Bay Distr., Mzuzu-Nkhata Bay road, c. 1000 m, fl. 19.vi.1986, *la Croix* 844 (K). S: Mulanje Mt., Chambe Plateau, 1800 m, fl. 4.ii.1979, *Blackmore, Brummitt & Banda* 423 (K; MAL). **Mozambique**. Z: Morrumbala, *Luja* 424 (BR). MS: Vallée du Revué, fl. 24.iv.1905, *Vasse* 222 (P).

Also in Zaire, Kenya, Tanzania, South Africa and Swaziland. Grassland, *Brachystegia* woodland; 1000–1940 m.

This variable species has been known as *Satyrium cheirophorum* Rolfe in East and Central Africa, but it is difficult to find any way in which it can be separated from the South African *Satyrium macrophyllum*. In Malawi, 2 different "forms" can be distinguished. Plants growing in montane grassland are smaller and more slender, usually less than 30 cm high, with deep carmine-red flowers blooming in February–March, while those occuring in woodland or grassland at lower altitudes are large and robust, with pink flowers and bloom during June and into July. However, details of the flowers seem to be identical.

7. **Satyrium crassicaule** Rendle in J. Bot. **33**: 295 (1895). —Rolfe in F.T.A. **7**: 271 (1898). —Kraenzlin, Orchid. Gen. Sp. **1**: 698 (1899). —Schlechter in Bot. Jahrb. Syst. **31**: 171 (1901) excl. syn. *S. nuttii* Rolfe. —Summerhayes in Kew Bull. **20**: 173 (1966); in F.W.T.A. ed. 2, **3**: 201 (1968); in F.T.E.A., Orchidaceae: 193 (1968). —Williamson, Orch. S. Centr. Africa: 92, t. 74–75 (1977). —Geerinck in Fl. Afr. Centr., Orchidaceae pt. 1: 216 (1984). —la Croix et al., Orch. Malawi: 135 (1991). TAB. **67**. Types from Uganda and Zaire.
    Satyrium niloticum Rendle in J. Bot. **33**: 296 (1895). —Rolfe in F.T.A. **7**: 271 (1898). Type from Kenya.
    Satyrium goetzenianum Kraenzl. in Bot. Jahrb. Syst. **24**: 506 (1898). —Rolfe in F.T.A. **7**: 574 (1898). Type from Zaire.
    Satyrium mystacinum Kraenzl. in Bot. Jahrb. Syst. **24**: 506 (1898). —Rolfe in F.T.A. **7**: 574 (1898). Type from Zaire.
    Satyrium fischeranum Kraenzl. in Bot. Jahrb. Syst. **24**: 507 (1898). —Rolfe in F.T.A. **7**: 573 (1898). Type from Tanzania.
    Satyrium kirkii Rolfe in F.T.A. **7**: 271 (1898) pro parte. Type: Malawi, without precise locality, *Buchanan* s.n. (K, lectotype, Summerhayes, 1966).
    Satyrium usambarae Kraenzl. in Bot. Jahrb. Syst. **33**: 56 (1902). Type from Tanzania.

Robust terrestrial herb to 1 m tall; rootstock short and stout, roots numerous. Leaves imbricate, 8–13 scattered along stem with the largest near the stem base, to 40 × 6 cm, lanceolate, semi-spreading, heavily veined; the upper leaves gradually decreasing in size becoming erect and sheath-like. Inflorescence c. 30 × 3.5 cm, cylindrical, densely many-flowered. Flowers pale pink. Ovary and pedicel c. 10 mm long, curved; bracts c. 25 mm long at base of inflorescence, reflexed, slightly pubescent. Sepals and petals projecting forwards and slightly decurved, joined to each other and lip in basal third. Median sepal 5.5–7 × 1.5–2.5 mm, oblanceolate or elliptical; lateral sepals similar but oblique and c. 3 mm wide. Petals similar to median sepal. Lip 5–6.5 mm long, 7–7.5 mm wide when spread out, very convex and hooded with a wide mouth; spurs 8–12 mm long, slender, parallel to ovary. Column 3–5 mm high; stigma c. 1 × 1.5–2 mm; rostellum mid-lobe spoon-shaped from a narrow base, much longer than side lobes.

Zambia. N: Lake Lusiwasi, 1575 m, fl. 7.i.1959, *van Zinderen Bakker* 922 (K). E: Lundazi Distr., Nyika Plateau, 2100 m, fl. 3.i.1959, *Richards* 10422 (K). **Malawi**. N: Nyika National Park, 12.5 km N of Chelinda Camp on way to Nganda, c. 2200 m, fl. 11.iii.1977, *Grosvenor & Renz* 1144 (K; SRGH). S: Zomba Mt., Chitinji Marsh, 1800 m, fl. 16.xii.1979, *Morris* 463 (K; MAL).
    Also in Nigeria, Cameroon, Zaire, Rwanda, Burundi, Ethiopia, Uganda, Kenya and Tanzania. Wet dambo, swamps and lake margins; 1300–2300 m.

8. **Satyrium hallackii** Bolus in J. Linn. Soc., Bot. **20**: 476 (1884). —Schlechter in Bot. Jahrb. Syst. **31**: 170 (1901). —Rolfe, F.C. **5**, **3**: 163 (1912). —Hall in Contrib. Bolus Herb., No. 10: 50 (1982). —Stewart et al., Wild Orch. South. Africa: 171 (1982). —la Croix et al., Orch. Malawi: 137 (1991). Type from South Africa.
    Satyrium foliosum var. *helonioides* Lindl., Gen. Sp. Orchid. Pl.: 337 (1838).

Subsp. **ocellatum** (Bolus) A.V. Hall in Contrib. Bolus Herb., No. 10: 54 (1982). Type from South Africa.
    Satyrium ocellatum Bolus, Icon. Orchid. Austro-Afr. **1**, 1: sub t. 23 (1893). —Grosvenor in Excelsa **6**: 85 (1976).
    Satyrium nutans Kraenzl. in Bot. Jahrb. Syst. **24**: 507 (1898). Type from South Africa.

Robust terrestrial herb 50–100 cm high. Stem leafy with 8–10 leaves, the lowermost sheathing, the next 2–4 leaves large, up to 28 × 7.5 cm, broadly lanceolate and borne near the stem base, the upper leaves decreasing in size and becoming bract-like. Inflorescence 10–24 × 4–5 cm, fairly densely many-flowered. Flowers white, sometimes flushed with pink, occasionally mauve-pink. Ovary and pedicel 10–14 mm long; bracts up to 25 mm long at base of inflorescence, reflexed, sometimes slightly

Tab. 67. SATYRIUM CRASSICAULE. 1, habit (×¼); 2, lower part of plant (×1); 3, inflorescence
(×1), 1–3 from *Richards* 14150; 4, flower, three-quarter front view (×3); 5, flower, back view
(×3); 6, sepals and petals (×3); 7, lip, side view (×3); 8, column, side view (×10); 9, column,
front view (×10), 4–9 from *Richards* 15857. Drawn by Heather Wood. From F.T.E.A.

pubescent. Sepals and petals projecting forwards, joined to each other and lip for almost half their length. Sepals 10–15 × 1–3 mm, oblong, rounded, the lateral sepals oblique and wider than the dorsal sepals; petals slightly shorter and c. 1 mm wide. Lip 8–13 mm long with a prominent dorsal crest, very convex and hooded, wide-mouthed, with the apex reflexed for c. 1 mm. Spurs 19–50(65) mm long, slender, tapering. Column c. 5 mm high.

Zimbabwe. N: Mazowe (Mazoe), 1300 m, fl. xii.1905, *Eyles* 233 (K). C: Harare, Chakoma Swamp, 1360 m, fl. 12.i.1951, *Greatrex* in *GHS* 30961 (K; SRGH). E: Chimanimani Distr., Rocklands, c. 13 km E of Chimanimani (Melsetter), fl. 23.i.1951, *Crook* 353 (K; PRE; SRGH). **Malawi**. S: Thyolo Distr., Conforzi Estate, c. 1000 m, fl. 14.xi.1982, *la Croix* 358 (K).
Also in South Africa. Marshy areas; 1000–1360 m.
Plants from Zimbabwe have large flowers with long spurs (33–65 mm). Specimens from Malawi are more like those from South Africa, with spurs 19–21 mm long.

9. **Satyrium sphaerocarpum** Lindl., Gen. Sp. Orchid. Pl.: 337 (1838). —Schlechter in Mem. Herb. Boissier No. 10: 31 (1900); in Bot. Jahrb. Syst. **31**: 167 (1901). —Rolfe, F.C. **5**, 3: 165 (1912). —Hall, Contrib. Bolus Herb., No. 10: 87 (1982). —Stewart et al., Wild Orch. South. Africa: 174 (1982). Type: Mozambique, Baía de Maputo (Delagoa Bay), *Forbes* s.n. (K, holotype).
   Satyrium militare Lindl., Gen. Sp. Orchid. Pl.: 342 (1838). Type from South Africa.

Terrestrial herb 20–40 cm high; tubers c. 2 cm long, ellipsoid. Stem leafy with c. 5 leaves. Lowermost leaves sheath-like, the next 2 leaves spreading or semi-erect, borne near the base of the stem, 5–12 × 2–4 cm, ovate, acute; upper leaves smaller and bract-like. Inflorescence 3–18 × 3–4 cm, fairly densely several- to many-flowered. Flowers whitish with purple blotches within, or pink and white. Ovary 7–9 mm long; bracts up to 20 mm long at base of inflorescence, reflexed. Sepals and petals projecting and slightly decurved, joined to each other and lip for about a third of their length. Median sepal 12–13 × 1–1.5 mm, oblong, obtuse, lateral sepals 12–15 × 2–2.5 mm, oblique. Petals 12–14 × c. 1 mm. Lip 13–14 mm long, 7–9 mm wide when spread out, wide-mouthed, with reflexed apical flap 1–2 mm long. Spurs 8–15 mm long, very slender from a relatively wide mouth. Column 9–11 mm long; stigma semi-orbicular, c. 1.5 mm high; rostellum less than 0.5 mm long, obtuse.

Mozambique. M: Baía de Maputo (Delagoa Bay), *Forbes* s.n. (K); Inhaca Island, Hlanginini Swamp, fl. October/November 1962, *Mogg* 30107 (K).
Also in South Africa. Coastal sandy soil; 0–200 m.

10. **Satyrium rhynchantoides** Schltr. in Bot. Jahrb. Syst. **53**: 529 (1915). —Summerhayes in F.T.E.A., Orchidaceae: 195 (1968). —Williamson, Orch. S. Centr. Africa: 93 (1977). —la Croix et al., Orch. Malawi: 139, fig. 83 (1991). Type from Tanzania.

Slender terrestrial herb 12–32 cm tall; tubers to 3.5 cm long, globose to ellipsoid, woolly. Stem leafy with 4–6 leaves. Lowermost 1–2 leaves sheath-like, the next 2–3 leaves borne near the stem base but not appressed to the ground, up to 7 × 2–3 cm, lanceolate, spreading, light green; the upper leaves bract-like. Inflorescence c. 8 cm long, laxly 3–25-flowered. Flowers pale pink or lilac or almost white. Ovary and pedicel c. 10 mm long, curved, slender; bracts erect, 10–15 mm long, lanceolate, acute. Sepals and petals spreading forwards or decurved, joined to each other and lip in basal third or quarter. Median sepal 7–9.5 × 1–2 mm, oblanceolate, obtuse; lateral sepals slightly shorter and c. 3 mm wide. Petals similar to median sepal but slightly shorter. Lip 7–9 mm long, 5–6 mm wide when spread out, very convex, wide-mouthed, the apex recurved. Spurs 6–8 mm long, slender, tapering, arched over ovary. Column 5 mm long; stigma 1 × 1.5 mm; rostellum with 3 equal teeth, the middle tooth bifid.

Zambia. N: Mpika Distr., 113 km from Mpika on Lusaka road, fl. 14.i.1975, *Brummitt & Polhill* 13791 (K). E: Lundazi Distr., Nyika National Park, Chowo Rocks, c. 2100 m, 1967, *G. Williamson* 243 (K). **Malawi**. N: Nyika National Park, Chowo Rocks, 2200 m, fl. 10.i.1974, *Pawek* 7946 (K; UC). C: Dedza Distr., Kalicelo (Kalichero) Hill, 1700 m, fl. 21.i.1979, *Robson & Jackson*

1290 (K). S: Zomba Mt., near Ngondola Forest Village, 1700 m, fl. 26.ii.1977, *Grosvenor & Renz* 966 (K; SRGH).

Also in Tanzania. Submontane grassland, high rainfall woodland, usually on rocky outcrops; 1700–2200 m.

11. **Satyrium monadenum** Schltr. in Bot. Jahrb. Syst. **53**: 527 (1915). —Summerhayes in F.T.E.A., Orchidaceae: 195 (1968). —Williamson, Orch. S. Centr. Africa: 94 (1977). —la Croix et al., Orch. Malawi: 137 (1991). Type from Tanzania.

Terrestrial herb 30–60 cm, high; tubers 1.5–2 × 1–2 cm, globose to ovoid, tomentose. Sterile shoot with 3(4) leaves, at various stages of development at flowering time, the lowest 2 sheath-like, the terminal one c. 5 cm long, lanceolate, acute. Flowering stem purplish in upper half with 6–7 loosely sheathing, imbricate leaves up to 7 cm long. Inflorescence 9–15 × 3 cm, densely many-flowered. Flowers rich carmine-red. Ovary and pedicel 10–12 mm long; bracts to 25 mm long at base of inflorescence, reflexed, tinged with red. Sepals and petals spreading and projecting forwards, joined to each other and lip for about a third of their length. Median sepal 12–14 × 2–2.5 mm, narrowly oblong, rounded; lateral sepals of similar length but broader and oblique. Petals similar to median sepal. Lip 10–12 mm long, 13–16 mm wide when spread out, very convex and hooded with a wide mouth, the apex slightly reflexed. Spurs 12–15 mm long, slender, parallel to ovary. Column 7–8 mm long, the stalk slender, curved; stigma 2 × 3 mm; rostellum shortly 3-lobed, the mid-lobe tooth-like and shorter than side lobes; viscidium single, constricted in middle.

Zambia. E: Nyika Plateau, 1967, *G. Williamson* 229 (K; SRGH). **Malawi**. N: Nyika National Park, wet montane grassland near Dam 3, 2300 m, fl. 6.i.1983, *la Croix* 403 (K).

Also in Tanzania. Seasonally wet montane grassland; usually over 2100 m.

This species sometimes forms large colonies.

12. **Satyrium sceptrum** Schltr. in Bot. Jahrb. Syst. **53**: 527 (1915). —Summerhayes in F.T.E.A., Orchidaceae: 197 (1968). —Geerinck in Fl. Afr. Centr., Orchidaceae pt. 1: 225 (1984). —la Croix et al., Orch. Malawi: 140 (1991). Type from Tanzania.
Satyrium acutirostrum Summerh. in Bull. Misc. Inform., Kew **1931**: 384 (1931); in Bot. Not. **1937**: 189 (1937); in F.T.E.A., Orchidaceae: 196 (1968). —Grosvenor in Excelsa **6**: 85 (1976). —Williamson, Orch. S. Centr. Africa: 94 (1977). Type from Zaire.

Robust terrestrial herb 40–90 cm high; tubers 1.5–5 cm long, ellipsoid to fusiform. Sterile shoot 3–4-leaved, the lowermost 1–2 leaves sheathing, the upper leaves up to 30 × 8 cm, lanceolate or elliptic. Flowering stem with c. 10 imbricate, sheathing leaves up to 16 cm long. Inflorescence 7–26 × 3–3.5 cm, densely many-flowered. Flowers orange, apricot-yellow or salmon-red. Ovary and pedicel 10–13 mm long; bracts reflexed, 20–30 mm long at base of inflorescence. Sepals and petals decurved, joined to each other and lip in basal quarter. Median sepal 7–11 × 2–3 mm, oblanceolate; lateral sepals longer, broader and oblique. Petals similar to median sepal. Lip thick-textured, with a wide mouth, 8–14 mm long, 10–12 mm wide when spread out, very convex and hooded with the apex reflexed; spurs 13–17 mm long, slender, parallel to ovary. Column 7–11 mm tall; stigma 2–3 × 3 mm; rostellum mid-lobe much longer than side lobes.

Zambia. E: Nyika Plateau, at Rest House, fl. iii.1967, *Williamson & Odgers* 267 (K; SRGH). **Zimbabwe**. E: Mt. Nyangani (Inyangani), c. 2100 m, fl. 14.ii.1931, *Norlindh & Weimarck* 5021 (K). **Malawi**. N: North Viphya, montane grassland near Chimaliro Forest, 1950 m, fl. 7.iii.1987, *la Croix* 1007 (K).

Also in Sudan, Zaire, Uganda, Kenya and Tanzania. Montane grassland; 1950–2300 m.

This species is closely related to *S. neglectum* Schltr., and Hall (1982) includes *S. sceptrum* in *S. neglectum* subsp. *woodii* (Schltr.) A.V. Hall. However, in the field the species are readily distinguished, not only by the colour but by the aspect of the flower, the lateral sepals in *S. sceptrum* being decurved while in *S. neglectum* they are spreading and lie beside the lip.

13. **Satyrium neglectum** Schltr. in Bot. Jahrb. Syst. **20**, Beibl. 50: 15 (1895). —Kraenzlin, Orchid. Gen. Sp. **1**: 706 (1900). —Schlechter in Bot. Jahrb. Syst. **31**: 157 (1901). —Rolfe in F.C. **5**, 3: 157 (1912). —Goodier & Phipps in Kirkia **1**: 53 (1961).—Moriarty, Wild Fls. Malawi: t. 30,

2 (1975). —Williamson, Orch. S. Centr. Africa: 94 (1977). —Stewart et al., Wild Orch. South. Africa: 163 (1982). —la Croix et al., Orch. Malawi: 139 (1991). Type from South Africa.
　　Satyrium neglectum subsp. *neglectum* —Hall in Contrib. Bolus Herb., No. 10: 86 (1982).

Robust terrestrial herb 25–80 cm high; tubers 2–3 × 1 cm, ellipsoid. Sterile shoot 2–5-leaved; the lowermost 1–2 leaves sheath-like; the upper leaves up to 27 × 8 cm, lanceolate or elliptic. Flowering stem sheathed by 6–12 sheathing leaves up to 13 cm long. Inflorescence 5–35 × 2–3 cm, densely many-flowered. Flowers pale to deep pink, rarely yellowish-white. Ovary and pedicel 10–12 mm long; bracts 20–40 mm long at base of inflorescence, reflexed, prominent, pink-tinged. Sepals and petals joined to each other and lip in basal third or quarter. Median sepal 5.5–8 × 1.5–2 mm, oblong, rounded, decurved; lateral sepals spreading and slightly twisted, oblique, slightly longer and broader than median sepal. Petals similar to median sepal, curled under. Lip 5–10 mm long, 5–8 mm wide when spread out, very convex and hooded, the apex recurved. Spurs 6–17 mm long, slender, parallel to ovary. Column 4.5–7 mm high; stigma 1–1.5 × 2 mm; rostellum mid-lobe spoon-shaped or orbicular with a narrow base, the side lobes much shorter and tooth-like.

Lip 6–9.5 mm long; spurs 13.5–17 mm long · var. *neglectum*
Lip 4.5–6.5 mm long; spurs 7–10.5 mm long · var. *brevicalcar*

Var. **neglectum** —Summerhayes in F.T.E.A., Orchidaceae: 198 (1968). —Grosvenor in Excelsa
　　6: 85 (1976).
　　Satyrium densum Rolfe in F.T.A. **7**: 270 (1898). —Schlechter in Bot. Jahrb. Syst. **31**: 195 (1901). —Summerhayes in Mem. N.Y. Bot. Gard. **9**, 1: 80 (1954). Type: Malawi, top of Mt. Zomba, *Buchanan* 303 (K, holotype).
　　Satyrium colliferum Schltr. in Bot. Jahrb. Syst. **53**: 528 (1915), pro parte, quoad descript. excl. *Stolz* 1188 & 2456. Type from Tanzania.

　　Zimbabwe. E: slopes of Mt. Nyangani (Inyangani) c. 2100 m, fl. 16.ii.1964, *Plowes* 2435 (K; SRGH). **Malawi**. N: South Viphya, Kawandama road, 1870 m, fl. 25.iii.1977, *Pawek* 12533 (K; MAL; MO). S: Zomba Plateau, 1720 m, fl. 15.iii.1970, *Brummitt* 9123 (K). **Mozambique**. MS: Beira Distr., Gorongosa Mt., Gogogo Summit Area, fl. iii.1972, *Tinley* 2430 (K; SRGH).
　　Also in Tanzania, South Africa, Lesotho and Swaziland. Montane grassland, occasionally in dambos; 1750–1900 m.

Var. **brevicalcar** Summerh. in Kew Bull. **20**: 174 (1966); in F.T.E.A., Orchidaceae: 198 (1968). Type from Tanzania.
　　Satyrium colliferum sensu Schlechter in Bot. Jahrb. Syst. **53**: 528 (1915), quoad *Stolz* 1188 & 2456.
　　Satyrium papyretorum Schltr. in Fries, Wiss. Ergebn. Schwed. Rhod.-Kongo-Exped.: 243, t. 18, fig. 2 (1916). Type: Zambia, Lake Bangweulu (Bangweolo), Kamindas, *R.E. Fries* 1006.

　　Zambia. N: Lake Bangweulu (Bangweolo), Kamindas, fl. 13.x.1911, *Fries* 1006 (K).
　　Also in Tanzania. Papyrus swamp.
　　The specimen cited above is the type of *S. papyretorum* Schltr. It agrees well with *S. neglectum* var. *brevicalcar*, apart from the flowers being fractionally smaller (lip 4.5 mm long, sepals and petals 4 mm long).

14. **Satyrium buchananii** Schltr. in Bot. Jahrb. Syst. **24**: 422 (1897). —Rolfe in in F.T.A. **7**: 572 (1898). —Schlechter in Bot. Jahrb. Syst. **53**: 526 (1915). —Eyles in Trans. Roy. Soc. S. Afr. **5**: 334 (1916). —Summerhayes in Bot. Not. **1937**: 190 (1937); in F.T.E.A., Orchidaceae: 199 (1968). —Moriarty, Wild Fls. Malawi: t. 30, 1 (1975). —Grosvenor in Excelsa **6**: 85 (1976). —Williamson, Orch. S. Centr. Africa: 94 (1977). —Geerinck in Fl. Afr. Centr., Orchidaceae pt. 1: 224 (1984). —la Croix et al., Orch. Malawi: 133 (1991). TAB. **68**. Type: Malawi, without precise locality, 1895, *Buchanan* s.n. (B†, holotype).
　　Satyrium longissimum Rolfe in F.T.A. **7**: 267 (1898). —Kraenzlin, Orchid. Gen. Sp. **1**: 720 (1900). Type from SE tropical Africa.
　　Satyrium nyassense Kraenzl. in Bot. Jahrb. Syst. **28**: 178 (1900); Orchid. Gen. Sp. **1**: 945 (1901). Type: Malawi, without precise locality, *Buchanan* 178 (B†, holotype).
　　Satyrium stolzianum Kraenzl. in Bot. Jahrb. Syst. **33**: 57 (1902). Type from Tanzania.
　　Satyrium kassnerianum Kraenzl. in Bot. Jahrb. Syst. **51**: 381 (1914). Type: Zambia, Katanino (Katinina) Hills, *Kassner* 2168 (B†, holotype; K).

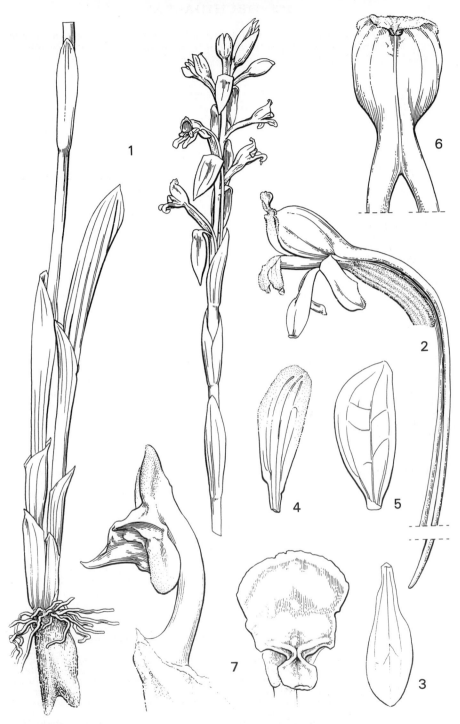

Tab. 68. SATYRIUM BUCHANANII. 1, habit ($\times\frac{2}{3}$), from *Corby* 273; 2, flower, side view ($\times$2), from *Cribb* 6012; 3, dorsal sepal ($\times$3); 4, petal ($\times$3); 5, lateral sepal ($\times$3); 6, lip, from above ($\times$3); 7, column, two views ($\times$6), 3–7 from *Richards* 10461. Drawn by Eleanor Catherine.

Terrestrial herb 30–70 cm high; tubers 1.5–4 × 1–2 cm, ovoid, tomentose. Sterile shoot 3–5-leaved, at various stages of development at flowering time; lowermost 2–3 leaves sheath-like, upper leaves up to 20 × 3.5 cm, lanceolate. Flowering stem with 5–8 sheathing leaves, sometimes imbricate, up to 8 cm long. Inflorescence 8–14 × 3–5 cm, laxly up to 20-flowered. Flowers white, the spurs green-tinged, with a strong, hyacinth-like scent. Ovary and pedicel 10–15 mm long, curved; bracts pinkish-brown, c. 30 mm long at base of inflorescence, reflexed. Sepals and petals deflexed, joined to each other and lip in basal quarter. Median sepal 10–13 × 1.5–3 mm, oblong, rounded at apex; lateral sepals slightly longer and wider, oblique. Petals similar to median sepal but usually slightly wider and with the margin ciliolate. Lip 10–15 mm long and of similar width when spread out, very convex and hooded with the apex recurved, wide-mouthed, hairy inside towards apex. Spurs 4.5–7.5 cm long, slender, pendent. Column 7.5–10 mm high; stigma 2–4 × 4–6 mm; rostellum c. 3 mm long, the mid-lobe spoon-shaped from a narrow base with the apex recurved, the side lobes much shorter, triangular.

Zambia. N: Mbala Distr., Mbala-Kasama road, 1500 m, fl. 19.i.1960, *Richards* 12441 (K). W: Mwinilunga Distr., SW of Dobeka Bridge, fl. 11.x.1937, *Milne-Redhead* 2704 (K). C: Chakwenga Headwaters, 100–129 km E of Lusaka, fl. 16.xi.1963, *E.A. Robinson* 5836 (K). **Zimbabwe**. N: Guruve (Sipolilo), Nyamunyeche Estate, fl. 18.i.1979, *Nyariri* 635 (K; SRGH). W: Matopos, 1500 m, fl. 22.vi.1920, *Borle* 51 (K; PRE). C: Harare, 1450 m, fl. xi.1919, *Eyles* 1914 (K; PRE; SRGH). E: Chimanimani Mts., E of Nat. Park Office, 1600 m, fl. 22.iii.1981, *Philcox, Leppard, Duri & Urayai* 9027 (K). **Malawi**. N: Mzimba Distr., Katoto, 8 km W of Mzuzu, 1300 m, fl. 16.i.1971, *Pawek* 4320 (K; MAL). C: Dedza Distr., Chongoni Forestry School, base of Chiwao Hill, 1650 m, fl. 4.ii.1959, *Robson* 1452 (K). S: Zomba Plateau, c. 1600 m, fl. 23.i.1966, *Jones* 8 (MAL).
Also in Zaire, Tanzania and Angola. Dambo and wet montane grassland; 1300–2300 m.

15. **Satyrium longicauda** Lindl., Gen. Sp. Orchid. Pl.: 337 (1838). —T. Durand & Schinz, Consp. Fl. Africa **5**: 96 (1895). —Bolus, Icon. Orchid. Austro-Afric. **1**,2: t. 70 (1896). —Kraenzlin, Orchid. Gen. Sp. **1**: 706 (1900). —Schlechter in Bot. Jahrb. Syst. **31**: 155 (1901). —Rolfe in F.C. **5**, 3: 158 (1912). —Summerhayes in Bot. Not. **1937**: 191 (1937). —Goodier & Phipps in Kirkia **1**: 53 (1961). —Summerhayes in F.T.E.A., Orchidaceae: 198 (1968). —Grosvenor in Excelsa **6**: 85 (1976).—Stewart et al., Wild Orch. South. Africa: 163 (1982). Type from South Africa.
Satyrium longicauda var. *longicauda* —Hall in Contrib. Bolus Herb. No. 10: 67 (1982).

Fairly slender terrestrial herb 10–65 cm high; tubers 1–3.5 cm long, ellipsoid or ovoid. Sterile shoot 3–5-leaved; the lowermost 1–3 leaves sheathing; upper leaves up to 11 × 7.5 cm, ovate. Flowering stem with 3–9 sheathing leaves c. 4 cm long. Inflorescence 6–8 × 2–2.5 cm, laxly up to 15-flowered. Flowers scented, white, with pale pink spots at apices of lip and lateral sepals, sometimes whole flower tinged with pink or purple. Ovary and pedicel c. 10 mm long; bracts reflexed, up to 35 mm long at base of inflorescence. Sepals and petals deflexed, joined to each other and lip in basal quarter. Median sepal 9–13 × 2–3 mm, oblong, obtuse; lateral sepals slightly longer and wider, oblique. Petals similar to median sepal but slightly shorter and wider, ciliolate at apex. Lip 9–12 mm long, of similar width when spread out, very convex and hooded, wide-mouthed with the apex reflexed. Spurs 3.3–4.3(5) cm, slender, tapering. Column 6.5–9 mm high; stigma 2–3 × 2.5–4 mm; rostellum 2–3 mm long, the mid-lobe ± orbicular from a narrow base, much longer than side lobes.

Zimbabwe. E: Nyanga Distr., foot of Mt. Inyangani, 2000 m, fl. 7.iii.1981, *Philcox, Leppard, Duri & Urayai* 8894 (K). **Malawi**. C: Dedza Mt., fl. iii.1971, *Westwood* 546A (K). **Mozambique**. MS: Chimanimani Mts., 1575 m, fl. 6.ii.1958, *Hall* 378 (K; SRGH).
Also in Tanzania, South Africa, Swaziland and Lesotho. Montane grassland; 1575–2000 m.
This species closely resembles *S. buchananii*, differing mainly in being a more slender plant with smaller flowers and shorter spurs. The leaves of the sterile shoot are generally broader and more fully expanded at flowering time than those of *S. buchananii*, and the sheaths on the fertile stem are tighter. Also, there seems to be a habitat difference, *S. longicauda* occurring in dry montane grassland and rocky hillsides while *S. buchananii* is a plant of bog and dambo. Some plants from Zimbabwe, e.g. *Fries, Norlindh & Weimarck* 2712 and *Gilliland* 1511, seem intermediate between the two, which could be explained either by hybridization, or by the two being one variable species. Should they prove to be conspecific, *S. longicauda* is the earlier name.

16. **Satyrium anomalum** Schltr. in Bot. Jahrb. Syst. **24**: 424 (1897). —Rolfe in F.T.A. **7**: 573 (1898). —Schlechter in Bot. Jahrb. Syst. **31**: 177 (1901). —Summerhayes in F.T.E.A., Orchidaceae: 205 (1968). —Moriarty, Wild Fls. Malawi: t. 30, 5 (1975). —Grosvenor in Excelsa **6**: 85 (1976). —Williamson, Orch. S. Centr. Africa: 96, t. 82 (1977). —la Croix et al., Orch. Malawi: 140 (1991). Type: Malawi, without precise locality, *Buchanan* s.n. (B†, holotype).
    Satyrium minax Rolfe in F.T.A. **7**: 268 (1898). Syntypes: Malawi, Blantyre, *Last* (K) and *Scott* (K).

Terrestrial herb 40–100 cm high; tubers to 5.5 cm long, globose, ellipsoid or fusiform. Sterile shoot 3–4-leaved; the lowermost 1–2 leaves sheathing; upper leaves 12–27 × 3–6 cm, elliptic or lanceolate. Flowering stem with c. 10 sheathing leaves up to 10 cm long. Inflorescence 12–48 × 2.5–5 cm, cylindrical, densely many-flowered. Flowers lilac with darker spurs or greenish-white; rhachis and ovary purple. Ovary 10 mm long; bracts green with purple edge, reflexed, to 30 mm long at base of inflorescence, lanceolate, acute. Sepals and petals curled under, joined for about half their length to each other and lip. Sepals 4.5–5.5 × 1–2 mm, oblanceolate, obtuse, the median narrower than the laterals. Petals 4–5 × 0.5 mm, ciliolate. Lip thick-textured with a narrow mouth, 5–7 mm long, ellipsoid, forming a hood; spurs 8–11 mm long curving upwards and diverging from each other. Column c. 4 mm high; stigma 1.5 × 1.5 mm; rostellum mid-lobe spoon-shaped from a narrow base, much longer than side lobes.

Zambia. N: Chinsali Distr., Ishiba Ngandu (Shiwa Ngandu), fl. ii.1969, *G. Williamson* 1579 (K). W: Mindolo, near Nkana, fl. 1.i.1936, *Brenan* 14 (K). C: Kabwe, in *Julbernardia paniculata* woodland, fl. 12.i.1961, *Morze* 40 (K). E: Mpangwe Hills, 1050 m, fl. 9.ii.1958, *Wright* 222 (K). S: 37 km NNE of Choma, 1180 m, fl. 3.i.1957, *E.A. Robinson* 2008 (K; SRGH). **Zimbabwe**. N: Guruve (Sipolilo) Distr., Nyamunyeche Estate, fl. 23.i.1979, *Nyariri* 666 (K; SRGH). C: Harare, 1500 m, fl. 16.i.1949, *Wild* 2722 (K; SRGH). E: Chimanimani Distr., Nyhodi Valley, fl. 20.ii.1951, *Crook* 387 (K; SRGH). **Malawi**. N: Rumphi Distr., c. 37 km N of Rumphi, 1250 m, fl. 22.ii.1978, *Pawek* 13816 (K; MAL; MO; SRGH; UC). C: Nkhota Kota (Nkhotakota) Escarpment, fl. 19.ii.1953, *Jackson* 1066 (K). S: Blantyre Distr., Ndirande Mt., 1450–1570 m, fl. 12.iii.1970, *Brummitt* 9042 (K).
    Also in Tanzania. In *Julbernardia* and *Brachystegia* woodland; 1050–1570 m.

17. **Satyrium confusum** Summerh. in Kew Bull. **20**: 176, fig. 5 (1966); in F.T.E.A., Orchidaceae: 202, fig. 35 (1968). —la Croix et al., Orch. Malawi: 141 (1991). TAB. **69**. Type from Tanzania.
    Satyrium kirkii sensu Rolfe in F.T.A. **7**: 271 (1898) pro parte, quoad specim. *Kirk*, non Rolfe sensu stricto.

Terrestrial herb 55–80 cm high; tubers 2–5 × 1–2 cm, ovoid or ellipsoid, tomentose. Sterile shoot 3–5-leaved; lowermost 1–2 leaves sheathing; upper leaves up to 20 × 3 cm, lanceolate or oblanceolate. Flowering shoot with 6–12 sheathing leaves up to 12 cm long. Inflorescence 10–24 × 1.5–2 cm, densely many-flowered. Flowers red, pink-red or reddish-yellow. Ovary and pedicel curved, 6–8 mm long; bracts up to 25 mm long at base of inflorescence, reflexed. Sepals and petals deflexed, joined to each other and lip in basal third. Median sepal 5–6 × 1 mm, oblong, rounded; lateral sepals similar but 2 mm wide and oblique. Petals similar to median sepal, the edges ciliolate. Lip 4.5–6 mm long, 7–8 mm wide when spread out, ellipsoid, forming hood, the apex slightly reflexed, thick-textured with a narrow mouth. Spurs 6–8 mm long, parallel to ovary. Column 4 mm long; stigma 1.5 × 2 mm; rostellum mid-lobe spoon-shaped from a narrow base, much longer than side lobes.

Malawi. C: Dedza Mt., at Radio Station, 2100 m, fl. 3.ii.1974, *Allen* 475 (K). S: Blantyre Distr., Mt. Soche, *Kirk* s.n. (K).
    Also in Tanzania. Open grassland, seasonal dambo; up to 2100 m.

18. **Satyrium elongatum** Rolfe in F.T.A. **7**: 268 (1898). —Summerhayes in Bot. Not. **1937**: 190 (1937); in F.T.E.A., Orchidaceae: 201 (1968). —Grosvenor in Excelsa **6**: 85 (1976). — Williamson, Orch. S. Centr. Africa: 95 (1977). Type: Zambia, Mbala Distr., Fwambo, *Carson* 60 (K, holotype).

Terrestrial herb 45–90 cm high; tubers to 4.5 × 2 cm, ovoid or ellipsoid, woolly. Sterile shoot 3–4-leaved; lowermost 1–2 leaves loosely sheathing; upper leaves to 17 × 3.5 cm, lanceolate or oblanceolate. Flowering shoot with 6–10 loosely sheathing

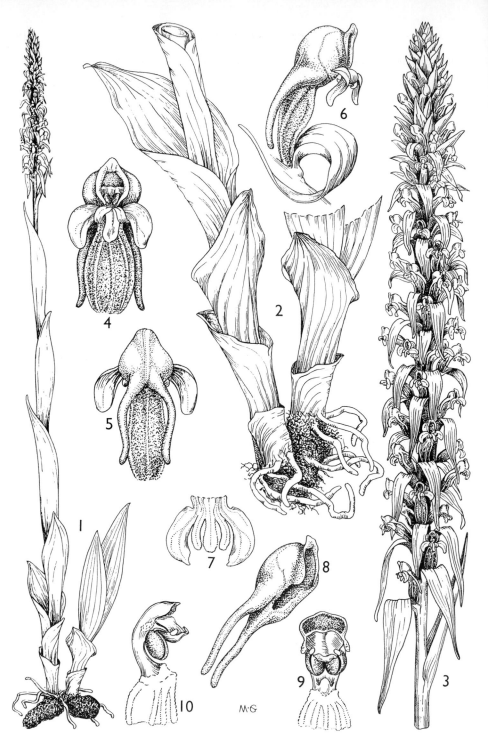

Tab. 69. SATYRIUM CONFUSUM. 1, habit ($\times\frac{2}{9}$); 2, base of plant ($\times$1); 3, inflorescence ($\times$1); 4, flower, front view ($\times$3); 5, flower, back view ($\times$3); 6, flower, side view ($\times$3); 7, sepals and petals ($\times$3); 8, lip, side view ($\times$3); 9, column, front view ($\times$4); 10, column, side view ($\times$4), 1–10 from *Polhill & Paulo* 1359. Drawn by Mary Grierson. From Kew Bull.

leaves up to 12 cm long. Inflorescence 15–30 × 1.5–2 cm, narrowly cylindrical, fairly laxly many-flowered, the rhachis visible between the flowers. Flowers greenish-white turning brown, rhachis and ovary pinkish. Ovary and pedicel 5–7 mm long; bracts up to 27 mm long at base of inflorescence, reflexed. Sepals and petals curled under, joined to each other and lip in basal third. Median sepal 4–6 × c. 1 mm, oblong, rounded; lateral sepals similar but about twice as wide and oblique. Petals 3.5–4.5 × c. 1 mm, the edge ciliolate. Lip 4.5–6 mm long, 5.5–7.5 mm wide when flattened, ellipsoid or obovoid with a narrow mouth, the apex slightly reflexed. Spurs 5–8 mm long, parallel to ovary. Column curved, 4 mm high; stigma 1.5–2 mm; rostellum mid-lobe broadly spoon-shaped from a narrow base, much longer than side lobes.

Zambia. N: Mbala Distr., Nakatali, 1660 m, fl. 4.ii.1951, *Bullock* 3652 (K). **Zimbabwe.** E: Nyanga (Inyanga), 1800 m, fl. 20.iii.1949, *Chase* 4063 (K; SRGH).
Also in Tanzania. Marshy ground, grassy slopes, open bush; 1500–1800 m.

19. **Satyrium riparium** Rchb.f. in Flora **48**: 183 (1865). —Rolfe in F.T.A. **7**: 267 (1898). —Kraenzlin, Orchid. Gen. Sp. **1**: 682 (1899) pro parte. —Schlechter in Bot. Jahrb. Syst. **31**: 179 (1901). —Summerhayes in F.T.E.A., Orchidaceae: 204 (1968). —Grosvenor in Excelsa **6**: 85 (1976). —Williamson, Orch. S. Centr. Africa: 96 t. 81 (1977). —Geerinck in Fl. Afr. Centr., Orchidaceae, pt. 1: 227 (1984). —la Croix et al., Orch. Malawi: 143 (1991). Type from Angola.

Terrestrial herb 40–80 cm high; tubers up to 4.5 cm long, ovoid or ellipsoid, villous. Sterile shoot 3–4-leaved; lowermost 1–2 leaves sheathing; upper leaves to 10 × 3.5 cm, lanceolate or elliptical. Flowering stem with 7–12 sheathing leaves up to 7 cm long. Inflorescence 11–32 × 2–3 cm, cylindrical, fairly laxly many-flowered. Flowers greenish-yellow, tinged with brown and purple. Ovary 7–8 mm long, curved, papillose; bracts 1–3 cm long at base of inflorescence, reflexed. Sepals and petals decurved, joined to each other and lip for about half their length. Median sepal 5–7.5 × 1–1.5 mm, oblong; lateral sepals of similar length, c. 2 mm wide, obliquely oblanceolate. Petals slightly shorter than sepals, puberulous. Lip fleshy, 5.5–7 mm long, ellipsoid, the apex reflexed, narrow-mouthed. Spurs slender, 18–25 mm long, straggling, ascending in young flowers but ± pendent in older flowers; often with an extra pair of vestigial spurs. Column curved, 5 mm long; rostellum mid-lobe spade-shaped from a narrow base, much longer than side lobes.

Zambia. N: Mpika Distr., c. 126 km from Kapiri Mposhi, 1200 m, fl. 2.xi.1962, *Richards* 16865 (K). W: Zambezi Rapids, fl. 28.x.1966, *Leach & Williamson* 13521 (K; SRGH). C: 131 km from Lusaka along Great East Road, c. 1120 m, fl. & fr. 16.xii.1972, *Strid* 2666 (K). E: Lundazi Distr., Nyika Plateau, Chowo Rocks, 2100 m, fl. 8.i.1982, *Dowsett-Lemaire* 200 (K). **Zimbabwe.** C: Marondera (Marandellas), c. 1350 m, fl. 23.xi.1962, *Moll* 329 (K; SRGH). **Malawi.** N: South Viphya, Lichelemu (Luchilemu) Dambo, 1300 m, fl. 4.i.1983, *la Croix* 388 (K). C: Lilongwe Distr., Dzalanyama Forest Reserve, Madzi Maiera Dambo, fl. 4.xii.1951, *Jackson* 708 (K).
Also in Zaire, Tanzania and Angola. Montane grassland, swampy grassland and dambo; 1100–2100 m.

20. **Satyrium coriophoroides** A. Rich. in Ann. Sci. Nat., sér. 2, **14**: 274, t. 18/ XI, 1–5 (1840); in Tent. Fl. Abyss. **2**: 298, t. 89 (1951). —Rolfe in F.T.A. **7**: 269 (1898). —Summerhayes in F.W.T.A. ed. 2, **3**: 201 (1968); in F.T.E.A., Orchidaceae: 201 (1968). —Geerinck in Fl. Afr. Centr., Orchidaceae pt. 1: 225 (1984). Type from Ethiopia.
Satyrium coriophoroides var. *sacculata* Rendle in J. Bot. **33**: 295 (1895). —Schlechter in Bot. Jahrb. Syst. **31**: 181 (1901). Type from Congo.
Satyrium sacculatum (Rendle) Rolfe in F.T.A. **7**: 266 (1898). —Schlechter in Bot. Jahrb. Syst. **53**: 534 (1915). —Summerhayes in F.T.E.A., Orchidaceae: 200 (1968). —Williamson, Orch. S. Centr. Africa: 94, t. 80 (1977). —la Croix et al., Orch. Malawi: 143 (1991). Type from Congo.
Satyrium stolzii Kraenzl. in Bot. Jahrb. Syst. **48**: 388 (1912). Type from Tanzania.

Terrestrial herb 40–80 cm high; tubers to 5.5 × 1.5 cm, ellipsoid. Sterile shoot 3–5-leaved; lowermost 1–3 leaves sheathing; upper leaves 6–15 × 2–4.5 cm, lanceolate or ovate. Flowering stem with 10–15 sheathing leaves up to 10 cm long. Inflorescence 8–30 × 1.5–2 cm, narrowly cylindrical, densely many-flowered. Flowers deep red, pink or white. Ovary 6–8 mm long; bracts up to 25 mm long at base of inflorescence, reflexed. Sepals and petals curled under, joined to each other and lip for one quarter to one third of their length. Median sepal 4.5–7 × c. 1 mm, oblong, rounded at apex; lateral sepals oblique and slightly longer, c. 2 mm wide. Petals 4–6.5 × 0.5–1 mm. Lip 5.5–7.5 mm long, globose, the apex slightly reflexed, fleshy with a narrow mouth. Spurs 8.5–17 mm long, parallel to ovary, sometimes with a pair of vestigial spurs c. 1 mm long in front of them. Column 3.5–5 mm long, curved; stigmas 1.5–3.5 mm; rostellum mid-lobe varying from slightly longer to much longer than side lobes.

Zambia. N: Mbala Distr., Ndundu, 1575 m, fl. 26.i.1952, *Richards* 636 (K). W: Mwinilunga Distr., E of Kewumbo R. (Kaoomba), fl. 22.xii.1937, *Milne-Redhead* 3782 (K). E: Lundazi Distr., Nyika Plateau, fl. ii.1968, *Williamson, Simon & Ball* 359 (K; SRGH). **Zimbabwe**. E: Nyanga (Inyanga), in vlei, 1850 m, fl. & fr. 22.ii.1946, *Wild* 952 (K; SRGH). **Malawi**. N: North Viphya, road to Uzumara near Mphompa, c. 1500 m, fl. 30.i.1986, *la Croix* 789 (K). C: Dedza Distr., Chongoni Mt., 2200 m, fl. 3.ii.1959, *Robson* 1430 (K).
   Widespread in tropical Africa. Grassland, and at edge of woodland, dambo; 1400–2300 m.
   S. sacculatum was separated from *S. coriophoroides* by having spurs 8.5–10.5 mm long, as opposed to 11–17 mm; by usually having a pair of additional spurs whereas *S. coriophoroides* never does; by rarely having white flowers rather than often white and by the mid-lobe of the rostellum being much longer, rather than slightly longer than the side lobes. The characters are not consistently linked, for example plants in Malawi which otherwise agree with *S. sacculatum* have spurs 13–15 mm long. There seems little reason to keep these species separate.

21. **Satyrium volkensii** Schltr. in Bot. Jahrb. Syst. **24**: 425 (1897). —Rolfe in F.T.A. **7**: 267 (1898). —Schlechter in Bot. Jahrb. Syst. **31**: 180 (1901), excl. syn. *S. chlorocorys*. —Summerhayes in Kew Bull. **8**: 577 (1954); in F.T.E.A., Orchidaceae: 202 (1968); in F.W.T.A. ed. 2, **3**: 201 (1968). —Grosvenor in Excelsa **6**: 86 (1976). —Williamson, Orch. S. Centr. Africa: 95 (1977). —Geerinck in Fl. Afr. Centr., Orchidaceae pt. 1: 228 (1984). —la Croix et al., Orch. Malawi: 146 (1991). Type from Tanzania.
      Satyrium brachypetalum sensu. Kraenzlin in Pflanzen Ost-Afr. **C**: 153 (1895) pro parte, non A. Rich.
      Satyrium trachypetalum Kraenzl. in Bot. Jahrb. Syst. **24**: 505 (1895); in Orchid. Gen. Sp. **1**: 683 (1899). —Rolfe in F.T.A. **7**: 574 (1898). Type from Tanzania.
      Satyrium leptopetalum Kraenzl. in Bot. Jahrb. Syst. **36**: 119 (1905). Type from Tanzania.
      Satyrium dizygoceras Summerh. in Bull. Misc. Inform., Kew **1932**: 508 (1932); in F.T.W.A. **2**: 417 (1936); in Bot. Not. **1937**: 190 (1937). Type from Kenya.

Terrestrial herb 45–85 cm tall; tubers to 5 cm long, ellipsoid, woolly. Sterile shoot 3–4-leaved; lowermost 1–2 leaves sheathing; upper leaves up to 18 × 5 cm, lanceolate or elliptic. Flowering stem with 7–13 sheathing leaves up to 10 cm long. Inflorescence 10–40 × 1.5–2.5 cm, narrowly cylindrical, fairly densely many-flowered. Flowers green or yellow-green, sometimes tinged with brown; scented. Ovary 6–10 mm long, glabrous or slightly papillate; bracts up to 35 mm long at base of inflorescence, reflexed. Sepals and petals curled under, joined to each other and lip in lower half or third. Median sepal 3.5–5 × 0.7 mm, oblanceolate, rounded at apex; lateral sepals slightly longer and about twice as wide, oblique. Petals similar to median sepal. Lip 4.5–5.5 mm long, ellipsoid, hooded, the apex reflexed, fleshy with a narrow mouth. Spurs 14–22 mm long, slender, pendent, sometimes with 2 extra rudimentary spurs less than 1 mm long below them. Column 3–5 mm long, curved; stigma 1–2 × 1–2 mm; rostellum mid-lobe spoon-shaped from a narrow base, much longer than side lobes.

Zambia. N: Mbala Distr., Kalambo Farm, 1500 m, fl. 8.i.1955, *Richards* 3941 (K). W: 10 km from Chingola on way to Solwezi, 1350 m, fl. 18.i.1975, *Brummitt, Chisumpa & Polhill* 13828 (K). E: Lundazi Distr., Nyika foothills, fl. xii.1966, *G. Williamson* 242 (K). **Zimbabwe**. E: Nyanga (Inyanga) Downs, c. 1950 m, fl. 29.i.1931, *Norlindh & Weimarck* 4675 (K). **Malawi**. N: South Viphya, Lichelemu (Luchilemu) Dambo, 1300 m, fl. 4.i.1983, *la Croix* 389 (K). C: Dedza Distr., Chencherere Hill, 1650 m, fl. 18.i.1959, *Robson* 1241 (K). S: Zomba Mt., near CCAP cottage, fl. 28.xii.1970, *Moriarty* 400 (K).
   Also in Nigeria, Cameroon, Zaire, Kenya and Tanzania. *Brachystegia* and mixed woodland, grassland with scattered bushes, montane grassland and dambo margins; 1200–2400 m.

22. **Satyrium chlorocorys** Rolfe in F.T.A. **7**: 268 (1898). —Kraenzlin, Orchid. Gen. Sp. **1**: 685 (1899), excl. syn. *S. volkensii*. —Summerhayes in Bot. Not. **1937**: 190 (1937); in Kew Bull. **4**: 433 (1949); in Mem. N.Y. Bot. Gard. **9**, 1: 80 (1954); in Kew Bull. **20**: 174 (1966); in

F.T.E.A., Orchidaceae: 206 (1968). —Grosvenor in Excelsa **6**: 85 (1976). —Williamson, Orch. S. Centr. Africa: 96, t. 83 (1977). —la Croix et al., Orch. Malawi: 141 (1991). Types from Tanzania.

Satyrium kraenzlinii Rolfe in F.T.A. **7**: 269 (1898). Type from Tanzania.

Satyrium brachypetalum sensu Kraenzlin, Orchid. Gen. Sp. **1**: 688 (1899), partim, non A. Rich.

Satyrium volkensii sensu Schltr. in Bot. Jahrb. Syst. **31**: 180 (1901), partim, non Schltr. sensu stricto.

Satyrium coriophoroides sensu Schltr. in Bot. Jahrb. Syst. **31**: 180 (1901), partim, non A. Rich.

Satyrium fallax Schltr. in Bot. Jahrb. Syst. **53**: 532 (1915). —Summerhayes in Bot. Not. **1937**: 190 (1937). Type from Tanzania.

Terrestrial herb 30–85 cm tall; tubers c. 3 × 1 cm, ellipsoid, tomentose. Stem leafy; leaves 5–8; the lowermost 1–2 leaves sheathing, the next 2(3) leaves spreading, borne near stem base but not appressed to ground, 5–24 × 3–9 cm, broadly lanceolate or ovate, pale glossy-green; the upper leaves erect, becoming bract-like. Inflorescence 6–34 × 1.5–4 cm, loosely or fairly densely many-flowered. Flowers green or yellow-green. Ovary 6–9 mm long; bracts up to 25 mm long at base of inflorescence, reflexed. Sepals and petals curled under, joined to each other and lip in basal half or third. Median sepal 4.5–6 mm long and less than 1 mm wide, oblanceolate, rounded; lateral sepals slightly longer and about twice as wide, oblique. Petals similar to median sepal but slightly shorter and narrower. Lip 4–6.5 mm long, the apex slightly reflexed, ellipsoid, hooded, fleshy with a small mouth. Spurs 8–20 mm long, slender, tapering. Column 4–4.5 mm high, curved; stigma 1.5 × 2–3 mm; rostellum mid-lobe spade-shaped from a narrow base, much longer than side lobes.

Zambia. E: Lundazi Distr., Nyika Plateau, fl. iv.1968, *G. Williamson* 938 (K; SRGH). **Zimbabwe**. E: Nyanga (Inyanga), Hill Fort, fl. 19.iv.1953, *Chase* 4949 (K; SRGH). **Malawi**. N: Nyika Plateau, Kasaramba Viewpoint, 2345 m, fl. 14.v.1970, *Brummitt* 10702 (K). S: Zomba Mt., near Chingwe's Hole, 1900 m, fl. 13.iii.1982, *la Croix* 296 (K). **Mozambique**. MS: Gorongosa Mt., Gogogo summit area, 1700–1868 m, fl. 12.iii.1972, *Tinley* 2434 (K; SRGH).

Also in Uganda and Tanzania. Montane grassland; 1450–2345 m.

23. **Satyrium microcorys** Schltr. in Bot. Jahrb. Syst. **53**: 533 (1915). —Summerhayes in Bot. Not. **1937**: 191 (1937); in F.T.E.A., Orchidaceae: 207 (1968). —Williamson, Orch. S. Centr. Africa: 96 (1977). —la Croix et al., Orch. Malawi: 142 (1991). Types from Tanzania.

Slender terrestrial herb 10–40 cm tall; tubers to 2 × 2 cm, ovoid, woolly. Stem leafy. Leaves 5–8; the lowermost 1–2 leaves sheathing; the next 2 leaves spreading, borne near stem base, 4–9 × 2–4 cm, elliptical or lanceolate; the upper leaves bract-like, spaced along stem. Inflorescence 4–12 × 1–1.5 cm, fairly densely several- to many-flowered. Flowers cream-yellow or yellow-green. Ovary 5 mm long; bracts to 15 mm long at base of inflorescence, reflexed. Sepals and petals curled under, joined to each other and lip in basal third. Median sepal 3–4 × 0.5 mm, oblong-oblanceolate, rounded; lateral sepals slightly longer and wider, oblique. Petals similar to median sepal but slightly shorter and narrower. Lip 3–5 mm long, ellipsoid, the apex slightly reflexed, fleshy with a narrow mouth. Spurs 3–4 mm long, bulbous at apex, parallel to ovary. Column c. 3 mm high, curved; stigma 1.5 × 2 mm; rostellum mid-lobe spade-shaped from a narrow base, much longer than the triangular side lobes.

Zambia. E: Nyika Plateau, fl. iii.1967, *G. Williamson* 290 (K; SRGH). **Malawi**. N: Nyika National Park, near Chelinda Bridge, 2300 m, fl. 3.iii.1987, *la Croix* 1002 (K; MAL); Chitipa Distr., Misuku Hills, fl. 10.v.1947, *Benson* 1262 (K).

Also in Tanzania. Montane grassland, usually amongst rocks in seepage areas; 1900–2300 m.

Specimens intermediate in appearance between *Satyrium microcorys* and *S. chlorocorys* are found where these two species occur together on the Nyika Plateau in Malawi. These intermediate specimens possess spurs which are c. 10 mm long, as in *S. chlorocorys*, but with bulbous tips as in *S. microcorys*. It seems possible that these could be hybrids.

24. **Satyrium shirense** Rolfe in F.T.A. **7**: 266 (1898). —Williamson, Orch. S. Centr. Africa: 96 (1977). —la Croix et al., Orch. Malawi: 144 (1991). Type: Malawi, Shire Highlands, *Buchanan* s.n. (K, holotype).

Slender terrestrial herb 13–45 cm tall. Leaves 6–7; the lowermost 1–2 leaves sheath-like; the next 2 leaves spreading, borne near the the stem base but not appressed to ground, 5–14.5 × 2–6 cm, broadly lanceolate to ovate, light green; the upper leaves bract-like, spaced along the stem. Inflorescence 3–15 × 1–2 cm, fairly densely several- to many-flowered. Flowers creamy-white. Ovary 6–8 mm long; bracts to 15 mm long at base of inflorescence, reflexed. Sepals and petals deflexed, the lateral sepals spirally twisted, joined to each other and lip in basal half. Median sepal 3–4 mm long, less than 1 mm wide, oblong-oblanceolate, rounded; lateral sepals slightly longer and almost twice as wide. Petals similar to median sepal but narrower. Lip 4–5 mm long, ± globose with a slight dorsal keel, fleshy with a narrow mouth. Spurs 8–12 mm long, slender, parallel to ovary; some flowers in an inflorescence have 2 vestigial spurs beside the main spurs. Column 3 mm long; rostellum very short, the apex reflexed, the side lobes represented by minute teeth.

Zambia. E: Lundazi Distr., Nyika Plateau, fl. iv.1968, *G. Williamson* 937 (K; SRGH). **Malawi**. N: Nyika National Park, junction of Chelinda Bridge and Kasaramba roads, c. 2300 m, fl. 4.iii.1977, *Pawek* 12444 (K; MAL; MO). C: Dedza Mt., seepage rocks near peak, 2160 m, fl. 23.ii.1985, *la Croix* 671 (K). S: Zomba Mt., rocky hill near Chingwe's Hole, c. 1900 m, fl. 23.iii.1983, *la Croix* 295 (K).

Not known elsewhere. Montane grassland, rocky hillsides and seepage slopes; 1750–2500 m.

In the herbarium, this species may be difficult to distinguish from small specimens of *S. chlorocorys* but in the field, they appear quite distinct. *S. shirense* is a smaller, more slender plant, consistently white-flowered; no intermediates between the two species have been seen. On Zomba Plateau in southern Malawi, *S. shirense* is common on rocky seepage slopes, replacing *S. chlorocorys* which is frequent in the montane grassland, but on the Nyika Plateau in northern Malawi, both species grow together in the montane grassland.

25. **Satyrium sphaeranthum** Schltr. in Bot. Jahrb. Syst. **53**: 532 (1915). —Summerhayes in F.T.E.A., Orchidaceae: 208 (1968). —Williamson, Orch. S. Centr. Africa: 96 (1977). —la Croix et al., Orch. Malawi: 145 (1991). Type from Tanzania.

Terrestrial herb 40–100 cm tall; tubers to 4.5 cm long, ellipsoid to fusiform, woolly. Stem leafy. Leaves 6–8; the lowermost 1–2 leaves sheath-like; the next 2 leaves large, spreading, borne near the stem base but not appressed to the ground, 8–30 × 5–13 cm, ovate or broadly lanceolate, funnel-shaped at base, light green; the upper leaves spaced up stem becoming bract-like. Inflorescence 13–30 × 1.5–2 cm, densely many-flowered. Flowers white or bright yellow. Ovary 5–7 mm long; bracts to 30 mm long at base of inflorescence, reflexed. Sepals and petals curled under, joined to each other in basal half and to basal quarter of lip. Median sepal 4.5–6 × c. 1 mm, oblong-oblanceolate, rounded; lateral sepals similar but wider and oblique. Petals similar to dorsal sepal but slightly smaller. Lip 5–7 mm long, ± globose, fleshy, the apex slightly reflexed, overhanging the narrow mouth. Spurs 10–12 mm long, slender, parallel to ovary, usually with an additional pair of rudimentary spurs below them. Column c. 4 mm high, curved; stigma 1–2 × 2–3 mm; rostellum mid-lobe broadly spoon-shaped from a narrow base, much longer than the tooth-like side lobes.

Zambia. E: Nyika Plateau, verge of evergreen forest, fl. iii.1967, *G. Williamson* 297 (K; SRGH). **Zimbabwe**. E: Mutare, 1575 m, fl. 19.iii.1964, *Chase* 8142 (K; SRGH). **Malawi**. N: Nyika National Park, Katizi Valley, c. 1900 m, fl. 24.iv.1982, *la Croix & Johnston-Stewart* 332 (K).

Also in Tanzania. Usually occurs in the transition zone between two vegetation types, e.g. between evergreen forest and montane grassland, and between thicket and *Brachystegia* woodland; 1280–2200 m.

In Malawi, yellow-flowered plants occur above 2000 m, while white-flowered plants occur at lower altitudes.

26. **Satyrium flavum** la Croix in Kew Bull. **48**, 2: 364 (1993). TAB. **70**. Type: Zimbabwe, Chimanimani Mts., *Wild* 3600 (K, holotype; SRGH).
   Satyrium parviflorum sensu Grosvenor in Excelsa **6**: 85 (1976).

Terrestrial herb 25–35 cm tall. Stem with 2 basal leaves appressed to the ground, and 4–6 bract-like, sheathing leaves. Basal leaves 3.5–7 × 2–5 cm, the larger ± orbicular, the smaller ovate; stem leaves 2–4 cm long. Inflorescence 7–9 × 1.5–2 cm, laxly 5–10-flowered. Ovary 8–10 mm long; bracts up to 15 × 9 mm at base of inflorescence, ovate,

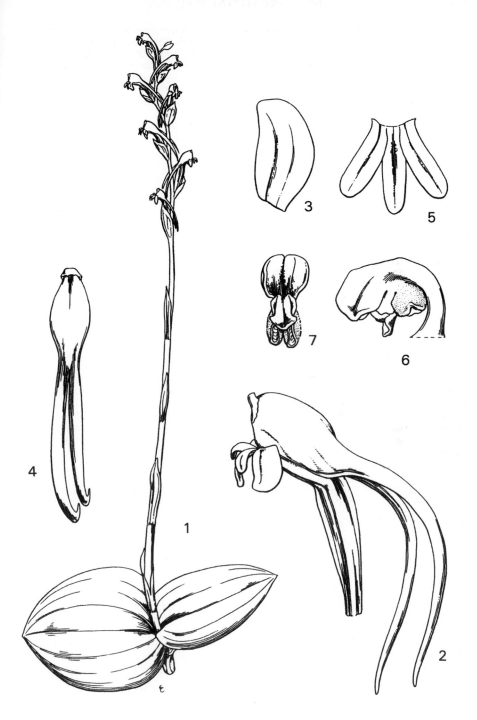

Tab. 70. SATYRIUM FLAVUM. 1, habit (×⅔); 2, flower, side view (×6); 3, lateral sepal (×8); 4, lip, from above (×4); 5, petals and dorsal sepal (×8); 6, column (×8); 7, column, front view (×8), 1–7 from *Wild* 3600. Drawn by Eleanor Catherine. From Kew Bull.

acute, not reflexed. Flowers yellow. Sepals and petals projecting then curved under, joined to each other and lip for about half their length. Intermediate sepal 6 × 0.6 mm, the free part 3 mm long, oblong-lanceolate, the apex rounded; lateral sepals similar but oblique and about twice as wide. Petals similar to intermediate sepal. Lip 6–8 mm long, of similar width when flattened, very convex, somewhat overhanging the narrow mouth, the apex reflexed; spurs 14–18 mm long, slender, arched. Column erect, 4 mm high with stalk 2 mm long; rostellum ± spade-shaped.

Zimbabwe. E: Chimanimani Mts., Mt. Peza, 2000 m, fl. 15.x.1950, *Wild* 3600 (K; SRGH); Nyanga (Inyanga), fl. 8.iv.1959, *Holmes* 0162 (K; SRGH).

Only known from the eastern highlands of Zimbabwe and adjacent areas of Mozambique. Amongst crags in grassland; 2000–2700 m.

This species has usually been identified as *Satyrium parviflorum* Sw. However, collectors describe the leaves as appressed to the ground, which is not the case with typical *S. parviflorum*. Also, the flowers are slightly larger, with longer spurs and are always described as yellow, and so it apparently forms a consistent and distinct entity.

27. **Satyrium amblyosaccos** Schltr. in Bot. Jahrb. Syst. **53**: 535 (1915). —Summerhayes in Kew Bull. **20**: 179 (1966); in F.T.E.A., Orchidaceae: 211 (1968). —Williamson, Orch. S. Centr. Africa: 97 (1977). —Geerinck in Fl. Afr. Centr., Orchidaceae pt. 1: 218 (1984). —la Croix et al., Orch. Malawi: 147 (1991). Type from Tanzania.
    Satyrium papillosum Schltr. in R.E. Fr., Wiss. Ergebn. Schwed. Rhod.-Kongo-Exped. **1**: 242, t. 18, fig. 1 (1916), nom. illegit., non Lindl. Type from Burundi.

Terrestrial herb 20–65 cm high; tubers to 3.5 cm long, ovoid or fusiform. Sterile shoot 3–4-leaved, the lower 2–3 leaves sheath-like, the upper leaves to 18 × 4 cm, lanceolate, acute. Flowering stem with 5–9 leaves, the largest erect, in the middle of the stem, 10–17 × 2.5–6.5 cm, ± lanceolate, loosely sheathing at base. Inflorescence 5–20 × 2–3 cm, fairly densely many-flowered. Flowers white, the sepals and petals tipped with green. Ovary 5–8 mm long, glabrous; bracts up to 30 × 7 mm long, prominent, usually reflexed towards base of inflorescence, spreading further up, but sometimes all reflexed. Sepals and petals deflexed, joined to each other in basal half and to lip in basal third. Median sepal 6–10 × 1.5–2 mm, oblong-oblanceolate, rounded; lateral sepals slightly shorter but wider. Petals slightly shorter than sepals, c. 1 mm wide, oblanceolate. Lip 6–9 mm long, very convex, with beak-like apex c. 2 mm long, wide-mouthed. Spurs 2–3 mm long and 2 mm in diameter, short and sac-like. Column 4–6 mm long, slender, hairy towards base; stigma concave, 1 × 1.5 mm; rostellum shortly 3-toothed.

Zambia. N: Mbala Distr., track to Goddard's Farm, 1500 m, fl. 5.ii.1957, *Richards* 8069 (K). W: Mwinilunga Distr., 80 km E of Mwinilunga, fl. 20.xii.1969, *Williamson & Simon* 1780 (K; SRGH). E: Nyika Plateau, fl. ii.1968, *G. Williamson* 358 (K). **Malawi**. N: Nyika National Park, road to Nganda, c. 3 km N of turn-off to Dembo Bridge, fl. 12.iii.1977, *Grosvenor & Renz* 1178 (K; SRGH). S: Zomba Plateau near Ku Chawe Inn, 1530 m, fl. 14.iii.1970, *Brummitt* 9179 (K).

Also in Zaire, Burundi and Tanzania. Montane, wet or dry grassland, marsh and *Brachystegia* woodland; 1200–2300 m.

28. **Satyrium trinerve** Lindl., Gen. Sp. Orchid. Pl.: 344 (1838). —Eyles in Trans. Roy. Soc. S. Afr. **5**: 334 (1916). —Grosvenor in Excelsa **6**: 86 (1976). —Hall in Contrib. Bolus Herb. No. 10: 63 (1982). —Stewart et al., Wild Orch. South. Africa: 148 (1982). —Geerinck in Fl. Afr. Centr., Orchidaceae, pt. 1: 219 (1984). —la Croix et al., Orch. Malawi: 150 (1991). TAB. **71**. Type from Madagascar.
    Satyrium atherstonei Rchb.f. in Flora **64**: 328 (1881). —Kraenzlin, Orchid. Gen. Sp. **1**: 660 (1899). —Rolfe in F.C. **5**, 3: 154 (1912). —Summerhayes in Kew Bull. **6**: 464 (1952). — Goodier & Phipps in Kirkia **1**: 53 (1961). —Summerhayes in F.T.E.A., Orchidaceae: 209 (1968); in F.W.T.A. ed. 2, **3**: 201 (1968). —Moriarty in Wild Fls. Malawi: t. 30, 3 (1975). — Williamson, Orch. S. Centr. Africa: 97 (1977). Type from South Africa.
    Satyrium zombense Rolfe in F.T.A. **7**: 273 (1898). Type: Malawi, Mt. Zomba, *Buchanan* 304 (K, holotype).
    Satyrium occultum Rolfe in F.T.A. **7**: 273 (1898). —Summerhayes in Bot. Not. **1937**: 191 (1937). Type: Malawi, without precise locality *Buchanan* 287 (K, holotype).
    Satyrium nuttii Rolfe in F.T.A. **7**: 273 (1898). Type: Zambia, Kambole, *Nutt* s.n. (K, holotype).
    Satyrium monopetalum Kraenzl., Orchid. Gen. Sp. **1**: 662 (1899). Type from South Africa.
    Satyrium triphyllum Kraenzl., Orchid. Gen. Sp. **1**: 660 (1899). Type from South Africa.

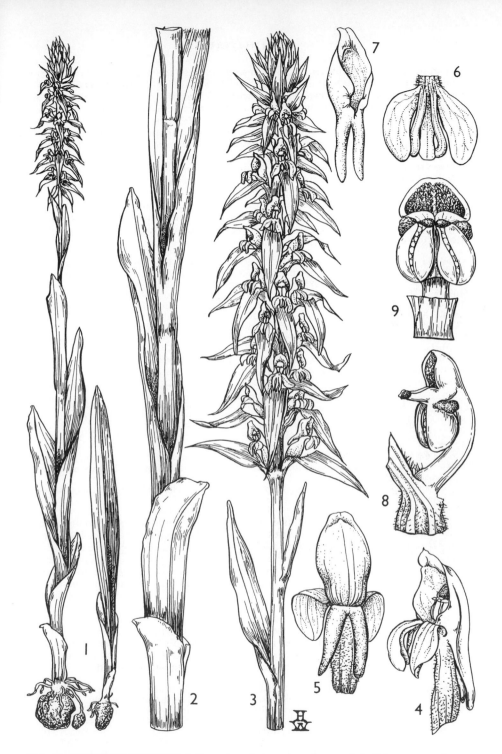

Tab. 71. SATYRIUM TRINERVE. 1, flowering plant and sterile stem ($\times\frac{1}{3}$); 2, lower part of stem ($\times$1); 3, inflorescence ($\times$1); 4, flower, three-quarter front view ($\times$3); 5, flower, back view ($\times$3); 6, sepals and petals ($\times$3); 7, lip ($\times$3); 8, column, side view ($\times$7); 9, column, front view ($\times$7), 1–9 from *Milne-Redhead & Taylor* 8307. Drawn by Heather Wood. From F.T.E.A.

Terrestrial herb 30–108 cm high; tubers to 4.5 cm long, ellipsoid to fusiform, woolly. Sterile shoot 4-leaved; the 2 lower leaves sheath-like; the upper leaves to 25 × 7.5 cm, lanceolate, acute. Flowering stem with 5–10 leaves, the largest in the middle, erect, loosely sheathing at base, 10–24 × 2.5–4 cm, ± lanceolate, acute, with prominent longitudinal veins. Inflorescence 5–27 × 2–3 cm, densely many-flowered, the rhachis hairy. Flowers white, the median sepal and petals sometimes yellow. Ovary 6–8 mm long, densely pubescent; bracts up to 55 mm long, lanceolate, acute, spreading or with some reflexed, untidy-looking, white towards apex. Sepals and petals ± deflexed, joined to each other in basal half and to lip for 1–2 mm. Median sepal 5–8 × 1 mm, oblanceolate, obtuse, it and the petals curved back; lateral sepals spreading, slightly longer and about twice as wide, obliquely oblong. Petals similar to median sepal but half as wide. Lip 6–8 mm long with upturned apex 2 mm long, very convex, with wide mouth 4–5 mm across. Spurs 4–7 mm long, 1.5–2 mm diameter, parallel to ovary. Column 3.5–5 mm high; stigma 1.5–2 mm high and wide; rostellum shortly and equally 3-lobed, the mid-lobe bifid.

Zambia. N: Mbala Distr., Lake Chila, fl. 11.xi.1949, *Bullock* 1411 (K). W: Mwinilunga Distr., Matonchi, Sinkabolo Dambo, fl. 22.xii.1969, *Williamson & Simon* 1812 (K; SRGH). C: Mkushi Distr., Fiwila, c. 1350 m, fl. 3.i.1958, *E.A. Robinson* 2584 (K; SRGH). **Zimbabwe**. C: Harare, Mukuvisi (Makabusi) River, c. 1500 m, fl. 10.i.1963, *Moll & Smith* 490 (K; SRGH). E: Nyanga (Inyanga), by Pungwe River, c. 1700 m, fl. 19.xii.1930, *Fries, Norlindh & Weimarck* 3841 (K). **Malawi**. N: Nyika National Park, by main road opposite old salt lick, 2300 m, fl. 17.ii.1976, *Phillips* 1229 (K). C: Dedza Distr., base of Kalicelo (Kalichero) Hill, 1700 m, fl. 21.i.1959, *Robson* 1293 (K). S: Zomba Plateau, by Chingwe's Hole, 1880 m, fl. 11.ii.1970, *Brummitt & Banda* 8503 (K; MAL). **Mozambique**. T: Moatize, Zóbuè, 900 m, fl. 11.i.1966, *Correia* 385 (K).

Widespread in tropical Africa, South Africa and Madagascar. Grassland, usually moist localities, dambo and *Brachystegia* woodland; 550–2300 m.

29. **Satyrium breve** Rolfe in F.T.A. **7**: 274 (1898). —Kraenzlin, Orchid. Gen. Sp. **1**: 661 (1899). —Schlechter in Bot. Jahrb. Syst. **53**: 535 (1915). —Goodier & Phipps in Kirkia **1**: 53 (1961). —Summerhayes in F.T.E.A., Orchidaceae: 211 (1968). —Moriarty, Wild Fls. Malawi: t. 30, 4 (1975). —Grosvenor in Excelsa **6**: 85 (1976). —Williamson, Orch. S. Centr. Africa: 98, t. 85, 86 (1977). —Geerinck in Fl. Afr. Centr., Orchidaceae pt. 1: 223 (1984). —la Croix et al., Orch. Malawi: 147 (1991). Type: Malawi, Shire Highlands, *Buchanan* 314 (K, holotype).

Satyrium paludosum sensu Schltr. in Bot. Jahrb. Syst. **31**: 183 (1901), pro parte, non Rchb.f. (1865).

Satyrium breve var. *minor* Summerh. in Bot. Not., **1937**: 189 (1937). Type: Zimbabwe, Mt. Inyanga, *Fries et al.* 3647 (LD, holotype).

Terrestrial herb 10–60 cm tall; tubers to 4 cm long, ovoid to fusiform, woolly. Sterile shoot 4-leaved; the lower 1–2 leaves sheath-like, the upper leaves to 31 × 2 cm, lanceolate, acute. Flowering stem with 5–8 leaves, the largest in the middle, ± erect, 8–26 × 2–5 cm, lanceolate. Inflorescence 2.5–14 × 2–5 cm, densely few- to many-flowered, pyramidal when young. Flowers pale to deep pink or red-purple, sometimes with darker streaks. Ovary 6–8 mm long, glabrous or papillose; bracts to 20 mm long, pubescent on upper surface, spreading or sometimes reflexed at base of inflorescence. Sepals and petals projecting forwards, joined to each other for a third to half of their length, and to the lip for a third or a quarter. Median sepal 7–16 × 1–2 mm, oblanceolate, rounded; lateral sepals slightly shorter but broader, oblique. Petals similar to median sepal but narrower. Lip 6–16 mm long, the apex acute and reflexed for c. 2 mm, very convex with a wide mouth. Spurs 2–5.5 mm long, 2 mm in diameter, blunt, parallel to ovary. Column 5–6.5 mm long; stigma c. 1.5 mm high and wide; rostellum equally 3-toothed.

Zambia. N: Mporokoso Distr., Kipoma Falls, c. 1550 m, fl. 10.xi.1967, *Simon & Williamson* 1093A (K; SRGH). W: Mwinilunga Distr., swamp c. 4 km from Kabompo Gorge, 1200 m, fl. 24.xi.1962, *Richards* 17498 (K). C: Rufunsa (Rufunza), on Great East Road, fl. 24.xi.1963, *Morze* 106a (K). E: Nyika Plateau, fl. xii.1966, *G. Williamson* 234 (K). **Zimbabwe**. C: Makoni Distr., Merryvale Farm, fl. 2.i.1976, *Best* 1245 (K; SRGH). E: Nyanga (Inyanga), Nyamziwa Falls, c. 1600 m, fl. 12.i.1951, *Chase* 3590 (K; SRGH). **Malawi**. N: Nyika National Park, by Dam 3, 2250 m, fl. 25.xi.1986, *la Croix* 863 (K). S: Zomba Plateau, Chitinji Marsh, 1800 m, fl. 6.xi.1979, *Morris* 60 (K; MAL). **Mozambique**. MS: Chimanimani Mts., between Skeleton Pass and Namadima, 1550 m, fl. 30.xi.1959, *Goodier & Phipps* 355 (K; SRGH).

Also in Zaire, Rwanda, Burundi and Tanzania. Marshy grassland; 1200–2300 m.

This species is very variable in plant and flower size, and has an unusually long flowering season. On Zomba Plateau in southern Malawi, plants with few-flowered inflorescences and larger, dark-coloured flowers and relatively long spurs (5 mm) are shorter in stature and flower earlier than plants with many-flowered inflorescences and smaller, paler, short-spurred flowers. It is possible that these could be distinguished at varietal level but detailed work is necessary.

30. **Satyrium compactum** Summerh. in Kew Bull. **20**: 179 (1966). —Grosvenor in Excelsa **6**: 85 (1976). —Williamson, Orch. S. Centr. Africa: 100 (1977). Type: Zambia, Kasama Distr., Mungwi, i.1961, *E.A. Robinson* 4257 (K, holotype).

Terrestrial herb 20–60 cm tall; tubers to 2 cm long, ellipsoid or globose, almost glabrous. Leaves 6–10; the lowermost 1–2 leaves sheath-like; stem leaves 6–14 × 0.7–1.3 cm, linear or lanceolate-linear, acute; the 3 uppermost leaves much smaller. Inflorescence 1–6 × 1–2 cm, shortly cylindrical, densely 4- to many-flowered. Flowers arched, spreading, white tinged green; rhachis shortly pubescent. Ovary c. 5 mm long, pubescent; bracts ± spreading, 5–12 mm long, lanceolate, acute. Sepals and petals decurved, adnate to each other and lip in lower quarter or third. Median sepal 3–5 × 1–1.5 mm, oblong-ligulate, obtuse; lateral sepals longer and c. 2 mm wide, oblique. Petals similar to median sepal but shorter and narrower. Lip almost horizontal, 4–5.5 mm long, ovate, apex obtuse or subacute. Spurs 1–2 mm long, c. 1 mm wide, parallel to ovary. Column 3–4 mm high, slightly curved; stigma 1–2 mm high, orbicular or transversely elliptical; rostellum 1 mm long, the apex shortly 3-lobed, the mid-lobe bidenticulate.

Zambia. N: Chinsali Distr., Lake Young, Ishiba Ngandu (Shiwa-Ngandu), 1350 m, fl. 15.i.1959, *Richards* 10677 (K). W: Mwinilunga Distr., fl. 1.xii.1958, *Holmes* 0110 (K; SRGH). **Zimbabwe**. C: Makoni Distr., c. 35 km N of Rusape, 1800 m, fl. 19.i.1976, *Best* 1259 (K; SRGH). E: Chimanimani Mts., 1800 m, fl. 30.i.1954, *Ball* 187 (K; SRGH).
Not known elsewhere. Wet grassland and dambo, often on peaty soil; 1350–1800 m.
This species resembles *S. trinerve* but differs from it in the shorter, more compact inflorescence, smaller bracts, and smaller flowers with shorter spurs.

31. **Satyrium oliganthum** Schltr. in Bot. Jahrb. Syst. **53**: 534 (1915). —Grosvenor in Excelsa **6**: 85 (1976). —Williamson, Orch. S. Centr. Africa: 100 (1977). —la Croix et al., Orch. Malawi: 148 (1991). Type from Angola.
   Satyrium paludosum var. *parvibracteatum* Schltr. in Warb., Kunene-Sambesi Exped. Baum: 209 (1903). Type from Angola.

Slender terrestrial herb 10–24(40) cm high; tubers to 2.5 cm long, ovoid, woolly. Leaves 5–6; the lowermost leaves sheath-like, the next 2 leaves borne near the stem base but not appressed to the ground, relatively large, spreading or ± erect, 2.5–5(11) × 1–2.5 cm, ovate, acute, sometimes with one withered at flowering time; the upper leaves bract-like, spaced along the stem. Inflorescence 4–8 × 2–3 cm, laxly or fairly densely 3–15-flowered. Flowers pale pink or mauve, the lip sometimes marked with purple. Ovary slender, 7–9 mm long; bracts 4–18 mm long, erect. Sepals and petals projecting forwards, joined to each other and lip in the basal one third. Median sepal 8–11 × 1–3 mm, oblanceolate, rounded; lateral sepals similar but wider and oblique. Petals 9–11 × c. 1 mm, oblanceolate. Lip 8–9 mm long with apical flap (not reflexed) 2–3 mm long, wide-mouthed. Spurs 4–8 mm long, parallel to ovary, quite stout but tapering to an obtuse apex. Column slender, 5.5 mm long; rostellum shallowly 3-toothed.

   Zambia. B: Mongu Distr., c. 50 km NE of Mongu, fl. 10.xi.1959, *Drummond & Cookson* 6310 (K; SRGH). N: Kasama Distr., Mungwi, fl. 6.xii.1960, *E.A. Robinson* 4142 (K). C: Kundalila Falls, fl. x.1968, *G. Williamson* 1098 (K). **Zimbabwe**. C: Marondera, Digglefold, 1500 m, fl. 8.ix.1951, *Corby* 740 (K; SRGH). E: Chimanimani Mts., 1660 m, fl. 21.ix.1953, *Ball* 61 (K; SRGH). **Malawi**. N: Khondowe "(Kondowe) to Karonga, 2000–6000 ft.", fl. vii.1896, *Whyte* s.n. (K). S: Mulanje Mt., Lichenya Plateau, 2100 m, fl. 21.ii.1983, *la Croix, Jenkins & Killick* 463 (K).
   Also in Tanzania and Angola. Montane grassland in tussocky grass amongst boulders and in dambos; 1120–2100 m.

32. **Satyrium paludosum** Rchb.f. in Flora **48**: 182 (1865). —Rolfe in F.T.A. **7**: 274 (1898). —Kraenzlin, Orchid. Gen. Sp. **1**: 662 (1899). —Schlechter in Bot. Jahrb. Syst. **31**: 183 (1901) pro parte. —Summerhayes in F.T.E.A., Orchidaceae: 212 (1968). —Grosvenor in Excelsa **6**: 85 (1976). —Williamson, Orch. S. Centr. Africa: 99 (1977). —Geerinck in Fl. Afr. Centr., Orchidaceae pt. 1: 222 (1984). —la Croix et al., Orch. Malawi: 149 (1991). Type from Angola.

Robust terrestrial herb 30–90 cm high; tubers to 2.5 cm long, ellipsoid, tomentose. Stem leafy; leaves 5–11, the lowermost 2–3 sheathing, the rest semi-erect, 12–27 × 1.5–5.5 cm, lanceolate, ribbed, decreasing in size towards stem apex. Inflorescence 3–6 × 2.5–4 cm, pyramidal, densely many-flowered. Flowers white, often with purple marks at base of petals and inside lip. Ovary 6–10 mm, very slightly pubescent; bracts to 26 mm long, erect, pale green. Sepals and petals projecting forwards, joined to each other in basal half or third and to lip in basal third or quarter. Median sepal 9–11 × 1.3–2.5 mm, ligulate; lateral sepals of similar length but about twice as wide, oblique. Petals similar to median sepal but slightly shorter and narrower. Lip 8.5–14 mm long, of similar length but about twice as wide, oblique. Petals similar to median sepal but slightly shorter and narrower. Lip 8.5–14 mm long, of similar width when spread out, very convex with a projecting apex, wide-mouthed, hairy inside at base. Spurs saccate, 1–2.5 mm long, sometimes almost absent. Column 5–6.5 mm high, slender; stigma c. 1.5 mm high and slightly wider; rostellum shortly 3-toothed.

Zambia. W: Mwinilunga Distr., R. Kasompe, fl. 8.x.1937, *Milne-Redhead* 2671 (K). C: Mkushi Dambo, fl. xii.1967, *Williamson & Simon* 618 (K). **Zimbabwe**. C: Harare, Cleveland Dam, 1480 m, fl. 24.xi.1945, *Wild* 420 (K; SRGH). E: Nyanga (Inyanga) Mts., 1800–2100 m, fl. xii.1899, *Cecil* 200 (K). **Malawi**. N: South Viphya, Lichelemu (Luchilemu) Dambo, 1300 m, fl. 4.i.1983, *la Croix* 392 (K). C: Dedza Distr., Chongoni Forest Reserve, fl. 16.xi.1970, *Salubeni* 1497 (MAL).

Also in Zaire, Kenya and Angola. Montane and swampy grassland; 1300–2100 m.

33. **Satyrium aberrans** Summerh. in Kew Bull. **20**: 182 (1966); in F.T.E.A., Orchidaceae: 212 (1968). —Grosvenor in Excelsa **6**: 85 (1976). TAB. **72**. Type: Zimbabwe, Chimanimani Mts., Mt. Peza, x.1950, *Wild* 3630 (K, holotype; SRGH).

Dwarf terrestrial herb 7–17 cm high; tubers to 2 cm long, ovoid, tomentose. Stem leafy. Leaves 3–5; the lowermost leaves sheathing; the next 2–3 leaves spreading, 2.5–6 × 1.5–3 cm, ovate or lanceolate; the uppermost 1–2 leaves bract-like. Inflorescence 2.5–4 × 1–1.5 cm, fairly densely several- to many-flowered. Flowers pale greenish-white, the bracts tinged purple, the stem also purple-brown. Ovary 6–7 mm long, papillose; bracts to 12 × 6 mm, spreading. Sepals and petals spreading forwards, joined to each other in basal half or third and to lip in basal third or quarter. Median sepal 4.5–6 × 1.5–2 mm, oblong; lateral sepals similar but broader and oblique. Petals 4.5 × 1–1.5 mm, lanceolate. Lip c. 4 mm long, very convex with a wide mouth (2.5 mm across), apex acute. Spurs 3.5–6 mm long, 1.5–2.5 mm diameter, straight, obtuse. Column 2.5–3 mm high; stigma transversely oblong, c. 1.5 mm wide; rostellum obscurely 3-lobed.

Zimbabwe. E: Chimanimani Mts., Mt. Peza, 1800 m, fl. 15.x.1950, *Wild* 3630 (K; SRGH); Chimanimani Main Range, 2100 m, fl. 9.iv.1955, *Ball* 546 (K; SRGH).

Also in Tanzania. Montane, in rock crevices; 1800–2100 m.

34. **Satyrium afromontanum** la Croix & P.J. Cribb in Kew Bull. **48**, 2: 361 (1993). Type: Malawi, Mulanje Mt., *Adamson* 426 (E, holotype).

　　Satyrium aberrans auct., non Summerhayes (1966), la Croix et al., Orch. Malawi: 151 (1991).

Dwarf terrestrial herb 2–10 cm high. Tubers 12–20 × 3–4 mm, ovoid or narrowly ellipsoid, somewhat tomentose. Leaves 4–6; the 2 lowermost leaves sheathing, the next 2 spreading, sub-opposite, near stem base but not appressed to the ground, the larger 2–3.2 × 1.4–2.2 cm, ovate, clasping the stem at base; scape sometimes with 1–2 loosely sheathing bract-like leaves, 12–18 × 8–10 mm. Inflorescence 1–2-flowered. Ovary 11–13 × 2–3 mm; constricted at apex; bracts loosely funnel-shaped, not reflexed, 7–10 × 9–10 mm, broadly ovate, acute, enfolding the ovary. Flowers white, the spurs green at the tips. Sepals and petals projecting forwards, intermediate sepal and petals joined to each other for half their length and to the lateral sepals and lip for a third of their length. Intermediate sepal 6–8 × 1–1.3 mm, oblong, the free part to 6.3 mm long; lateral sepals 8–9 × 3.5–4.3 mm, free part to 6.2 mm long, obliquely ovate. Petals similar to intermediate sepal but slightly shorter. Lip very convex, 7–8 mm long and c. 8 mm wide, the apex reflexed; spurs 5–6 × 1–1.3 mm, the apices usually rounded, lying along the ovary. Column bent over, the stalk fairly slender, c. 1 mm long, 0.5 mm diameter; fertile part 3.5 mm long.

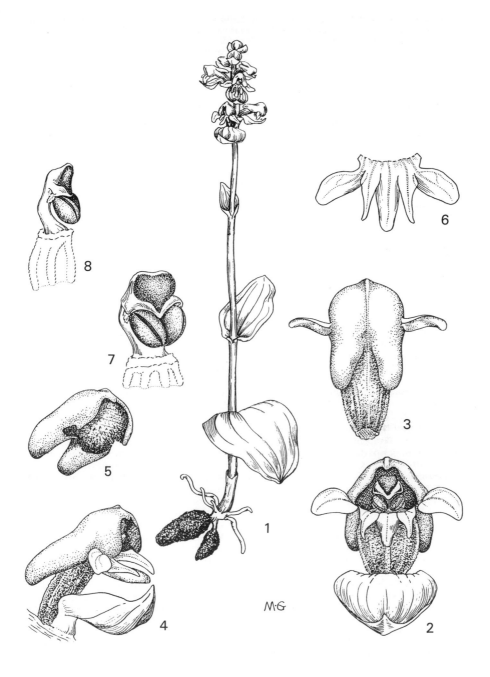

Tab. 72. SATYRIUM ABERRANS. 1, habit (×1); 2, flower, front view (×4); 3, flower, back view (×4); 4, flower, side view (×4); 5, lip with spurs (×4); 6, sepals and petals (×10); 7, column, front view (×6); 8, column, side view (×6), 1–8 from *Richards* 14303. Drawn by Mary Grierson.

Malawi. S: Mulanje Mt., Madzeka Plateau, 2400 m, fl. 6.iii.1984, *Royle* in *la Croix* 566 (K); Mulanje Mt., in damp situation, fl. 1897, *Adamson* 426 (E).

Endemic. In shelter beneath rocks forming colonies, in damp places; 2100–2400 m.

This dwarf species resembles *Satyrium comptum* Summerh. but differs in the inflorescences being consistently 1–2-flowered, in the spurs lying along the ovary and not diverging from it, and in the loose bracts that enclose the ovary and which are never reflexed.

35. **Satyrium ecalcaratum** Schltr. in R.E. Fr., Wiss. Ergebn. Schwed. Rhod.-Kongo-Exped. **1**: 242 (1916). —Summerhayes in F.T.E.A., Orchidaceae: 214 (1968). —Geerinck in Fl. Afr. Centr., Orchidaceae pt. 1: 222 (1984). —la Croix et al., Orch. Malawi: 152 (1991). Type from Zaire.

Terrestrial herb 10–35 cm high; tubers to 16 mm long, ellipsoid or fusiform, tomentose. Stem leafy; leaves 4–6, the lowermost sheath-like, the next 2 borne near the stem base, spreading, up to 6.5 × 3 cm, ovate, acute; the upper leaves smaller and bract-like. Inflorescence 2–6 × 1.5–2.5 cm, ± pyramidal, densely c. 15-flowered. Flowers deep pink. Ovary 5–7 mm long; bracts erect, c. 15 mm long. Sepals and petals projecting, joined to each other and lip for 1–1.5 mm at base. Median sepal 4–5 × 1.6–1.8 mm, oblong-lanceolate, obtuse; lateral sepals slightly longer and wider, oblique. Petals c. 4 × 1.5 mm, lanceolate. Lip 4–5 mm long, c. 5.5 mm wide when spread out, wide-mouthed; spurs absent. Column 2–3 mm long, curved; stigma c. 1 mm high and wide; rostellum obscurely 3-lobed.

Malawi. S: Mulanje Mt., under rocks, 2400–2575 m, fl. & fr. v.1965, *Morris* 170 (K). Also in Zaire, Rwanda, Uganda and Tanzania. Under rocks in humus; c. 2500 m. Known in the Flora Zambesiaca area from only 1 collection.

24. **CORYCIUM** Sw.

Corycium Sw. in Kongl. Vetensk. Acad. Nya Handl. **21**: 220 (1800). —Rolfe in F.C. **5**(3): 281 (1913).

Terrestrial herb with undivided sometimes stalked tubers. Stem erect, leafy; leaves radical and cauline, linear or oblong, suberect or sometimes spreading, flat or sometimes crisped. Inflorescence terminal, many-flowered, often densely spicate; bracts lanceolate, shorter or longer than the flowers. Flowers small to medium-sized, subglobose. Dorsal sepal erect, narrow, concave. Lateral sepals usually united, narrowed at the base and connate into a concave limb. Petals obliquely falcate, curved and united with the margins of the dorsal sepal to form a hood, usually somewhat contracted in front, concave or obliquely saccate at the base. Lip ascending along the face of the column and adnate to it between the stigmatic lobes and rostellum arms, narrowed into a claw below, dilated into a reflexed transversely lunate or 2-lobed limb above, limb more rarely oblong or lanceolate; produced above the junction with the column into a large reflexed or erect variously-shaped fleshy appendage. Column short, dilated, produced in front into 2 horizontal arms, holding at their extreme ends the glands of the pollinia; anther cells more or less distant, placed in front of or sometimes behind the arms of the rostellum, erect or ascending, glands of the pollinia uppermost; stigma posticous, pulvinate, lunate or 2-lobed, or with 2 lateral and distant stigmas.

A genus of 15 species, the majority confined to South Africa, but with the range of one species extending to Tanzania and another species to Malawi.

Corycium dracomontanum Parkman & Schelpe in Contrib. Bolus Herb., No. 10: 158 (1982). —Stewart et al., Wild Orch. South. Africa: 195 (1982). TAB. **73**. Type from South Africa.

An erect glabrous herb, 15–54 cm tall, drying black. Leaves erect spreading, up to 17 × 1.5 cm, lanceolate, acuminate, those above tightly sheathing the stem. Inflorescence densely many-flowered, cylindrical, spicate; bracts 1–2 cm long,

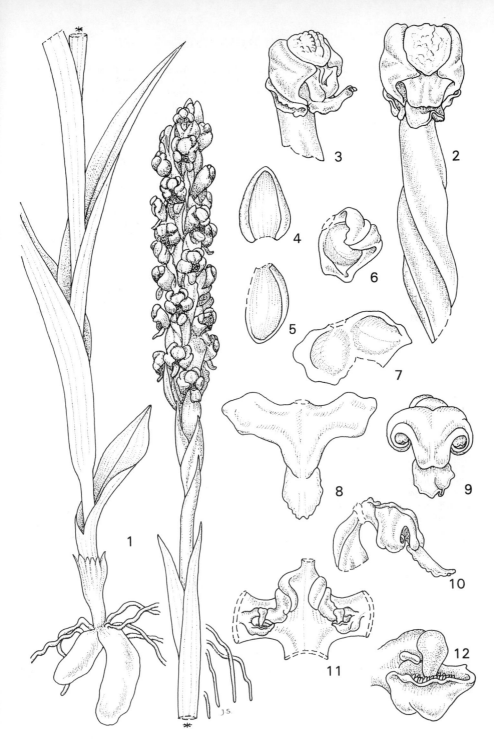

Tab. 73. CORYCIUM DRACOMONTANUM. 1, habit (×1); 2, flower, front view (×4); 3, flower, side view, ovary removed (×4); 4, dorsal sepal (×4); 5, lateral sepal (×4); 6, petal (×4); 7, petal, opened out (×4); 8, column and lip, opened out (×4); 9, column and lip, front view (×4); 10, column and lip, side view (×4); 11, column from below, with segments unfurled (×4); 12, close-up of pollinarium (×14), 1–12 from *H. Kurweil* 1492. Drawn by Judi Stone.

lanceolate, acuminate, as long as or longer than the flowers. Flowers c. 0.5 cm across, globose, green turning black later, lip and petals green tinged with purple, the lip-appendage bright green; ovary 1–1.2 cm long. Dorsal sepal concave, 0.4–0.5 cm long, orbicular, adnate to the petals to form a hood. Lateral sepals spreading, saccate, 0.4–0.5 cm long, connate for half their length. Petals concave, enclosing the lip appendage, 0.4 cm long, suborbicular, acute. Lip sharply reflexed, pointing downwards, apparently 3-lobed when flattened; mid-lobe 0.4–0.5 cm long, tapering, obtuse to acute, undulate or flat and crenate; the side lobes (lip-appendage) divaricate, deflexed, 0.5 cm tall, oblong or somewhat rounded, dome-shaped with a ridge above, bipartite. Column 0.9–1.2 mm long.

Malawi. S: Mt. Mulanje, Sombani Plateau, 2080 m, 21.ii.1991, *Kurzweil* 1492 (NBG).
Also in South Africa (Eastern Cape Province, Natal, Orange Free State and Transvaal), Transkei and Lesotho. Terrestrial in grassland, often near water.
This species is very closely allied to *C. nigrescens* Sond. which is widespread in South Africa and Lesotho, and has also been collected in southern Tanzania (Kitulo Plateau). The latter differs in having slightly smaller, predominantly purple-brown flowers that turn black with age, connate lateral sepals, and a distinct lip and lip-appendage. The distribution of *C. nigrescens* suggests that it may also occur in the Flora Zambesiaca region.

25. DISPERIS Sw.

Disperis Sw. in Kongl. Vetensk. Acad. Nya Handl. **21**: 218 (1800).
—Schlechter in Bull. Herb. Boissier **6**: 911 (1898).

Erect terrestrial epiphytic or lithophytic herb, usually small, with small tubers. Stem with 1–several cataphylls at the base; foliage leaves 1–few, opposite or alternate, sometimes absent. Flowers small, solitary or in racemes, white, yellow, green, pink, lilac or magenta-purple. Bracts leaf-like. Dorsal sepal adnate to the petals to form an open or elongate hood; lateral sepals sometimes partly joined, each with a sac-like spur near the inner margin (spur lacking in 1 or 2 species). Petals variously shaped, sometimes auriculate at the base. Lip greatly modified, its claw joined to the face of the column and ascending above it, variously curving into the spur if present, often dilated into a smooth or papillate limb and usually bearing a simple or 2-lobed appendage which varies greatly in shape from species to species. Column erect; rostellum large, 2-lobed, produced in front into 2 rigid cartilaginous arms with the viscidia at their apices. Anther loculi distinct, parallel; pollinia-granules secund, in double row on the edges of the flattened caudicles; staminodes present in some species. Stigma 2-lobed, the lobes on either side of the adnate claw of the lip. Capsule cylindrical or ovoid, ribbed, usually developing rapidly.

A genus of about 75–80 species in tropical and South Africa, Madagascar, the Mascarene Islands and India eastwards to New Guinea. Several species are very small and obscurely coloured and it seems likely that they are under-recorded.

1. Dorsal sepal and petals joined to form an open or boat-shaped hood, but not a distinct spur · 2
   – Dorsal sepal and petals joined to form a spur-like structure, stout or slender · · · · · · · 13
2. Leaves alternate · 3
   – Leaves opposite · 11
3. Stem and sometimes the ovary hairy; leaf and bract ciliate · · · · · · · · · · · · · · · · · · · 4
   – Plant glabrous; leaf and bract not ciliate · 5
4. Plant dwarf, 2–4 cm tall; ovary papillate-pubescent; lip with a ligulate claw and a reflexed, triangular lamina; spurs not pointing forwards, shortly conical, very much shorter than the sepals · 18. *pusilla*
   – Plants 7–15 cm tall; ovary glabrous; lip not as above; spurs of lateral sepals pointing forwards and almost as long as the sepals · 14. *macowanii*
5. Hood very concave, as broad as long · 6
   – Hood shallow, longer than broad · 7
6. Lateral sepals ± free; their spurs 2 mm long, curved · · · · · · · · · · · · · · · · 20. *thorncroftii*
   – Lateral sepals joined at base for a quarter to half of their length; their spurs very short, c. 0.5 mm long · 11. *katangensis*

7. Limb of lip folded, ovate-oblong; appendage broadly oblong, apiculate, keeled · · · · · · · ·
· 9. *concinna*
 − Limb of lip flat or fleshy, not folded; appendage 2-lobed · · · · · · · · · · · · · · · · · · · 8
8. Limb of lip fleshy, oblong-cuneate or hoof-shaped · · · · · · · · · · · · · · 2. *reichenbachiana*
 − Limb of lip flat, bearing a papillate crest or keels · 9
9. Petals half-cordate at the base; hood distinctly cordate at base · · · · · · · · · · 21. *togoensis*
 − Petals tapering or narrowly clawed at the base; hood rounded or narrowed at base · · · 10
10. Appendage lobes glabrous; flowers mauve · 16. *mozambicensis*
 − Appendage lobes papillose; flowers bicoloured, in the Flora Zambesiaca area usually yellow
and white but sometimes mauve and white · 3. *johnstonii*
11. Leaves with white or pink longitudinal venation on the upper surface; lip with a 2-lobed,
papillate appendage near the junction of the claw with the column · · · · · · 12. *leuconeura*
 − Leaves plain green on the upper surface, the venation not white or pink; lip with 2
appendage lobes, the lobes long and bifid · 12
12. Limb and appendages of lip papillose-hairy · 22. *virginalis*
 − Limb and appendages of lip glabrous · 10. *dicerochila*
13. Leaves alternate or single · 14
 − Leaves opposite or sub-opposite · 18
14. Leaves much reduced, less than 1 cm long · 15
 − Leaves 1–4.5 cm long · 16
15. Apex of lip appendage clavate and hairy · 17. *parvifolia*
 − Apex of lip appendage 2-lobed and glabrous · 8. *breviloba*
16. Plants epiphytic (in the Flora Zambesiaca area); lip with an erect, terete appendage where
the limb is bent just below the apex · 1. *kilimanjarica*
 − Plants terrestrial; lip not as above · 17
17. Leaf single; lateral sepals 8–9 mm long · 13. *lindleyana*
 − Leaves 2–3; lateral sepals c. 3 mm long · 15. *micrantha*
18. Spur relatively short and broad, or shortly cylindric and slightly incurved; lip appendage
deeply 2-lobed · 19
 − Spur narrowly cylindrical, 9–20 mm long; lip appendage shallowly 2-lobed · · · · · · · · 20
19. Lateral sepals 5–6 mm long, aristate at apex; sacs 2–2.5 mm long · · · · · · · · 19. *thomensis*
 − Lateral sepals 7–9 mm long, not aristate; sacs c. 1.5 mm long · · · · 6. *aphylla* subsp. *bifolia*
20. Spur 10–12 mm long; lip reaching apex of spur where it is recurved; appendage 9 mm long,
almost reaching mouth of spur · 4. *nemorosa*
 − Spur 12–20 mm long; lip not reaching end of spur; appendages very short not nearly
reaching mouth of spur · 21
21. Flowers bright rose-pink; apex of spur bifid; lip appendage 3-lobed, with mid-lobe entire
and side lobes lacinate · 7. *bifida*
 − Flowers white; apex of spur not bifid; lip appendage with 2 fimbriate lobes · 5. *anthoceros*

1. **Disperis kilimanjarica** Rendle in J. Linn. Soc., Bot. **30**: 400, t. 32/8–10 (1895). —Schlechter in
Bull. Herb. Boissier **6**: 937 (1898). —Verdcourt in F.T.E.A., Orchidaceae: 225 (1968). —
Williamson, Orch. S. Centr. Africa: 103 (1977). —Geerinck in Fl. Afr. Centr., Orchidaceae pt.
1: 237 (1984). —la Croix et al., Orch. Malawi: 157 (1991). Tab. **74**, fig. **1**. Type from Tanzania.

Epiphytic or terrestrial herb 5–12 cm tall; tuber 10–15 × 5–10 mm, ovoid. Leaves
2, alternate, 1.5–3.5 × 1.2 cm, ovate, cordate at base. Inflorescence single-flowered.
Flowers white tinged with green and sometimes with mauve. Ovary and pedicel
10–11 mm long; bracts leaf-like, 10–15 × 7 mm, ovate. Dorsal sepal 8–16 mm long,
joined to petals to form a blunt spur. Lateral sepals 7–12 × 3–5 mm, obliquely ovate,
apiculate, each with a sac-like spur 1–2 mm long. Petals 8 × 4 mm, oblong, the free
margin irregular. Lip 6 mm long, curved back into spur, the linear claw expanded
into a triangular blade reflexed at the tip, with an oblong appendage 3 mm long
arising at the point of reflexion, papillate at tip, almost reaching apex of the spur.

Zambia. E: Nyika Plateau, Chowo Forest, fl. iv.1976, *G. Williamson* 887 (K). **Malawi**. N: Nyika
National Park, Zovochipolo, 2210 m, fl. 8.iv.1984, *la Croix* 601 (K); South Viphya, Kawandama
Forest, 1750 m, fl. 15.v.1982, *Dowsett-Lemaire* 721 (K).
Also in Rwanda, Burundi, Uganda, Kenya and Tanzania. Always epiphytic in the Flora
Zambesiaca area, occurring in montane or submontane forest on mossy branches and trunks
of trees; 1750–2250 m.

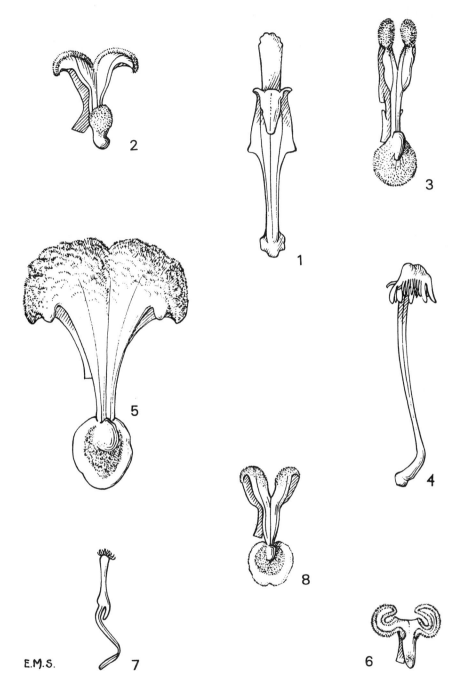

Tab. 74. DISPERIS. —lips of various species (all ×6). 1, D. KILIMANJARICA; 2, D. REICHENBACHIANA; 3, D. JOHNSTONII; 4, D. ANTHOCEROS; 5, D. KATANGENSIS var. KATANGENSIS; 6, D. LEUCONEURA; 7, D. PARVIFOLIA; 8, D. TOGOENSIS. Drawn by Margaret Stones. From F.T.E.A.

2. **Disperis reichenbachiana** Rchb.f. in Flora **48**: 180 (1865). —Rolfe in F.T.A. **7**: 290 (1898). —Verdcourt in F.T.E.A., Orchidaceae: 221 (1968). —Grosvenor in Excelsa **6**: 80 (1976). —Williamson, Orch. S. Centr. Africa: 103 (1977). —Geerinck in Fl. Afr. Centr., Orchidaceae pt. 1: 236 (1984). —la Croix et al., Orch. Malawi: 160 (1991). Tab. **74**, fig. **2**. Type from Angola.

Slender terrestrial herb 7–25 cm tall; tuber 5–20 × 5–10 mm, ellipsoid or globose. Leaves 2–4, alternate, dark green with white veins above, purple below, 1.7 × 1–3 cm, ovate or lanceolate-ovate, cordate at base. Immature plants have 2 opposite leaves. Inflorescence 1–4-flowered. Flowers mauve-pink, the petals paler mauve with a dark comma-shaped mark; the lip whitish, yellow towards apex. Ovary and pedicel 13–15 mm long; bracts leaf-like, 0.5–2 cm long. Dorsal sepal 7–8 × 1–2 mm, joined to petals to form shallow hood 6–8.5 mm wide. Lateral sepals 9–10.5 × 3.5–5.5 mm, obliquely elliptical, joined for c. 2 mm at base, each with shallow sac c. 1 mm deep. Petals 6–10 × 2–3.5 mm, obliquely elliptical. Lip 6.5–8 mm long, the claw bent back on itself and bearing there a 2-lobed, papillose appendage; the claw ending in a fleshy, papillose, oblong appendage joined by its middle to the end of the claw.

Zambia. N: Kasama Distr., Chishimba (Chisimba) Falls, fl. 13.ii.1961, *E.A. Robinson* 4371 (K). W: Solwezi, fl. 22.i.1963, *van Rensburg* 1253 (K). **Zimbabwe**. C: Goromonzi Distr., Binga Swamp Forest, fl. 8.ii.1976, *Grosvenor* 855 (K; SRGH). E: Chimanimani Distr., near Nyambaba Falls, 800 m, fl. 29.xii.1954, *Ball* 445 (K; SRGH). **Malawi**. N: Nkhata Bay Distr., Kaningina Forest Reserve, 1100 m, fl. 7.i.1987, *la Croix* 928 (K; MAL). C: Lilongwe Distr., Dzalanyama Forest Reserve, 1600 m, fl. 17.iii.1984, *Dowsett-Lemaire* 1113 (K).

Also in Central African Republic, Burundi, Kenya, Uganda, Tanzania and Angola. In shade of riverine or submontane forest, growing in leaf litter; 800–1600 m.

3. **Disperis johnstonii** Rolfe in F.T.A. **7**: 291 (1898). —Summerhayes in Hooker's Icon. Pl. **33**: t. 3269 (1935); in F.W.T.A. ed. 2, **3**: 203 (1968). —Verdcourt in F.T.E.A., Orchidaceae: 219 (1968). —Grosvenor in Excelsa **6**: 80 (1976). —Williamson, Orch. S. Centr. Africa: 103 (1977). —Stewart in Fl. Pl. S. Africa **46**: t. 1828 (1981). —la Croix et al., Orch. Malawi: 160 (1991). TAB. **74**, fig. 3; TAB. **75**. Type from Tanzania.
   Disperis stolzii Schltr. in Bot. Jahrb. Syst. **53**: 548 (1915); in Fedde, Repert. Spec. Nov. Regni Veg., Beih. 68: t. 45/177 (1932). Type from Tanzania.

Small terrestrial herb 4–15 cm tall; tubers 7–15 mm long, ovoid or globose, hairy. Leaves 2–3, alternate, green with white veins above, purple below, sessile, 2–3 × 1.5–2 cm, ovate, clasping stem at base. Inflorescence 1–5-flowered. Hood yellow with darker marks inside (hood pale purple in east African specimens); lateral sepals white or very pale mauve. Ovary and pedicel 10–12 mm long; bracts leaf-like, of similar length. Dorsal sepal 8–11 × c. 1 mm, joined to petals to form an open hood 9–10 mm wide. Lateral sepals 10–12 × 5.5–7 mm, joined at the base for about a third of their length, projecting forwards, obliquely semicircular, each with small sac c. 1 mm long. Petals c. 10 × 2–3 mm, narrowly elliptical, falcate. Lip 4–6.5 mm long, the claw bent back on itself near the base and at that point bearing a 2-lobed appendage with papillose lobes; claw ending in an ovate or almost round blade with a papillose protuberance in centre.

Zambia. E: Chipata Distr., between Katete and Naviruli, 1060 m, fl. 2.iii.1973, *Kornas* 3377 (K). **Zimbabwe**. N: Guruve Distr., c. 40 km E of Guruve (Sipolilo), fl. 15.iii.1960, *Drummond* 6796 (K; SRGH). **Malawi**. N: Mzimba Distr., 5 km W of Mzuzu at Katoto, 1350 m, fl. 21.vi.1975, *Pawek* 9795 (K; MAL; MO; SRGH; UC). C: Lilongwe Distr., Dzalanyama Forest Reserve, below Choulongwe (Chaulongwe) Falls, 1220–1280 m, fl. 28.iii.1970, *Brummitt* 9464 (K). S: Mangochi Distr., slopes of Uzuzu Hill, 1075 m, fl. 21.ii.1982, *Brummitt, Polhill & Patel* 16016 (K; MAL).

Also in Nigeria, Cameroon and Tanzania. *Brachystegia* woodland, forest patches, usually in shelter of rocks; 1050–1350 m.

4. **Disperis nemorosa** Rendle in J. Bot. **33**: 297 (1895). —Schlechter in Bull. Herb. Boissier **6**: 950 (1898). —Rolfe in F.T.A. **7**: 292 (1898). —Verdcourt in F.T.E.A., Orchidaceae: 227 (1968). —Williamson, Orch. S. Centr. Africa: 100 (1977). —la Croix et al., Orch. Malawi: 159 (1991). TAB. **76**. Type from Uganda.
   Disperis centrocorys Schltr. in Bot. Jahrb. Syst. **53**: 549 (1915); in Fedde, Repert. Spec. Nov. Regni Veg., Beih., 68: t. 43/170 (1932). Type from Tanzania.

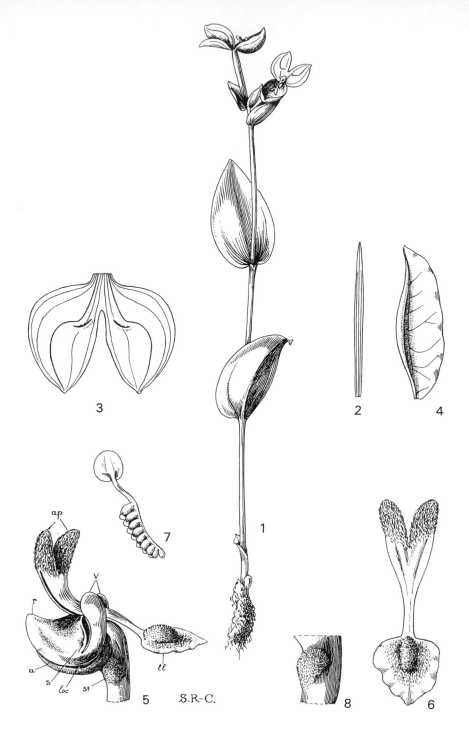

Tab. 75. DISPERIS JOHNSTONII. 1, habit (×1); 2, dorsal sepal (×4); 3, lateral sepals (×4); 4, petal (×4); 5, lip and column, lateral view (×12); 6, lip, front view (×12); 7, pollinium (×16); 8, staminode (×16); a, anther; ap, appendage of lip; ll, lamina of lip; loc, anther loculus; r, mid-lobe of rostellum; s, stigma; st, staminode; v, viscidia, 1–8 from the type *Johnston* s.n. Drawn by Stella Ross-Craig. From Hooker's Icon. Pl.

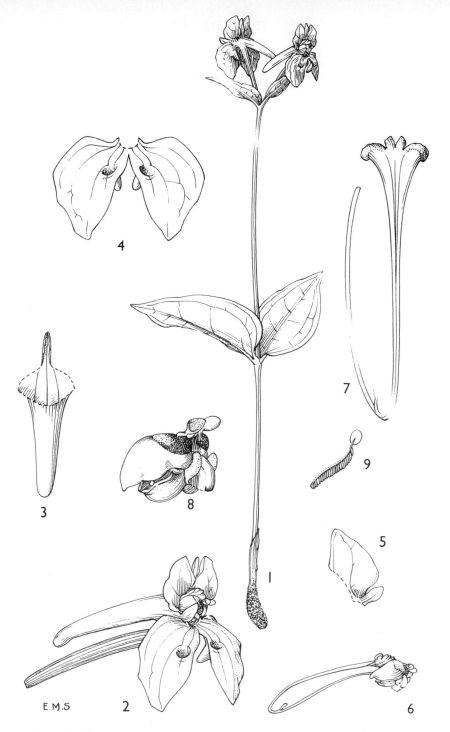

Tab. 76. DISPERIS NEMOROSA. 1, habit (×1); 2, flower (×3); 3, dorsal sepal (×3); 4, lateral sepals (×3); 5, petal (×3); 6, lip and column (×3); 7, lip (×6); 8, column, side view (×8), 1–8 from *Bally* 7874; 9, pollinarium (×8), from *Polhill & Paulo* 1630. Drawn by Margaret Stones. From F.T.E.A.

Terrestrial herb 10–30 cm tall; tubers 1–2.5 × 0.8–1 cm, ovoid or globose. Leaves 2, subopposite, glossy bright green, sessile or with a very short petiole, 2.5–6.5 × 2–4 cm, ovate, sheathing at base. Inflorescence 1–4-flowered. Flowers white or bright magenta-pink with lilac spur, sweetly violet-scented in the evening. Ovary and pedicel 15–20 mm long; bracts leaf-like, of similar length. Dorsal sepal joined with petals to form a slender spur c. 10 mm long from top of ovary to apex of spur, 12 mm long from apex of spur to front margin of dorsal sepal. Lateral sepals projecting forwards in same plane as the spur, 6–8 × 4–5 mm, obliquely oblong, obtuse, almost free, each with small sac 1–1.5 mm long. Petals adnate to dorsal sepal, auriculate at base. Lip with a slender claw 9–12 mm long, reaching the spur apex, then doubled back for 1.5–2 mm, with a long appendage arising at the point of reflexion and reaching the open end of the spur. Appendage linear-oblanceolate, growing wider towards bifid apex, with pair of obtuse lobes c. 1 mm long just before the apex.

Zambia. E: Nyika Plateau, Chowo Forest, c. 2000 m, fl. 8.iv.1984, *la Croix* 600 (K). **Malawi**. N: Nyika National Park, Kasyaula Forest, 2000 m, fl. 22.ii.1982, *Dowsett-Lemaire* 351 (spirit) (K).
Also in Uganda, Kenya and Tanzania. Deep shade in forest, often near streams; c. 2000 m.

5. **Disperis anthoceros** Rchb.f., Otia Bot. Hamburg. **2**: 103 (1881). —Schlechter in Bull. Herb. Boissier **6**: 951 (1898). —Rolfe in F.T.A. **7**: 292 (1898); in F.C. **5**, 3: 311 (1913). —Pole-Evans in Fl. Pl. S. Afr., t. 327 (1929). —Summerhayes in F.W.T.A. ed. 2, **3**: 205 (1968). —Verdcourt in F.T.E.A., Orchidaceae: 229 (1968). —Morris, Epiphyt. Orch. Malawi: 32 (1970). —Williamson, Orch. S. Centr. Africa: 100, t. 87 (1977). —Stewart, Wild. Orch. S. Afr.: 198 (1982). —Geerinck, Fl. Afr. Centr., Orchidaceae pt. 1: 230 (1984). —la Croix et al., Orch. Malawi: 153 (1991). Tab. **74**, fig. 4. Type from Ethiopia.
Disperis hamadryas Schltr. in Ann. Transvaal Mus. **10**: 252 (1924). Type from South Africa.

Terrestrial herb 6–30 cm tall; tubers 1–1.5 × 1 cm, ovoid or globose, woolly. Leaves 2, opposite, almost sessile, mid to dark green, sometimes tinged with purple below, 2.5–5.5 × 2.5–3.7 cm, ovate, acute, sheathing at base. Inflorescence 1–3(7)-flowered; flowers white with green or purplish marks, the hood green-veined inside. Ovary and pedicel 13–17 mm long; bracts 9–17 mm long, leaf-like. Dorsal sepal joined to petals to form a slender spur 12–20 mm long; lateral sepals 7–10 × 3–5 mm, obliquely obovate, joined for about half their length, each with a conical, sac-like spur 1–1.5 mm long. Petals falcate, auriculate at the base. Lip not reaching the spur apex, doubling back for 1–1.5 mm; the appendage consisting of 2 fimbriate lobes c. 1.5 mm long; lip claw 8–15 mm long.

Zambia. W: Kitwe, fl. 2.ii.1970, *Fanshawe* 10744 (K; NDO). C: c. 11 km E of Lusaka, 1270 m, fl. 10.ii.1958, *King* 419 (K). E: Nyika National Park, Manyenjere Forest, 2050 m, fl. 3.ii.1982, *Dowsett-Lemaire* 343a (K). S: Livingstone Distr., Victoria Falls, fl. iii.1969, *G. Williamson* 1531 (K; SRGH). **Zimbabwe**. N: Makonde Distr., Umboe, Mhangura (Mangula), fl. 14.ii.1968, *Wild* 7690 (K; SRGH). C: Harare Distr., Calgary Farm, fl. 16.ii.1977, *Grosvenor & Renz* 893 (K; SRGH). E: Mutare Distr., edge of Banti Forest, fl. 21.iii.1981, *Philcox, Leppard, Duri & Urayai* 9038 (K). **Malawi**. N: Nkhata Bay Distr., Choma Livestock Centre near Mzuzu, 1300 m, fl. 9.ii.1986, *la Croix* 799 (K; MAL). C: Ntchisi Forest Reserve, 1600 m, fl. 20.iii.1984, *Dowsett-Lemaire* 1125 (K). S: Zomba Distr., Mpita Estate, 1150 m, fl. 3.iii.1984, *la Croix* 558 (K).
Also in Nigeria, Zaire, Rwanda, Burundi, Ethiopia, Sudan, Uganda, Kenya, Tanzania and South Africa (Transvaal and Natal). In leaf litter in shade of evergreen forest, sometimes under *Pinus* or *Cupressus* in plantations; 1100–2300 m.

6. **Disperis aphylla** Kraenzl. in Bull. Soc. Roy. Bot. Belgique **38**: 71 (1900). —Verdcourt in F.T.E.A., Orchidaceae: 226 (1968). —Williamson, Orch. S. Centr. Africa: 102, t. 89 (1977). —Geerinck in Fl. Afr. Centr., Orchidaceae pt. 1: 233 (1984). —la Croix et al., Orch. Malawi: 154 (1991). Type from Zaire.

Terrestrial herb 6–16 cm tall, the stem greenish-white with purple streaks; tubers 10–20 × 5–10 mm, ovoid or globose. Leaves 2 (1 or absent in subsp. *aphylla*), opposite or subopposite, purple on the undersurface, 2–3 × 1.5–2.5 cm, ovate; margins undulate. Inflorescence 1–5-flowered; flowers white or pale lilac-pink, the sepals darker towards the base. Ovary and pedicel purplish, 10–14 mm long; bracts leaf-like, 9–13 × 3–5 mm. Dorsal sepal 9–10.5 mm long from apex to top of spur and 5–6 mm

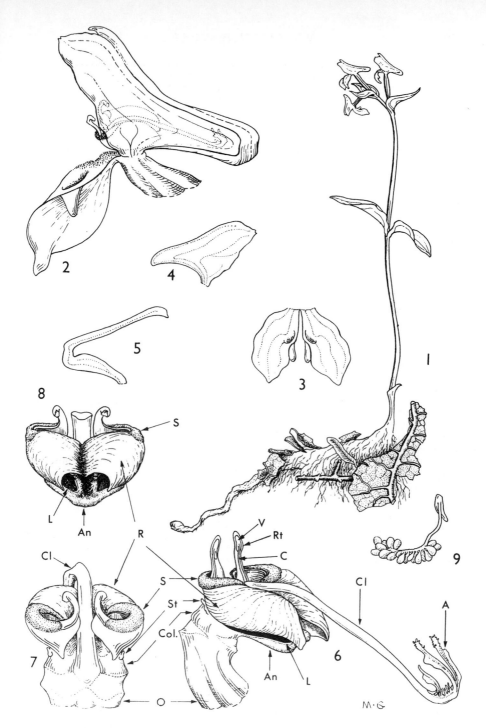

Tab. 77. DISPERIS APHYLLA subsp. BIFOLIA. 1, habit (×1), from *Verdcourt* 213; 2, flower, side view (×6); 3, lateral sepals (×4); 4, petal (×4); 5, dorsal sepal, side view (×4); 6, lip and column, three-quarter rear view (An, anther; R, rostellum; Rt, tip of rostellum; S, stigma; St, staminode; Col, column; O, ovary; L, locules; Cl, free part of claw of lip; A, appendage; V, viscidium; C, caudicle) (×12); 7, lip and column, front view (×12); 8, lip and column, rear view, claw cut off (×12); 9, pollinium (×14), 2–9 from *Verdcourt* 3028. Drawn by Mary Grierson.

long from top of spur to apex of ovary, adnate to petals to form a fairly slender spur 3–4 mm long. Lateral sepals free, 7–9 × 2–3 mm, obliquely ovate, slightly curved, each with a sac-like spur c. 1.5 mm long. Petals c. 7 × 3 mm, triangular. Lip with a linear claw 2–4.5 mm long, reflexed near the apex; limb short and narrow and just before the apex bearing 2 linear, 2-lobed, papillate appendages c. 2 mm long.

Subsp. **bifolia** Verdc. in Kew Bull. **22**: 96 (1968); in F.T.E.A., Orchidaceae: 227 (1968). —la Croix et al., Orch. Malawi: 154 (1991). TAB. **77**. Type from Tanzania.

Leaves 2, opposite or subopposite. Lip with a linear claw 4.5 mm long.

Zambia. E: Nyika Plateau, Chowo Forest, fl. iv.1968, *G. Williamson* 888 (K). **Malawi**. N: Nyika National Park, Kasyaula Forest, 2000 m, fl. 22.ii.1982, *Dowsett-Lemaire* 344 (K).

Also in Tanzania. In leaf litter in deep shade of evergreen forest; c. 2000 m.

Plants from the Nyika Plateau (Malawi and Zambia) have larger leaves and longer lateral sepals than the Tanzanian specimens.

Subsp. *aphylla* occurs in Cameroon, Zaire, Uganda, Kenya and Angola, and may be distinguished by its leafless stems and by a lip claw 2 mm long.

7. **Disperis bifida** P.J. Cribb in Kew Bull. **40**, 2: 403 (1985). —la Croix et al., Orch. Malawi: 155 (1991). TAB. **78**. Type: Zambia, Nyika Plateau, between Rest House and main road junction, *Dowsett-Lemaire* 347 (K, holotype).

Terrestrial herb 15–30 cm tall. Leaves 2, opposite, dark green, 2.5–4.5 × 1.5–3.5 cm, ovate, acute, cordate at the base. Inflorescence 1–4-flowered; flowers bright rose-pink with 2 rows of carmine spots inside the hood, spur and underside of lateral sepals paler. Ovary and pedicel 20 mm long; bracts leaf-like, 10 × 7 mm. Dorsal sepal together with the petals forming a hood, extending into a slender spur c. 1 mm in diameter, bifid at the apex, 14–17 mm long from apex of dorsal sepal to tip of spur, 7–9 mm long from tip of spur to apex of ovary. Lateral sepals deflexed or projecting, 10 × 5 mm, obliquely semi-orbicular, each with a short spur c. 1 mm long about halfway along their length. Petals 8–10 × 5 mm, lobed and erose on the front margin. Lip c. 11 mm long, linear, extending into the hood but not reaching the top, recurved before the apex; appendage 3-lobed, the mid-lobe short and entire, the side lobes lacinate.

Zambia. E: Nyika Plateau, between Rest House and main road junction, 2180 m, fl. 23.ii.1982, *Dowsett-Lemaire* 347 (K).

Endemic. In leaf litter in deep shade of evergreen forest; 2180 m.

This species is remarkable for its limited distribution; so far it is known from only one small forest patch on the Zambian side of the Nyika Plateau. There, it apparently hybridizes with *D. dicerochila* as plants occur which are intermediate between the two.

D. bifida × D. dicerochila

Spurs short, c. 2 mm, conical; lips 6–11 mm long, linear, bent over in the apical 2–3 mm, and bearing yellow and white fimbriae.

Zambia. E: Nyika Plateau, in forest patch, 2180 m, fl. 15.ii.1987, *la Croix* 973 (K; MAL).

8. **Disperis breviloba** Verdc. in Kew Bull. **32**: 9 (1977). —la Croix et al., Orch. Malawi: 155 (1991). TAB. **79**. Type: Zambia, Kundalila Falls, *Williamson & Gassner* 2344 (K, holotype). *Disperis parvifolia* Schltr., sensu Williamson, Orch. S. Centr. Africa: 101, t. 88 (1977).

Terrestrial herb 8–16 cm tall, the entire plant usually purple-red or pink, sometimes greenish; tuber 7 mm long, ellipsoid. Stem erect, purple-brown at the apex, purplish at the base, glabrous or slightly papillate, leafless apart from 2–3 sheath-like basal cataphylls 5–12 mm long. Inflorescence 1–2(3)-flowered. Spur green-brown at the apex, lilac at the base; sepals and petals pinkish, the sepals red inside. Pedicel 3–4 mm long; ovary 6 mm long, green-brown, slightly papillose; bracts 4.5–8 × 3–4 mm, green. Dorsal sepal joined to the petals to form an obtuse, conical spur 7 mm long and 3.7 mm wide at the base. Lateral sepals free, 3.5–4 × 1.5–2.2 mm, irregularly oblong with

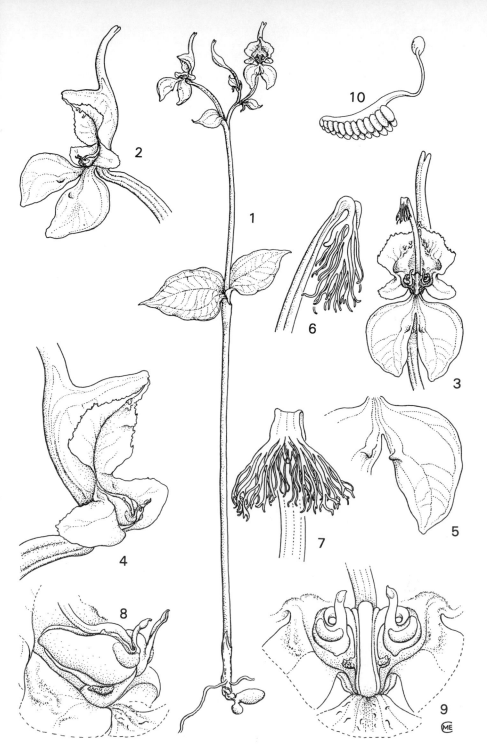

Tab. 78. DISPERIS BIFIDA. 1, habit (×⅔); 2, flower, side view (×2); 3, flower, front view, with lip removed from hood (×2); 4, flower, side view (×4); 5, lateral sepals (×3); 6 & 7, lip apex (×8); 8 & 9, column, side and front views (×8); 10, pollinium (×8); 1–10 from *Dowsett-Lemaire* 347. Drawn by M.E. Church. From Kew Bull.

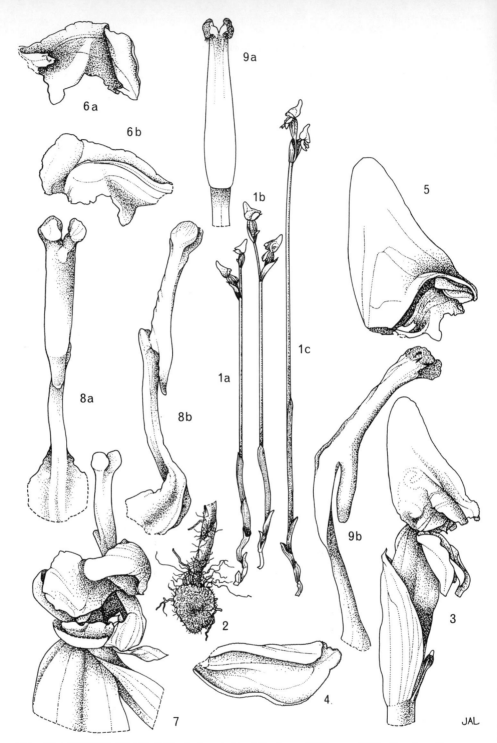

Tab. 79. DISPERIS BREVILOBA. 1a,b,c, habit (×⅔), from *Williamson* 2344; 2, tuber, from *Williamson* 362; 3, flower, side view (×4.8); 4, lateral sepal (×10); 5, spur formed of dorsal sepal and petals (×6.6); 6a,b, petals (×10); 7, column and lip (×10); 8a, lip, three-quarter side view (×15); 8b, lip, lateral view (×15), 3–8 from *Williamson* 2344; 9a, lip of variant, front view (×16); 9b, lip of variant, lateral view (×18), from *Williamson* 362. Drawn by J.A. Langhorne. From Kew Bull.

the outer edge rounded, lacking spurs. Petals 3–3.5 × 1.5–2 mm, irregularly oblong. Lip white, claw 2.5 × 0.2–0.4 mm, lip apex produced for 0.4 mm beyond the point of attachment of the appendage; appendage 5 × 0.4 mm, linear, translucent, shortly 2-lobed at the apex with lobes 0.3–0.6 mm long, rugulose. Anther 1.7 mm long; rostellum mid-lobe 1.5–2 × 1.5–2 mm, side arms 2 × 1.5 mm, incurved.

Zambia. B: above east bank of Kabompo River, 1200 m, fl. xii.1969, *Williamson & Simon* 1767 (SRGH). C: Kundalila Falls, fl. 20.xii.1974, *Williamson & Gassner* 2344 (K). **Malawi**. N: Nyika Plateau, c. 10 km from Chelinda, 2340 m, fl. ii.1968, *Williamson, Ball & Simon* 362 (K; SRGH).
Not known elsewhere. *Brachystegia* woodland, open dambo and montane grassland, usually in shallow soil over rocks; 1200–2340 m.

9. **Disperis concinna** Schltr. in Bot. Jahrb. Syst. **20**, Beibl. 50: 43 (1895). —Rolfe in F.C. **5**: 299 (1913). —Stewart et al., Wild Orch. S. Afr.: 202 (1982). Type from South Africa.
   Disperis sp. no. 1, Grosvenor in Excelsa **6**: 80 (1976).

Slender terrestrial herb 10–25 cm tall. Leaves 1–4, alternate, erect, sessile, up to 2 cm long, elliptic or lanceolate. Inflorescence 1–4-flowered; flowers greenish-white with purple or rose-pink veins. Ovary 10 mm long; bracts 13 × 6 mm, leaf-like. Dorsal sepal 8 × 2.5–3 mm, forming a sac-like hood with the petals; hood 5 mm wide across the mouth. Lateral sepals spreading, 5–7 × 1.5–2 mm, obliquely oblanceolate, apiculate, each with a conical, sac-like spur c. 1.5 mm long. Petals c. 8 × 2.5 mm, obliquely falcate-lanceolate. Lip claw narrowly ovate-oblong, obtuse; limb conduplicate, c. 2 mm long; appendage longer than limb and broadly oblong, apiculate, keeled. Rostellum large, ovate, acute; side arms 3 mm long, oblong.

Zimbabwe. E: Nyanga (Inyanga), 1800 m, fl. ii.1957, *Payne* 77 (K; SRGH).
Also in South Africa (Transvaal and Natal). Moist submontane grassland; c. 1800 m.

10. **Disperis dicerochila** Summerh. in Hooker's Icon. Pl. **33**: t. 3272 (1935). —Verdcourt in F.T.E.A., Orchidaceae: 223 (1968). —Morris, Epiphyt. Orch. Malawi: 33 (1970). —Moriarty, Wild Fls. Malawi: t. 27, 3 (1975). —Grosvenor in Excelsa **6**: 80 (1976). —Williamson, Orch. S. Centr. Africa: 103 (1977). —Geerinck in Fl. Afr. Centr., Orchidaceae pt. 1: 232 (1984). —la Croix et al., Orch. Malawi: 156 (1991). TAB. **80**. Type from Uganda.

Terrestrial herb 7–24 cm tall; stem purplish; tubers 1–3 cm long, ellipsoid or globose, densely hairy. Leaves 2, opposite, 1.5–5.5 × 1–3.5 cm, ovate, acute, usually cordate at base, dark green, sometimes purplish beneath. Inflorescence 1–3-flowered; flowers pale to deep mauve-pink, occasionally almost white, with purple marks inside hood. Ovary and pedicel c. 12 mm long; bracts leaf-like, of similar length. Dorsal sepal 7–11 mm long, linear-lanceolate, joined to petals to form an open hood. Lateral sepals ± free, 7–10 × 4–8 mm, obliquely ovate, acute, each with a sac-like spur 1.5–2 mm long. Petals 7–10 × 3.5–4 mm, lobed on free margin. Lip 7–9 mm long; claw linear; limb short, reflexed at the apex with 2 appendages at the point of reflexion; appendages 2-lobed, the upper lobe long with a papillate apex, the lower lobe short and recurved.

Zambia. E: Nyika Plateau, forest patch near Zambian Rest-house, 2200 m, fl. 15.ii.1987, *la Croix* 974 (K; MAL). **Zimbabwe**. E: Nyanga (Inyanga), Romneydale, 1800 m, fl. 20.iii.1949, *Chase* 4054 (K; SRGH). **Malawi**. N: South Viphya, Chikangawa, 1750 m, fl. 7.iii.1987, *Cornelius* in *la Croix* 1006 (K; MAL). C: Dedza Mt., just below peak, 2150 m, fl. 23.ii.1985, *la Croix & Jenkins* 672 (K). S: Mulanje Mt., Chambe Plateau, in pine plantation, fl. 11.ii.1979, *Blackmore, Brummitt & Banda* 326 (K; MAL).
Also in Ethiopia, Zaire, Rwanda, Burundi, Uganda, Kenya and Tanzania. Forming colonies in leaf litter or on mossy logs in evergreen forest, sometimes in pine and cypress plantations; 1700–2300 m.
Dowsett-Lemaire 345 from the Nyika Plateau in Malawi is anomalous in having alternate leaves, but the flowers are correct for *D. dicerochila*.

11. **Disperis katangensis** Summerh. in Bull. Misc. Inform., Kew **1931**: 384 (1931); in Hooker's Icon. Pl. **33**: t. 3271 (1935). —Verdcourt in F.T.E.A., Orchidaceae: 219 (1968). —Williamson, Orch. S. Centr. Africa: 103, t. 91 (1977). —Geerinck in Fl. Afr. Centr., Orchidaceae pt. 1: 234 (1984). TAB. **81**. Type from Zaire.

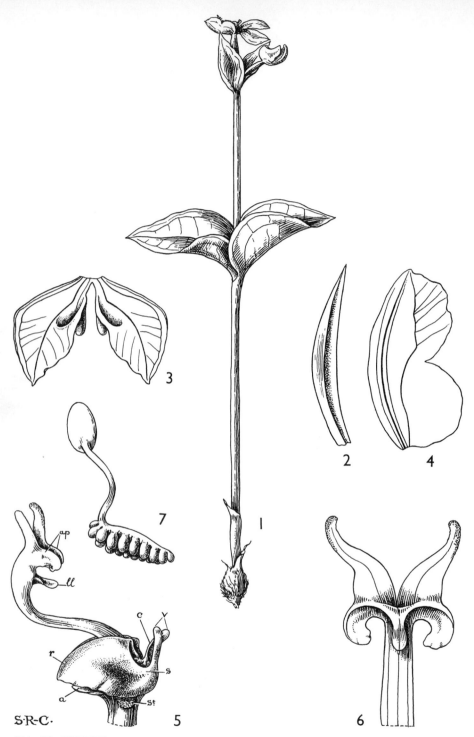

S·R·C·

Tab. 80. DISPERIS DICEROCHILA. 1, habit (×1), from *Napier* 635; 2, dorsal sepal (×6); 3, lateral sepals (×4); 4, petal (×6); 5, column and lip, side view (a, anther; ap, appendages of lip; c, caudicle; ll, apex of lip; r, mid-lobe of rostellum; s, stigma; st, staminode; v, viscidia) (×8); 6, lip-apex, opened out (×12); 7, pollinarium (×16), 2–7 from *Eggeling* 1382. Drawn by Stella Ross-Craig. From F.T.E.A.

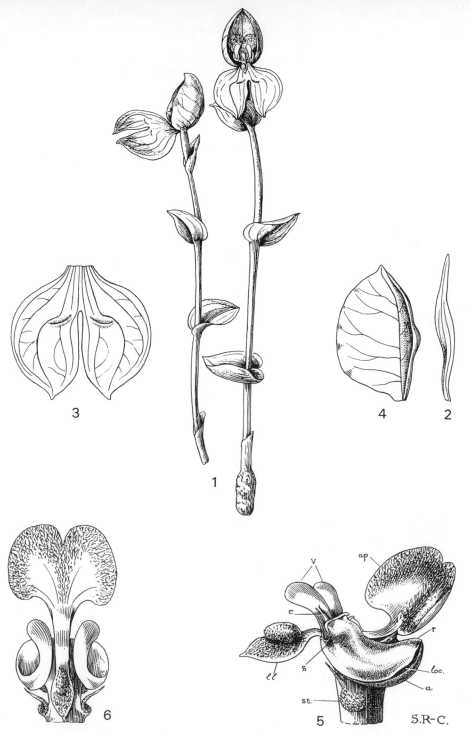

Tab. 81. DISPERIS KATANGENSIS. 1, flowering plants (×1); 2, dorsal sepal (×2); 3, lateral sepals (×2); 4, petal (×2); 5, lip and column, lateral view (×8); 6, lip and column, front view (×6); a, anther; ap, appendage of lip; c, caudicle; ll, lamina of lip; loc, anther loculus; r, mid-lobe of rostellum; s, stigma; st, staminode; v, viscidia, 1–6 from the type *von Hirschberg* 168. Drawn by Stella Ross-Craig. From Hooker's Icon. Pl.

Terrestrial herb 5–20 cm tall; stem greenish above, beetroot-coloured towards the base; tuber c. 1 cm long, globose. Leaves 2, alternate, 1–3 × 0.8–2 cm, ovate, cordate at the base, deep dull green with whitish veins on upper surface, deep purple below. Inflorescence 1–5-flowered. Flowers pale mauve with yellow and purple marks inside hood; lip yellow with white claw and appendages. Ovary 10–13 mm long; bracts leaf-like, 6–20 mm long. Dorsal sepal 7–17 mm long, linear, adnate to petals to form an open hood 7–15 mm long, 5.5–11 mm deep and 8–18 mm wide across mouth. Lateral sepals 9–15 × 6–11 mm, obliquely elliptic, apiculate, joined for quarter to one third of their length, each with a sac-like spur up to 0.5 mm long. Petals 7–17 × 7–11 mm, elliptic. Lip 7.5–10 mm long, the claw sharply bent back near the base and then dilated into an appendage; appendage densely papillate, reniform and emarginate or deeply 2-lobed; the claw ending in a spathulate, elliptic or almost circular limb with a papillate crest.

Lateral sepals more than 12 mm long; hood 13–14 mm long, c. 15 mm wide across mouth; lip appendage broadly reniform · var. *katangensis*
Lateral sepals 9–11 mm long; hood 7–8 mm long, 8–10 mm wide across mouth; lip appendage deeply 2-lobed · var. *minor*

Var. **katangensis** Tab. **74**, fig. **5**.

Zambia. N: Mbala Distr., Kawimbe Rocks, Old Sumbawanga Road, 1740 m, fl. 31.xii.1962, *Richards* 17086 (K). W: 4 km SW of Ndola, in forest reserve, 1200 m, fl. 13.i.1953, *Draper* 23 (K). Also in Zaire, Tanzania and Angola. *Brachystegia* woodland; 1500–1800 m.

Var. **minor** Verdc. in Kew Bull. **22**: 93 (1968). Type: Zambia, Mwinilunga Distr., *Milne-Redhead* 3540 (K, holotype; BR; SRGH).

Differs from the typical variety in having smaller flowers and smaller leaves (5–12 × 5–11 mm) and a deeply 2-lobed lip appendage.

Zambia. W: Mwinilunga Distr., NE of Dobeka Bridge, fl. 7.xii.1937, *Milne-Redhead* 3540 (K; BR; SRGH).
Endemic. *Cryptosepalum* woodland on sand; c. 1400 m.

12 **Disperis leuconeura** Schltr. in Bot. Jahrb. Syst. **53**: 549 (1915); in Fedde, Repert. Spec. Nov. Regni Veg., Beih. 68: t. 44/173 (1932). —Verdcourt in F.T.E.A., Orchidaceae: 222 (1968). —Morris, Epiphyt. Orch. Malawi: 34 (1970). —Grosvenor in Excelsa **6**: 80 (1976). — Williamson, Orch. S. Centr. Africa: 104 (1977). —la Croix et al., Orch. Malawi: 157 (1991). Tab. **74**, fig. **6**. Type from Tanzania.

Slender terrestrial herb 4–16 cm tall; tubers c. 1 × 1 cm, ± globose, hairy. Leaves 2, opposite, sheathing at the base, 2–3.5 × 1.4–1.8 cm, ovate, upper surface dark green with silvery-pink veins, lower surface maroon-red. Inflorescence 1–3-flowered; flowers mauve-pink or purple-pink. Ovary purplish, 9–14 mm long. Bracts leaf-like, 6–7 × 4 mm. Dorsal sepal c. 6 mm long, linear-lanceolate, joined with petals to form an open hood c. 5 mm tall, 6 mm wide across the mouth. Lateral sepals free, spreading, 4–5 × 2–4 mm, obliquely ovate, apiculate, the apex sometimes curved, each with small sac 1–1.5 mm long. Petals 5.5 mm long, the free margin undulate. Lip 4 mm long; claw bent back near base, appendage papillose with 2 upward-curved lobes; limb linear, glabrous. Rostellum arms c. 2.5 mm long, curved, thin.

Zambia. S: Victoria Falls, fl. iii.1969, *G. Williamson* 419 (K). **Zimbabwe**. E: Chimanimani, Haroni-Makuripini, high evergreen forest, 450 m, fl. 13.i.1969, *Biegel* 2824 (K; SRGH). **Malawi**. S: Mulanje Distr., Esperanza Tea Estate, 750 m, fl. 21.xii.1985, *Johnston-Stewart* 487 (K). Also in Tanzania. Evergreen and riverine forest, deep shade in leaf litter; 450–800 m.

13. **Disperis lindleyana** Rchb.f. in Flora **48**: 181 (1865). —Grosvenor in Excelsa **6**: 80 (1976). — Stewart et al., Wild Orch. South. Africa: 205 (1982). Type from South Africa.

Terrestrial herb 13–30 cm tall; tubers c. 15 × 10 mm, ovoid, woolly. Leaf 1, about halfway up the stem, 1.7–7 × 1.5–4 cm, ovate, cordate at the base. Inflorescence 1–5-flowered. Flowers white tinged with green, with violet markings, aging to creamy-yellow. Ovary 20 mm long; bracts leaf-like, 10–15 mm long. Dorsal sepal 15 mm long, joined to petals to form a conical, sac-like hood 8–10 mm long from top of ovary to apex of hood. Lateral sepals spreading, 8–10 × 5–8 mm, obliquely ovate, acute, each with a sac-like spur 2 mm long. Petals 8–9 mm long, falcate-oblong, the outer margin undulate. Lip broadly unguiculate, reflexed; limb 8 mm long, broadly linear, obtuse, bearing an appendage at the point of reflexion; appendage obovate-oblong, fleshy and papillose at the apex, shorter than the limb.

Zimbabwe. E: Mutare Distr., Himalaya Range, c. 2000 m, fl. 17.xii.1955, *Ball* 553 (K; SRGH); Nyanga (Inyanga), Rhodes Estate, fl. 23.xii.1962, *West* 4375 (K; SRGH).
Also in South Africa and Swaziland. Evergreen forest, in forest-floor leaf litter, also in pine plantations; 1650–2000 m.

14. **Disperis macowanii** Bolus in J. Linn. Soc., Bot. **22**: 77 (1885). —Verdcourt in Kew Bull. **30**, 4: 603 (1976). —Stewart et al., Wild Orch. South. Africa: 198 (1982). —la Croix et al., Orch. Malawi: 158 (1991). Type from South Africa.

Small terrestrial herb 7–15 cm tall, the stem sparsely hairy. Leaves 2, alternate, 14–16 × 6–10 mm, ovate, cordate. Inflorescence 1–2-flowered. Flowers white to cream with green and purple spotting on the petals. Ovary 8–10 mm long, bright green; bracts leaf-like, of similar length. Dorsal sepal joined to petals to form blunt, conical hood 5–6 mm high. Lateral sepals free, projecting forwards or deflexed, 4–5 × 2 mm, obliquely ovate, acute; spur 1.5–2 mm long, conical, pointing forwards and almost reaching end of the sepal. Petals 6 mm long, ± falcate, obscurely lobed towards base on outer margin. Lip with claw 2.5 mm long widening into an elliptic dilation 1 mm wide before narrowing into an apical part 1 mm long; appendage 5–5.5 × 1 mm, linear, slightly curved, obtusely folded longitudinally, minutely crenulate on margins, the apex papillate and slightly 2-lobed with a tooth on each side and thus appearing 4-lobed.

Malawi. N: Nyika National Park, road to Chikomanankazi, 2225 m, fl. 5.iii.1982, *Dowsett-Lemaire* 350 (K).
Also in Tanzania, South Africa and Swaziland. In the Flora Zambesiaca area, known only from the above collection. Montane grassland, amongst rocks and under shrubs; c. 2000 m.

15. **Disperis micrantha** Lindl., Gen. Sp. Orchid. Pl.: 370 (1839). —Rolfe in F.C. **5**, 3: 308 (1913). —Bolus, Icon. Orchid. Austro-Afric. **3**: t. 96 (1913). —Pole Evans in Fl. Pl. S. Afr. **9**: plate 330 (1929). —Grosvenor in Excelsa **6**: 80 (1976). —Stewart et al., Wild Orch. South. Africa: 207 (1982). Type from South Africa.

Slender terrestrial herb 9–21 cm tall. Leaves 2–3, alternate, 2–4.5 × 1.5–3 cm, ovate, clasping stem at base. Inflorescence 1–6-flowered, usually 2-flowered. Flowers white or greenish-white, sometimes veined and tipped with wine-red. Ovary 8–10 mm long; bracts leaf-like, 13–15 × 5–7 mm. Dorsal sepal 3–6 mm long, forming a saccate hood with the petals. Lateral sepals spreading, 3–4 × 2 mm, obliquely ovate-oblong, acuminate, producing a short, broadly conical sac-like spur above the middle. Petals 3–6 mm long, obliquely falcate-ovate, acute. Lip narrowly unguiculate; limb reflexed with 2 short, diverging lobes; appendage broadly oblong, obtuse, papillose, longer than the limb. Column short; rostellum broadly rounded, obtuse, concave; arms 1 mm long, cartilaginous, twisted.

Zimbabwe. E: Chimanimani Mts., near Bridal Veil Falls, c. 1350 m, fl. 10.iii.1955, *Ball* 519 (K; SRGH).
Also in South Africa and Swaziland. Riverine forest; c. 1350 m.

16. **Disperis mozambicensis** Schltr. in Bot. Jahrb. Syst. **24**: 428 (1897). —Rolfe in F.T.A. **7**: 575 (1898). Type: Mozambique, Beira, Pungwe R, April 1895, *Schlechter* s.n. (B†, tracing of flower dissection at K).

Dwarf terrestrial herb 7–13 cm tall. Leaves 3, alternate, the lowermost 1–2 × 0.7–1 cm, oblong or ovate-lanceolate, acute or subacute. Inflorescence few-flowered; flowers pale rose-pink. Ovary 9 mm long; bracts leaf-like, shorter than the ovary. Dorsal sepal reflexed, ± horizontal, narrowly linear, joined to petals to form an open hood 12–14 × 10 mm. Lateral sepals 10 × 6 mm, sub-falcate, oblique, narrow at base, joined for quarter of their length, with very short, obtuse sac-like spurs. Petals 13 × 5 mm, concave, lanceolate-falcate, sub-obtuse, the outer margin slightly wavy. Lip erect; claw 3 mm long, narrowly linear, suddenly bent back and at point of reflexion bearing a bifurcate appendage; the rest of the claw narrowly linear, ending in a suborbicular limb 2 mm in diameter, densely papillose in the middle of the upper surface and with a fleshy, longitudinal keel 1 mm high.

Mozambique. MS: Beira, R. Pungwe, fl. iv.1895, *Schlechter* s.n. (tracing of flower dissection at K). Endemic. Amongst bushes on bank of river.

In his description, Schlechter says that this species is allied to *D. reichenbachiana*, but the lip appendage resembles that of *D. johnstonii*.

17. **Disperis parvifolia** Schltr. in Bot. Jahrb. Syst. **53**: 547 (1915); in Fedde, Repert. Spec. Nov. Regni Veg., Beih. 68: t. 44/174 (1932). —Summerhayes in F.W.T.A. ed. 2, **3**: 203 (1968). — Verdcourt in F.T.E.A., Orchidaceae: 226 (1968). —la Croix et al., Orch. Malawi: 159 (1991). Tab. **74**, fig. **7**. Type from Tanzania.

Very slender terrestrial herb 4–11 cm tall; tubers less than 1 cm in diameter, ± globose, hairy. Leaves 1–2 near base of plant, c. 4 × 4 mm, more or less sheathing, dark green. Inflorescence 1-flowered. Spur green with magenta apex, lateral sepals magenta-purple. Ovary 10 × 3.5 mm, bright green, most conspicuous; bract 6–8 mm long, tightly enclosing lower part of the ovary. Dorsal sepal forming a conical spur 6–6.5 mm long. Lateral sepals free, 2.5–4.5 × 2 mm, obliquely elliptical, concave, with very small, obscure sacs. Petals adnate to the dorsal sepal at mouth of spur, 3.5–4 × 2 mm, obliquely oblong, somewhat lobed on free margin. Lip 4–4.5 mm long including the appendage; limb reflexed, linear, enlarged and hairy at the apex; appendage arising from the limb at the point of reflexion, 1.5 mm long, linear.

Malawi. N: Nyika National Park, Mbuzinandi road, c. 1675 m, fl. 9.i.1983, *la Croix* 413 (K). Also in Cameroon and Tanzania. Sub-montane grassland; c. 1675 m.

In a specimen observed in Malawi, seed was being shed from the ovary while the plant was still in flower.

18. **Disperis pusilla** Verdc. in Kew Bull. **22**: 94, fig. 1 (1968); in F.T.E.A., Orchidaceae: 223 (1968). —Williamson, Orch. S. Centr. Africa: 105 (1977). —Geerinck in Fl. Afr. Centr., Orchidaceae pt. 1: 238 (1984). TAB. **82**. Type: Zambia, Mbala Distr., Kalambo R., 2.4 km above Sansia Falls, *Richards* 10373 (K, holotype; EA).

Dwarf terrestrial herb 2.5–4 cm high; tubers ellipsoid or globose, 5–14 × 3–7.5 mm, woolly. Stem densely covered with papillae-like hairs. Leaf 1, basal, 6–8 × 6–10 mm, broadly ovate or orbicular, sessile, sheathing at base, densely papillose on upper surface and edge. Inflorescence 1-flowered. Flowers deep yellow, almost orange, or dull yellow-brown, sometimes purple-tinged. Ovary 8 mm long, pubescent; bract leaf-like, c. 7 × 3 mm. Dorsal sepal 8 × 1.8 mm, linear-lanceolate, forming open hood with petals, c. 7.5 mm across mouth. Lateral sepals 6 × 3.5 mm, free, semi-orbicular, with sacs 0.8 mm long. Petals 8 × 2.8 mm, oblong-linear, falcate. Lip claw bent back on itself shortly above column and expanded into a papillate, triangular limb 3.5 mm long, 2 mm wide, produced apically into a narrow tip.

Zambia. N: Mbala Distr., in bog 2.4 km above Sansia Falls, Kalambo River, 1740 m, fl. 29.xii.1958, *Richards* 10373 (K; EA); c. 5 km from Mbala, 1650 m, fl. 20.xii.1962, *Richards* 17074 (K). Also in Zaire, Kenya and Tanzania. River banks and marshy ground, deciduous woodland; 1650–1750 m.

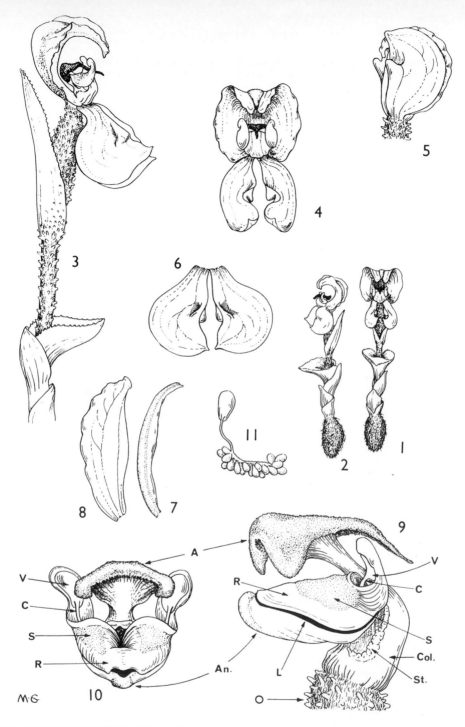

Tab. 82. DISPERIS PUSILLA. 1, habit, front view (×3); 2, habit, side view (×3); 3, habit, side view (×6); 4, flower, front view (×6); 5, flower in bud (×6); 6, lateral sepals (×6); 7, dorsal sepal, side view (×6); 8, petal (×6); 9, lip and column, side view (×21); 10, lip and column, front view (A, appendage; Col, column; An, anther; S, stigma; St, staminode; L, loculus; C, caudicle; V, viscidium; R, rostellum; O, ovary) (×21); 11, pollinium (×21), 1–11 from *Richards* 7159. Drawn by Mary Grierson.

19. **Disperis thomensis** Summerh. in Bull. Misc. Inform., Kew **1937**: 458 (1937). —Verdcourt in Kew Bull. **30**: 606 (1976). —Williamson, Orch. S. Centr. Africa: 103 (1977). —Geerinck in Fl. Afr. Centr., Orchidaceae pt. 1: 233 (1984). —la Croix et al., Orch. Malawi: 160 (1991). Type from São Tomé.

Terrestrial herb 9–15 cm tall; tuber c. 1 cm long ± cylindrical, woolly. Leaves 2, opposite, dark green, shortly petiolate, sheathing at the base, 2–5 × 1.3–3.5 cm, ovate, shortly acuminate, the margins undulate. Inflorescence 1–4-flowered; flowers white marked with green and purple, held horizontally. Ovary 12–17 mm long; bracts leaf-like, of similar length. Dorsal sepal lanceolate, acute, concave, bent above the middle, joined to petals to form a hood c. 5 mm long. Lateral sepals free, 5–6 × 2.5–3 mm, obliquely obovate, aristate at the apex, with incurved, conical spur 2–2.5 mm long near inner edge. Petals 5.3–6.5 mm long, narrowly oblong, the front edge with a rounded lobe at base, the posterior edge with a suberect median lobe 1.5 mm long; apex shortly bifurcate, the anterior fork longer than the posterior. Lip 4.5–7 mm long, the apex bent abruptly inwards and shortly apiculate, the front face joined to the apex of 2 papillose appendages; appendages 6 mm long, 2-lobed with one lobe straight, the other recurved.

Zambia. W: Mwinilunga Distr., West Lunga River, 8 km N of Mwinilunga, 1300 m, fl. 23.i.1975, *Brummitt, Chisumpa & Polhill* 14014 (K). E: Nyika Plateau, Chowo Forest, 2150 m, fl. 16.iii.1982, *Dowsett-Lemaire* 353 (K). **Malawi**. N: South Viphya, Chikangawa, 1750 m, fl. 25.iv.1982, *la Croix & Johnston-Stewart* 333 (K).

Also in Ivory Coast, Ghana, Bioko (Fernando Po), São Tomé, Angola, Nigeria, Guinea, Sierra Leone, Liberia, Zaire and Tanzania. In evergreen and riverine forests, in leaf litter in deep shade and rock faces; 1300–2150 m.

20. **Disperis thorncroftii** Schltr. in Bot. Jahrb. Syst. **20**, Beibl. 50: 19 (1895). —L. Bolus in Fl. Pl. Africa **25**: t. 963 (1945). —Grosvenor in Excelsa **6**: 80 (1976). —Stewart et al., Wild Orch. South. Africa: 208 (1982). Type from South Africa.

Terrestrial herb 14–27 cm tall; tubers 15–20 mm long, obovoid, woolly. Leaves 2, alternate, the lower much larger than the upper; lower leaf 3–5 × 2–3 cm, ovate, cordate at the base, often marbled with white on the upper surface, dark purple beneath. Inflorescence 1–3-flowered; flowers white or purple-pink with green spots, almost always erect. Ovary c. 15 mm long; bracts leaf-like, 9–11 mm long. Dorsal sepal 7–8 mm long, acuminate, erect then bent over, joined with the petals to form an open hood c. 8 mm wide. Lateral sepals free, 6–8 mm long, obliquely ovate, acuminate or apiculate, each with a curved, saccate spur 2 mm long. Petals 7–8 mm long with green tubercles on the upper half, shortly clawed, obliquely ovate with a crenulate outer margin, acuminate. Lip narrowly unguiculate; limb c. 2 mm long, ovate, acuminate, reflexed; appendage erect, linear, slightly curved at the apex, almost as long as the limb. Rostellum almost orbicular; arms c. 4 mm long, dilated at the apex.

Zimbabwe. E: Mutare, Himalaya Range, 2000 m, fl. 17.xii.1955, *Ball* 554 (K; SRGH); Chimanimani Mts., Charter Forest Estate, 1700 m, fl. 13.xii.1964, *Ball* 1054 (K; SRGH).
Also in South Africa. Montane, evergreen forest, on forest floor; 1700–2000 m.

21. **Disperis togoensis** Schltr. in Bot. Jahrb. Syst. **38**: 2, fig. 1 (1905); in Fedde, Repert. Spec. Nov. Regni Veg., Beih. 68: t. 44/176 (1932). —Summerhayes in F.W.T.A. ed. 2, **3**: 203 (1968). — Verdcourt in F.T.E.A., Orchidaceae: 221 (1968). —Geerinck in Fl. Afr. Centr., Orchidaceae pt. 1: 237 (1984). Tab. **74**, fig. **8**. Type from Togo.
    Disperis cordata Summerh. in Bull. Misc. Inform., Kew **1933**: 252 (1933) non Sw., nom illegit. Type from Cameroon.
    Disperis cardiopetala Summerh. in Hooker's Icon. Pl. 33, t. 3270 (1935); in F.W.T.A. **2**: 418 (1936). Type as for *D. cordata* Summerh.
    Disperis nr. cardiopetala Summerh. —Morris, Epiphyt. Orch. Malawi: 32 (1970).

Terrestrial herb 5–15 cm tall; tubers 7–15 × 5–7 mm, ovoid or ellipsoid, woolly. Leaves 1–3, alternate, sessile, 0.6–3.5 × 0.5–2 cm, broadly ovate, cordate, dark green with white veins on upper surface, purple beneath. Inflorescence 3–10-flowered; flowers pink or pale mauve-pink with darker lines. Ovary 9–10 mm long; bracts leaf-like, c. 6 mm long. Dorsal sepal 7–8.5 × 0.5–1 mm, linear, cordate at the base, joined to the petals to form an open hood. Lateral sepals 8–11 × 4–5 mm, obliquely ovate, joined for about a quarter to half their length, each with a small sac c. 0.5 mm long. Petals 7–10 × 2–3.5 mm, obliquely ovate, half-cordate at the base. Lip 4.5–5 mm long, the claw bent back, with a 2-lobed appendage at the point of reflexion, the appendage lobes diverging and papillate; claw ending in a ± round lamina 1–2 mm in diameter with a papillate keel near the middle.

Malawi. C: Nkhota Kota (Nkhotakota) Escarpment, fl. 19.ii.1953, *Jackson* 1070 (K). **Mozambique**. N: Niassa Prov., Lizombe, c. 110 km NE of Lichinga, 1000–1500 m, *B. Pettersson* 222 (K).
Also in West Africa, Zaire, Uganda and Tanzania. Grassland and open woodland; 1000–1500 m.

22. **Disperis virginalis** Schltr. in Bot. Jahrb. Syst. **24**: 431 (1897). —Rolfe in F.C. **5**, 3: 311 (1913). —Grosvenor in Excelsa **6**: 80 (1976). —Stewart et al., Wild Orch. South. Africa: 197 (1982). Type from South Africa.
Disperis nelsonii Rolfe in F.C. **5**, 3: 311 (1913). Type from South Africa.

Slender terrestrial herb 8–25 cm tall. Leaves 2, ± opposite, 2–7.5 × 1.5–4.5 cm, ovate, with petiole c. 3 mm long. Inflorescence 1–3-flowered; flowers pale to deep pink or white, with purple dots within. Ovary 12–20 mm long; bracts leaf-like, of similar length. Dorsal sepal 10–13 mm long, linear, concave, joined with the petals to form an open hood. Lateral sepals joined near the base, 8–10 × 4–4.5 mm, obliquely obovate-oblong, subacute or apiculate, each with a slender, acute sac-like spur 1.5–2.5 mm long directed toward the sepal apex. Petals projecting beyond the dorsal sepal, 11–14 mm long, obliquely rhomboid-falcate, acute, with a rounded lobe on the lower half of the free margin. Lip erect; claw c. 6 mm long, unguiculate; limb minute, reflexed; appendage about 4 times as long as the limb, recurved, dilated into 4 spreading lobes or 2-lobed at the apex and the lobes shortly 2-lobed. Limb and appendage densely papillose-hairy on the inside. Rostellum broadly ovate-orbicular, shortly bifid with reflexed margin; arms 2–3 mm long.

Zimbabwe. E: Mutare, Himalaya Range, Engwa, 1950 m, fl. 2.iii.1954, *Wild* 4441 (K; SRGH); Nyanga (Inyanga), in old pine plantations, 1800 m, fl. 11.ii.1955, *Ball* 500 (K; SRGH).
Also in South Africa. Floor of montane evergreen forest and pine plantations, and on mossy rocks under *Podocarpus*; 1500–1950 m.
Similar to *D. dicerochila* Summerh. but differing in having limb and appendages of lip papillose hairy.

26. **VANILLA** Mill.

Vanilla Mill., Gard. Dict., abr. ed. 4 (1754). —Rolfe in J. Linn. Soc., Bot. **32**: 439–478 (1896).

Lianes with aerial or clinging adventitious roots, rooting in the soil where the stem touches the ground. Leaves absent, or if present, sessile to shortly petiolate, fleshy. Inflorescences axillary or terminal, few- to many-flowered. Flowers large; white, yellow or green, often marked with purple or yellow on the lip. Sepals and petals similar, free. Lip usually larger than the sepals and petals, adnate to column for part of its length and forming a funnel; disk variously appendaged with lamellae or hairs. Column elongate, curved, auriculate; anther attached to margin of clinandrium, incumbent, subglobose, operculate; stigma transverse, bifid at the apex, situated under the rostellum. Rostellum broad, membranous, articulated at the base, usually deflexed. Capsule long, cylindrical, unilocular, dehiscent; seeds black, relatively large.

A genus of about 100 species in the tropics and subtropics of the Old and New World. Two species occur in the Flora Zambesiaca area.
Vanilla flavouring is extracted from the pods of *V. planifolia* Andr., a Mexican species which is cultivated in many parts of the world, particularly Madagascar.

1. Plant apparently leafless · 1. *roscheri*
2. Plant with distinct leaves · 2. *polylepis*

1. **Vanilla roscheri** Rchb.f. in Linnaea **41**: 65 (1877). —Rolfe in F.T.A. **7**: 179 (1897). —Cribb & P.F. Hunt in F.T.E.A., Orchidaceae: 258 (1984). Type from Zanzibar.

Climbing and scrambling herb with a stem of indefinite length, lacking green leaves. Roots short, arising at the nodes. Stem c. 10 mm in diameter, terete but with 2 shallow channels on each side; internodes up to 15 cm long; nodes with brown vestigial leaves up to 3 cm long. Inflorescences terminal or axillary, unbranched, up to 30 cm long, many-flowered. Flowers white flushed with pink, strongly and sweetly scented; lip salmon-pink or yellow in the throat. Ovary and pedicel short and erect at anthesis, becoming pendulous in fruit. Dorsal sepal up to 8 × 2.5 cm, lanceolate-oblong, apiculate; lateral sepals similar but slightly narrower. Petals up to 8 × 3.8 cm, elliptic-oblong or ovate, apiculate. Lip to 8 × 4.5 cm, funnel-shaped, the edges adnate to the column for 2 cm at the base; disk with 2 rows of laciniate crests up to 4 mm high, and a small crest up to 15 mm long composed of 2 rows of digitate lamellae arising between the main crests at the base. Column up to 2.5 cm long. Capsule to 17.5 cm long, 7.5 mm wide.

Mozambique. MS: Beira Distr., Dondo Forest, 60 m, fl. 9.xii.1962, *Ball* 1000 (K; SRGH). GI: Inhambane, Pomene, 1 km from airport, fl. 24.ix.1980, *Jansen, de Koning & Zunguze* 7492 (K; WAG). M: Maputo, N shore of Bay, fl. 8.xii.1944, *Danitree* s.n. (K; PRE).
Also in Kenya, Tanzania and South Africa. Coastal bush and forest, and in grassy fields with scattered trees; 1–100 m.
This species is closely related to, and possibly conspecific with, *V. phalaenopsis* Rchb.f. from the Seychelles.

2. **Vanilla polylepis** Summerh. in Bot. Mus. Leafl. Harv. Univ. **14**: 219 (1951). —P.F. Hunt in Kew Bull. **29**: 425 (1974). —Grosvenor in Excelsa **6**: 86 (1976). —Williamson, Orch. S. Centr. Africa: 146 (1977). —Ball, S. Afr. Epiph. Orch.: 234 (1978). —Cribb & P.F. Hunt in F.T.E.A., Orchidaceae: 261 (1984). —Geerinck in Fl. Afr. Centr., Orchidaceae pt. 1: 243 (1984). —la Croix et. al., Orch. Malawi: 162 (1991). TAB. **83**. Type: Zambia, Mwinilunga Distr., Matonchi River, 23.x.1937, *Milne-Redhead* 2930 (K, holotype).

Climbing and trailing liana up to 8 m long; roots adventitious, arising at the nodes, aerial or clinging to substrate, rooting in soil where they touch the ground. Stem succulent, channelled, bright green, 10–12 mm in diameter; internodes 12–15 mm long. Leaves up to 24 × 6.5 cm, fleshy, lanceolate, oblanceolate or ovate, acute. Inflorescence axillary or pseudo-terminal, up to 20-flowered, with 1–3 flowers open at a time. Flowers white or greenish-white, lip yellow towards the base, usually purple at the apex, or pure white. Ovary and pedicel 5.5–6 cm long. Sepals 3.6–6 × 0.5–2 cm, oblanceolate, fleshy; petals similar, but slightly broader and less fleshy with a keel on the outer surface, terminating in a projecting point c. 2 mm long. Lip 4.5–6 cm long, 2.5 cm wide, obscurely 3-lobed, funnel-shaped, adnate to column at base, undulate on the margin; disk (below apex of column) with crest of up to 12 rows of transverse, parallel, branched scales up to 2 mm long; basal 3 cm of lip with papillae. Column 3–4.5 cm long, the basal 2–3 cm adnate to lip. Capsule up to 15 × 1.5 cm.

Zambia. N: Mbala Distr., Lunzua Hydro-Electric Station, 1500 m, fl. 16.xi.1961, *Richards* 15355 (K). W: Mwinilunga Distr., Matonchi River, fl. 23.x.1937, *Milne-Redhead* 2930 (K). C: Mkushi Distr., Chiwefwe, 1200 m, fl. xi.1934, *Stevenson* 1653 (K). **Zimbabwe.** E: Chimanimani Range, by pool in Bundi Gorge, fl. 22.xi.1959, *Ball* 844 (K; SRGH). **Malawi.** N: South Viphya, Luchilemu Valley, 1250 m, fl. 15.xi.1986, *la Croix* 880 (K).
Also in Zaire, Kenya and Angola. Evergreen fringing forest, mushitu and scrubby woodland by river; 1200–1500 m

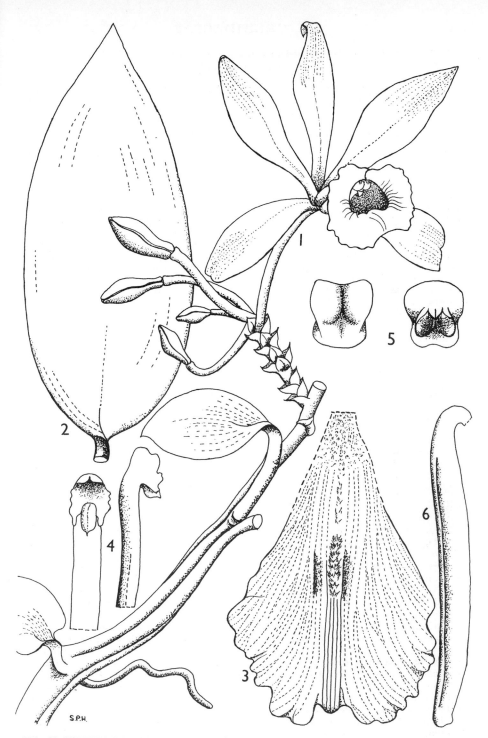

Tab. 83. VANILLA POLYLEPIS. 1, habit (×⅔), from *Napper* slide 30893; 2, leaf (×⅔); 3, lip (×1⅓), 2 & 3 from *Morze* 126A; 4, column, front and side view (×2); 5, anther, back and front views (×4), 4 & 5 from *Polhill & Lucas* 12103; 6, fruit (×⅔), from *Ball* 934. Drawn by Susan Hillier. From F.T.E.A.

27. **NERVILIA** Commer. ex Gaudich.
by Börge Pettersson

Nervilia Commer. ex Gaudich. in Freycinet, Voyage Monde, Bot.: t. 35 (1827),
421 (1829) nom. conserv. —Summerhayes in F.W.T.A., ed. 2, **3**: 206 (1968). —
Cribb in F.T.E.A., Orchidaceae: 267 (1984). —Geerinck in Fl. Afr. Centr.,
Orchidaceae pt. 1: 251 (1984). —Pettersson in Lindleyana **4**: 33 (1989); in
Nordic J. Bot. **9**: 487 (1990); in Orchid Monogr. **5** (1991).
Pogonia Juss. sect. *Nervilia* (Comm. ex Gaudich.) Lindl., Gen. Sp. Orchid. Pl.:
414 (1840). —Rolfe in F.T.A. **7**: 186 (1897).

Terrestrial tuberous herb in which the leaves are produced after the flowers
(hysteranthous); tubers subterranean subspherical, rhizomatous with 2–7
internodes, bearing short straight roots and a short ascending stem 1–15 cm long.
Foliar leaf solitary, erect or prostrate, plicate, non-articulate, elliptic to reniform or
almost circular, the upper surface sometimes pubescent and/or with various silvery
patterns, the lower surface often purple. Inflorescence erect, 1–many-flowered,
racemose; scape 2–60 cm long. Flowers resupinate or (in some 1-flowered species)
erect; tepals except the lip similar, green or brownish-green; lip spurless or shortly
spurred, 3-lobed to almost entire, often more or less papillate or pubescent and
variously marked with red; column long, curved or almost straight, more or less
slender towards the base; clinandrium a deep apical cavity embracing a large part of
the anther; stigma ventral, elliptic to almost square, viscidium diffuse; anther
incumbent, hinged; pollinia 2, bipartite, sectile.

A genus of about 60 species in tropical and subtropical parts of Africa, Asia and Australia. 13
species are recorded from mainland Africa, of which 11 are known to occur in the Flora
Zambesiaca area.

Schlechter recognised 4 sections: *Nervilia, Linervia, Vinerlia* and *Kyimbilaea* [Bot. Jahrb. Syst.
45: 400 (1911); **53**: 551 (1915)]. Sect. *Vinerlia*, as here understood, is not represented in Africa
(see Pettersson in Lindleyana **4**: 40, (1989)). Sect. *Kyimbilaea* is here included in sect. *Linervia*
(see Pettersson in Orchid Monogr. **5**: 28 (1991)). Species 1–6 belong to sect. *Linervia*, and
species 7–11 belong to sect. *Nervilia*.

A. *Key to flowering specimens*

1. Lip with spur · 2
   – Lip without spur · 3
2. Lip white with purple markings; scent of camphor · · · · · · · · · · · · · · · · · · 4. *pectinata*
   – Lip mainly lilac, purple or red; scent sweet · 3. *stolziana*
3. Flower single · 4
   – Flowers 2–several · 9
4. Lip mainly white, erect, trumpet-shaped · 5
   – Lip with strong red or purple markings, horizontal, not trumpet-shaped · · · · · · · · · · 6
5. Lip with a tuft of spaghetti-like outgrowths; flower very small, sepals mostly less than 13 mm
   long · 1. *petraea*
   – Lip with thin, hairlike outgrowths (sometimes lacking) and 3 rows of blunt tubercles or
   elongated, tapering emergences; flower larger, sepals mostly more than 13 mm long · · · ·
   · 2. *crociformis*
6. Mid-lobe of lip recurved · 10. *ballii*
   – Mid-lobe of lip not recurved · 7
7. Flower cleistogamous · 6. *gassneri*
   – Flower not cleistogamous · 8
8. Lip mid-lobe ± flat with a convex central ridge · 6. *gassneri*
   – Lip mid-lobe with erect or in-rolled edges · 5. *adolphi*
9. Inflorescence 4–12-flowered · 10
   – Inflorescence usually 2–3-flowered, secund · 11
10. Flowers yellow-green with purple or green veining; column 10–15 mm long · 7. *bicarinata*
    – Flowers yellow with red veining; column 15–19 mm long, at least in lowest flower · · · · ·
    · 8. *renschiana*

11. Flowers small, sepals less than 25 mm long; column 7–11 mm long · · · · · · · · · 9. *kotschyi*
–   Flowers large, sepals more than 25 mm long; column 16–25 mm long · · · · · 11. *shirensis*

B. *Key to vegetative specimens*

1. Leaf pubescent · 2
–   Leaf glabrous · 4
2. Leaf polygonal in outline, thin, held a few centimeters above the ground · · · 4. *pectinata*
–   Leaf reniform to orbicular (margin may be scalloped at the ends of the main veins), rather thick, prostrate · 3
3. Leaf small, less than 3 cm across, purple beneath, reticulated with silver on upper surface (silvery pattern may be absent) · 3. *stolziana*
–   Leaf larger, 4.5–14 cm across, green on both surfaces · · · · · · · · · · · · · · · · · 2. *crociformis*
4. Leaf base attenuate to truncate · 5
–   Leaf base cordate · 7
5   Leaf olive-green with silvery veins on upper surface, purple beneath · · · · · · · · · 10. *ballii*
–   Leaf green on both surfaces · 6
6. Leaf more than 4 cm wide · 11. *shirensis*
–   Leaf less than 4 cm wide · 9. *kotschyi*
7. Petiole more than 10 cm long; leaf held well above the ground · · · · · · · · · · · 7. *bicarinata*
–   Petiole less than 6 cm; leaf often ± prostrate · 8
8. Leaf pleated with ragged keels running along the ridges of the pleats · · · · · · 9. *kotschyi*
–   Leaf without such keels · 9
9. Leaf large, more than 9 cm across, very pleated · 8. *renschiana*
–   Leaf smaller, less than 9 cm across, not pleated · 10
10. Leaf very small, less than 2 cm across, with silvery rays (silvery pattern may be absent), purple beneath · 1. *petraea*
–   Leaf larger · 11
11. Leaf less than 5 cm wide · 12
–   Leaf more than 5 cm wide · 13
12. Leaf with a triangular apex, not strongly prostrate, never dark purple beneath · 6. *gassneri*
–   Leaf ± rounded at apex, often strongly prostrate and dark purple beneath · · · · 5. *adolphi*
13. Leaf spade-shaped, slightly longer than wide, thick · · · · · · · · · · · · · · · · · · · 9. *kotschyi*
–   Leaf reniform, cordate to orbicular, wider than long, thin · · · · · · · · · · · · · · · · · · 14
14. Leaf broadly cordate with a distint apical point, not strongly prostrate · · · · · · 5. *adolphi*
–   Leaf reniform to almost orbicular without a distinct apical point, often strongly prostrate
· 2. *crociformis*

1. **Nervilia petraea** (Afzel. ex Sw.) Summerh. in Bot. Mus. Leafl. (Harv. Univ.) **11**: 249 (1945). —Grosvenor in Excelsa **6**: 84 (1976). —Williamson, Orch. S. Centr. Africa: 140, t. 76 fig. 3 (1977). —Cribb in F.T.E.A., Orchidaceae: 272, t. 54 fig. 4 (1984). —Pettersson in Lindleyana **4**: 38 (1989); in Nordic J. Bot. **9**: 495 (1990); in Acta Univ. Ups., Comp. Sum. Upps. Diss. Fac. Sci. **281**: 19 (1990); in Orchid Monogr. **5**: 43 (1991). —la Croix et al., Orch. Malawi: 166, t. 106 (1991). Type from Sierra Leone.
  Arethusa petraea Afzel. ex Sw. in Neues J. Bot. **1**, 1: 62 (1805).
  Arethusa simplex Thouars, Hist. Orchid. [Fl. Iles Austr. Afr.]: tab. gen., fig 24 (1822). Type from Mauritius.
  Stellorkis aplostellis Thouars, Hist. Orchid. [Fl. Iles Austr. Afr.]: tab. gen., fig 24 (1822), (as "*Stellorchis*"). Type as for *Arethusa simplex.*
  Pogonia thouarsii sensu Rolfe in F.T.A. **7**: 187 (1897) non Blume.
  Nervilia simplex (Thouars) Schltr. in Bot. Jahrb. Syst. **45**: 410 (1911). —H. Perrier in Fl. Madag., Orch. 1: 209 (1939).
  Nervilia afzelii Schltr. in Bot. Jahrb. Syst. **45**: 402 (1911). Type as for *N. petraea.*

Erect terrestrial herb 2–8.5(10.5) cm tall, glabrous except for lip and subterranean parts. Tuber 3–8 mm in diameter, subspherical or ovoid, (2)3(4)-noded. Leaf solitary, appearing after flowering, prostrate, 6–15 × 8–25 mm, cordate to almost orbicular, sometimes obscurely angled, glabrous, dark olive-green, often with silvery rays, purple beneath; veins 7–9(10); petiole up to 10 mm long, sulcate with 1–2 sheathing cataphylls. Scape erect, terete, 1-flowered, with (1)2–4 sheathing cataphylls, lengthening to c. 10 cm when fruiting; bract 0–1.5 mm long. Flower more or less erect. Sepals and petals spreading, closing again after about 8 hours anthesis or earlier if pollinated, subequal, linear-ligulate, acute, brownish-green; lateral sepals 8–15.5 × 3.5 mm, oblique. Lip to 9–15.5 × 8 mm, obovate-cuneate, 3-lobed, white with a yellow centre; lateral lobes oblong-triangular, obtuse; mid-lobe fimbriate, suborbicular, with numerous subulate appendages. Column 4–9 mm long, clavate, glabrous; pollinia c. 1.5 mm long; ovary c. 3 mm long; capsule c. 9 mm long.

Zambia. N: Mbala Distr., Kalambo R., c. 25 km above Sansia Falls, 1470 m, 29.xii.1958, *Richards* 10372 (K). C: 35 km E of Lusaka, fl. & st. 6.xii.1974, *Williamson & Gassner* 2336 (K). **Zimbabwe**. C: Harare Distr., Denda, 1300 m, fl. 29.xi.1948; & st. ii.1948, *Greatrex in GHS* 50267 (K; SRGH). **Malawi**. N: Nkhata Bay Distr., Kalwe Forest Reserve, near Nkhata Bay, 540 m, st. 2.iv.1989, *Pettersson, Englund, Kustvall & Ranaivo* VJRB 106 (UPS). C: Dedza Distr., Chongoni, st. 18.iv.1982, *la Croix & Johnston-Stewart* 336 (K). S: Zomba Distr., Malosa above Secondary School, c. 900 m, fr. & st. 14.xii.1984, *Pettersson* 353 (UPS). **Mozambique**. N: Lizombe, c. 110 km NE of Lichinga, st. 16.ii.1982, *Pettersson* 211 (LMU; UPS).

Also in West Africa, Central African Republic, Gabon, Congo, East Africa, Zaire, Madagascar and Mauritius. *Brachystegia* woodland, grassland, dambos and pine plantations; 900–1470 m.

2. **Nervilia crociformis** (Zoll. & Moritzi) Seidenf. in Dansk Bot. Arkiv **32**, 2: 151 (1978). — Pettersson in Lindleyana **4**: 37 (1989); in Nordic J. Bot. **9**: 494 (1990); in Acta Univ. Ups., Comp. Sum. Upps. Diss. Fac. Sci. **281**: 19 (1990); in Orchid Monogr. **5**: 44 (1991). Type from Indonesia (Java).
    Bolborchis crociformis Zoll. & Moritzi in Moritzi, Syst. Verz. Pfl. Zoll.: 89 (1846).
    Nervilia humilis Schltr. in Bot. Jahrb. Syst. **53**: 551 (1915); in Fedde, Repert. Spec. Nov. Regni Veg., Beih. **68**: t. 47 fig. 185 (1932). —Grosvenor in Excelsa **6**: 84 (1976). — Williamson, Orch. S. Centr. Africa: 140, t. 75, Pl. 120 (1977); in J. S. African Bot. **45**: 467 (1979). —Cribb in F.T.E.A., Orchidaceae: 271, t. 54 fig. 3 (1984). —la Croix et al., Orch. Malawi: 165, t. 104 (1991). Type from Tanzania.
    Nervilia reniformis Schltr. in Bot. Jahrb. Syst. **53**: 551 (1915); in Fedde, Repert. Spec. Nov. Regni Veg., Beih. **68**: t. 47 fig. 187 (1932). —Saunders in Soc. Malawi J. **33**, 2: 15 (1980). —Cribb in F.T.E.A., Orchidaceae: 272, t. 54 fig. 6 (1984). Type from Tanzania.
    Nervilia erosa P.J. Cribb in Kew Bull. **32**: 155, t. 10 (1977). Type: Zambia, 75 km S of Mwinilunga, *Williamson & Drummond* 1679 (K, holotype; SRGH).
    Nervilia sp. (*Williamson & Drummond* 1679) Williamson, Orch. S. Centr. Africa: 140, t. 76 fig. 2 (1977).

Erect terrestrial herb 1.5–10 cm tall, glabrous except for leaf, lip and subterraneanan parts. Tuber 0.5–1.4 cm in diameter, subspherical or ovoid, (2)3–6-noded. Leaf solitary, appearing after flowering, prostrate or nearly so, 1.5–9 × 3.5–14 cm, reniform to almost orbicular, usually densely pubescent on the upper surface, but glabrous clones or clones with hairs only near the leaf margin not uncommon; veins (5)7–16; petiole c. 0–6 cm long, sulcate with 1–3 sheathing cataphylls. Scape erect, terete, 1-flowered, with 2–3 sheathing cataphylls, lengthening to c. 20 cm when fruiting. Flower ± erect, fragrant; bract 1–2 mm long. Sepals and petals spreading, closing again after about 8 hours anthesis or earlier if pollinated, subequal, linear-ligulate, acute, brownish-green; lateral sepals oblique, 12–19 × 3.5 mm; petals slightly shorter and narrower. Lip oblong-cuneate, 12–18 × 9–11 mm, 3-lobed, white with a yellow centre and often with a faint lilac or purple tinge or markings in the apical area; inner surface ± covered with thin hair-like outgrowths and a few thicker formations arranged ± in three rows along the centre of the mid-lobe and ranging from low tubercles to long, acute, tapering emergences of different appearance; lateral lobes obtusely triangular or rounded, sometimes less prominent or absent; mid-lobe ovate-triangular to ovate, subacute to obtuse; margin crenulate to undulate-fimbriate. Column almost straight, (5.5)7–9 mm long, clavate, glabrous; pollinia yellow with a few purplish massulae at lower end, (0.8)1.5–2 mm long; ovary 4–5 mm long.

Zambia. W: Solwezi Distr., Kansanshi Mine, fl. & st. cult. 13.xi.1962, *Holmes* 0385 (K). C: Lusaka Distr., Chisamba Forest Reserve, st. 30.xii.1963 & fl.? 20.xi.1960, *Morze* 152 (K). E: Nyika Plateau, near Chowo and around Kasoma, 2000–2200 m, st. i.1982, *Dowsett-Lemaire* 280 (K). **Zimbabwe**. N: Makonde Distr., top of Great Dyke Pass on Harare-Sinoia Rd., st. 9.i.1977, *Grosvenor* 876A (SRGH). C: Harare Distr., Rumani, Chekujibire Kopje, st. iii.1945 & iv.1948, & fl. 30.xi.1948, *Greatrex in GHS* 228708 (PRE; SRGH). E: Chimanimani Distr., near Nyabamba Falls, 750 m, fl. 19.xi.1954 & st. iii.1954, *Ball* 409 (K; SRGH). **Malawi**. N: Mzimba Distr., Mzuzu, Marymount, 1350 m, st. 24.iv.1974, *Pawek* 8560 (K; MAL; MO). C: Dedza Distr., Chongoni

Forestry School, lower slope Chiwao Hill, st. 28.ii.1981, *Patel & Hargreaves* 799 (MAL). S: Thyolo Distr., Nchima Tea Estate, 950 m, fl. 20.xi.1982 & st. 13.iii.1983, *la Croix* 362 (K; MAL). **Mozambique**. N: Lichinga, 1356 m, st. 1.ii.1982 & fl. cult., *Pettersson* 151 (LMU; UPS).

Widespread in Africa from Cape Verde Is., Senegal and Ethiopia to South Africa (Transvaal) and Madagascar, also in Asia and Australia. *Brachystegia* woodland, evergreen forest, grassland, riverine forest and pine plantations; 650–2000 m.

3. **Nervilia stolziana** Schltr. in Bot. Jahrb. Syst. **53**: 550 (1915). —Williamson in Orch. Review **1976**: 380, t. 9 (1976); Orch. S. Centr. Africa: 142, t. 76 fig. 6, Pl. 122 (1977); in Proc. 10th World Orch. Conf.: 87 (1982). —Cribb in F.T.E.A., Orchidaceae: 269, t. 54 fig. 5 (1984). — Pettersson in Lindleyana **4**: 40 (1989); in Nordic J. Bot. **9**: 496 (1990); in Acta Univ. Ups., Comp. Sum. Upps. Diss. Fac. Sci. **281**: 20 (1990); in Orchid Monogr. **5**: 48 (1991). —la Croix et al., Orch. Malawi: 166, t. 107 (1991). TAB. **84**. Type from Tanzania.

Nervilia fuerstenbergiana sensu Williamson, Orch. S. Centr. Africa: 145 (1977) non Schlechter.

Erect terrestrial herb 2.7–13.5 cm tall, glabrous except for leaf, lip and subterranean parts. Tuber 6–9 mm in diameter, subspherical or ovoid, 3–4-noded. Leaf solitary, appearing after flowering, prostrate, 1.4–3.0(3.9) × 2.0–3.5(4.2) cm, reniform to orbicular, obtuse, covered with short stiff hairs on the upper surface and often with a silvery reticulate pattern, usually purple beneath; veins(5)7–11; petiole 5–7 mm long, sulcate with 1–2 sheathing cataphylls. Scape erect, terete, 1-flowered, with 2–3 sheathing cataphylls, lengthening to 16–23 cm when fruiting. Flowers ± horizontal; bracts 0.5–3 mm long. Sepals and petals spreading, closing again after about 8 hours anthesis or after pollination, subequal, linear-lanceolate, acute, reddish-green; lateral sepals slightly oblique, 12–16 × 2 mm; petals slightly shorter, to 14 mm long. Lip 14–23(including spur) × 8–12.5 mm, elliptic to elliptic-cuneate, spurred, lilac, mauve or red with white markings and yellow centre, with numerous long papillae on inner surface; veins with several anostomoses; spur straight, short, often slightly bifid, 3–4.2 × 2.6 mm. Column almost straight, 6–7 mm long, clavate, glabrous; pollinia yellow, c. 2.5 mm long.

Zambia. W: S of Kitwe on Baluba Stream, fl. 6.xii.1974 & st. cult. 22.i.1975, *Williamson & Gassner* 2333 (K). C: Lusaka Distr., Chisamba Forest Reserve, st. 8.ii.1964, *Morze* 161 (K). **Malawi**. N: Nyika Plateau, Chisanga Falls, fl. 20.xii.1969, *Ball* 1199 (SRGH); Mzimba Distr., Mzuzu, Katoto, 1350 m, st. 14.iv.1974, *Pawek* 8331 (K); Mzimba Distr., Viphya, Mtangatanga Forest, c. 1500 m, fl. 2.xii.1961, *Chapman* 1502 (SRGH). S: Zomba Distr., Malosa above Secondary School, c. 1000 m, fl. 14.xi.1984, *Pettersson* 302 (BR; K; LISC; LMU; MAL; NHT; SRGH; UPS) & st. 14.xii.1984, *Pettersson* 352 (K; MAL; UPS). **Mozambique**. MS: Sussundenga Distr., 15 km E of Zimbabwean border, 15 km E of Chimanimani summit, 900 m, st. ii.1961, *Ball* 909 (K; SRGH).

Also in Zaire and Tanzania. *Brachystegia* woodland, submontane and riverine vegetation and in pine and *Eucalyptus* plantations ; 900–1800 m.

4. **Nervilia pectinata** P.J. Cribb in Kew Bull. **32**: 153, fig. 9 (1977). —Williamson in J. S. African Bot. **45**: 465 (1979). —Cribb in F.T.E.A., Orchidaceae: 268, fig. 54(2), fig. 56 (1–7) (1984). —Dowsett-Lemaire in Bull. Jard. Bot. Nat. Belg. **55**: 319, 377 (1985); **58**: 96 (1988). — Pettersson in Lindleyana **4**: 40 (1989); in Acta Univ. Ups., Comp. Sum Upps. Diss. Fac. Sci. **281**: 21 (1990); in Orchid Monogr. **5**: 49 (1991). —la Croix et al., Orch. Malawi: 165, t. 105 (1991). TAB. **85**, fig. A. Type: Zimbabwe, Mutare Distr., Vumba Mts., on 'Thordale', *Ball* 410 (K, holotype; SRGH).

Nervila sp. no. 5 (*Ball* 1010), Grosvenor in Excelsa **6**: 84 (1976).

Nervilia sp. (*Williamson* 2102), Williamson, Orch. S. Centr. Africa: 146 (1977).

Erect terrestrial herb 5–14 cm tall, glabrous except for leaf, lip and subterranean parts. Tuber 7–10 × 7 mm, subspherical or ovoid. Leaf solitary, appearing after flowering (sometimes sterile specimens with fully developed leaves can be seen together with flowering specimens), horizontal but not prostrate, (1.5)2.9–5.0 × (1.9)3.5–5.5 cm, ± polygonal or star-shaped, the upper surface sparsely covered with short hairs (very occasionally glabrous), hairs absent from the main veins but following the small veins radially, dark green above, pale green or sometimes purple beneath; veins 5–7; petiole 2–6 cm long, sulcate with 1–2 sheathing cataphylls. Scape erect, terete, 1-flowered, with 2–3 sheathing cataphylls, lengthening to 9–19 cm when fruiting. Flowers more or less horizontal, fragrant with a camphor scent. Sepals and

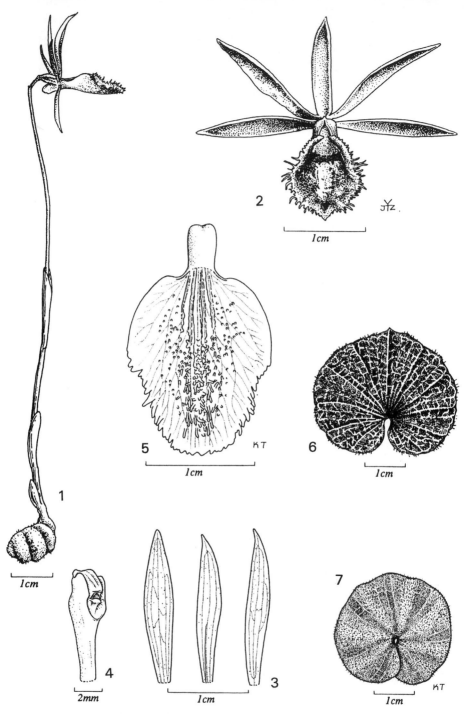

Tab. 84. NERVILIA STOLZIANA. 1, flowering plant; 2, flower, 1 & 2 from *Pettersson* 320; 3, from left to right: dorsal sepal, petal, lateral sepal; 4, column; 5, lip, 3–5 from *Pettersson* 302; 6, leaf with silvery pattern, from *Pettersson* 352 (fresh, cult.); 7, leaf without silvery pattern, *Pettersson, Englund, Kustvall & Ranaivo* 70. Drawn by Yue Jingzhu and Kerstin Thunberg. From Nervilia (Orchidaceae) in Africa.

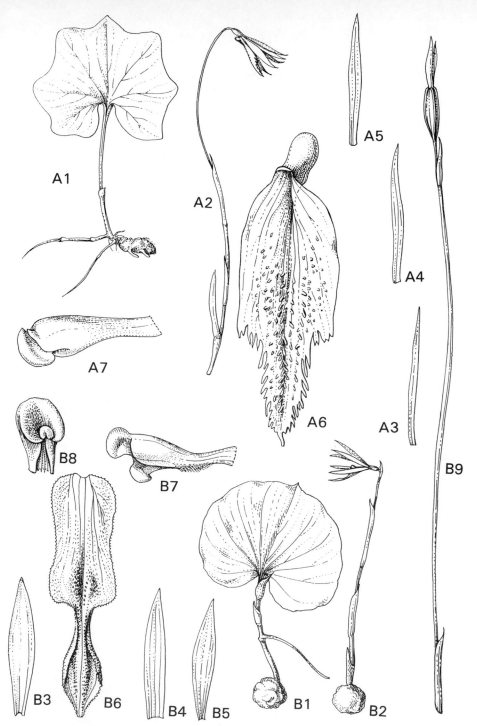

Tab. 85. A. —NERVILIA PECTINATA. A1, plant in leaf (×⅔); A2, flowering scape (×⅔); A3, dorsal
sepal (×2); A4, lateral sepal (×2); A5, petal (×2); A6, lip (×4); A7, column (×6), A1–A7 from
Ball 410. B. —NERVILIA ADOLPHI var. ADOLPHI. B1, plant in leaf (×⅔); B2, plant in flower
(×⅔); B3, dorsal sepal (×2); B4, lateral sepal (×2); B5, petal (×2); B6, lip (×4); B7, column (×4);
B8, detail of column-tip (×4), B1–B8 from *Morze* 100B and *Milne-Redhead & Taylor* 7849; B9,
fruiting scape (×⅔), from *Bowling* 38101. Drawn by Loura Mason. From F.T.E.A.

petals spreading, closing again after about 8 hours anthesis, subequal, lanceolate, acute, pale green lined with purple; lateral sepals slightly oblique, 10.5–22 × 3 mm; petals slightly shorter and 2 mm wide, lanceolate, acute. Lip spurred, 10–19(including spur) × 8.5–10 mm, elliptic, ± obscurely 3-lobed in middle, white with purple dots and yellow towards base, shortly papillate on inner surface, veins with no or very few anastomoses, margins pectinate in apical half; spur straight, short, rounded, 1.5–3 × 1.2 mm. Column almost straight, 4.5–6 mm, clavate, glabrous; pollinia 1.5–2 mm long.

Zambia. E: Lundazi Distr., Nyika Plateau, Chowo Forest, fl. & st. xii.1971, *G. Williamson* 2102 (K). **Zimbabwe**. E: Mutare Distr., Vumba Mts., on 'Thordale', st. iv.1952 & fl. cult. 26.xi.1954, *Ball* 410 (K; SRGH). **Malawi**. N: Nyika Plateau, from Kasyaula to Zovochipolo and Nyika Juniperus Forest Reserve, 2000–2200 m, st. ii. 1982, *Dowsett-Lemaire* 342 (K). S: Blantyre Distr., Mt. Soche near summit, 1510 m, fl., fr. & st. 10.xii.1984, *Pettersson* 343 (K; MAL; UPS). **Mozambique**. MS: Sussundenga Distr., E of Chimanimani, 15 km S of Martin's Falls, c. 1000 m, fl. & st. cult. 17.xi.1963, *Ball* 1010 (SRGH).

Also in Tanzania. Evergreen montane forest and cypress plantations; 990–2220 m.

5. **Nervilia adolphi** Schltr. in Bot. Jahrb. Syst. **53**: 552 (1915); in Fedde, Repert. Spec. Nov. Regni Veg., Beih. 68: t. 46 fig. 182 (1932). —Grosvenor in Excelsa **6**: 84 (1976). —Williamson, Orch. S. Centr. Africa: 141, t. 76 fig. 4 (1977); in J. S. African Bot. **45**: 467 (1979). — Saunders in Soc. Malawi J. **33**, 2: 15 (1980). —Cribb in F.T.E.A., Orchidaceae: 269, fig. 54(1), fig. 56(8–16) (1984). —Pettersson in Lindleyana **4**: 38 (1989); in Acta Univ. Ups., Comp. Sum. Upps. Diss. Fac. Sci. **281**: 21 (1990); in Orchid Monogr. **5**: 51 (1991). —la Croix et al., Orch. Malawi: 164, t. 103 (1991). Type from Tanzania.
    Nervilia sp. no. 1 (*Ball* 715), Grosvenor in Excelsa **6**: 84 (1976).

Erect terrestrial herb 2–10(13) cm tall, glabrous except for subterranean parts. Tuber 6–15 mm in diameter, subspherical or ovoid, (2)3(4)-noded. Leaf solitary, appearing after flowering, prostrate to suberect (blade up to 5 cm above litter surface), glabrous, 1.8–5.5 × 2–6.5 cm, broadly cordate or reniform, subacute to obtuse, dark olive-green above and purple beneath, or shiny green above, sometimes tessellate with silver, and green or ± purple beneath; veins (5)7–9; petiole c. 0–5 cm long, sulcate with 1–2 sheathing cataphylls. Scape erect, terete, 1-flowered, with 2 sheathing cataphylls, lengthening to 8.6–21 cm when fruiting; bract 1.5–5.5(10.5) mm long. Flowers small, ± horizontal, fragrant. Sepals and petals spreading, closing again after about 8 hours anthesis, subequal, ligulate, acute, brownish-green; lateral sepals slightly oblique, 10–22 × 2.5–3.5 mm; petals slightly narrower than sepals. Lip 9.5–20 × 4–7 mm, oblong, 3 -lobed, white to pale pink with purple-violet shortly papillate markings; lateral lobes short, obtuse, erect, tightly clasping the column; mid-lobe (=epichile) 3–9.5 mm long, ovate, with a convex central ridge and an incurved margin. Column almost straight, 6–10 mm long, clavate, pubescent on ventral side; pollinia yellow, 2–3 mm long; ovary 2–7 mm long; capsule 7.5–10 × 4–5 mm.

Var. **adolphi** TAB. **85,** fig. B.

Leaf uniformly green, or olive-green above and green or purple beneath.

Zambia. W: Solwezi, fl. 20.x.1959, *Holmes* 0180 (K; SRGH). C: Lusaka Distr., Chisamba Forest Reserve, fl. 16.xi.1963 & st. 6.ii.1963, *Morze* 100B (K). **Zimbabwe**. E: Chimanimani Distr., Welgelegen by R. Haroni, 1050 m, fl. 19.xi.1954, fr. & st. iii.1954, *Ball* 407 (K; SRGH). **Malawi**. N: Chitipa Distr., Misuku Hills, path to Mugesse Forest Reserve on the south, st. 15.iii.1977, *Grosvenor & Renz* 1235A (K; SRGH); Nkhata Bay Distr., 9 km S of Chikangawa, 1750 m, fl. 19.xi.1978, *Phillips* 4255 (K; MAL; MO; SRGH; WAG; Z). C: Ntchisi Forest Reserve, 1400 m, st. 21.iii.1984, *Dowsett-Lemaire* 1129 (K). S: Thyolo Distr., Bvumbwe, 1200 m, st. 22.ii.1982, *la Croix* 270 (K; MAL); Zomba Distr., Malosa above Secondary School, c. 1000 m, fl. 14.xi.1984, *Pettersson* 306 (UPS). **Mozambique**. N: Lichinga, 1356 m, st. 4.ii.1982, *Pettersson* 159 (LMU; UPS).

Also in West Africa, Equatorial Guinea, Tanzania, Uganda and Zaire. High rainfall woodland, submontane grassland, riverine fringe-forest and pine plantations; 900–1750 m.

Var. **seposita** N. Hallé & Toilliez in Adansonia, Ser. 2, **11**: 460 (1971). —Cribb in F.T.E.A., Orchidaceae: 271 (1984). —Pettersson in Acta Univ. Ups., Comp. Sum. Upps. Diss. Fac. Sci. **281**: 22 (1990); in Orchid Monogr. **5**: 53 (1991). Type from Ivory Coast.

Leaf ± prostrate, tessellate with silver on the upper surface, purplish-grey beneath.

Zambia. C: Lusaka Distr., Chisamba Forest Reserve, Kamaila, fl. 24.ii.1964, *Morze* 100B/2 (K). Also in West Africa, Zaire and Uganda.

Hallé & Toilliez circumscribed var. *seposita* as always having a short lip, with the mid-lobe about one-third of the total lip length. However, while most short-lipped West African specimens have tessellate leaves, short-lipped specimens with green leaves and long-lipped specimens with tessellate leaves also occur. *La Croix* 879 from Malawi is an example of a green-leaved specimen with a short lip. Because of the apparent unreliability of the lip-length character var. *seposita* is here distinguished on leaf characters.

6. **Nervilia gassneri** Börge Pett. in Nordic J. Bot. **9**: 492 (1990); in Acta Univ. Ups., Comp. Sum. Upps. Diss. Fac. Sci. **281**: 22 (1990); in Orchid Monogr. **5**: 55 (1991). TAB. **86**. Type: Malawi, Zomba Plateau, 1530 m, 15.xii.1984, *Pettersson & Gassner* 359 (UPS, holotype; BR; K; LISC; LMU; MAL; NHT; SRGH).
  Nervilia sp. (*Renny* 250), Renny in S. Afr. Orch. J. Jun.–Jul. 1972: 16, fig. B–G (1972).
  Nervilia sp., van Ede in Proc. 8th World Orch. Conf.: 233 (1975).
  Nervilia sp. no. 2 (*Ball* 427), Grosvenor in Excelsa **6**: 84 (1976).
  Nervilia adolphi sensu Stewart et al., Wild Orch. S. Afr.: 213 pro parte excl. photo (1982). —McMurtry in Proc. 10th World Orch. Conf.: 83 (1982). —Gibbs Russell et al. in Mem. Bot. Survey S. Afr. **48**: 43 (1984) non Schlechter (1915).
  Nervilia fuerstenbergiana sensu Cribb in F.T.E.A., Orchidaceae: 271 (1984) non Schlechter (1911).
  Nervilia "gassneri" sensu Pettersson in Lindleyana **4**: 38 (1989), nom. nud.

Erect terrestrial herb 2.5–10 cm tall, glabrous except for subterranean parts. Tuber 7–13 × 8–9 × 6–9 mm, subspherical or ovoid. Leaf solitary, appearing after flowering (sterile specimens sometimes develop leaves while the fertile ones are flowering), prostrate or nearly so (blade up to 2 cm above litter surface), glabrous, 1.4–5.5 × 1.2–4.5 cm, cordate, subacute to obtuse, apical part triangular with sides straight or often slightly concave; veins (5)7, the middle one often with a more prominent groove on the upper surface; petiole 1–6 cm long, sulcate with 1–2 sheathing cataphylls. Scape erect, terete, 1-flowered, with (1)2 sheathing cataphylls, lengthening to (4.5)6–20 cm when fruiting; bract 1–2.6 × 0.5 mm. Flowers small, c. 22 × 18 mm, ± horizontal, faintly scented. Sepals and petals spreading, subequal, ligulate, acute, reddish-green with maroon longitudinal flecks; lateral sepals slightly oblique, 9.5–14 × 2 mm; petals slightly shorter than sepals, narrower towards the base and with fainter colours. Lip 9–13 × 5–6 mm, oblong, 3-lobed, divided in the middle by a narrow waist into a hypochile and epichile of subequal length, white to pale-pink with cerise, shortly papillate markings; lateral lobes short, obtuse, erect, enclosing the column; mid-lobe (= epichile) ovate, flat with a convex central ridge, almost as wide as the hypochile. In the cleistogamous flowers the perianth parts are often very reduced in size. Column almost straight, 4.5–6 × 1.3 mm, clavate, pubescent on the ventral side; pollinia creamy-white to pinkish-grey 1.5 × 0.6 mm, sometimes sticking together, forming a single compound pollinium, massulae scale-like; capsule 9–12 × 4.5 mm.

Zimbabwe. E: Chimanimani Distr., near Bridal Veil Falls, 1350 m, fl., fr. & st. 9.xii.1954, *Ball* 427 (K; SRGH). **Malawi**. S: Zomba Plateau, 1530 m, fl. & st. 15.xii.1984, *Pettersson & Gassner* 359 (BR; K; LISC; LMU; MAL; NHT; SRGH; UPS).
  Also in Tanzania and South Africa (Transvaal). Riverine forest and submontane cypress plantations; 1150–1530 m.

The following material differs from typical *N. gassneri* in having very acute, sometimes horn-like, side lobes to the lip, and tooth-like projections along the centre of the mid-lobe. It was treated as *Nervilia sp.* (*Williamson & Odgers* 181) by Williamson in Orch. S. Centr. Africa: 141, t. 76 fig. 5 (1977), who intended to describe it as a new species (pers. comm.) but the material turned out to be inadequate; a situation which still prevails. It is here provisionally put under *N. gassneri* but may prove to represent a distinct taxon when more material becomes available.

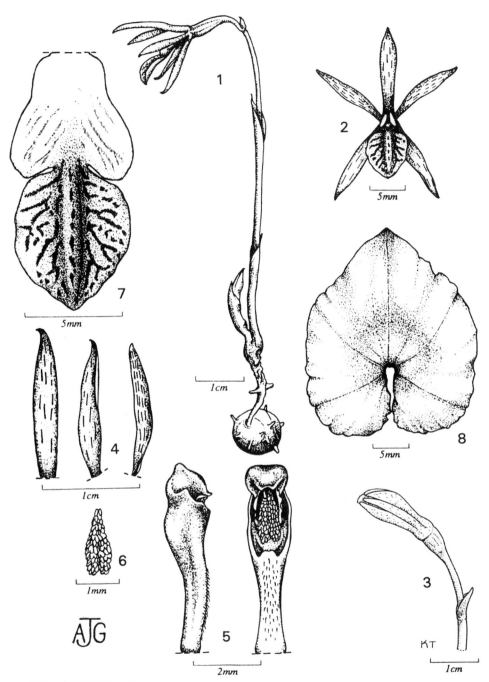

Tab. 86. NERVILIA GASSNERI. 1, flowering plant; 2, flower, 1 & 2 from *Pettersson & Gassner* 359; 3, cleistogamous flower, from *Pettersson, Hedrén & Kibuwa* 89; 4, from left to right: dorsal sepal, lateral sepal, petal; 5, column, side and front view; 6, pollinia; 7, lip; 8, young leaf, 4–8 from *Pettersson & Gassner* 359. Drawn by Anthony Gassner and Kerstin Thunberg. From Nervilia (Orchidaceae) in Africa.

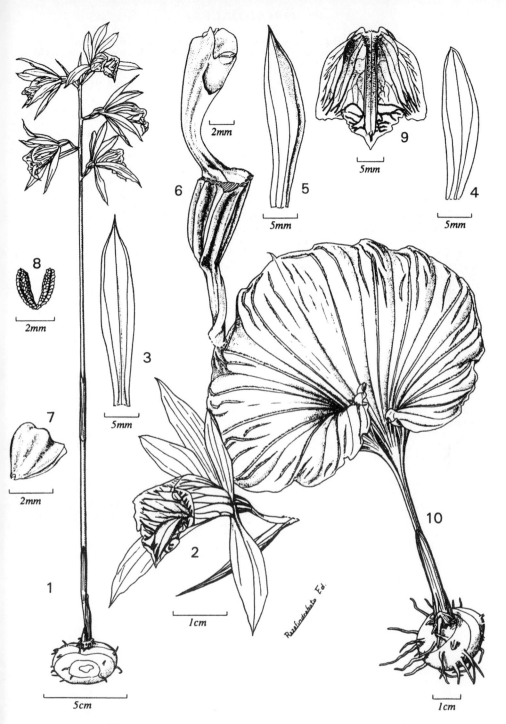

Tab. 87. NERVILIA BICARINATA. 1, flowering plant; 2, flower; 3, dorsal sepal; 4, petal; 5, lateral sepal; 6, column (with anther removed) and ovary; 7, anther; 8, pollinia; 9, lip; 10, plant in leaf, 1–10 from *Morat* 3117 (drawn from flowering plant in cultivation). From Nervilia (Orchidaceae) in Africa.

Zambia. W: a tributary of Baluba Stream 6 km N of Fisenge Interpass, fl. (fragment) & st. xii.1964, *Williamson & Odgers* 181 (K); C: Kabwe (Broken Hill) Forest Reserve, Mpima, at old Mkushi Rd., fl. 13.xi.1962, *Morze* 100A (K).

Brachystegia woodland and pine plantation.

Pettersson 161 (LMU; UPS) from Mozambique (Lichinga) is sterile but probably also belongs here.

7. **Nervilia bicarinata** (Blume) Schltr. in Bot. Jahrb. Syst. **45**: 405 (1911). —Pettersson in Lindleyana **4**: 37 (1989); in Nordic J. Bot. **9**: 487 (1990); in Acta Univ. Ups., Comp. Sum. Upps. Diss. Fac. Sci. **281**: 22 (1990); in Orchid Monogr. **5**: 56 (1991). —la Croix et al., Orch. Malawi: 168, t. 109 (1991). TAB. **87**. Type from Madagascar.
   Pogonia bicarinata Blume, Coll. Orchid. Archip. Ind. Japon.: 152, t. 60 fig. 1 A–D (1859).
   Pogonia umbrosa Rchb.f. in Flora **50**: 102 (1867). Types from São Tomé & Príncipe.
   Nervilia umbrosa (Rchb.f.) Schltr., Westafr. Kautschuk-Exp.: 274 (1900). —Williamson, Orch. S. Centr. Africa: 145, t. 77 fig. 12 (1977). —Cribb in F.T.E.A., Orchidaceae: 277, t. 55 fig. 3, t. 57 (1984).
   Nervilia renschiana sensu H. Perrier in Fl. Madag., Orch. 1: 203 (1939) non Rchb.f.

Erect terrestrial herb 17–75 cm tall, glabrous except for lip and subterranean parts. Tuber 1–2 × 1–3 × 1–4.7 cm, subspherical or ovoid, 3–7-noded. Leaf solitary, appearing after flowering, ± horizontal, held well above ground, very large, up to 22.5 cm long and 26.5 cm wide, orbicular (reniform when flattened), cordate at base, apiculate, glabrous, 10–30-veined, heavily pleated; pleats keeled with keels sometimes white and always somewhat ragged near the margin; petiole (2)5–26 cm long, sulcate with 1–2 sheathing cataphylls when young. Scape erect, terete, laxly (1)4–10(12)-flowered, with 2–5 sheathing cataphylls, the uppermost often bract-like and up to 4 cm long. Flowers evenly spaced along the rhachis, 2–4 cm apart; bracts 10–27 × 0.9–2.5 mm, filiform to lanceolate. Sepals and petals subequal, ligulate-lanceolate, acute, greenish; lateral sepals slightly oblique, 17–31 × 0.9–4 mm; petals slightly shorter than sepals. Lip greenish-white with purple or green venation, 20–31 × 17–25 mm, ovate, obscurely 3-lobed, bearing 2 parallel fleshy pubescent ridges running from base of lip to base of mid-lobe; side lobes short, oblong to shortly triangular, obtuse, erect, enclosing the column; mid-lobe triangular to ovate, acute, recurved. Column curved, 10–15(16) mm long, clavate, glabrous; anther front rim with two protruding teeth; pollina yellow, 2.5–3.2 mm long; stigmatic surface 4–5 mm in diameter. Capsule c. 13 mm long.

Zambia. N: Mbala, fl., *Glover* 13 (BR). C: c. 11 km W of Lusaka, fl. & st. xi. 1969, *G. Williamson* 437 (BOL; K; LISC; MO; P; PRE; SRGH). S: Livingstone Distr., below Victoria Falls, 900 m, fl. xi.1950 & st. i.1951, *Glen* in *GHS* 44515 (K; PRE; SRGH). **Zimbabwe**. W: Hwange Distr., Victoria Falls Rain Forest, 880 m, fl. 6.xii.1977 & st. 3.i.1978, *Langman* 21 (SRGH). **Malawi**. N: Nkhata Bay Distr., foot of Kandoli Mts., near Lumpheza R., 620 m, st. 28.ii.1987, *la Croix* 990 (K). S: Thyolo Distr., Mwalanthunzi, 1100 m, fl. & st. 29.xi.1981, *la Croix* 222 (K).

Widespread in Africa from Senegal and Ethiopia to Angola and South Africa (Natal), and in Madagascar, the Comoro Is. and the Mascarenes. Also in Yemen and Oman. Riverine and waterfall-spray forest, and in *Syzygium* thicket; 600–1100 m.

8. **Nervilia renschiana** (Rchb.f.) Schltr. in Bot. Jahrb. Syst. **45**: 404 (1911); in Fedde, Repert. Spec. Nov. Regni Veg., Beih. 33: 119 (1925). —Pettersson in Orch. Res. Newsl. **4**: 7, Fig. 2 (1984); in Taxon **35**: 549, fig. 1 A–C (1986); in Lindleyana **4**: 37 (1989); in Nordic J. Bot. **9**: 492 (1990); in Acta Univ. Ups., Comp. Sum. Upps. Diss. Fac. Sci. **281**: 23 (1990); in Orchid Monogr. **5**: 59 (1991). —la Croix et al., Orch. Malawi: 169, t. 111 (1991). TAB. **88**. Type from Madagascar.
   Pogonia renschiana Rchb.f. in Bot. Zeit. **39**, 28: 449 (1881).
   Nervilia insolata Jum. & H. Perrier in Ann. Fac. Sci. Marseille. **21**, 2: 192 (1912). —H. Perrier in Fl. Madag., Orch. 1: 202 (1939). —Cribb in F.T.E.A., Orchidaceae: 277, t. 55 fig. 5 (1984). Type from Madagascar.
   Nervilia umbrosa sensu Schelpe in J. S. African Bot. **24**: 391 (1976) non (Rchb.f.) Schltr. (1901).
   Pogonia sp., Schinz in Akad. Wiss. Wien. Math.-Naturwiss. Kl., Denkschr. **78**: 42 (1905). —Gomes e Sousa, Pl. Menyharth.: 62 (1936).
   Nervilia sp. no. 6 (*Wild* 4194), Grosvenor in Excelsa **6**: 84 (1976).
   Nervilia sp. (*Drummond* 5417), Williamson, Orch. S. Centr. Africa: 143, t. 77 fig. 10 (1977).

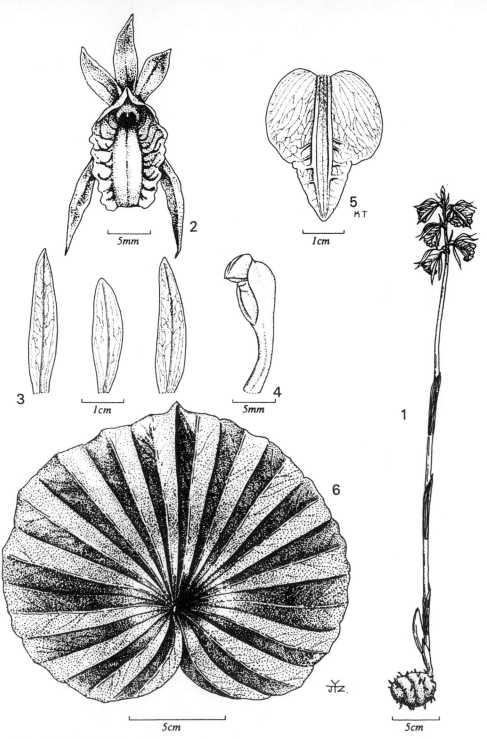

Tab. 88. NERVILIA RENSCHIANA. 1, flowering plant, from *Johnston-Stewart* 312; 2, flower; 3, from left to right: dorsal sepal, petal, lateral sepal; 4, column; 5, lip, 3–5 from *Milne-Redhead & Taylor* 7852; 6, leaf, from *Johnson-Stewart* 312. Drawn by Yue Jingzhu. From Nervilia (Orchidaceae) in Africa.

Erect terrestrial herb 25–80 cm tall, glabrous except for lip and subterraneaen parts. Tuber 3–3.5 cm in diameter, subspherical or ovoid. Leaf solitary, appearing after flowering, large, prostrate, (5)9–17 cm long and (6)11–19 cm wide, reniform to almost orbicular, deeply cordate at base, apiculate, glabrous, 11–20-veined, heavily pleated (especially when young); lateral veins strongly curved towards the leaf margin; petiole 0–1.5(5) cm long, sulcate with 1–2 sheathing cataphylls when young. Scape erect, thick, yellowish-brown to purple, laxly to rather densely 3–8-flowered, with 2–4 sheathing cataphylls, the uppermost rarely bract-like and up to c. 4 cm long. Flowers large, yellowish tinged with brown; bracts 14–35 × 1–4.2 mm, filamentous to lanceolate. Sepals and petals subequal, ligulate-lanceolate, acute, light green to yellowish-green; lateral sepals slightly oblique, 24–27 × 4–5.5 mm. Lip yellowish with reddish veins, 24–38 × 17.5–26 mm, ovate, obscurely 3-lobed, bearing 2 parallel fleshy pubescent ridges from base of lip to base of mid-lobe; side lobes semicircular to oblong, rounded, erect, enclosing the column; mid-lobe ovate, subacute, recurved. Column curved, 13.5–19 mm (in lowest flower 15–19 mm), clavate, glabrous; anther front rim with two protruding teeth.

Zambia. S: Mazabuka Distr., tributary of Zambezi, c. 5 km SW of Chirundu (Otto Beit Bridge), st. 1.ii.1958, *Drummond* 5417 (BR; K; PRE; SRGH). **Zimbabwe**. N: Sebungwe Distr., near R. Kariyangwa, Lubu Flygate, 900 m, fl. & fr. 16.xi.1956, *Lovemore* 505 (K; SRGH). E: Mutare Distr., Burma Valley, Manyera Farm, 800 m, fl. 14.xi.1961, *Chase* 7555 (SRGH). **Malawi**. N: Karonga Distr., hill N of Mwakashunguti, 700 m, fl. 18.xii.1978, *Hargreaves* 564 (MAL); North Rukuru R., 1760 m, st. 31.iii.1989, *Pettersson, Englund, Kustvall & Ranaivo* VJRB 82 (UPS). S: Thyolo Distr., Kumadzi Tea Estate, 600 m, fl. & st. 15.xi.1984, *Johnston-Stewart* 312 (K). **Mozambique**. Z: Mocuba Distr., Lugela, Namagoa Tea Plantations, 200 km inland from Quelimane, 60–120 m, fl. & st. xi–xii.1943, *Faulkner* 77 (K; PRE; SRGH). T: Songo, junto a estrada do planalto do Songo para a Maroeira, lado direito a p. de 2 km do Posto de Controle, fl. 12.xi.1978, *Correia & Marques* 22 (K; LMU); Boruma, Chimambe, st. xii.1890, *Menyhárth* 549 (K; WU; Z).

Also in Burundi, Tanzania, Zaire, South Africa (Natal) and Madagascar. *Brachystegia* woodland and riverine forest fringe, often on termite mounds; 100–900 m.

9. **Nervilia kotschyi** (Rchb.f.) Schltr. in Bot. Jahrb. Syst. **45**: 404 (1911). —Summerhayes in Bull. Misc. Inform., Kew **1937**: 459 (1937). —Jacobsen in Kirkia **9**, 1: 155 (1973) pro parte excl. specim. *Jacobsen* 2986. —Grosvenor in Excelsa **6**: 84 (1976). —Williamson, Orch. S. Centr. Africa: 144, t. 77 fig. 11, pl. 123 (1977); in J. S. African Bot. **45**: 467 (1979). —Cribb in F.T.E.A., Orchidaceae: 273, t. 55 fig. 4 (1984). —Pettersson in Lindleyana **4**: 37 (1989); in Nordic J. Bot. **9**: 489 (1990); in Acta Univ. Ups., Comp. Sum. Upps. Diss. Fac. Sci. **281**: 23 (1990); in Orchid Monogr. **5**: 60 (1991). —la Croix et al., Orch. Malawi: 168, t. 110 (1991). Type from Sudan.
    Pogonia kotschyi Rchb.f. in Öst. Bot. Zeitschr. **14**: 338 (1864). —Rolfe in F.T.A. **7**: 187 (1987).
    Nervilia sp. Saunders in Soc. Malawi J. **33**, 2: 38, 40 (1980) pro parte.

Erect terrestrial herb 8–42 cm tall, glabrous except for lip and subterranean parts. Tuber up to 14–32 mm in diameter, subspherical or ovoid, (3)4–6-noded. Leaf solitary, appearing after flowering, prostrate or erect (var. *purpurata*), (2.3)3–17 cm long and (1.6)2.5–16 cm wide, broadly cordate or sometimes ovate, acute to apiculate, glabrous, 6–25-veined, often dark olive-green above and purple beneath, sometimes with silvery lines along the veins on the upper surface, usually heavily pleated and raggedly keeled on the pleats, keels often purple, if keels absent then leaves green on both surfaces and not pleated (*N. sakoae*); petiole c. 0–6 cm long, sulcate with 1–2 sheathing cataphylls when young. Scape erect, terete, (1)2–8(12)-flowered, with 2–4 sheathing cataphylls, the uppermost sometimes bract-like and up to 25 mm long. Flowers greenish with purple veins; bracts 7–25 × 1–2 mm linear, acuminate. Sepals and petals subequal, linear-lanceolate, acute, greenish; lateral sepals slightly oblique, (8)12–26 × 3–3.5 mm; petals slightly shorter than sepals. Lip greenish-white with purple venation, 10–19 × 7–12 mm, elliptic, obscurely 3-lobed, bearing 2 parallel fleshy pubescent ridges running from base of lip to base of mid-lobe; side lobes shortly triangular, acute or subacute, erect, enclosing the column; mid-lobe ovate-triangular or triangular often with an irregular callus in the middle, acute or subacute, recurved. Column curved, 7–11.5 mm long, clavate, glabrous; anther front rim with two protruding teeth; pollinia yellow, c. 2 mm long; ovary c. 4 mm long; capsule 8–13 mm long.

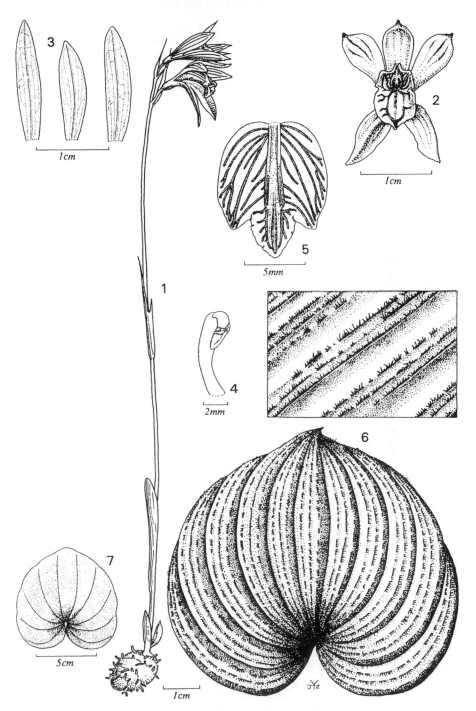

Tab. 89. NERVILIA KOTSCHYI var. KOTSCHYI. 1, flowering plant; 2, flower; 3, from left to right: dorsal sepal, petal, lateral sepal; 4, column; 5, lip; 6, leaf (much-fringed form), 1–6 from *Pettersson* 164; 7, leaf (smooth-leaved form), from *Pettersson* 148.   Drawn by Yue Jingzhu. From Nervilia (Orchidaceae) in Africa.

Var. **kotschyi** —Pettersson in Nordic J. Bot. **9**: 489 (1990); in Acta Univ. Ups., Comp. Sum. Upps. Diss. Fac. Sci. **281**: 24 (1990); in Orchid Monogr. **5**: 62 (1991). TAB. **89**.
  Nervilia sakoae Jum. & H. Perrier in Ann. Fac. Sci. Marseille **21**, 2: 194 (1912) (as "*saokae*"). Type from Madagascar.
  Nervilia diantha Schltr. in Bot. Jahrb. Syst. **53**: 553 (1915); in Fedde, Repert. Spec. Nov. Regni Veg., Beih. 68: t. 46 fig. 183 (1932). Type from Tanzania.

Leaf prostrate, up to 16 cm wide, broadly cordate, often heavily pleated, 10–14-veined, mostly with 1(4) ragged irregularly interrupted keels or fringes running along the pleats; fringes and underside of leaf often dark purple; upper surface sometimes with silvery lines along the veins; petiole very short. Inflorescence usually 2-flowered in the Flora Zambesiaca area. Column 7–9 mm long.

Zambia. W: c. 67 km W of Chingola on Solwezi road., fl. 21.x.1969 & st. cult., *Williamson & Drummond* 1626 (SRGH). C: Kamaila, Chisamba Forest Reserve, fl. 1.ii.1964, *Morze* 100C (K). E: Chipata, ICRAF Field Station, st. 24.i.1989, *Palmberg* s.n. (UPS). S: Mazabuka Distr., Munali Pass, 64 km SW of Lusaka, fl. & st. xi.–xii.1971, *G. Williamson* 2082 (SRGH). **Zimbabwe**. N: Makonde Distr., c. 19 km E of Banket, Airey's Pass, Makwadzi R., near Sutton Mine, Great Dyke, st. 14.i.1977 & fl. cult. 6–10.xi.1978, *Grosvenor* 877 (SRGH). W: Hwange Distr., Victoria Falls, 880 m, st. 20.i.1951, *Glen* in *GHS* 228,907 (SRGH). C: Harare Distr., Enterprise, Denda Farm, 1300 m, fl. xi.1946 & st. ii.1947, *Greatrex* in *GHS* 15538 (K; SRGH). E: Chimanimani Distr., Welgelegen by R. Haroni, 900m, st. iii.1954 & fl. 19.xi.1954, *Ball* 408 (K; SRGH). **Malawi**. N: Mzimba Distr., Mzuzu, Katoto, 1350 m, st. 14.iv.1974, *Pawek* 8330 (K; MO); and 5 km W of Mzuzu at Katoto, 1350 m, fl. 19.xii.1975, *Pawek* 10401 (K; MAL; MO; SRGH; UC). S: Zomba Distr., Thondwe Area, Mbala Estate, 1070 m, fl. 13.xi.1984 & st., *Pettersson* 301 (K; MAL; UPS). **Mozambique**. N: Lichinga, 1356 m, st. & fl. cult. 6.ii.1982, *Pettersson* 164 (LMU; UPS). Z: Milange, Serra do Chiperone, ascent to chief Marrega, NW slope, 770 m, st. 24.i.1972, *Correia & Marques* 2288 (LMU). MS: Chimoio, Tete road., st. 6.i.1948, *Mendonça* 3606 (LISC).
Widespread from Senegal, Mali, Ethiopia and Sudan southwards to the Flora Zambesiaca area, also in Madagascar. *Brachystegia* woodland, riverine forests, thickets submontane forests and pine plantations; 700–1500 m.
The leaf upper surface is sometimes totally smooth, and not keeled. Such collections are known from Zambia (C) and Mozambique (N). However, there is a continuous variation smooth and heavily keeled leaves.

Var. **purpurata** (Rchb.f. & Sonder) Börge Pett. in Nordic J. Bot. **9**: 489 (1990); in Acta Univ. Ups., Comp. Sum. Upps. Diss. Fac. Sci. **281**: 24 (1990); in Orchid Monogr. **5**: 63 (1991). Type from South Africa.
  Pogonia purpurata Rchb.f. & Sonder in Flora **48**: 184 (1865).
  Nervilia purpurata (Rchb.f. & Sonder) Schltr. in Warb., Kunene-Sambesi Exped. Baum: 210 (1903). —Grosvenor in Excelsa **6**: 84 (1976). —Williamson, Orch. S. Centr. Africa: 143, t. 77 fig. 8 (1977). —Cribb in F.T.E.A., Orchidaceae: 273 (1984).
  Nervilia dalbergiae Jum. & H. Perrier in Ann. Fac. Sci. Marseille **21**, 2: 196 (1912). Type from Madagascar.

Leaf erect, 6–13 × 2.5–4 cm, elliptic to lanceolate, plicate, 6–10-veined; ridge of pleats keeled, edge of keels smooth; petiole 3–6 cm long. Inflorescence mostly 3-flowered in the Flora Zambesiaca area. Column 8–11 mm long.

Zambia. B: 30 km W of Katima Mulilo on S side of Zambezi R., c. 260 km W of Livingstone, st. i.1973 & fl. cult. xii.1973, *G. Williamson* 2288 (K). C: 10 km W of Lusaka, fl. xii.1973 & st. i.1974, *G. Williamson* 2286 (EA; K). **Zimbabwe**. C: Harare (Salisbury City), Highlands, Salisbury Commonage, 1350 m, fl. & st. 12.xi.1948, *Greatrex* 119 (K; PRE; SRGH). **Malawi**. C: Kasungu Nat. Park, near Lifupa Lodge, 1000 m, st. 25.iii.1989, *Pettersson, Englund, Kustvall & Ranaivo* VJRB 16 (K; MAL; UPS). **Mozambique**. N: Msumba near Metangula, shore of Lake Nyasa, 475 m, fl. & st. 7.i.1937, *Seddon* 1 (K).
Also in Ethiopia, Sudan, Tanzania, South Africa (Transvaal) and Madagascar. Grassland and dambos; 475–1350 m.
Flowering specimens are difficult to determine to variety. The following are specimens for which the material is inadequate for determination at infraspecific level. However, *Junod* 394 because of its geographical locality probably belongs to var. *purpurata*.
Zambia. N: Kasama Distr., 28 km SE of Kasama, fl. & fr. 22.xi.1960, *E.A. Robinson* 4102 (K; M; SRGH). **Mozambique**. MS: Moribane, at the border, fl. 17.xi.1942, *Salbany* 84 (LISC). M: Libombo Marsh, Rikatla, fl. & fr. 1918 or 1919, *Junod* 394 (BOL; G; K; PRE).

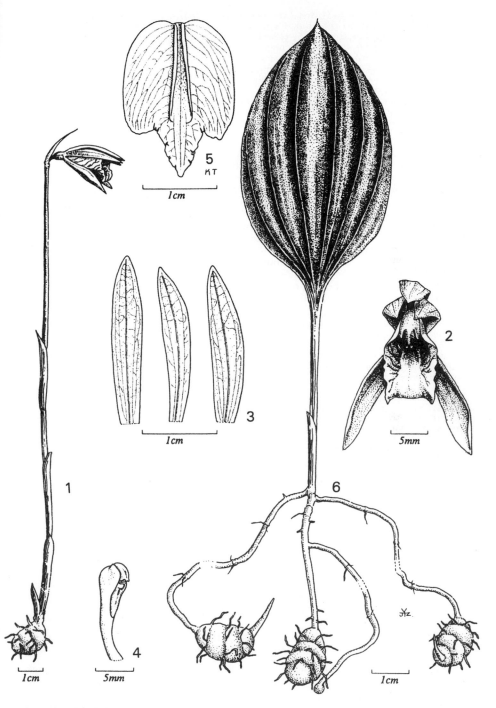

Tab. 90. NERVILIA BALLII. 1, flowering plant; 2, flower; 3, from left to right: dorsal sepal, petal, lateral sepal; 4, column; 5, lip; 6, plant in leaf, 1–6 from *Pettersson* 299. Drawn by Yue Jingzhu. From Nervilia (Orchidaceae) in Africa.

10. **Nervilia ballii** G. Will. in Pl. Syst. Evol. **134**: 66, t. 7 (1980). —Saunders in Soc. Malawi J. **33**, 2: 15 (1980). —Cribb in F.T.E.A., Orchidaceae: 275, [fig. 54 (7) as "*N. purpurata*"] (1984). —Pettersson in Lindleyana **4**: 37 (1989); in Acta Univ. Ups., Comp. Sum. Upps. Diss. Fac. Sci. **281**: 24 (1990); in Orchid Monogr. **5**: 64 (1991). —la Croix et al., Orch. Malawi: 167, t. 108 (1991). TAB. **90**. Type: Zimbabwe, Mutare Distr., Vumba Foothills below Richard's Homestead, *Ball* 585 (SRGH, holotype).

Nervilia kotschyi sensu Jacobsen in Kirkia **9**, 1: 155 (1973) pro parte quoad specim. *Jacobsen* 2986 non (Rchb.f.) Schltr.

Nervilia sp. (761), Moriarty, Wild. Fl. Malawi: 57, t. 29 fig. 3 (1975).

Nervilia sp. no. 3 (*Wild* 4174), Grosvenor in Excelsa **6**: 84 (1976).

Nervilia sp. (*Williamson & Drummond* 1639), Williamson, Orch. S. Centr. Africa: 143, t. 76 fig. 7 (1977).

Erect terrestrial herb 11–30 cm tall, glabrous except for lip and subterranean parts. Tuber 15–17 × 15 × 10–15 mm in diameter, subspherical or ovoid, (2)3–5-noded. Leaf solitary, appearing after flowering, erect, 5–6 × 3–4 cm, ovate, acute, glabrous, heavily pleated, dark olive-green above with silvery lines along the veins, dark purple beneath; petiole c. 6 cm long, sulcate with 1–2 sheathing cataphylls when young. Scape erect, 1(2)-flowered, with 3–4 sheathing cataphylls, the uppermost sometimes bract-like and up to 4 cm long. Flower reddish tinged on the outside; bracts 1–2, 14–23 × 0.6–1.6 mm, filamentous to lanceolate. Sepals and petals subequal, lanceolate, acute, greenish with red veins and a general reddish tinge to the outside; lateral sepals slightly oblique, 19.5–25 × 3–4.5 mm; petals slightly shorter than sepals. Lip cream or pale yellow with a prominent red or purple venation, 18–25 × 14–18 mm, ovate to oblong-ovate, obscurely 3-lobed, bearing 2 parallel fleshy pubescent ridges from base of lip to base of mid-lobe; side lobes rounded or oblong, obtuse, erect, enclosing the column; mid-lobe ovate, reflexed. Column curved, 9–13.5 mm long, clavate, glabrous; anther front rim with two protruding teeth; anther purple; pollinia yellow, c. 2.5 mm long; ovary 4.5–5.5 mm long.

Zambia. W: near East Lumwana R., fl. & st. 22.x.1969, *Williamson & Drummond* 1639 (SRGH). C: 130 km E of Lusaka, fl. 4.xii.1974 & st. cult.? 22.i.1975, *Williamson & Gassner* 2332 (K). E: 10 km N of Chipata, 1150 m, fl. 24.xi.1961, *King* 469 (BOL; SRGH). S: Mazabuka Distr., 64 km S of Lusaka, fl. 11.xi.1971 [1970?] & st. xii.1971, *G. Williamson* 2078 (K; SRGH). **Zimbabwe**. N: c. 10.5 km from Gokwe on the road to Sare/Nemangwe, near Marope R., st. 24.ii.1964, *Bingham* 1147 (SRGH); Hurungwe Distr., near Kanyanga, tributary of Msukwe R., 900 m, fl. 18.xi.1952, *Wild* 4174 (K; MO; PRE; SRGH). C: Harare Distr., Rumani, 1300 m, fl. xi.1946 & st. iii.1946, *Greatrex* 112 (K; PRE; SRGH). E: Mutare Distr., Burma Valley, left bank of Nyamakari R., Manyera Farm, 650 m, st. 19.i.1959, *Chase* 7157 (SRGH); Burma Valley, Manyera Farm, 650 m, fl. 30.xi.1964, *Chase* 8185 (K; SRGH). **Malawi**. N: Rumphi Distr., Nyika Approach, c. 5 km after turn-off to Thazima, 1500 m, st. 17.i.1987, *I. & E. la Croix* 938 (K). S: Zomba Distr., Thondwe Area, Mbala Estate, 1070 m, fl. 13.xi.1984 & st. 12.xii.1984, *Pettersson* 299 (K; MAL; UPS). **Mozambique**. N: Near Serra Jeci, along road to lime quarry, st. 7.ii.1982, *Pettersson* 189 (LMU; UPS).

Also in Tanzania. Woodland, grassland, grassland-scrub, *Syzygium* thicket and pine or *Gmelina* plantations; 600–1500 m.

11. **Nervilia shirensis** (Rolfe) Schltr. in Bot. Jahrb. Syst. **45**: 403 (1911); **53**: 553 (1915). —Jacobsen in Kirkia **9**, 1: 155 (1973) pro parte. —Grosvenor in Excelsa **6**: 84 (1976). —Williamson, Orch. S. Centr. Africa: 143 (1977). —Saunders in Soc. Malawi J. **33**, 2: 15, 38, 40 (1980). —Cribb in F.T.E.A., Orchidaceae: 279, fig. 55,1 (1984). —Pettersson in Lindleyana **4**: 34 (1989); in Acta Univ. Ups., Comp. Sum. Upps. Diss. Fac. Sci. **281**: 25 (1990); in Orchid Monogr. **5**: 65 (1991). —la Croix et al., Orch. Malawi: 170, t. 112, pl. 52 (1991). TAB. **91**. Type: Malawi, Shire Highlands, *Buchanan* 317 (K, holotype).

Pogonia shirensis Rolfe in F.T.A. **7**: 187 (1897). —Binns, First Check-list Herb. Fl. Malawi: 75 (1968).

Pogonia buchananii Rolfe in F.T.A. **7**: 187 (1897). —Binns, First Check-list Herb. Fl. Malawi: 75 (1968). Type: Malawi, (Blantyre or Chiradzulu Distr. [fide Binns]), *Buchanan* 1342 (K, holotype).

Nervilia buchananii (Rolfe) Schltr. in Bot. Jahrb. Syst. **45**: 403 (1911); Die Orchideen: 105 (1915); in Bot. Jahrb. Syst. **53**: 553 (1915). —Fries, Wiss. Ergebn. Schwed. Rhod.-Kongo-Exped.: 243 (1916).

Pogonia purpurata sensu Brain in Rhod. Agric. J. **1939**: 827, t. 65 (1939) non Rchb.f. & Sonder.

Erect terrestrial herb (10)15–44 cm tall, glabrous except for lip and subterranean parts. Tuber 1.5–3 × 1.5–2.8 × 0.8–2 cm, subspherical or ovoid, 4–7-noded. Leaf

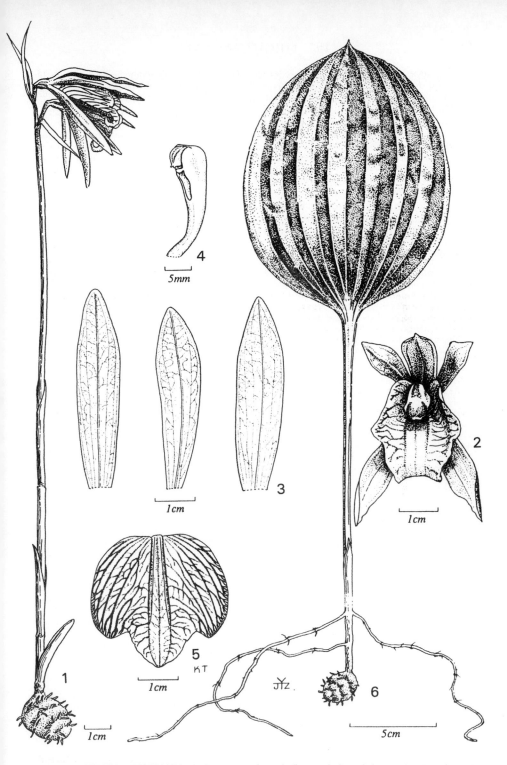

Tab. 91. NERVILIA SHIRENSIS. 1, flowering plant; 2, flower; 3, from left to right: dorsal sepal, petal, lateral sepal; 4, column; 5, lip; 6, plant in leaf, 1–6 from *Pettersson* 300. Drawn by Yue Jingzhu and Kerstin Thunberg. From Nervilia (Orchidaceae) in Africa.

solitary, appearing after flowering, erect to suberect, large, 6–16 cm long and 3.5–11 cm wide, ovate to almost round, glabrous, pleated; apex obtuse to acute or slightly acuminate; base attenuate to truncate or slightly cordate; veins 6–10; petiole 5.5–17 cm long, sulcate with 1–5 sheathing cataphylls when young. Scape erect, terete, (1)2–3-flowered, with 2–4 sheathing cataphylls. Flowers large, more or less secund; bracts 22–31 mm long, linear, setose, acute. Sepals and petals subequal, linear-lanceolate, acute, greenish-yellow with red veins and a general reddish tinge to the outside; lateral sepals slightly oblique, 25–42(45) × 5.2–7.5 mm; petals slightly shorter than sepals. Lip cream or pale yellow with prominent red or purple venation, (19)24–40 × 22–31 mm, ovate, 3-lobed, bearing 2 parallel fleshy pubescent ridges running from base of lip to base of mid-lobe; side lobes oblong, rounded to obtuse, erect, enclosing the column; mid-lobe ovate, obtuse, reflexed. Column curved, 16–25 mm long, clavate, glabrous; anther front rim with 2 protruding teeth; pollinia yellow, 3.9–5.2 mm; stigmatic surface 5–8 mm in diameter. Capsule 14–19 × 12 mm.

Zambia. N: Mbala Distr., Kaka R. Gorge, 1950 m, fl. 14.xi.1956, *Richards* 6984 (K). W: Kabompo Distr., 57 km N of Manyinga, fl. 24.x.1969, *Williamson & Drummond* 1680 (SRGH); Solwezi Distr., Kansanshi Mine, st. 13.xi.1962, *Holmes* 0382 (K). C: Kapiri Mposhi, fl. & fr. 15.xi.1963, *Fanshawe* 8140 (K); 13 km W of Lusaka, st. 9.ii.1975, *Williamson & Gassner* 2370 (K). E: Katete, St Francis' Hospital, 1050 m, fl. 29.xi.1955, *Wright* 46A (K). S: Livingstone Distr., fl. vi.1955, *Seale* 15 (K). **Zimbabwe**. N: Mazowe Distr., between Amandas and Marodzi R., 3 km N of Amandas on Portlock Rufaro Rd., fl. 9.xii.1975 & st. i.1976, *Maratos* in *GHS* 242,318 (K; SRGH). C: Highlands, 1350 m, fl. 3.xi.1948, *Greatrex* 70 (B; K; L; PRE; SRGH); Highlands, st. ii.1946, *Greatrex* in *GHS* 22945 (K; SRGH). E; Mutare, Commonage, West base of Cross Hill, 1050–1200 m, st. 26.ii.1956, *Chase* 5999 (BM; K; SRGH); Mutare, Commonage, NE of old railway water supply intake, E of Morningside, 1120 m, fl. 12.xi.1961, *Chase* 7554 (EA; K; LISC; PRE; SRGH). **Malawi**. N: Mzimba Distr., 11 km S of Euthini (Eutini), 11 km W of Rukuru R., 1250 m, st. 31.i.1976, *Pawek* 10781B (K); Nkhata Bay Distr., near Chikangawa, 1650 m, fl. 26.xi.1978, *Phillips* 4285 (K; MAL; MO; SRGH; WAG; Z). C: Dedza Distr., near Bembeke, Ngonoonda-Bembeke Rd., fl. 15.xi.1967, *Salubeni* 892 (K; MAL; SRGH). S: Zomba Distr., Thondwe Area, Mbala Estate, 1070 m, fl. 13.xi.1984 & st. 12.xii.1984, *Pettersson* 300 (K; LISC; MAL; SRGH; UPS). **Mozambique**. N: Near Lake Malawi (Nyasa), c. 480 m, fl. 1902, *Johnson* 523 (K); Lichinga, 1356 m, st. 3.ii.1982, *Pettersson* 152 (K; LISC; LMU; MAL; SRGH; UPS). Z: Near Milange, at km 5, fl. immat. 13.xi.1942, *Mendonça* 1250 (LISC).

Also in Tanzania, Zaire and Angola. *Brachystegia* woodland, riverine forest, grassland, and in pine, cypress or *Gmelina* plantations; 480–1950 m.

28. **DIDYMOPLEXIS** Griff.

Didymoplexis Griff. in Calcutta J. Nat. Hist. **4**: 383, t. 17 (1844).
Apetalon Wight, Ic. Pl. Ind. Or. **5**: 22, t. 1758 (1851).

Leafless, saprophytic herb lacking chlorophyll; tubers fleshy, horizontal. Stems erect, unbranched; inflorescence terminal, racemose, 1- to many-flowered. Flowers small; tepals adnate at base forming a short tube; lip free from tepals, adnate to column foot. Column long, wider at apex, with ± slender arms; stigma near apex; anther declinate. Pollinia 4, lacking caudicle. Pedicel elongating rapidly as capsule matures.

A genus of about 20 species, mostly Asiatic, with one in tropical and South Africa and one in Madagascar.

Didymoplexis africana Summerh. in Kew Bull. **6**: 465 (1952); in Hooker's Icon. Pl. 36, t. 3563 (1956). —P.F. Hunt in F.T.E.A., Orchidaceae: 263 (1984). —la Croix et al., Orch. Malawi: 171 (1991). TAB. **92**. Type from Tanzania.
Gastrodia sp. no. 1 (*Ball* 1422), Grosvenor in Excelsa **6**: 82 (1976).

Leafless, glabrous saprophytic herb; tubers up to 4 × 0.7 cm, fusiform. Scape slender, 8–21 cm high. Stem with 2–3 sheathing cataphylls c. 2 mm long near base. Inflorescence densely 1- to many-flowered. Flowers spreading or erect, white or cream flushed with pink. Ovary and pedicel 10 mm long at anthesis. Tepals ± connate for about half their length forming a bilabiate tube. Dorsal sepal erect, 15 × 3.5–4 mm,

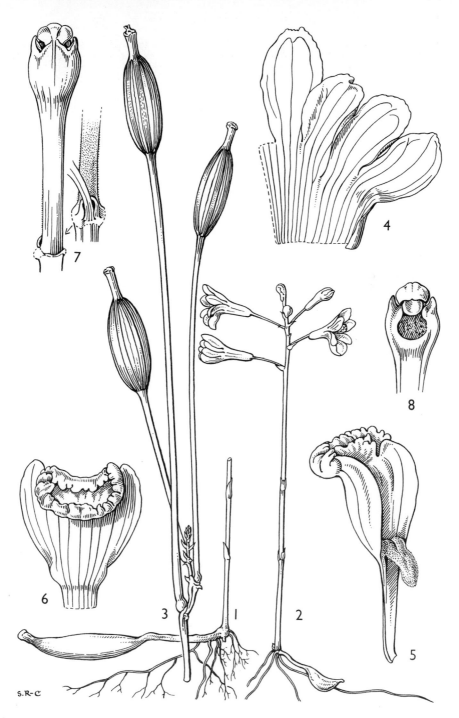

Tab. 92. DIDYMOPLEXIS AFRICANA. 1, base of scape and tuber (×1), from *Zimmermann* s.n.; 2, habit (×1); 3, infructescence (×1); 4, tepals (dorsal sepal, one petal, both lateral sepals) (×4); 5, lip, side view (×6); 6, lip, partially flattened, underside (×6); 7, column, from back and front showing attachment of lip (×6); 8, tip of column, front view (×6), 2–8 from *Moreau* 269. Drawn by Stella Ross-Craig. From F.T.E.A.

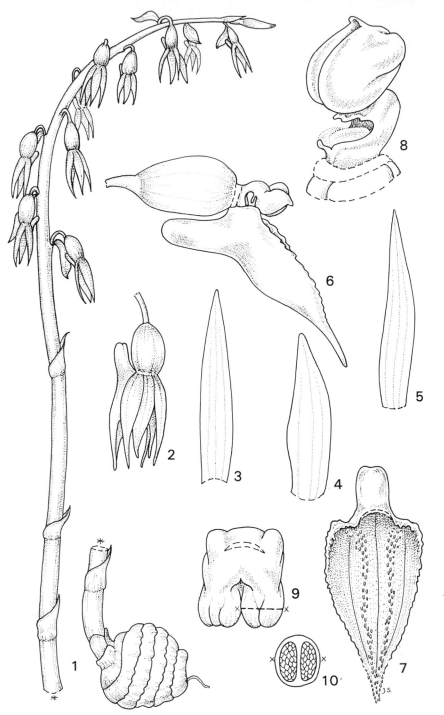

Tab. 93. EPIPOGIUM ROSEUM. 1, habit (×1), from *Friis & Vollesen* 1139; 2, flower (×2); 3, dorsal sepal (×4); 4, petal (×4); 5, lateral sepal (×4); 6, ovary, column and lip, side view (×4); 7, lip and spur, front view (×4); 8, column, side view (×8); 9, anther cap, ventral view (×8); 10, cross-section through anther cap between crosses, showing pollinia (×8), 2–10 from *la Croix* 74 (in K spirit coll. 42348). Drawn by Judi Stone.

subspathulate, joined to petals for c. 7.5 mm. Lateral sepals 10.5–12 × 3.25–3.5 mm, subspathulate, joined to each other for 7.5 mm and to petals for 4.5 mm. All sepals ± verrucose outside. Petals 10.5–12 × 3.5–4.5 mm. Lip free, 10–11 × 5.5 mm, clawed at base then flabellate-orbicular, erose-denticulate at margin with a transverse fleshy callus below apex and a reflexed, ligule-like callus at apex of claw. Column c. 10 mm long, straight, with tridentate apex. Anther hemispherical; pollinia 4, ovoid. Rostellum fleshy.

Zimbabwe. E: Mutare Distr., Vumba, Shigodora Farm, map ref. VP 710 864, fr. 24.iv.1976, *Ball* 1422 (SRGH). Malawi. S: Mulanje Distr., Ruo Gorge, near Lujeri Power Station, c. 760 m, fr. 2.ix.1973, *Ball* 1333 (K; SRGH). Mozambique. MS: Border Farm, near Penhalonga, 24.v.1959, *Ball* 812 (SRGH).

Also in ?Ghana and Tanzania. In deep shade in leaf litter of submontane evergreen forest; up to c. 800 m.

All the specimens from the Flora Zambesiaca area are in fruit and so cannot be assigned to this species with absolute certainty.

29. EPIPOGIUM Borkh.

Epipogium Borkh., Tent. Disp. Pl. German.: 139 (1792).

Leafless, saprophytic herb lacking chlorophyll; rhizome tuberous. Scape erect with a few basal sheaths; inflorescence terminal. Tepals free, subequal; lip spurred, concave with verrucose crests. Column and anther short; pollinia 2, sectile, each with a slender, curved caudicle and a single broad viscidium. Stigma broad, prominent; rostellum absent. Capsule ovoid, pendulous, developing quickly.

A genus of 2 species, one in temperate Eurasia and one in the Old World tropics.

Epipogium roseum (D. Don) Lindl. in J. Linn. Soc., Bot. 1: 177 (1857). —Summerhayes in F.W.T.A. ed. 2, 3: 207 (1968). —Geerinck in Fl. Afr. Centr., Orchidaceae pt. 1: 262 (1984). —Hunt in F.T.E.A., Orchidaceae: 238 (1984). —la Croix et al., Orch. Malawi: 172 (1991). TAB. 93. Type from Nepal.

Limodorum roseum D. Don, Prodr. Fl. Nepal: 30 (1825).
Epipogium nutans Rchb.f. in Bonplandia 5: 36 (1857). —Rolfe in F.T.A. 7: 188 (1897). Type from Java.
Epipogium africanum Schltr. in Bot. Jahrb. Syst. 45: 399 (1911). Type from Cameroon.

Leafless, saprophytic herb; tuberous rhizome up to 5 × 3.5 cm, ovoid, horizontal, villous, spongy. Stem 18–60 cm tall, hollow, pinkish-white with red streaks, with c. 6 bract-like cataphylls. Inflorescence 12–20 cm long, laxly several- to many-flowered. Pedicel 7 mm long, ovary 4 mm long at anthesis, soon enlarging; bracts 12–15 mm long, ovate. Flowers drooping, not opening wide, off-white to dull pink, with purple or dull red spots. Sepals and petals 9–12 × c. 2 mm, lanceolate, acute. Lip c. 10 × 5 mm, ovate, with 2 verrucose crests running along its length; spur blunt, 3–4 × 1–1.5 mm. Column shorter than anther; stigma at base of column.

Malawi. S: Thyolo Distr., Naming'omba Forest, 1100 m, fl. 11.xii.1980, *Johnston-Stewart* 14 (K).
Also in Ghana, Nigeria, Cameroon, Bioko, Annobon, Zaire and Angola, and in Indo-Malaysia to Australia and Vanuatu.

Deep shade in forest, in leaf-litter; c. 1100 m.

The life cycle of this species is very short, taking less than a week from the time the shoot appears above ground to dehiscence of the fruit.

Tab. 94. CALANTHE SYLVATICA. 1, pseudobulbs (×⅔), from *Cribb & Grey-Wilson* 10706; 2, leaf (×⅔), from *Cribb & Grey-Wilson* 10610; 3, inflorescence (×⅔); 4, dorsal sepal (×2); 5, lateral sepal (×2); 6, petal (×2); 7, lip, showing variation (×2), 3–7 from *Cribb et al.* 11480; 8, lip, showing variation (×2), from *Cribb & Grey-Wilson* 10706; 9, column and spur (×2); 10, clinandrium, front view (×2); 11, anther cap, front and back views (×4); 12, pollinia (×6), 9–12 from *Cribb et al.* 11480. Drawn by Susan Hillier. From F.T.E.A.

30. CALANTHE R. Br.

Calanthe R. Br. in Bot. Reg. 7, t. 573 (1821) nom conserv. —Rolfe in F.T.A. 7: 45 (1897).

Large or medium-sized terrestrial herb; pseudobulbs leafy and often obscure, with 1 to several nodes. Leaves deciduous or persistent, petiolate, plicate. Inflorescence erect, racemose, few- to many-flowered. Flowers usually large and showy; tepals spreading; lip fused to the column at base, usually 3-lobed, spurred, often with a warty callus at the base. Column short; pollinia 8, clavate, viscidium elliptic or lanceolate; rostellum bifid.

A genus of over 200 species, widespread from Asia to Australasia, a few occurring in Madagascar, with one species each in Africa, Central America and the West Indies.

Calanthe sylvatica (Thouars) Lindl., Gen. Sp. Orchid. Pl.: 250 (1833). —Rchb.f. in Walpers, Ann. Bot. **6**: 914 (1864). —Schltr. in Fedde, Repert. Spec. Nov. Regni Veg., Beih. **33**: 165 (1925). —Geerinck in Fl. Afr. Centr., Orchidaceae pt. 1: 264 (1984). —Cribb in F.T.E.A., Orchidaceae: 282 (1984). —la Croix et al., Orch. Malawi: 173 (1991). TAB. **94**. Type from Mascarene Islands.
    Centrosis sylvatica Thouars, Hist. Orchid. [Fl. Iles Austr. Afr.]: tab. gen., figs. 35, 36 (1822).
    Calanthe natalensis (Rchb.f.) Rchb.f. in Bonplandia **4**: 322 (1856). —Eyles in Trans. Roy. Soc. S. Afr. **5**: 335 (1916). —Wild in Clark, Victoria Falls Handb.: 140 (1952). —Grosvenor in Excelsa **6**: 79 (1976). —Williamson, Orch. S. Centr. Africa: 131 (1977). Type from South Africa (Natal).
    Calanthe corymbosa Lindl. in J. Linn. Soc., Bot. **6**: 129 (1862). —Rolfe in F.T.A. **7**: 46 (1897). —Summerhayes in F.W.T.A. ed. 2, **3**: 226 (1968). —Grosvenor in Excelsa **6**: 79 (1976). —Williamson, Orch. S. Centr. Africa: 130 (1977). Type from Bioko (Fernando Po).
    Calanthe sanderiana Rolfe in Gard. Chron., Ser. 3: 396 (1892). —Eyles in Trans. Roy. Soc. S. Afr. **5**: 335 (1916). —Morris, Epiphyt. Orch. Malawi: 36 (1975).
    Calanthe delphinioides Kraenzl. in Bot. Jahrb. Syst. **17**: 55 (1893). —Rolfe in F.T.A. **7**: 46 (1897). Type from Cameroon.
    Calanthe volkensii Rolfe in F.T.A. **7**: 46 (1897). Type from Tanzania.
    Calanthe neglecta Schltr. in Bot. Jahrb. Syst. **53**: 570 (1915). Type from Tanzania.
    Calanthe stolzii Schltr. in Bot. Jahrb. Syst. **53**: 569 (1915). Type from Tanzania.
    Calanthe schliebenii Mansfeld in Notizbl. Bot. Gart. Berlin **11**: 808 (1933). Type from Tanzania.

Large terrestrial herb up to 70 cm tall. Pseudobulbs rather obscure, 2–5 × 1.5 cm, conical, with 4–5 nodes. Leaves 3–5 ± rosulate, suberect to spreading; leaf up to 35 × 12 cm, lanceolate, acute, plicate, dark green; petiole 5–25 cm long. Peduncle slightly pubescent with 2 sheath-like leaves; inflorescence densely several- to many-flowered. Flowers white or purple with orange callus. Pedicel and ovary each 10–15 mm long; bracts 1–3 cm long. Sepals 9–37 × 4–14 mm, lanceolate, acute, the lateral sepals usually slightly longer than the dorsal ones and oblique. Petals 8–25 × 4–14 mm, elliptic or oblanceolate. Lip 3-lobed at the base, 11–15 × 6–25 mm, with 3 ridges at the base. Side lobes auriculate or crescent-shaped, 5–7 mm long; mid-lobe flabellate to obovate, emarginate, sometimes with a small tooth in the sinus. Spur slender, straight or sigmoid, 1–4 cm long.

Zambia. N: Chinsali Distr., Ishiba Ngandu (Shiwa Ngandu), 1500 m, fl. 8.vi.1956, *E.A. Robinson* 1630 (K). W: Kitwe, Ichimpi, fl. 26.iii.1968, *Mutimushi* 2583 (K; NDO). E: Lundazi Distr., Nyika Plateau, Chowo Forest, c. 2000 m, fl. *Laing* (SRGH). S: Victoria Falls, fl. 22.ii.1961, *Morze* 21 (K). **Zimbabwe**. W: Victoria Falls, 900 m, fl. 9.ii.1912, *Rogers* 5697 (K; PRE). C: Makoni Distr., c. 20 km from Rusape, Chiduku CL (TTL), fl. ii.1967, *Masterson* in *GHS* 188, 878 (K; SRGH). E: Mutare Distr., N Vumba Mts., 1120 m, fl. 19.iv.1959, *Chase* 7102 (K). **Malawi**. N: Chitipa Distr., Misuku Hills, Mugesse (Mughese) Forest, c. 1700 m, fl. 10.iv.1984, *la Croix* 614 (K). C: Lilongwe Distr., Dzalanyama Forest Reserve, 1580 m, fl.iii.1984, *Dowsett-Lemaire* 1115 (K). S: Thyolo Distr., Mwalanthunzi Forest, 1100 m, fl. 31.iii.1984, *la Croix* 578 (K). **Mozambique**. MS: SE slopes of Mt. Gorongosa, 1100 m, fl. 25.vii.1970, *Müller & Gordon* 1456 (K; SRGH).
    Widespread in tropical Africa, South Africa, Madagascar, Mascarene Is., and possibly tropical Asia. In shade in evergreen and riverine forest, often near streams and in swampy areas; 900–1700 m.

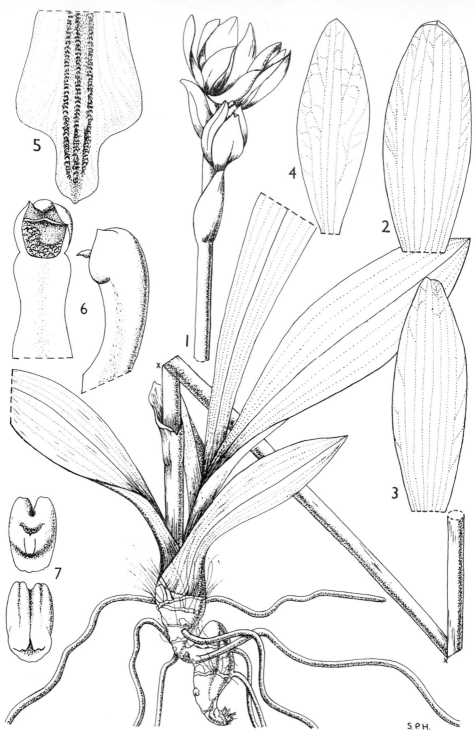

Tab. 95. PHAIUS OCCIDENTALIS. 1, habit (×$\frac{2}{3}$), from *Morze* 145 and *Greenway* 7078; 2, dorsal sepal (×2); 3, lateral sepal (×2); 4, petal (×2); 5, lip (×2); 6, column, front and side views (×4); 7, anther cap, back and front views (×4), 2–7 from *E.A. Robinson* 5831. Drawn by Susan Hillier. From F.T.E.A.

This species is very variable, particularly in size of flower. Populations in any one area tend to have flowers of similar size and colour, for example plants from the southern and central regions of Malawi have white, relatively small flowers, while those from the northern region are larger and always purple-flowered.

31. PHAIUS Lour.

Phaius Lour., Fl. Cochinch.: 529 (1790).
Gastorkis Thouars in Nouv. Bull. Sci. Soc. Philom. Paris 1: 317 (1809).
Tankervillia Link, Handb. 1: 251 (1829).

Terrestrial (rarely epiphytic) herb with fibrous roots, sometimes with pseudobulbs. Leaves arising from a pseudobulb or spaced along the stem, petiolate, plicate. Inflorescence 1- to few-flowered. Sepals and petals free, subequal; lip erect, entire or 3-lobed, spurred or saccate, with longitudinal ridges on the upper surface. Column long, clavate; anther operculate, incumbent, 2-celled. Pollinia 8 in 2 groups of 4 with caudicles but no viscidia. Stigma near apex of column.

About 50 species in tropical regions of the Old World, with 2 in mainland Africa.

Phaius occidentalis Schltr. in Warb., Kunene-Sambesi-Exped. Baum: 211 (1903). —Grosvenor in Excelsa **6**: 84 (1976). —Williamson, Orch, S. Centr. Africa: 132 (1977). —Geerinck in Fl. Afr. Centr., Orchidaceae pt. 1: 266 (1984). —Hunt in F.T.E.A., Orchidaceae: 281 (1984). —la Croix et al., Orch. Malawi: 174 (1991). TAB. **95**. Type from Angola.

Glabrous terrestrial herb 50–70 cm tall, the whole plant turning blue-black when damaged or dried. Roots many, arising from a pseudobulb-like structure, fleshy, 2–3 mm in diameter. Leaves 5–6 ± rosulate; leaf c. 16 × 5 cm, elliptic-lanceolate to obovate, plicate; petiole up to 5.5 cm long. Inflorescence densely 5–6-flowered; flowers erect, white, the lip spotted with brownish-pink at base, the disk yellow. Ovary and pedicel 13 mm long; bracts 25–30 mm long. Dorsal sepal 25–30 × 6–10 mm, lanceolate, obtuse, cucullate at apex; lateral sepals 28–30 × 6.5–8 mm, obliquely elliptic-lanceolate. Petals c. 27 × 7 mm, lanceolate, acute, keeled. Lip up to 28 × 16 mm, 3-lobed, with a blunt, sac-like spur c. 3 mm long; mid-lobe up to 10 mm long with 3 raised hairy keels running from near apex to base of lip; side lobes rounded. Column 10 mm long, slightly hairy. Capsule hairy.

Zambia. N: Kasama Distr., Mungwi, fl. 20.xi.1960, *E.A. Robinson* 4009 (K). W: Mwinilunga Distr., NE of Dobeka Bridge, fl. 8.xi.1937, *Milne-Redhead* 3151 (K). C: Chakwenga Headwaters, 100–129 km E of Lusaka, fl. 16.xi.1963, *E.A. Robinson* 5831 (K). **Zimbabwe**. C: Harare, in vlei, fl. 23.xi.1957, *Greatrex* s.n. (K). **Malawi**. N: South Viphya, bog near Nthungwa forest, fl. 8.xi.1968, *Salubeni* 1224 (MAL).
Also in Zaire, Uganda, Tanzania and Angola. Seasonally wet, swampy grassland or dambos; 1150–1500 m.

32. LIPARIS Rich.

Liparis Rich. in Orch. Eur. Annot.: 21, 30, 38 (1817); in Mém. Mus. Hist. Nat. 4: 43, 52, 60 (1818), nom. conserv.
Leptorchis Thouars in Nouv. Bull. Sci. Soc. Philom. Paris 1: 317 (1809), as "*Leptorkis*"; orth. mut. Thouars, Hist. Orchid. [Fl. Iles Austr. Afr.]: t. 25 (1822).

Epiphytic, lithophytic or terrestrial herb; stems usually ± swollen at the base to form pseudobulbs. Leaves 1 to several, broad, plicate, thin-textured and unjointed, or narrow, rigid and articulated at the base. Inflorescence terminal or subumbellate, usually erect, racemose, few- to many-flowered. Flowers usually yellow, yellow-green or purplish. Tepals spreading or reflexed, petals often linear. Lip simple, usually much larger than the sepals and petals, 2-lobed or 3-lobed, with entire, dentate or crenate margins, usually with 2 calli at the base. Column fairly long, arched, terete or somewhat winged, with or without a column-foot; pollinia 4, in 2 pairs.

Tab. 96. LIPARIS BOWKERI. 1, habit (×1), from *Greatrex* in *GHS* 50258; 2, flower, side view (×2); 3, flower, front view (×2); 4, cross-section of stem (×4); 5, dorsal sepal (×3); 6, lateral sepal (×3); 7, petal (×3); 8, lip (×3); 9, column and lip, side view (×3); 10, column, back view (×6); 11, column, side view (×6); 12, column, front view (×6); 13, pollinia (×6), 2–13 from *Hepper* 7382 (in K spirit coll. 52255). Drawn by Judi Stone.

A genus of about 250 species, mostly tropical but also in temperate regions of Old and New World.

1. Foliage leaf 1 · 2
 − Foliage leaves 2 to several · 3
2. Plant terrestrial; leaf prostrate, cordate, arising from an underground tuber · · · · · · · · · ·
   · 6. *mulindana*
 − Plant epiphytic; leaf erect, ± oblanceolate, arising from a pseudobulb · · · · · · 2. *caespitosa*
3. Lip 3-lobed, mid-lobe much longer than the side lobes · · · · · · · · · · · · · · · 10. *tridens*
 − Lip entire or 2-lobed, sometimes with a small apiculus · 4
4. Leaves cordate at the base, with a distinct petiole; inflorescence subumbellate; lip 2-lobed, deeply toothed · 4. *latialata*
 − Leaves not cordate; petiole obscure; inflorescence racemose · · · · · · · · · · · · · · · · 5
5. Margin of lip dentate · 6
 − Margin of lip entire · 7
6. Plant very small, c. 5 cm tall; dorsal sepal 4–5 mm long · · · · · · · · · · 3. *chimanimaniensis*
 − Plant 20–25 cm tall; dorsal sepal 10 mm long; capsule winged at apex · · · 5. *molendinacea*
7. Plant epiphytic · 8
 − Plant terrestrial · 9
8. Dorsal sepal c. 5 mm long; column less than 1.5 mm long · · · · · · · · · · · · · · 8. *nyikana*
 − Dorsal sepal 8–11 mm long; column 3–3.6 mm long · · · · · · · · · · · · · · · · · · · 1. *bowkeri*
9. Plant less than 8 cm tall including inflorescence; pseudobulbs white, subterranean, less than 1 cm in diameter; lip as long as broad · 9. *rungweensis*
 − Plants over 8 cm tall, usually much taller; pseudobulbs green, not subterranean; lip broader than, or as broad as, long · 10
10. Dorsal sepal 5–6.5 mm long; lip less than 5 × 5 mm, emarginate, usually maroon · · · · · ·
   · 7. *nervosa*
 − Dorsal sepal 8–11 mm long; lip 5–7 × 6–8 mm, obtuse or rounded at apex, green, yellow or buff · 1. *bowkeri*

1. **Liparis bowkeri** Harv., Thes. Cap. **2**: 6, t. 109 (1863). —Ridley in J. Linn. Soc., Bot. **22**: 270 (1886). —Bolus, Icon. Orchid. Austro-Afric. **1**, 1: t. 2 (1893). —Rolfe in F.T.A. **7**: 21 (1897). —Stewart et al., Wild Orch. South. Africa: 216 (1982). —Cribb & Bowden in F.T.E.A., Orchidaceae: 299 (1984). —Geerinck in Fl. Afr. Centr., Orchidaceae pt. 1: 276 (1984). — la Croix et al., Orch. Malawi: 175 (1991). TAB. **96**. Type from South Africa.
   Liparis ruwenzoriensis Rolfe in F.T.A. **7**: 20 (1897). Type from Zaire.
   Liparis neglecta Schltr. in Bot. Jahrb. Syst. **53**: 561 (1915). —Summerhayes in Kew Bull. **6**: 468 (1952). —Morris, Epiphyt. Orch. Malawi: 37 (1970). —Moriarty, Wild Fls. Malawi: t. 29, 5 (1976). —Grosvenor in Excelsa **6**: 84 (1976). —Williamson, Orch. S. Centr. Africa: 113 (1977). Type from Tanzania.
   Liparis purseglovei Summerh. in Kew Bull. **8**: 132 (1953). Type from Uganda.

Terrestrial, lithophytic or epiphytic herb, 8–40 cm tall. Pseudobulbs to 7 × 1.5 cm, conical, 2–5-leaved. Leaves 5–14 × 1–7.5 cm, lanceolate or ovate, acute, plicate, thin-textured. Inflorescence laxly c. 25-flowered; peduncle with 1–several triangular bracts. Flowers green to yellow-green, or clear yellow, turning buff-orange with age; lip with a metallic grey or orange central line. Pedicel and ovary 7–11 mm long; bracts 10–15 mm long, lanceolate, acuminate. Dorsal sepal erect, 8–11 × 1–2.5 mm, linear-lanceolate; lateral sepals 6–9 × 2.5–4 mm, obliquely elliptic, falcate, obtuse, lying under the lip. Petals 5–12 × c. 0.5 mm, linear, acute, pointing downwards and outwards. Lip 5–7 × 6–8 mm, orbicular or transversely oblong, auriculate at base, the margin entire, with a bifid callus in the throat. Column 3–4 mm long, arched to form almost a right angle.

Zambia. N: Mbala Distr., Lake Chila outflow, 1600 m, fl. 8.iii.1950, *Bullock* (K). E: Nyika Plateau, in forest on rocks, fl. xii.1966, *G. Williamson* 240 (K). **Zimbabwe**. C: Shurugwe Distr., Dunraven Falls, 1200 m, fl. 17.ii.1964, *Mitchell* 721 (K; SRGH). E: Chimanimani Distr., Fairview, 1750 m, fl. 16.i.1955, *Greatrex* in *GHS* 50258 (K; SRGH). S: Bikita Distr., S slope of Mt. Horzi, c. 1100 m, fl. 11.v.1969, *Pope* 153 (K; SRGH). **Malawi**. N: South Viphya, Mtangatanga Forest Reserve, 1650 m, fl. 31.i.1987, *Cornelius* in *la Croix* 952 (K; MAL). S: Blantyre Distr., Ndirande Mt., 1360 m, fl. 30.i.1955, *Morris* 107 (K).
   Also in Zaire, Rwanda, Burundi, Uganda, Kenya, Tanzania and South Africa. Riverine forest, high rainfall *Brachystegia* woodland, under rocks and on mossy rocks and fallen trunks in shade; 1100-2200 m.
   Plants of this species are very variable in size, those growing lithophytically or epiphytically usually being larger than terrestrial specimens.

2. **Liparis caespitosa** (Thouars) Lindl. in Bot. Reg. **11**: sub t. 882 (1825). —Summerhayes in Kew Bull. **6**: 468 (1952). —Holttum in Fl. Malaya **1**: 205 (1964). —Cribb & Bowden in F.T.E.A., Orchidaceae: 293 (1984). —la Croix et al., Orch. Malawi: 176 (1991). Type from Réunion.

 Epidendrum cespitosum Lam., Encycl. **1**: 187 (1783). Type from Réunion.

 Malaxis caespitosa Thouars, Hist. Orchid. [Fl. Iles Austr. Afr.]: tab. gen., figs. 90 (1822), but not based on *Epidendrum cespitosum* Lam.

 Leptorchis caespitosa (Thouars) O. Kuntze, Rev. Gen. Pl. **2**: 671 (1891).

 Cestichis caespitosa (Thouars) Ames, Orchid. **2**: 132 (1908).

Small epiphytic herb; pseudobulbs 13–25 × 10 mm, ovoid, smooth and green when young, later yellowish and wrinkled, 1-leafed at apex, pseudobulbs set close together on the rhizome to form dense mats. Leaf stiff, erect, 6–10 × 1–1.5 cm, oblanceolate, usually light green, jointed at base. Inflorescence terminal, c. 10 cm long, densely 20–40-flowered. Flowers non-resupinate, greenish-yellow to creamy-yellow, very small. Ovary and pedicel 5 mm long; bracts acuminate, 3 mm long. Sepals ovate, acute, reflexed; dorsal sepal 2.5 × 0.5 mm. Lip 2.5 mm long, less than 2 mm wide, oblong, ± 4-lobed, apiculate, with 2 small calli at the base.

 Malawi. S: Zomba Mt., near Chingwe's Hole, 1900 m, fl. 6.ii.1982, *Johnston-Stewart in la Croix 5* (K). **Mozambique**. Z: Serra do Gurué, c. 1280 m, fl. 24.vii.1979, *de Koning 7440* (K; LMU).

 Also in Uganda, Tanzania, Madagascar, Réunion, and extending from Sri Lanka to NE India and to the Philippines, New Guinea, Solomon Is. and Fiji. Epiphytic on trunks and lower branches of trees in evergreen forest; 1280–1900 m.

3. **Liparis chimanimaniensis** G. Will. in Pl. Syst. Evol. **142**: 153, fig. 3 (1983). Type: Zimbabwe, Chimanimani Mts., *Hall 410* (SRGH, holotype).

 Liparis sp. no. 1 (*Hall 410*), Grosvenor in Excelsa **6**: 84 (1976).

Dwarf terrestrial herb, usually less than 5 cm tall; pseudobulbs 7 mm long, ellipsoid. Leaves 2, sometimes 1, c. 3.5 × 1 cm, narrowly elliptical or lanceolate, erect. Peduncle without bracts; inflorescence laxly c. 4-flowered. Flowers semi-erect, green, the lip brown with a yellow margin. Ovary and pedicel slender, 5 mm long. Dorsal sepal 4–5 × 0.5–1 mm, linear, curved; lateral sepals spreading, 4–5 × 0.8–1.5 mm, obliquely lanceolate, subacute. Petals 5–6 × 0.3–0.7 mm, narrowly linear, reflexed. Lip 3–4 × 1.5–2 mm, stiff, porrect, with 2 prominent veins running the length of the upper surface, and with 2 calli 0.5 mm high at the base; margin of epichile minutely serrate. Column curved, 2–3 mm long.

 Zimbabwe. E: Chimanimani Mts., west of Point 71, fl. ii.1958, *Hall 410* (SRGH). Endemic. Well drained rocky slopes in montane zone; c. 7000 ft.

4. **Liparis latialata** Mansf. in Notizbl. Bot. Gart. Berlin **12**: 703 (1935). —Grosvenor in Excelsa **6**: 84 (1976). —Williamson, Orch. S. Centr. Africa: 115, t. 93 (1977). —Cribb & Bowden in F.T.E.A., Orchidaceae: 296 (1984). Type from Tanzania.

Terrestrial herb to 12 cm tall. Pseudobulbs up to 9 cm long and 3 mm in diameter, with several sheathing leaves and 2–3 foliage leaves. Foliage leaves 3–6 × 2.3–5 cm, ovate, cordate, dark green, the margin sometimes undulate; petiole short, c. 1 cm long. Inflorescence subumbellate, 3–8-flowered, shorter than or equal in length to the foliage leaves. Sepals and petals plum-purple coloured; lip yellow or white, purple veined. Pedicel and ovary 9–12 mm long; bracts 3–4 mm long. Dorsal sepal 7–10 × 1–2 mm, linear-lanceolate; lateral sepals 11–15 × 5–6 mm, joined for three quarters of their length. Petals 7–10 mm long, less than 1 mm wide, ± filiform. Lip fleshy, with a small 2-lobed callus at the base, up to 11 × 12 mm, fan-shaped, emarginate with a small apiculus, the apical edge undulate and dentate. Column 2.5–3.5 mm long, curved, winged.

 Zambia. W: Ndola, mushitu of Mwekera stream, fl. 7.xii.1963, *Morze 137a* (K). **Zimbabwe**. E: Chimanimani, Mutzingazi, 1060 m, fl. 14.i.1956, *Ball 556* (K; SRGH).

 Also in Zaire and Tanzania. Riverine and swamp forest, leaf mould in marshy ground; 1060–1120 m.

5. **Liparis molendinacea** G.Will. in Pl. Syst. Evol. **142**: 149, fig. 1 (1983). Type: Zambia, Kawamba Distr., Kalungwishi River above Lumangwe Falls, *G. Williamson* 1244 (SRGH, holotype).

Terrestrial herb c. 25 cm tall; pseudobulbs pyriform, c. 8 cm high, 1.5 cm in diameter at the base. Leaves c. 4, up to 8 × 3 cm, ovate or elliptic, plicate, the margins undulate, the apex somewhat recurved. Peduncle with several ovate, acuminate bracts c. 10 × 5 mm long. Inflorescence laxly c. 12-flowered. Flowers semi-erect, green, the lip with a shiny yellow callus. Ovary and pedicel to 2 cm long; bracts leafy. Dorsal sepal c. 10 × 1.2 mm, linear, curved, obtuse; lateral sepals porrect, c. 10 × 4 mm, obliquely ovate, acute. Petals porrect, c. 12 × 1 mm, linear, obtuse. Lip ± 4-lobed, 10–12 × 4–6 mm, rectangular or oblong, constricted in the middle; hypochile porrect with a 2-lobed basal callus 0.8 mm high and a yellow longitudinal crest; side lobes erect, c. 6 × 6 mm, oblong, rounded; epichile bent down at right angles, somewhat 2-lobed at the apex with a central apiculus, the margins dentate. Column curved, 6 mm long. Capsule ellipsoid, 6-winged with triangular extensions at the apex.

Zambia. N: Kawamba Distr., bank of Kalungwishi River above Lumangwe Falls, fl. xii.1968, *G. Williamson* 1244 (SRGH).
Endemic. Swamp forest, in humus on forest floor; c. 1300 m.

6. **Liparis mulindana** Schltr. in Bot. Jahrb. Syst. **53**: 560 (1915). —Grosvenor in Excelsa **6**: 84 (1976). —Williamson, Orch. S. Centr. Africa: 115 (1977). —Cribb & Bowden in F.T.E.A., Orchidaceae: 292 (1984). —Geerinck in Fl. Afr. Centr., Orchidaceae pt. 1: 279 (1984). — la Croix et al., Orch. Malawi: 176 (1991). Type from Tanzania.

Dwarf terrestrial herb with a short, fleshy rhizome. Leaf 1, appressed to the ground, up to 6 × 5 cm, ovate or ± orbicular, cordate at the base with 8–9 deeply impressed veins, dark olive-green with a reddish tinge above, deep purple beneath. Inflorescence to 15 cm long, 4–8-flowered; peduncle 8–12 cm long, reddish-green with 2–3 narrow, acuminate, reddish bracts c. 5 mm long. Flowers dull reddish, green towards the centre. Pedicel and ovary 5–8 mm long, winged; bracts 4–5 mm long, narrowly triangular, acuminate, purple. Dorsal sepal erect, 6–10 × 1–1.5 mm, linear, fleshy; lateral sepals 6–7 × 3–4 mm, obovate, falcate, joined at first but later separating, projecting forwards, the tips curled down, lying below the lip. Petals 6–11 mm long, less than 1 mm wide, linear, fleshy, deflexed or spreading. Lip 8 × 4–5 mm, geniculate, constricted after a broadly auriculate base then ± heart-shaped, concave with a fleshy disk in the centre; slightly emarginate at the apex. Column 4–6 mm long, geniculate, winged towards the apex.

Zambia. N: Mbala Distr., Chilongwelo, 1480 m, fl. 10.i.1952, *Richards* 433 (K). W: Solwezi Boma, fl. 25.i.1962, *Holmes* 0331 (K). **Zimbabwe**. N: Hurungwe Distr., 8 km E of Kapiri Hill near Mwami (Miami), fl. 11.ii.1969, *James* s.n. (K; SRGH). **Malawi**. N: Nyika National Park c. 5 km N of Thazima, 1650 m, fl. 16.ii.1987, *la Croix* 963 (K). S: Mangochi Distr., Phirilongwe Hill, 1100 m, fl. 17.iii.1985, *Johnston-Stewart* 402 (K).
Also in Zaire, Burundi and Tanzania. *Brachystegia* woodland, often on stony hill sides; 1100–1650 m.

7. **Liparis nervosa** (Thunb.) Lindl., Gen. Sp. Orchid. Pl.: 26 (1830). —Moriarty, Wild Fls. Malawi: t. 29, 4 (1975). —Grosvenor in Excelsa **6**: 84 (1976). —Williamson, Orch. S. Centr. Africa: 114 (1977). —Cribb & Bowden in F.T.E.A., Orchidaceae: 298 (1094). —Geerinck in Fl. Afr. Centr., Orchidaceae pt. 1: 275 (1984). —la Croix et al., Orch. Malawi: 177 (1991). TAB. **97**. Type from Japan.
Ophrys nervosa Thunb., Fl. Jap.: 27 (1784).
Liparis guineensis Lindl. in Bot. Reg. **20**: t. 1671 (1834). —Rolfe in F.T.A. **7**: 20 (1897). — Summerhayes in F.W.T.A., ed. 2, **3**: 214 (1968). —Morris, Epiphyt. Orch. Malawi: 36 (1970). Type from Sierra Leone.
Liparis elata var. *rufina* Ridl. in J. Linn. Soc., Bot. **22**: 260 (1886). Type from Nigeria.
Liparis rufina (Ridl.) Rolfe in F.T.A. **7**: 19 (1897). —Summerhayes in F.W.T.A., ed. 2, **3**: 214, fig. 389 (1968). —Morris, Epiphyt. Orch. Malawi: 39 (1970).
Liparis nyassana Schltr. in Bot. Jahrb. Syst. **53**: 560 (1915). Type from Tanzania.
For further synonymy see Garay & Sweet, Orchids of the S Ryukyu Islands (1974).

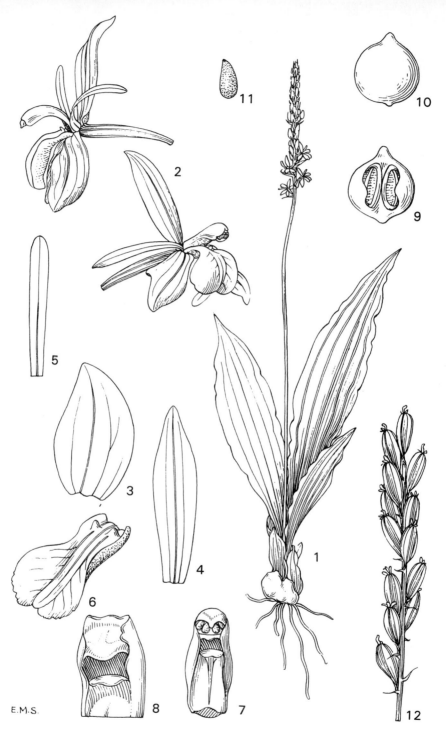

Tab. 97. LIPARIS NERVOSA. 1, habit (×$\frac{1}{2}$), from *Linder* 345; 2, flower (×6); 3, lateral sepal (×9); 4, dorsal sepal (×9); 5, petal (×9); 6, lip (×9); 7, column, from below (×9); 8, column, anther cap and pollinia removed (×20); 9, anther cap, ventral view (×20); 10, anther cap, dorsal view (×20); 11, pollinium (×20); 12, fruiting spike (×1), 11 & 12 from *Hepburn*. Drawn by Margaret Stones. From F.W.T.A.

Terrestrial herb to 60 cm tall. Pseudobulbs up to 4 × 2.5 cm, ovoid, covered with leaf sheaths, partly underground. Leaves 2–3, petiolate with a sheathing base, the lamina to 35 × 9 cm, lanceolate to ovate, acute, ribbed, light green. Peduncle 5-angled with several sheaths; inflorescence densely many-flowered, rhachis to 15 cm long. Flowers small, c. 6 mm in diameter, green or yellow-green, the lip deep purple or green-purple. Pedicel and ovary 10–12 mm long, erect; bracts to 7 mm long. Dorsal sepal 5.3–6.3 × 1–2 mm, linear, fleshy, erect or reflexed; lateral sepals 3–5 × 2–3 mm, oblong, falcate, often rolled up, lying below lip. Petals 4.5–6 mm long, narrowly linear, reflexed. Lip 2.5–4 × 2.5–4.5 mm, ± orbicular, auriculate at the base. Column arched, 2.5–4 mm long with lateral wings towards the apex and an apical wing around the anther.

Zambia. N: Kasama, falls at Power Station, fl. 8.i.1961, *Holmes* 0300 (K). W: Mwinilunga Distr., 10 km E of Mwinilunga, 1300 m, fl. 23.i.1975, *Brummitt, Chisumpa & Polhill* 14038 (K). C: c. 10 km E of Lusaka, 1270 m, fl. 16.i.1956, *King* 267 (K). **Zimbabwe**. C: Marondera, Digglefold, fl. 20.i.1949, *Corby* 354 (K; SRGH). E: Mutare Distr., Vumba Mts., 1120 m, fl. 22.i.1955, *Greatrex* in *GHS* 50096 (K; SRGH). **Malawi**. N: Mzimba Distr., Mzuzu, Lunyangwa Research Stn., 1330 m, fl. 28.i.1987, *la Croix* 946 (K; MAL). C: Ntchisi Forest Reserve, in *Widdringtonia* plantation, 1600 m, fl. 3.ii.1983, *Dowsett-Lemaire* 617 (K). S: Thyolo Distr., Mindali Estate, 1050 m, fl. 19.i.1982, *la Croix* 256 (K).

Widespread in tropical Africa, and also in India and Japan and extending to the Philippines, and in Costa Rica, the West Indies, Brazil, Colombia, Venezuela and Paraguay. *Brachystegia* woodland, riverine vegetation, high altitude or submontane grassland and in cypress and other plantations; 500–1675 m.

Plants of this species are variable in size, those growing in open grassland are usually much smaller and have suberect more strongly pleated leaves than those growing in woodland.

8. **Liparis nyikana** G. Will. in Pl. Syst. Evol. **142**: 151, fig. 2 (1983). —la Croix et al., Orch. Malawi: 178 (1991). Type: Malawi, Nyika Plateau, *Holmes* 0233 (SRGH, holotype; K).

Dwarf epiphytic herb to 12 cm tall. Pseudobulbs 12–28 × 4–8 mm, conical, green, set close together on the rhizome. Leaves 3–6, lowermost 2–3 sheathing, the rest semi-erect, 6–8 × 2–4 cm, ovate, acute, somewhat ribbed. Inflorescence c. 5 cm long, rather laxly up to 6-flowered; peduncle with 1 bract c. 1 mm long. Flowers ± erect, very small, 3 mm in diameter, greenish-yellow turning straw-coloured with age, lip with a reddish central line. Pedicel 2 mm long, ovary 5 mm long; bracts 3 mm long, ovate. Dorsal sepal 4–5 × 0.8 mm, linear, erect; lateral sepals 2–3 × 1–3 mm, united for most of their length, projecting forwards and lying under the lip. Petals 3–5 mm long, narrowly linear, almost filiform. Lip c. 3 × 3 mm, quadrate, curving back, with 2 calli at the base and a thickened central line; margin undulate. Column arched, 1.5–2 mm long.

Malawi. N: Nyika National Park, on tree fern trunks in gully forest, *Holmes* 0233 (K; SRGH); Nyika National Park, Zovochipolo, 2225 m, fl. & fr. 16.ii.1987, *la Croix* 978 (K).

Also in Tanzania. Evergreen forest, on mossy trunks and branches of forest trees, usually near streams; 2000–2350 m.

9. **Liparis rungweensis** Schltr. in Bot. Jahrb. Syst. **53**: 562 (1915). —Cribb & Bowden in F.T.E.A., Orchidaceae: 297 (1984). —la Croix et al., Orch. Malawi: 179 (1991). Type from Tanzania. *Liparis rupicola* Schltr. in Bot. Jahrb. Syst. **53**: 563 (1915). Types from Tanzania.

Very small terrestrial herb; pseudobulbs subterranean, c. 10 × 8 mm. Leaves 2, erect, arising at ground level, 25–35 × 15–25 mm, ovate to oblong-ligulate, coriaceous. Inflorescence erect, 4–6.5 cm long, c. 10-flowered. Flowers green. Ovary and pedicel 5–7 mm long; bracts of similar length. Dorsal sepal 7 × 2 mm, lanceolate, acute, auricled at the base, the margin rolled back; lateral sepals 4 × 2 mm, falcate curving forwards under the lip. Petals 7 × 1 mm, linear, deflexed, margins revolute. Lip entire, 6 × 4 mm, geniculate, with a fleshy 2-lobed callus at the base; epichile tapering to an acute apex which is reflexed, so that lip appears 2-lobed. Column 5 mm long, curved.

Zambia. E: Nyika Plateau, near Zambian Rest-house, 2100 m, fl. 11.iii.1982, *Dowsett-Lemaire* 360 (K).

Also in southern Tanzania. Submontane grassland, in tall grass and scrub; c. 2100 m.

10. **Liparis tridens** Kraenzl. in Bot. Jahrb. Syst. **28**: 162 (1900). —Summerhayes in F.W.T.A. ed. 2, **3**: 214 (1968). —Cribb & Bowden in F.T.E.A., Orchidaceae: 295 (1984). —la Croix et al., Orch. Malawi: 179 (1991). Type from Cameroon.

Dwarf epiphytic herb forming small clumps. Pseudobulbs up to 13 × 7 mm, ovoid, with about 7 distichous leaves. Largest leaf at apex of pseudobulb, up to 7.5 × 1.5 cm, oblong-lanceolate, keeled below, the margins undulate. Inflorescence 5–10 cm long, fairly densely up to 20-flowered. Flowers yellow-green. Dorsal sepal 3 × 1 mm, ovate, acute; lateral sepals 3.5 × 1.5–2 mm, ovate, acute. Petals 3 × 0.5 mm, linear. Lip 3-lobed near the base, with a basal callus; side lobes triangular, mid-lobe 5 × 0.75 mm, linear, acuminate, sharply curved up about half-way. Column 1.5 mm long, with 2 rounded wings in the basal half and 2 rounded wings projecting forwards at the apex.

Malawi. S: Mulanje Distr., Mchese Mt., fl. i.1982, *Johnston-Stewart*, sight record only.

Also in Ivory Coast, Nigeria, Cameroon, Bioko, Uganda and Tanzania. Epiphytic on mossy branches in riverine forest; c. 900–1800 m.

INCOMPLETELY KNOWN SPECIES

Liparis hemipilioides Schltr. in Bot. Jahrb. Syst. **26**: 341 (1899). Type: Mozambique, Companhia de Mozambique area, "25 mile Station", c. 30 m, fl. iv.1898, *Schlechter* s.n. (B†, holotype).

Terrestrial herb 15–20 cm tall; tubers 1.5–2 cm long, ovoid. Leaf 1, basal, ovate, cordate, glabrous, dark green. Scape slender, 3–5-flowered, with many small lanceolate sheaths. Sepals and petals olive-green, lip violet. Dorsal sepal 7 mm long, linear, obtuse, margins revolute; lateral sepals 6 × 5 mm, broadly oblong, connate. Petals 7 mm long, erect to spreading. Lip almost panduriform, narrowing above a broad auriculate base with thickened margins, then expanding into a ±quadrangular-oblong very obtuse lobe, ± emarginate at the apex; lobe 5 × 3 mm, with a small, medial, low, 2-lobed rugulose callus. Column slender, 4 mm long, somewhat incurved towards the apex.

Mozambique. MS: "Companhia de Mozambique" area, in primeval forest by Dondo, "25 mile Station", c. 30 m, fl. iv.1898, *Schlechter* s.n. (B†).

This species appears to be very close to *Liparis mulindana* Schltr. If these taxa prove to be synonymous then *Liparis hemipilioides* Schltr. is the earlier name.

33. **MALAXIS** Sol. ex Sw.

Malaxis Sol. ex Sw., Prodr.: 119 (1788); in Kongl. Vetensk. Acad. Nya Handl. **21**: 233 (1800). —Bentham, Gen. Pl. **3**: 493 (1883).
Microstylis Nutt., Gen. N. Amer. Pl. **2**: 196 (1818). —Bentham, Gen. Pl. **3**: 494 (1883).

Dwarf terrestrial, lithophytic or epiphytic herb with creeping rhizomes or fusiform tuberous roots. Stems leafy, ± swollen at the base; leaves thin-textured, plicate. Inflorescence erect, terminal, racemose or subumbellate, few- to many-flowered. Flowers small, resupinate in the Flora Zambesiaca area, green, yellow-green, buff, orange or purple. Sepals and petals free, similar or dissimilar, lip larger than tepals, entire or lobed, often ± auriculate at the base; often with dentate margins. Column short; stigma ventral; anther terminal; pollinia ovoid, 4 in 2 pairs joined at the base.

A genus of about 300 species, cosmopolitan in distribution but with most in tropical Asia. Seven species are known from Africa, 5 occurring in the Flora Zambesiaca area.

1. Inflorescence very short, subumbellate · 2. *maclaudii*
– Inflorescence a cylindrical raceme · 2
2. Stem very short; leaves ± basal, sub-opposite · 3
– Stem 3–15 cm long; leaves borne on upper stem (± in the middle of the flowering plant) · 4
3. Raceme c. 10 mm in diameter; flowers c. 6 mm in diameter; front margin of lip denticulate · 4. *schliebenii*
– Raceme to 5 mm in diameter; flowers less than 4 mm in diameter; front margin of lip entire · 1. *katangensis*
4. Raceme less than 7 mm in diameter; lip with a central pubescent cushion · · 3. *prorepens*
– Raceme over 7 mm in diameter; lip with 2 lateral, lunate pubescent cushions · 5. *weberbaueriana*

1. **Malaxis katangensis** Summerh. in Bot. Mus. Leafl. Harv. Univ. **14**: 221 (1951). —Williamson, Orch. S. Centr. Africa: 111 (1977). —Cribb in Kew Bull. **32**: 739 (1978); in F.T.E.A., Orchidaceae: 288 (1984). —Geerinck in Fl. Afr. Centr., Orchidaceae pt. 1: 273 (1984). — la Croix et al., Orch. Malawi: 180 (1991). Type from Zaire.

Dwarf terrestrial herb 3–15 cm tall; rhizome creeping, but cigar-shaped and tuber-like below the flowering stems. Leaves 2–3 at base of stem, the lowermost sheath-like, the others sub-opposite, to 6 × 4 cm, ovate or elliptic, light green, plicate. Inflorescence 2–14 cm long, narrowly cylindrical, densely many-flowered; flowers very small, yellow-green, turning buff-orange with age. Pedicel and ovary 2 mm long; bracts 2–3.5 mm long. Dorsal sepal erect, 1.7–2.5 × 1 mm, lanceolate; lateral sepals 1.3–2.2 × 1–1.5 mm, obliquely ovate. Petals 1.3–2.3 × 0.3–0.6 mm, linear-lanceolate, the margins glandular. Lip 1.3–1.8 × 1.5–1.8 mm, obovate, auriculate at the base, with a central pad of hairs; margins smooth. Column 0.4–0.7 mm long.

Var. **katangensis**

Dorsal sepal 2–2.5 mm long, lateral sepals 1.5–2.2 mm long; petals 1.8–2.3 mm long; lip 1.5–1.8 mm long. Column 0.7 mm long.

Zambia. W: Mwinilunga Distr., 104 km E of Mwinilunga, fl. xii.1969, *G. Williamson* 1865 (K; SRGH). **Malawi**. N: Mzimba Distr., Kaningina Forest Reserve near Mzuzu, 1300 m, fl. & fr. 22.iii.1986, *la Croix* 825 (K). S: Zomba Mt., near stables, 1300 m, fl. 22.i.1983, *la Croix & Gassner* 428 (K).
Also in Sierra Leone, Nigeria, Zaire and Tanzania. *Brachystegia* woodland; 1300–1600 m.

Var. **pygmaea** (Summerh.) P.J. Cribb in Kew Bull. **32**: 739 (1978). Type: Zambia, N of Mwinilunga, fl. 26.i.1938, *Milne-Redhead* 4359 (K, holotype).
Malaxis pygmaea Summerh. in Bot. Mus. Leafl. Harv. Univ. **14**: 223 (1951).

Dorsal sepal 1.7–1.9 mm long, lateral sepals 1.3 mm long; petals 1.3–1.5 mm long; lip 1.3 mm long. Column 0.4 mm long. Differs from the typical variety in the smaller size of the flower.

Zambia. W: Mwinilunga Distr., just N of Mwinilunga, fl. 26.i.1938, *Milne-Redhead* 4359 (K, holotype).
Endemic. Woodland; c. 1400 m.

2. **Malaxis maclaudii** (Finet) Summerh. in Bull. Misc. Inform., Kew **1934**: 208 (1934); in F.W.T.A. ed. 2, **3**: 211 (1968). —Cribb in Kew Bull. **32**: 738 (1978). —Geerinck in Fl. Afr. Centr., Orchidaceae pt. 1: 272 (1984). Type from Guinée.
Microstylis maclaudii Finet in Bull. Soc. Bot. France **54**: 533 (1907).
Malaxis hirschbergii Summerh. in Bot. Mus. Leafl. Harv. Univ. **14**: 225 (1951). — Williamson, Orch. S. Centr. Africa: 110 (1977). Type from Zaire.

Terrestrial herb to 15 cm tall, with creeping rhizome. Leaves 2–3; bases sheathing, petiolate up to 2 cm long; blade 5–7 × 2.5–3.5 cm, ovate or ovate-lanceolate. Inflorescence subumbellate, densely several-flowered. Flowers mauve or lilac, scented. Pedicel and ovary 8 mm long; bracts 4 mm long. Sepals 5–6 × 2 mm, ovate; petals 5 × 1.5–2.5 mm, the margins ciliolate. Lip 4–5 × 4.5–6 mm, flabellate, the margins fimbriate, with 2 longitudinal basal calli and 1 or more tooth-like projections in the centre. Column c. 1.5 mm long.

Zambia. W: Kitwe Distr., Mwekera, swamp forest at S Mutundu Stream, fl. 6.i.1981, *Liheshi* 3 (K; NDO).
Also in Guinée, Sierra Leone, Liberia, Ghana, Nigeria, Cameroon, Bioko, Central African Republic, Zaire and Gabon. Swamp forest; c. 1300 m.

3. **Malaxis prorepens** (Kraenzl.) Summerh. in Bull. Misc. Inform., Kew **1934**: 208 (1934); in F.W.T.A. ed. 2, **3**: 211 (1968). —Cribb in Kew Bull. **32**: 740 (1978); in F.T.E.A., Orchidaceae: 288 (1984). Type from Sierra Leone.
  Microstylis prorepens Kraenzl. in Bot. Jahrb. Syst. **17**: 48 (1893).
  Microstylis katochilos Schltr. in Bot. Jahrb. Syst. **38**: 5 (1905). Type: Mozambique, 40 km from Beira, *Schlechter* 13249 (B†, holotype; K, drawing of holotype).
  Malaxis katochilos (Schltr.) Summerh. in Kew Bull. **6**: 465 (1952).

Terrestrial herb 15–30 cm tall; rhizome creeping; flowering shoots erect. Leaves usually 3, around middle of stem; bases sheathing, petiolate, 2.5–3 cm long; blades suberect or spreading, 5–6 × 3–4 cm, ovate, thin-textured. Inflorescence 4–10 cm × 3–6 mm, narrowly cylindrical, fairly densely many-flowered. Flowers reddish-purple, sometimes green. Dorsal sepal 2–2.5 × 0.7–0.9 mm, oblong-elliptic, rounded; lateral sepals 1.8–2 × 1–1.3 mm, obliquely oblong-elliptic, rounded. Petals 1.8–2.2 × 0.4–0.6 mm, linear-lanceolate; lip 2–2.3 × 1.8–2 mm, obovate, auriculate at base, with a central white cushion of hairs. Column 0.5–0.6 mm long.

Zambia. W: Kitwe Distr., Chati Forest Reserve, between Kalulushi and Lufwanyama Rivers, 1265 m, fl. 13.i.1962, *Baker* 134 (K; SRGH). **Mozambique**. MS: 40 km N of Beira, 60 m, fl. 10.iv.1898, *Schlechter* 13249 (B; K).
Also in West Africa, from Sierra Leone to Nigeria, and in Ethiopia, Uganda and Tanzania. Riverine vegetation and rain forest, in deep shade; 60–1265 m.

4. **Malaxis schliebenii** (Mansf.) Summerh. in Kew Bull. **6**: 465 (1952). —Williamson, Orch. S. Centr. Africa: 110 (1977). —Cribb in Kew Bull. **32**: 739 (1978); in F.T.E.A., Orchidaceae: 286 (1984). —la Croix et al., Orch. Malawi: 180 (1991). TAB. **98**. Type from Tanzania.
  Microstylis schliebenii Mansf. in Notizbl. Bot. Gart. Berlin **11**: 808 (1933).

Small terrestrial herb 5–10 cm tall; rhizome creeping, but swollen to form cigar-shaped tubers below the erect flowering stems. Leaves 2–3, the lowermost sheathing, the others sub-opposite at or near the stem base, spreading, up to 6 × 4 cm, ovate, pleated, light green. Inflorescence c. 12-flowered, up to 9 cm long and c. 1 cm in diameter, cylindrical. Flowers c. 6 mm in diameter, translucent yellow-green. Dorsal sepal 3–4 × 1–1.5 mm, lanceolate; apex rounded, erect or slightly recurved. Lateral sepals spreading, 3.5–4 × 3–4 mm, flabellate, emarginate, with 2 fleshy ridges in the centre, the basal margins thickened, the apical margin denticulate. Column 1 mm long.

Zambia. W: Ndola Distr., c. 12 km on Mufulira Road, fl. 5.i.1960, *Holmes* 0185 (K; SRGH). **Malawi**. N: South Viphya, Luchilemu Valley, 1250 m, fl. 12.ii.1986, *la Croix* 805 (K). S: Thyolo Distr., Bvumbwe, 1150 m, fl. 22.ii.1982, *la Croix* 269 (K).
Also in Zaire and Tanzania. Woodland, often in large colonies; 600–1300 m.

5. **Malaxis weberbaueriana** (Kraenzl.) Summerh. in Bull. Misc. Inform., Kew **1934**: 208 (1934); in F.W.T.A. ed. 2, **3**: 211 (1968). —Cribb in Kew Bull. **32**: 740 (1978); in F.T.E.A., Orchidaceae: 289 (1984). —Geerinck in Fl. Afr. Centr., Orchidaceae pt. 1: 274 (1984). —la Croix et al., Orch. Malawi: 181 (1991). Type from Cameroon.
  Microstylis weberbaueriana Kraenzl. in Orchis **2**: 128 (1908).
  Microstylis stolzii Schltr. in Bot. Jahrb. Syst. **53**: 559 (1915). Type from Tanzania.
  Malaxis stolzii (Schltr.) Summerh. in Kew Bull. **2**: 126 (1948). —Grosvenor in Excelsa **6**: 84 (1976). —Williamson, Orch. S. Centr. Africa: 110, t. 92 (1977).

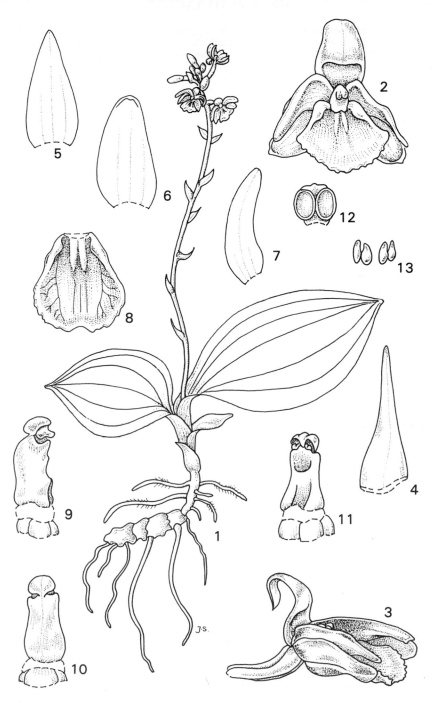

Tab. 98. MALAXIS SCHLIEBENII. 1, habit (×1), from *la Croix* 269; 2, flower, front view (×6); 3, flower, side view (×6); 4, bract (×6); 5, dorsal sepal (×6); 6, lateral sepal (×6); 7, petal (×6); 8, lip (×6); 9, column, side view (×14); 10, column, back view (×14); 11, column, front view (×14); 12, anther cap (×20); 13, pollinia (×20), 2–12 from *Johnston-Stewart* 44 (in K spirit coll. 45158). Drawn by Judi Stone.

Terrestrial herb up to 35 cm tall; rhizome creeping, swollen to form cigar-shaped tubers below the erect stems. Leaves usually 3, up to 6 × 3 cm, ovate, ribbed, thin-textured. Inflorescence up to 17 cm long, laxly several- to many-flowered. Flowers small, maroon-purple, occasionally yellow-green. Pedicel and ovary 3–3.5 mm long; bracts 1.5 mm long. Dorsal sepal 2–3.3 × 1.5–1.7 mm, ovate, obtuse; lateral sepals slightly smaller, oblique, the apex more rounded. Petals 2–3 × 0.3–0.6 mm, linear-lanceolate. Lip c. 2 × 2 mm, suborbicular to quadrate, with 2 lunate, pubescent calli towards the sides. Column 0.6–1 mm long.

Zambia. N: Kawambwa, fl. xii.1967, *G. Williamson* 671 (K). W: Mwinilunga Distr., by R. Luao, fl. 27.xii.1937, *Milne-Redhead* 3837 (K). **Zimbabwe**. E: Chimanimani Mts., near Nyabamba Falls, 900 m, fl. 27.i.1955, *Ball* 458 (K; SRGH). **Malawi**. N: South Viphya, near Nkhalapya, 1650 m, fl. 21.ii.1987, *la Croix* 980 (K; MAL). S: Thyolo Distr., Mandimwe Estate, 1050 m, fl. 17.ii.1980, *la Croix* 8 (K). **Mozambique**. Z: Gurué, c. 1200 m, fr. xii.1979, *Schäfer* 6895 (K; LMU).

Also in Cameroon, Bioko, Zaire, Kenya and Tanzania. Riverine forest, dense woodland and *Syzygium* swamp, in deep shade; 900–1750 m.

34. **OBERONIA** Lindl.

Oberonia Lindl., Gen. Sp. Orch. Pl.: 15 (1830); Fol. Orch. 8, *Oberonia*: 1–8 (1859).—Seidenfaden in Dansk. Bot. Arkiv. **25**, 3: 7–125 (1968), nom. conserv.

Epiphytic herb, rarely lithophytic. Leaves distichous, terete or bilaterally flattened, fleshy, sometimes jointed. Inflorescence terminal, cylindrical, usually densely many-flowered. Flowers very small, non-resupinate, green, yellow or orange. Sepals and petals free, subsimilar, the petals slightly smaller; lip entire or lobed, ecallose and lacking a spur. Column very short; pollinia 4, pyriform, waxy, free.

A genus of 80–100 species, mostly Asiatic but extending into Australasia; one species in mainland Africa, Madagascar, the Comoros and Mascarene Islands.

Oberonia disticha (Lam.) Schltr. in Fedde, Repert. Spec. Nov. Regni Veg., Beih. 33: 132 (1925). —H. Perrier in Fl. Madag., Orch. 1: 153 (1939). —Moriarty, Wild Fls. Malawi: t. 28, 3 (1975). —Grosvenor in Excelsa **6**: 84 (1976). —Ball, South. Afr. Epiph. Orchids: 146 (1978). —Stewart et al., Wild Orch. South. Africa: 215 (1982). —Cribb in F.T.E.A., Orchidaceae: 291 (1984). —Geerinck in Fl. Afr. Centr., Orchidaceae pt. 1: 282 (1984). —la Croix et al., Orch. Malawi: 181 (1991). TAB. **99**. Type from Réunion.
    Epidendrum distichum Lam., Encycl. **1**: 189 (1783).

Small epiphytic herb, usually pendent. Roots arising at base of plant, less than 1 mm in diameter. Stems clustered, few to many, 2–15 cm long. Leaves several, distichous, usually imbricate, bilaterally flattened, succulent, light green, 2–5 × 0.5–1 cm, lanceolate, acute or acuminate, decreasing in size towards the stem apex. Inflorescence racemose, densely many-flowered, 4–10 cm long, cylindrical, tapering. Flowers less than 2 mm in diameter, yellow-ochre to orange. Sepals c. 0.7 × 0.5 mm, ovate, obtuse; petals 0.5 × 0.2 mm, elliptic, obtuse; lip c. 1 × 0.7 mm, oblong-pandurate, deflexed. Column c. 3 mm long.

Zimbabwe. E: Chimanimani Mts., Makurupini Forest, 430 m, fl. 13.i.1969, *Bisset* 56 (K; SRGH). **Malawi**. S: Thyolo Distr., Namikweya Estate, 1200 m, fl. 9.ii.1984, *la Croix* 526 (K; MAL).
    Also in Cameroon, Zaire, Uganda, Tanzania, South Africa (Transvaal), Madagascar, and the Comoro and Mascarene Islands. Evergreen forest or high rainfall *Brachystegia* woodland, epiphytic on tree trunks, 430–1200 m.
    This species will almost certainly also occur in the adjacent Mozambique Makurupini forest.

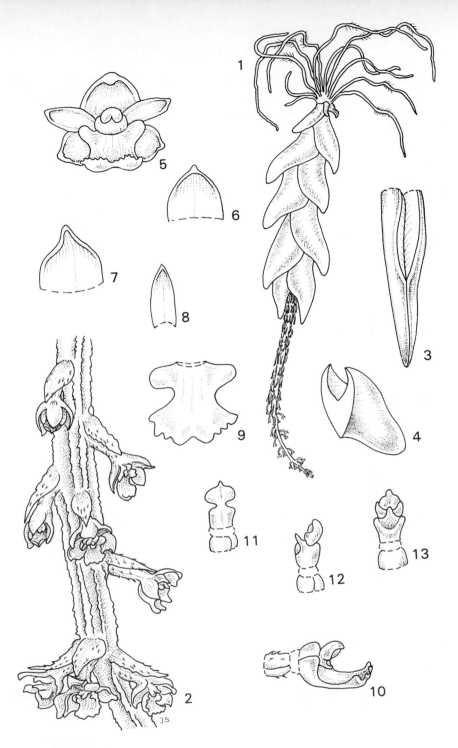

Tab. 99. OBERONIA DISTICHA. 1, habit (×1); 2, lower portion of inflorescence (×8); 3, tip of leaf, from above (×3); 4, cross-section of leaf (×3); 5, flower (×20); 6, dorsal sepal (×20); 7, lateral sepal (×20); 8, petal (×20); 9, lip (×20); 10, column and lip, side view (×20); 11, column, back view (×20); 12, column, side view (×20); 13, column, front view (×20), 1–13 from *Du Puy et al.* M796. Drawn by Judi Stone.

35. **STOLZIA** Schltr.

Stolzia Schltr. in Bot. Jahrb. Syst. **53**: 564 (1915). —Mansfeld in Notizbl. Bot.
Gart. Berlin **11**: 1061 (1934). —Summerhayes in Kew Bull. **8**: 140 (1953); in
Kew Bull. **17**: 557 (1964). —Cribb in Kew Bull. **33**: 79–89 (1978).

Dwarf, mat-forming epiphytic herb, rarely lithophytic; stems creeping, bearing
pseudobulbs. Pseudobulbs asymmetrical ovoid, fusiform or clavate, 1–2-leaved at the
apex. Leaves fleshy or leathery, spreading or erect. Inflorescence terminal on the
pseudobulb, erect, 1- to many-flowered. Flowers ± secund, somewhat campanulate,
green, yellow, orange, brown or red, sometimes striped. Lateral sepals joined at the
base, forming a mentum with the column-foot. Lip entire, recurved, V-shaped in
cross-section, not spurred. Column-foot at least three times as long as the column;
pollinia 8, comprising 4 large and 4 small pollinia; stigma concave, with a flap-shaped
rostellum in front.

A genus of about 15 species, all from tropical Africa.

1. Pseudobulbs 1-leafed · 1. *compacta*
– Pseudobulbs 2-leaved · 2
2. Pseudobulbs ovoid; inflorescence several-flowered · · · · · · · · · · · · · · · · · · · 2. *nyassana*
– Pseudobulbs rhizome-like, only swollen below the leaves; inflorescence 1-flowered · · · 3
3. Leaves fleshy; flowers resupinate, striped yellow and red-brown; dorsal sepal recurved
   towards the apex · 3. *repens*
– Leaves thin-textured; flowers non-resupinate, orange-brown; dorsal sepal not recurved · ·
· 4. *williamsonii*

1. **Stolzia compacta** P.J. Cribb in Kew Bull. **32**: 159 (1977); in Kew Bull. **33**: 81 (1978); in
   F.T.E.A., Orchidaceae: 328 (1984). —la Croix et al., Orch. Malawi: 183 (1991). Type:
   Malawi, Nyika Plateau, Kasaramba Forest, fl. ii.1968, *Williamson, Ball & Simon* 371 (K,
   holotype).

Creeping epiphytic herb forming chains or mats. Pseudobulbs set close together,
up to 1 cm high and 1.5 cm wide, obliquely ovoid, pale green, 1-leafed at the apex.
Leaf c. 5–9 × 1 cm, oblanceolate, erect, light green, thin-textured. Inflorescence 1-
flowered, much shorter than the leaf. Flowers resupinate, campanulate, rather fleshy,
yellow, yellow-green, purple or brick-red. Dorsal sepal 7–19 × 2.3–3.5 mm, lanceolate,
acuminate; lateral sepals similar but oblique, joined in the basal half to each other
and the column-foot to form a saccate mentum c. 2 mm high. Petals 4–5 × 1–1.5 mm,
linear-lanceolate, acute. Lip 3.5–4 × 2–2.5 mm, ovate or elliptic-ovate, rounded, with
obscure, verrucose basal callus. Column less than 1 mm long; column-foot incurved,
2–3 mm long.

Flowers greenish-yellow · subsp. *compacta*
Flowers brick-red or purple · subsp. *purpurata*

Subsp. **compacta**

Flowers yellow-green, dorsal sepal 9–10 mm long; plants larger than in subsp.
purpurata.

Malawi. N: Nyika Plateau, Kasaramba Forest, fl. ii.1968, *Williamson, Ball & Simon* 371 (K).
Endemic. Submontane evergreen forest, on mossy trunks and branches of trees; 2150–2350 m.

Subsp. **purpurata** P.J. Cribb in Kew Bull. **33**: 83 (1978). TAB. **100**. Type: Zimbabwe, Mutare
Distr., Himalaya Range, Engwa, fl. 9.xi.1954, *Wild* 4627 (K, holotype).
   Stolzia sp., Morris, Epiphyt. Orch. Malawi: 76 (1970).
   Stolzia sp. no. 1 (*Wild* 4627), Grosvenor in Excelsa **6**: 86 (1976).

Flowers dull purple or brick-red, slightly smaller than those of subsp. *compacta*;
whole plant also slightly smaller.

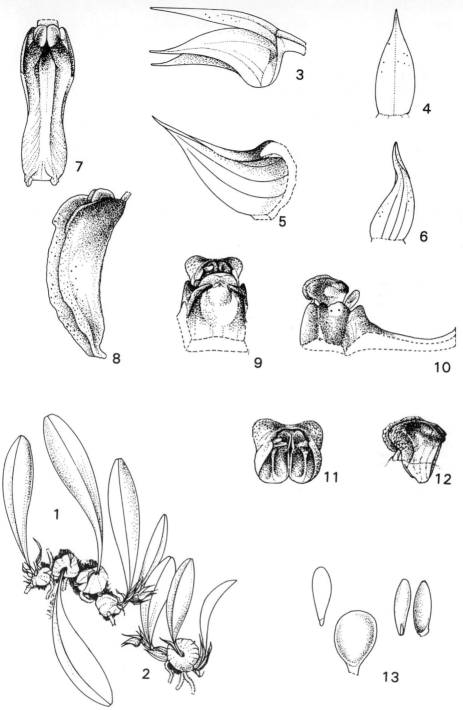

Tab. 100. STOLZIA COMPACTA subsp. PURPURATA. 1, habit (×⅔), from *Wild* 4627; 2, habit (×⅔), from *Greatrex* in *GHS* 29829; 3, flower, side view (×3½); 4, dorsal sepal (×4); 5, lateral sepal (×5); 6, petal (×6); 7, lip, front view (×12); 8, lip, side view (×12); 9, column, front view (×14); 10, column, side view (×14); 11, anther cap, front view (×20); 12, anther cap, side view (×20); 13, pollinia, large and small (×54), 3–13 from *Wild* 4627. Drawn by J.A. Langhorne. From Kew Bull.

Zimbabwe. E: Mutare Distr., Himalaya Range, Engwa, fl. 9.xi.1954, *Wild* 4627 (K). **Malawi**. S: Mulanje Mt., Lichenya Plateau, st. v.1965, *Morris* 158 (K); Mulanje Mt., Ruo Gorge, 1600 m, fl. in cult. 16.v.1984, *la Croix* 656 (K).

Not known elsewhere. Montane, submontane and riverine forest, epiphytic on mossy branches of trees; 1600–2000 m.

Subsp. *iringana* P.J. Cribb occurs in southern Tanzania, and differs from the other subspecies in its smaller size being less than 4 cm tall with pseudobulbs c. 5 mm in diameter.

2. **Stolzia nyassana** Schltr. in Bot. Jahrb. Syst. **53**: 565 (1915). —Morris, Epiphyt. Orch. Malawi: 76 (1970). —Cribb in Kew Bull. **33**: 85 (1978); in F.T.E.A., Orchidaceae: 328 (1984). —la Croix et al., Orch. Malawi: 184 (1991). Type from Tanzania.

Creeping epiphytic herb. Pseudobulbs c. 10 × 13 mm, obliquely ovoid, set close together forming mats, 2-leaved at the apex. Leaves slightly fleshy, 6–7 × 1–1.3 cm, oblong-elliptic, rounded and slightly 2-lobed at the apex, light green. Inflorescence erect, c. 6 cm long, several-flowered. Flowers secund, greenish-yellow. Dorsal sepal 5 × 2.5 mm, ovate; lateral sepals similar but oblique and slightly broader, joined in the basal third and forming a saccate mentum with the column-foot. Petals 4.5 × 1.5 mm, narrowly lanceolate, acute. Lip fleshy, 3 × 0.5 mm, tongue-like, V-shaped in cross-section. Column 1.5 mm long; column-foot 2 mm long.

Malawi. N: Misuku Hills, Matipa Peak, exposed site at edge of forest, *Dowsett-Lemaire* 437 (K). Also in Tanzania. Submontane forest, epiphytic on mossy branches of trees; c. 2000 m.

3. **Stolzia repens** (Rolfe) Summerh. in Kew Bull. **8**: 141 (1953); in F.W.T.A. ed. 2, **3**: 226 (1968). —Morris, Epiphyt. Orch. Malawi: 75 (1970). —Williamson, Orch. S. Centr. Africa: 128, t. 105 (1977). —Ball, Epiph. Orch. S. Afr.: 212 (1978). —Cribb in F.T.E.A., Orchidaceae: 330 (1984). —la Croix et al., Orch. Malawi: 184 (1991). Type from Uganda.
   Polystachya repens Rolfe in Bull. Misc. Inform, Kew **1912**: 132 (1912).

Dwarf, creeping epiphytic herb forming mats. Pseudobulbs prostrate, 2–3 cm long, swollen to c. 3 mm in diameter at one end, 2-leaved. Leaves fleshy, 5–14 × 5–7 mm, suborbicular, obovate or elliptic. Inflorescence very short, 1-flowered. Flowers yellow with red-brown stripes or red-brown with yellow stripes. Dorsal sepal to 7 × 2.5 mm, lanceolate, acute or obtuse; lateral sepals similar but oblique and slightly shorter, joined at the base and forming a saccate mentum with the column-foot. Petals 5 × 1.5 mm, lanceolate-falcate. Lip fleshy, 2 × 1 mm, tongue-like. Column less than 1 mm, long; column-foot 2 mm long.

Var. **repens**

Flowers yellow with red-brown stripes at the base of dorsal sepal and petals. Dorsal sepal up to 7 × 2.5 mm, lanceolate, acute; lateral sepals lanceolate, acute. Petals up to 5 × 1.5 mm, lanceolate-falcate, acute.

Zambia. W: Zambezi Rapids, fl. 26.x.1966, *Leach & Williamson* 13563 (K; SRGH). E: Nyika Plateau, Manyenjere Forest, 2070 m, st. 20.ii.1982, *Dowsett-Lemaire* 365 (K). **Zimbabwe**. C: Wedza, 1660 m, st. 27.ii.1964, *Wild* 6350 (K; SRGH). E: Mutare Distr., Vumba Mts., 1500 m, fl. iii.1949, *Wild* 2811 (K; SRGH). **Malawi**. N: Nyika National Park, Chisanga Falls, 1700–1800 m, st. 12.iii.1982, *Dowsett-Lemaire* 366 (K). S: Blantyre Distr., Ndirande Mt., 1600 m, fl. 14.xii.1980, *la Croix* 72 (K).
   Also in Ghana, Nigeria, Cameroon, Zaire, Ethiopia, Uganda, Kenya and Tanzania. High rainfall woodland and evergreen forest; 1050–2200 m.

Var. **obtusa** G. Will. in J. S. African Bot. **46**: 333 (1980). Type: Zimbabwe, Mutare Distr., N & NE faces of Castle Beacon, fl. i.1976, *Ball* 1398 (SRGH, holotype).
   Stolzia sp. no. 2 (*Ball* 1398), Grosvenor in Excelsa **6**: 86 (1976).

Flowers red-brown with 2 yellow stripes at the base of dorsal sepal and petals. Dorsal sepal 5 × 1.5 mm, oblong, obtuse, projecting forwards; lateral sepals 5 × 1.2 mm, deflexed, falcate-lanceolate, obtuse; mentum saccate, somewhat bent forwards. Petals spreading, 4 × 1 mm, oblong, obtuse. Lip c. 2 mm long.

Zimbabwe. E: Mutare Distr., N & NE faces of Castle Beacon, fl. i.1976, *Ball* 1398 (SRGH). Malawi. S: Mulanje Mt., Lichenya Plateau near Rest-house, 2000 m, fl. 29.xii.1982, *Jenkins* in *la Croix* 381 (K). Not known elsewhere. Montane and submontane evergreen forest, epiphytic on mossy branches; c. 2000 m.

4. **Stolzia williamsonii** P.J. Cribb in Kew Bull. **33**: 88 (1978). —la Croix et al., Orch. Malawi: 185 (1991). Type: Malawi, Nyika Plateau, Kasaramba Forest, fl. ii.1968, *Williamson, Ball & Simon* 370 (K, holotype).

Dwarf, creeping epiphytic herb. Pseudobulbs prostrate, elongate, 2–4 cm long, swollen at one end, 2-leaved. Leaves 10–30 × 5–8 mm, elliptic or obovate, rounded at the apex, thin-textured, articulated c. 1 mm above apex of the pseudobulb, one leaf of the pair usually larger than the other. Flowers solitary; peduncle c. 3 mm long; bract 1.5 mm long, ovate, acute. Flowers red-brown, non-resupinate. Dorsal sepal 8.5 × 3 mm, concave, elliptic, obtuse, not recurved towards the apex; lateral sepals similar but oblique and slightly shorter, united in the basal third, forming a saccate mentum 2–2.5 mm high. Petals 7 × 2 mm, adnate to the dorsal sepal, ovate-elliptic, obtuse, papillose. Lip recurved, 3.5 × 1.2 mm, lanceolate-ligulate, acute or acuminate, papillose with an erose margin. Column 0.5 mm long; column-foot 2 mm long.

Malawi. N: Nyika Plateau, Kasaramba Forest, fl. ii.1968, *Williamson, Ball & Simon* 370 (K). Also in Tanzania. Submontane forest, forming mats on mossy branches of trees; c. 2350 m.

36. **CHASEELLA** Summerh.

Chaseella Summerh. in Kirkia **1**: 88 (1961).

Dwarf epiphytic herb. Pseudobulbs small, spaced out along a slender creeping rhizome, 6–12-leaved. Leaves needle-like. Flowers small, resupinate. Pollinia 2.

A monotypic genus confined to tropical Africa.

Chaseella pseudohydra Summerh. in Kirkia **1**: 89 (1961). —Grosvenor in Excelsa **6**: 79 (1976). —Ball in S. Afr. Epiph. Orch.: 94 (1978). —Cribb in F.T.E.A., Orchidaceae: 323 (1984). — Vermeulen in Orchid Monogr. **2**: 164 (1987). TAB. **101**. Type: Zimbabwe, Mutare Distr., Honde Gorge, *Chase* in *GHS* 22738 (K, holotype; SRGH).

Creeping epiphytic herb; rhizome slender, c. 1 mm in diameter. Pseudobulbs 4–8 × 3–7 mm, orbicular to ovoid, 0.5–2.5 cm apart, 6–12-leaved at the apex. Leaves fleshy, 2–12 × 0.5–1 mm, linear, acute, resembling pine needles. Inflorescence 1(2)-flowered; peduncle c. 2 mm long. Flowers brick-red or pinkish-orange, the petals straw-coloured, somewhat campanulate. Pedicel and ovary 3 mm long, with some short, black hairs. Dorsal sepal c. 4 × 3 mm, ovate; lateral sepals c. 4 × 3–4 mm, obliquely ovate, acute, recurved at the apex, forming a rounded mentum 3.5 mm high. Petals c. 1.8 × 1.2 mm, obliquely elliptical. Lip mobile, somewhat recurved, 2–2.5 × 2–2.5 mm, broadly ovate, rounded at the apex, slightly cordate at the base, ± channelled. Column c. 2 mm long, including c. 1 mm long stelida; column-foot c. 1.5 mm long. Pollinia 2, triangular.

Zimbabwe. E: Mutare Distr., Honde Gorge, fl. ii.1949, *Chase* in *GHS* 22738 (K; SRGH). Also in Kenya. Evergreen and riverine forest, epiphytic.

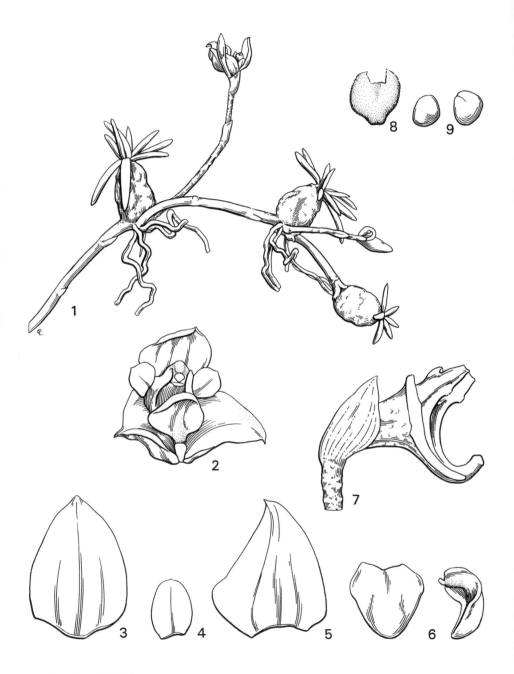

Tab. 101. CHASEELLA PSEUDOHYDRA. 1, habit (×2); 2, flower (×6); 3, dorsal sepal (×8); 4, petal (×8); 5, lateral sepal (×8); 6, lip (×8); 7, column (×8); 8, anther cap (×16); 9, pollinia (×16), 1–9 from *Chase* in *GHS* 22738. Drawn by Eleanor Catherine.

37. **BULBOPHYLLUM** Thouars

Bulbophyllum Thouars, Hist. Orchid. [Fl. Iles Austr. Afr.]: tab. gen., figs. 93–97 (1822), nom. conserv. —Vermeulen in Orchid Monogr. **2**: 1–300 (1987).
Cirrhopetalum Lindl. in Bot. Reg. **10**: sub t. 832 (1824).
Megaclinium Lindl. in Bot. Reg. **12**: t. 989 (1826).

Epiphytic or lithophytic herb with sympodial growth. Rhizomes creeping, often woody, sometimes branched. Pseudobulbs each comprising 1 internode (representing the main axis of a sympodial branch), clustered or spaced out along the rhizome, 1–2(3)-leaved at the apex. Leaves mostly coriaceous or fleshy, rarely thin-textured. Inflorescences arising from base of the pseudobulb, racemose or rarely umbellate, (1)few- to many-flowered. Rhachis sometimes swollen or flattened; flowers white, cream, yellow, green, orange or purple, ± fleshy. Sepals usually free and subequal, the lateral sepals united at the base to the column foot to form an obscure mentum. Petals usually smaller than the sepals. Lip often much smaller than the sepals, hinged to end of the column foot, usually motile, often fleshy and tongue-like, sometimes fringed with long or short hairs. Column short, usually with apical lateral extensions (stelidia), and often winged. Pollinia 2 or 4 in 2 pairs.

A very large, pantropical genus of over 1000 species, with about 70 species in tropical and South Africa.

1. Pseudobulbs 1-leafed · 2
 – Pseudobulbs 2(3)-leaved · 10
2. Inflorescence subumbellate; lateral sepals more than twice as long as the dorsal sepal · · ·
 · 21. *longiflorum*
 – Inflorescence racemose; lateral sepals less than twice as long as dorsal sepal · · · · · · · · 3
3. Pseudobulbs of mature plants well spaced out, 2–6 cm apart on the rhizome · · · · · · · · 4
 – Pseudobulbs clustered on rhizome, less than 2 cm apart · · · · · · · · · · · · · · · · · · 7
4. Inflorescence arched, less than 3 cm long · 9. *humblotii*
 – Inflorescence over 5 cm long · 5
5. Rhachis and bracts with fine dark hairs; bracts overlapping, longer than the flowers · · · ·
 · 16. *lupulinum*
 – Rhachis and bracts glabrous; bracts not overlapping, shorter than the flowers · · · · · · · 6
6. Inflorescence usually more than 20 cm long; underside of lip with large, irregular warts ·
 · 19. *encephalodes*
 – Inflorescence usually less than 20 cm long; underside of lip without such warts · · · · · · · ·
 · 20. *unifoliatum*
7. Rhachis somewhat swollen, wider than the peduncle · · · · · · · · · · · · · · · · · 14. *ballii*
 – Rhachis not swollen · 8
8. Pseudobulbs c. 1 cm high; rhachis wiry, zigzag, flowers well spaced out · · · 6. *intertextum*
 – Pseudobulbs usually 2–3 cm high; rhachis straight, densely many-flowered · · · · · · · · · 9
9. Inflorescence erect, secund; pseudobulbs smooth, shiny, green or yellowish, very rarely purple · 8. *expallidum*
 – Inflorescence arched or pendent, not secund; pseudobulbs with pitted surface, maroon or brownish · 7. *josephi*
10. Rhachis bilaterally flattened, leaf-like or fleshy · 11
 – Rhachis terete, fleshy or wiry but not bilaterally flattened · · · · · · · · · · · · · · · · · 14
11. Margin of lip dentate or pectinate in lower half; inflorescence usually more than 25 cm long
 · 12
 – Margin of lip entire; inflorescence usually less than 25 cm long · · · · · · · · · · · · · · 13
12. Lateral sepals cuspidate to caudate · 10. *maximum*
 – Lateral sepals acute or acuminate · 11. *injoloense*
13. Dorsal sepal erect, lateral sepals spreading from base; petals narrowly linear, subclavate at apex · 12. *sandersonii*
 – Dorsal sepal porrect, lateral sepals straight, or recurved at apex only; petals linear-oblong
 · 13. *scaberulum*
14. Rhachis ± fleshy, wider than the peduncle · 15
 – Rhachis not fleshy, not noticeably wider than the peduncle · · · · · · · · · · · · · · · · 18
15. Pseudobulbs spaced out on rhizome, usually more than 2 cm apart; flowers distichous, yellow or orange-red · 16

– Pseudobulbs clustered on rhizome, less than 2 cm apart; flowers not distichous, greenish or purplish ·· 17

16. Lip recurved, swollen towards the base and apex, the 2 parts separated by a transverse furrow on the underside; sepals usually yellow or orange with red-purple stripes ······ ··· 18. *oreonastes*

– Lip not or very slightly recurved, without transverse furrow on the underside; sepals yellow or orange, without stripes ································ 17. *fuscum*

17. Inflorescence dense, arched or spreading; pseudobulbs squat, as broad as, or broader than high ·· 15. *elliotii*

– Inflorescence lax, ± erect; pseudobulbs ovoid, taller than broad ············ 14. *ballii*

18. Lip fringed with hairs ··· 19

– Lip glabrous or papillose but not hairy ··· 21

19. Pseudobulbs cylindrical or narrowly conical, more than 5 times as long as wide ········ ·· 1. *cochleatum*

– Pseudobulbs orbicular, ovoid or ellipsoid, not more than 4 times as long as wide ···· 20

20. Pseudobulbs greenish-yellow flushed with purple, widely spaced on rhachis; peduncle and rhachis ± equal in length ······································· 2. *gravidum*

– Pseudobulbs dark purple-red, clustered and mat-forming; rhachis about half as long as peduncle ······································· 3. *rugosibulbum*

21. Rhachis and bracts with fine dark hairs; bracts overlapping, longer than the flowers ···· ··· 16. *lupulinum*

– Rhachis and bracts glabrous; bracts shorter than the flowers ·················· 22

22. Lip upper surface and margin densely papillose ···················· 4. *bavonis*

– Lip upper surface and margin glabrous ····························· 5. *stolzii*

---

1. **Bulbophyllum cochleatum** Lindl. in J. Linn. Soc., Bot. **6**: 125 (1862). —Rolfe in F.T.A. **7**: 28 (1897). —Summerhayes in Kew Bull. **8**: 144 (1953); **14**: 138 (1960); in F.W.T.A. ed. 2, **3**: 236 (1968). —Williamson, Orch. S. Centr. Africa: 136 (1977). —la Croix et al., Malawi Orch. **1**: 50 (1983). —Cribb in F.T.E.A., Orchidaceae: 315 (1984). —Vermeulen in Orchid Monogr. **2**: 41 (1987). —la Croix et al., Orch. Malawi: 194 (1991). —Geerinck in Fl. Afr. Centr., Orchidaceae pt. 2: 260 (1992). Type from Bioko.

Erect, epiphytic herb with a stout branching rhizome. Pseudobulbs borne 1–3 cm apart on a woody rhizome, 2-leaved, 3–11 × 0.4–1.3 cm, narrowly conical to almost cylindrical, smooth, green, usually with 4 purple vertical lines. Leaves 7–23 × 0.5–1.3 cm, linear or linear-lanceolate, dark green. Inflorescence 9–21 cm long, densely many-flowered; peduncle with 3 scarious sheaths; rhachis terete; bract distichous, boat-shaped, c. 8 mm long, green in bud but turning brown when flowers open. Flowers tinged green with purple or mauve, with dark purple lip. Ovary and pedicel 2.5 mm long. Dorsal sepal 3–7.5 × 1.5–3 mm, ovate, acute; lateral sepals 2–7 × 1–3 mm, lanceolate, acute, reflexed near the base. Petals 1–2.5 × 0.2–0.8 mm, linear, parallel to the column. Lip 1.5–5 × 0.3–1.5 mm, linear, fleshy at the base, long ciliate on the margin in the apical half. Column 1–2 mm long; stelidia c. 1 mm long, long acuminate.

Zambia. W: Mwinilunga Distr., 10 km SE of Matonchi, fl. 17.ii.1975, *Williamson & Gassner* 2416 (K; SRGH). **Malawi**. N: North Viphya, Uzumara Forest, 1850 m, fl. 4.xi.1986, *la Croix* 870 (K; MAL). Also in W Africa, Sudan, Rwanda, Uganda, Kenya and Tanzania. Riverine or submontane evergreen forest.

---

2. **Bulbophyllum gravidum** Lindl. in J. Linn. Soc., Bot. **6**: 126 (1862). —Rolfe in F.T.A. **7**: 27 (1897). —Summerhayes in F.W.T.A. ed. 2, **3**: 236 (1968). —Morris, Epiphyt. Orch. Malawi: 51 (1970). —la Croix et al., Malawi Orch. **1**: 50 (1983); Orch. Malawi: 195 (1991). Type from Bioko.
　　Bulbophyllum monticolum Hook.f. in J. Linn. Soc., Bot. **7**: 219 (1864). —T. Durand & Schinz, Consp. Fl. Africa **5**: 12 (1895).
　　Bulbophyllum cochleatum var. *gravidum* (Lindl.) J.J Verm. in Orchid Monogr. **2**: 42 (1987) pro parte excl. B. *rugosibulbum*. —Geerinck in Fl. Afr. Centr., Orchidaceae pt. 2: 321 (1992).

Epiphytic or lithophytic herb; pseudobulbs borne 3–4 cm apart on a woody rhizome, 2(3)-leaved, 2–3.5 × 1–1.5 cm, ellipsoid or ovoid, 4–5-angled with knobbly edges, greenish-yellow flushed with purple. Leaves suberect, 5–12 × 0.8–1.3 cm, lanceolate or oblanceolate. Inflorescence erect, up to 22 cm long; peduncle 5–12 cm long, with 2–3 tubular sheathing bracts; rhachis 6–10 cm long, not swollen, densely many-flowered. Flowers distichous; sepals mustard-yellow, petals pinkish, lip purple. Ovary and pedicel 1.5 mm long; bracts 5 × 3 mm. Dorsal sepal c. 4 × 2.5 mm, ovate, acute; lateral sepals 4 × 1 mm, lanceolate, acute, curling back. Petals 1.5 mm long, linear. Lip 3 × 0.5 mm, linear, the margin hairy. Column 1 mm long; stelidia 0.5 mm long.

Zambia. W: Mwinilunga Distr., c. 1 km S of Matonchi Farm, fl. 3.xi.1937, *Milne-Redhead* 3074 (K). **Malawi**. N: Nkhata Bay Distr., 8 km E of Mzuzu, fl. *Pawek* 3699 (K). S: Zomba Distr., Malosa Mt., 1200 m, fl. 30.i.1983, *Johnston-Stewart* in *la Croix* 429 (K).
Also in W Africa and Tanzania. *Brachystegia* woodland or open grassland, often lithophytic; 1200–1400 m.
Vermeulen, in his revision of continental African *Bulbophyllinae* (Orchid Monogr. **2**: 1987), treated *B. gravidum* as a variety of *B. cochleatum*, and included *B. rugosibulbum* in that variety. All three species have distinct ecological preferences, *B. cochleatum* occurring in evergreen forest, in dense shade; *B. gravidum* in open woodland, sometimes grassland, usually growing on rocks, while *B. rugosibulbum* grows in dense woodland, usually at higher altitudes, forming mats on trunks and lower branches. When grown together over a period of 7 years, each species retained its distinct character. In the herbarium, pseudobulbs shrink, but in the field they are readily distinguished and can easily be keyed out, so we propose to keep them separate here.

3. **Bulbophyllum rugosibulbum** Summerh. in Kew Bull. **14**: 138 (1960); in Hooker's Icon. Pl. 37: t. 3670 (1969). —Williamson, Orch. S. Centr. Africa: 138, t. 119 (1977). —la Croix et al., Malawi Orch. **1**: 55 (1983). —Cribb in F.T.E.A., Orchidaceae: 315 (1984). —la Croix et al., Orch. Malawi: 196 (1991). TAB. **102**. Type: Zambia, Solwezi, cult. Kew, *Holmes* 05 (K, holotype).
  Bulbophyllum cochleatum var. *gravidum* (Lindl.) J.J.Verm. in Orchid Monogr. **2**: 43 (1987) pro parte quoad syn. *B. rugosibulbum* excl. typ. *B. gravidum*. —Geerinck in Fl. Afr. Centr., Orchidaceae pt. 2: 321 (1992) as "*rugosilabium*".

Epiphytic herb; pseudobulbs borne close together on the rhizome, forming mats, 2-leaved, 10–20 × 8–15 mm, ± globose to ovoid, purple-red, wrinkled. Leaves 35–45 × 4–6 mm, linear, grass-like, deciduous in dry season. Peduncle 3.5–5 cm long; rhachis 1.5–2 cm long, up to 16-flowered. Flowers ± secund, deep maroon-purple. Ovary and pedicel c. 1 mm long; bracts 2–4 mm long, about half as long as the flowers. Dorsal sepal 3–4 × 1.5–2.5 mm, ovate, acute; lateral sepals 2–4.2 × 1–2 mm, lanceolate, curling back. Petals 1–1.5 mm long, linear. Sepals 2–3 × 0.4–0.5 mm, motile, fringed with hairs on the upper half. Column and stelidia each c. 1 mm long.

Zambia. W: Solwezi Distr., fl. 20.viii.1959, *Holmes* 05 (K). **Malawi**. N: Nyika Nat. Park, near Thazima, c. 1700 m, fl. 8.i.1987, *la Croix* 929 (K). S: Mulanje Mt., path to Chambe Plateau, c. 1250 m, fl. 28.xii.1986, *Jenkins* in *la Croix* 918 (K).
Also in southern Tanzania. *Brachystegia* woodland, forming mats on trunks and lower branches; 1250–1800 m.

4. **Bulbophyllum bavonis** J.J. Verm. in Blumea **29**: 589 (1984); in Orchid Monogr. **2**: 50 (1987). —la Croix et al., Orch. Malawi: 187 (1991). Type: Malawi, Misuku Hills, Mugesse (Mughese) Forest, *Dowsett-Lemaire* 674 (K, holotype).

Epiphytic herb. Pseudobulbs borne 6–7 cm apart on a wiry rhizome 1 mm in diameter, 2-leaved, 12–30 × 8–12 mm, ovoid, 4-angled, yellow or reddish. Leaves c. 3 × 1 cm, oblong or lanceolate, obtuse. Inflorescence arching or pendent, to about 10 cm long, laxly 4–15-flowered; flowers distichous. Sepals white with purple mid-line, petals white, lip yellow. Pedicel and ovary 2–3 mm long; bracts 3 mm long. Sepals 6 × 2.5 mm, oblong or elliptic; dorsal sepal ± erect, finely papillose in the apical half; lateral sepals oblique and slightly recurved. Petals 2 × 1 mm, obovate, finely papillose towards the apex. Lip fleshy, 3–4 × 1.5 mm, oblong, obtuse, papillose on the margins, recurved near the base. Column 2.3–3 mm long; stelidia 1.2–1.6 mm long.

Malawi. N: Misuku Hills, Mugesse (Mughese) Forest, 1700 m, fl. 3.iii.1983, *Dowsett-Lemaire* 674 (K); Nyika, east escarpment, Mwenembwe Forest, c. 2350 m, fl. 8.vi.1983, *Dowsett-Lemaire* 776 (K).
Also in Tanzania. Evergreen forest; 1600–2350 m.

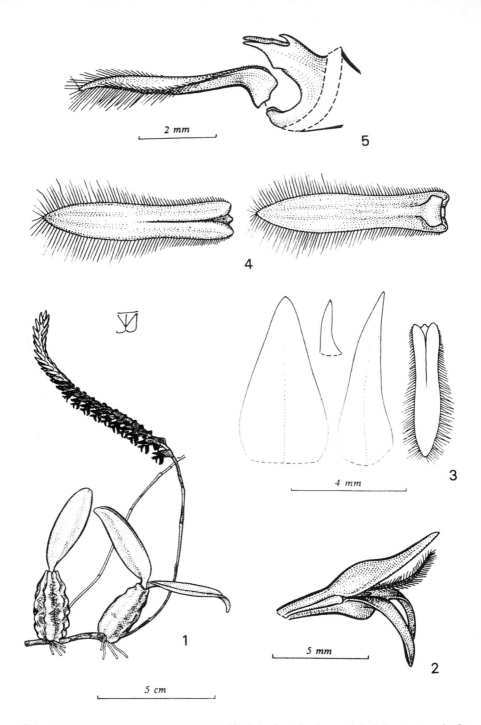

Tab. 102. BULBOPHYLLUM RUGOSIBULBUM. 1, plant (back pseudobulb has lost one leaf);
2, flower; 3, flower analysis, from left to right: dorsal sepal, petal, lateral sepal, lip; 4, lip,
left: adaxially, right: abaxially, 1–4 from *Milne-Redhead & Taylor* 8912; 5, column and lip,
lateral view, from *Milne-Redhead & Taylor* 8912. Drawn by J.J. Vermeulen. From Orchid
Monographs.

5. **Bulbophyllum stolzii** Schltr. in Bot. Jahrb. Syst. **53**: 571 (1915). —Summerhayes in Kew Bull. **8**: 581 (1954). —Morris, Epiphyt. Orch. Malawi: 51 (1970). —la Croix et al., Malawi Orch. **1**: 58 (1983). —Cribb in F.T.E.A., Orchidaceae: 313 (1984). —Vermeulen in Orchid Monogr. **2**: 52, fig.22 (1987). —la Croix et al., Orch. Malawi: 192 (1991). TAB. **103**. Type from Tanzania.

Small epiphytic herb with a creeping rhizome less than 1 mm in diameter. Pseudobulbs borne up to 11 cm apart on the rhizome, 2-leaved, 1–2.5 cm × 8–10 mm, ± globose to ovoid, reddish-yellow, smooth when young but rugose when old. Leaves 2–6 × 0.2–1 cm, elliptic, sometimes slightly 2-lobed at the apex. Inflorescence 3–9 cm long, erect at first then bending over, up to c. 16-flowered; peduncle very slender with c. 4 sheaths; rhachis slightly zigzag. Flowers distichous, with a slightly unpleasant smell. Sepals greenish-white with purple margins and mid-line; petals white; lip yellow. Dorsal sepal 4–6 × 2–4 mm, ovate; lateral sepals 4–5 × 2–2.5 mm, obliquely ovate, somewhat spreading. Petals 2–2.5 × 1 mm, obovate, obtuse. Lip fleshy, 2–3.5 × 1–1.5 mm, recurved about half-way, glabrous. Column 1–2 mm long, including stelidia almost 1 mm long.

Malawi. N: Mafinga Mts., Jembya Forest, 1850 m, *Dowsett-Lemaire* 524 (K). S: Zomba Mt., by Chagwa Dam, c. 1700 m, fl. 13.v.1984, *la Croix* 630 (K); Mulanje Mountain, Chambe Plateau, 1980 m, fl. 14.v.1986, *J.D. & E.G. Chapman* 7556 (K; MO).
Also in Tanzania. Evergreen montane forest and swamp forest, growing with ferns and mosses on trunks and branches of *Syzygium cordatum* and other trees; 1700–2300 m.

6. **Bulbophyllum intertextum** Lindl. in J. Linn. Soc., Bot. **6**: 127 (1862). —Rolfe in F.T.A. **7**: 29 (1897). —Summerhayes in F.W.T.A. ed. 2, **3**: 234 (1968). —Morris, Epiphyt. Orch. Malawi: 49 (1970). —Grosvenor in Excelsa **6**: 79 (1976). —Williamson, Orch. S. Centr. Africa: 138 (1977). —Ball, S. Afr. Epiph. Orch.: 76 (1978). —la Croix et al., Malawi Orch. **1**: 46 (1983). —Cribb in F.T.E.A., Orchidaceae: 309 (1984). —Vermeulen in Orchid Monogr. **2**: 54 (1987). —la Croix et al., Orch. Malawi: 192 (1991). —Geerinck in Fl. Afr. Centr., Orchidaceae pt. 2: 328 (1992). Type from Nigeria.
Bulbophyllum usambarae Kraenzl. in Bot. Jahrb. Syst. **34**: 58 (1904). Type from Tanzania.
Bulbophyllum intertextum var. *parvilabium* G. Will. in Pl. Syst. Evol. **134**: 62 (1980). Type: Zambia, Zambezi Rapids, Kalene Hill, Dec. 1968, *Williamson & Simon* 442 (K, holotype; SRGH) non *Bulbophyllum parvilabium* Schltr. (1911), nec Schltr. (1919).

Dwarf epiphytic herb. Pseudobulbs usually set close together on a slender rhizome, 1-leafed, up to 12 × 10 mm, ovoid or ± orbicular, shiny green tinged with purple-maroon. Leaves 15–50 × 7–10 mm, narrowly oblong, acute or obtuse. Inflorescence erect, 5–10(30) cm long, c. 12-flowered; peduncle and rhachis wiry. Flowers greenish-white, the sepals tipped with purple-pink. Pedicel and ovary 2 mm long. Dorsal sepal 3–6 × 1–2 mm, ovate, acute; lateral sepals 3–6 × 0.8–3 mm, acute or acuminate, slightly recurved. Petals c. 2 × 1 mm, oblong-elliptic, acute or rounded. Lip fleshy, recurved, 1.2–4 × 0.7–2 mm, ovate, rounded at the apex, auriculate at the base; margin hairy; upper surface with median slit. Column 1–2 mm long, including stelidia 0.3–1 mm long.

Zambia. W: Mwinilunga Distr., Sinkabolo Dambo, fl. 16.ii.1975, *Williamson & Gassner* 2406 (K; SRGH). **Zimbabwe**. E: Nyanga (Inyanga), by Nyaningwa River, fl. 23.ii.1975, *Ball* 1308 (K; SRGH). **Malawi**. N: Mzimba Distr., Mzuzu, Marymount Secondary School, 1320 m, fl. 29.v.1975, *Pawek* 9668 (K; UC). S: Mulanje Mt., Ruo Gorge, near Lujeri, c. 1000 m, *Morris* 95 (K).
Also in West Africa, Ethiopia, Kenya, Tanzania, Angola and the Seychelles. Evergreen and riverine forest, forming clumps on trunks and branches, usually embedded in moss, rarely in *Brachystegia* woodland; 1000–1900 m.

7. **Bulbophyllum josephi** (Kuntze) Summerh. in Bot. Mus. Leafl. Harv. Univ. **11**: 250 (1945); in F.W.T.A. ed. 2, **3**: 234 (1968). —Vermeulen in Orchid Monogr. **2**: 67 (1987) as "*josephii*". —la Croix et al., Orch. Malawi: 190 (1991). —Geerinck in Fl. Afr. Centr., Orchidaceae pt. 2: 335 (1992). Type from Cameroon.
Phyllorchis josephi Kuntze, Revis. Gen. Pl. **2**: 676 (1891).
Bulbophyllum mahonii Rolfe in Bull. Misc. Inform., Kew **1906**: 32 (1906). —Morris, Epiphyt. Orch. Malawi: 47 (1970). —Grosvenor in Excelsa **6**: 79 (1976). —Williamson, Orch. S. Centr. Africa: 138 (1977). —Ball, S. Afr. Epiph. Orch.: 72 (1978). —Cribb in F.T.E.A., Orchidaceae: 310 (1984). Type: Malawi, Mulanje Mt., c. 1200 m, fl. v.1899, *Mahon* s.n. (K, holotype).
Bulbophyllum amanicum Kraenzl. in Bot. Jahrb. Syst. **51**: 382 (1914). Type from Tanzania.

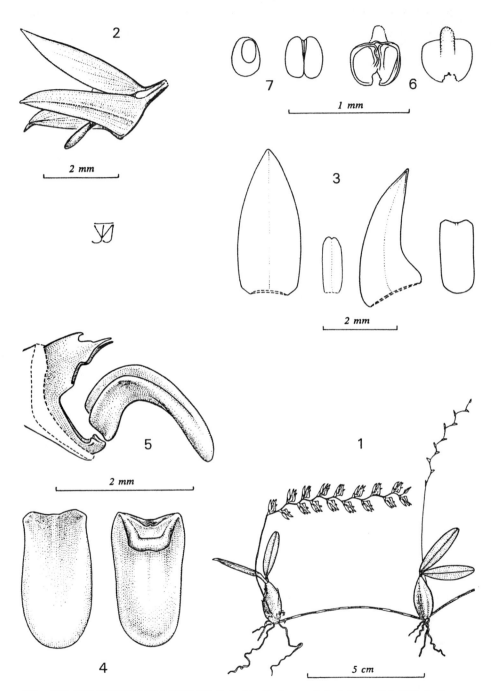

Tab. 103. BULBOPHYLLUM STOLZII. 1, habit; 2, flower, 1 & 2 from *Stolz* 767; 3, flower analysis, from left to right: dorsal sepal, petal, lateral sepal, lip; 4, lip, left: adaxially, right: abaxially; 5, column and lip, lateral view, 3–5 from *Cribb & Grey-Wilson* 10818; 6, anther, left: adaxially, right: abaxially, from *Stolz* 767; 7, pollinia, left: single pair, right: both pairs, from *Cribb & Grey-Wilson* 10818. Drawn by J.J. Vermeulen. From Orchid Monographs.

Bulbophyllum winkleri Schltr. in Orchis **8**: 132, t. 3/8–15 (1914), non Schltr. (1906). Type from East Africa.

Bulbophyllum schlechteri De Wild. in Pl. Bequaert. **1**: 120 (1921). Type as for *B. winkleri.*

Bulbophyllum sennii Chiov. in Atti R. Acad. Ital., Mem. Cl. Sci. Fis. **11**: 57 (1940). Type from Ethiopia.

Bulbophyllum josephi var. *mahonii* (Rolfe) J.J. Verm. in Bull. Jard. Bot. Belg. **56**: 232 (1986) as "*josephii*".

Epiphytic herb. Pseudobulbs set close together on the rhizome, 1-leafed, up to 30 × 15 mm, ovoid, slightly flattened, sometimes obscurely angled, dark maroon-red or greenish-brown, the surface pitted like orange peel. Leaf erect, stiff, 8–13 × 1.5–2 cm, elliptical, minutely 2-lobed at the apex. Inflorescence arching or pendent; peduncle c. 5 cm long, rhachis 12–20 cm long, fairly densely many-flowered, the flowers ± secund, not opening widely. Flowers white, the sepals sometimes purple or orange-vermilion at the tips, the lip white or yellow; sometimes with a strong fruity scent, sometimes scentless. Pedicel and ovary 3–4 mm long, bracts 4–5 mm long, whitish or brownish. Dorsal sepal 5–7 × 1–2 mm, narrowly triangular, acute, lateral sepals slightly larger and broader. Petals 2–3 × 1 mm, oblong, minutely papillose. Lip 2–3 × 1–1.5 mm, recurved, ovate, glabrous or papillose. Column c. 1 mm long, stelidia c. 1 mm long.

Zambia. W: Mwinilunga Distr., fl. iv.1969, *G. Williamson* 22 (K; SRGH). **Zimbabwe**. E: Vumba Mts., 1500 m, fl. iii.1949, *Wild* 2804 (K; SRGH). **Malawi**. N: Nyika Nat. Park, top of path to Chisanga Falls, 1900 m, fl. 23.iv.1982, *la Croix* 337 (K). S: Blantyre Distr., Mt. Soche, c. 1500 m, fl. 16.iii.1965, *Morris* 61 (K). **Mozambique**. Z: Gurué, c. 1200 m, fl. viii.1979, *Schäfer* 6884 (K; LMU). MS: Chimanimani Mts., c. 1500 m, fl. 12.iv.1967, *Grosvenor* 383 (K; SRGH).

Widespread in tropical Africa. *Brachystegia* woodland and submontane evergreen forest; 1050–1900 m.

This species is named after Sir Joseph Hooker.

8. **Bulbophyllum expallidum** J.J. Verm. in Bull. Soc. Roy. Bot. Belgique **54**: 146 (1984); in Orchid Monogr. **2**: 72, fig. 37, Pl. 5c (1987). —la Croix et al., Orch. Malawi: 188 (1991). — Geerinck in Fl. Afr. Centr., Orchidaceae pt. 2: 336 (1992). Type: Malawi, Mzuzu; cult. Hort. Leiden 22269 (L, holotype).

Bulbophyllum buntingii sensu Williamson, Orch. S. Centr. Africa: 138 (1977).

Bulbophyllum acutisepalum sensu Cribb in F.T.E.A., Orchidaceae: 310 (1984).

Epiphytic herb. Pseudobulbs set close together on the rhizome, 1-leafed, up to 3 × 2.5 cm, ovoid, obscurely 4–6-angled, glossy yellow, rarely purple-tinged. Leaf erect, stiff, up to 12 × 3 cm, oblong. Inflorescence erect or suberect, to 35 cm long, many-flowered, the flowers ± secund. Flowers creamy-white, faintly scented, not opening widely. Pedicel and ovary 3–4 mm long; bracts 2–5 mm long, whitish. Dorsal sepal c. 10 × 2 mm, lanceolate, acuminate, slightly recurved; lateral sepals slightly longer, otherwise similar. Petals 2.5 × 1.5 mm, ovate, acute. Lip 2–4 × 0.5 mm, fleshy, channelled, with purple line. Column 1–1.8 mm long, including stelidia; stelidia 0.4–0.7 mm long.

Zambia. N: Kasama, Malale Rocks, 1275 m, fl. 1.iii.1960, *Richards* 12670 (K). W: Solwezi Distr., Chifubwa, 40 km on Kyushi Road, fl. 5.i.1961, *Holmes* 010 (K). C: Mkushi Distr., Irumi Hills, fl. 9.iii.1963, *Morze* 35 (K). **Malawi**. N: South Viphya, Lusangadzi, near Mzuzu, c. 1350 m, fl. 2.i.1982, *la Croix* 249 (K).

Also in Zaire and Tanzania. In mixed woodland, forming clumps on trunks and lower branches of trees, common where it occurs, often growing with *B. josephi*, rarely lithophytic; 1275–1900 m.

9. **Bulbophyllum humblotii** Rolfe in J. Linn. Soc., Bot. **29**: 50 (1891). —H. Perrier in Fl. Madag., Orch. 1: 447 (1939). —Grosvenor in Excelsa **6**: 79 (1976). —Ball, S. Afr. Epiph. Orch.: 80 (1978). —la Croix et al, Malawi Orch. **1**: 44 (1983). —Cribb in F.T.E.A., Orchidaceae: 321 (1984). —Vermeulen in Orchid Monogr. **2**: 79, fig. 43 (1987). —la Croix et al., Orch. Malawi: 189 (1991). Types from Madagascar.

Bulbophyllum linguiforme P.J. Cribb in Kew Bull. **32**: 157 (1977). Type: Zimbabwe, Chimanimani Mts., *Ball* 316 (K, holotype; SRGH).

Bulbophyllum sp. (*Morris* 27), Morris, Epiphyt. Orch. Malawi: 47 (1970).

Dwarf epiphytic herb. Pseudobulbs set (0.5)1–3 cm apart on the rhizome, 1-leafed, c. 10 × 9 mm, ± globose, glossy green or yellow-green. Leaf fleshy, erect, 2–3(5) ×

0.6–1 cm, oblong, obtuse. Inflorescence 15–35 mm long, first erect then arching, fairly densely 4–10-flowered. Flowers creamy-white with yellow lip, not opening widely. Pedicel and ovary 0.5–1.5 mm long; bracts c. 2 mm long, whitish. Dorsal sepal c. 4–5 × 3 mm, ovate, apiculate, concave; lateral sepals 4–6 × 2–3 mm, triangular, acute. Petals c. 2 × 1 mm, oblong. Lip fleshy, 2.5 × 1–1.5 mm, tongue-like, geniculate at base. Column c. 1.5 mm long; stelidia c. 1 mm long.

Zimbabwe. E: Chimanimani Mts., Haroni Valley, 1060 m, fl. 11.iv.1954, *Ball* 316 (K; SRGH). **Malawi**. S: Mulanje Distr., Limbuli Estate, 800 m, fl. 26.vii.1985, *la Croix & Spurrier* 711 (K; MAL).
Also in Tanzania, Seychelles and Madagascar. In *Brachystegia/Uapaca* woodland, a high-level epiphyte, usually on *Uapaca spp.*; 640–1250 m.

10. **Bulbophyllum maximum** (Lindl.) Rchb.f. in Walp., Ann. Bot. Syst. **6**: 259 (1861) as "*Bolbophyllum*". —Summerhayes in Kew Bull. **12**: 120 (1957); in F.W.T.A. ed. 2, **3**: 241 (1968). —la Croix et al, Malawi Orch. **1**: 83 (1983). —Cribb in F.T.E.A., Orchidaceae: 317 (1984). —Vermeulen in Orchid Monogr. **2**: 97 (1987). —la Croix et al., Orch. Malawi: 202 (1991). —Geerinck in Fl. Afr. Centr., Orchidaceae pt. 2: 344 (1992). Type from Nigeria.
    Megaclinium maximum Lindl., Gen. Sp. Orchid. Pl.: 47 (1830).
    Megaclinium oxypterum Lindl. in Bot. Reg. **25**: Misc. Not. 14 (1839). —Rolfe in F.T.A. **7**: 38 (1897). Type from Sierra Leone.
    Bulbophyllum oxypterum (Lindl.) Rchb.f. in Walp., Ann. Bot. Syst. **6**: 258 (1861). — Summerhayes in Kew Bull. **12**: 121 (1957); in F.W.T.A. ed. 2, **3**: 241 (1968). —Goodier & Phipps in Kirkia **1**: 54 (1961). —Morris, Epiphyt. Orch. Malawi: 41 (1970). —Grosvenor in Excelsa **6**: 79 (1976). —Williamson, Orch. S. Centr. Africa 132 (1977). —Ball, S. Afr. Epiph. Orch.: 82 (1978).
    Megaclinium platyrhachis Rolfe in F.T.A. **7**: 43 (1897). Type: Malawi, hort. *O'Brien s.n.* (K, holotype).
    Bulbophyllum platyrhachis (Rolfe) Schltr., Bot. Jahrb. Syst. **38**: 15 (1905). —Morris, Epiphyt. Orch. Malawi: 40 (1970). —la Croix et al., Malawi Orch. **1**: 55 (1983). —Cribb in F.T.E.A., Orchidaceae: 319 (1984).
    Megaclinium oxypterum var. *mozambicense* Finet, Not. Syst. **1**: 169 (1910). Type: Mozambique, Aringa, 11.ii.1905, *le Testu* 668 (BM; P, holotype)
    Bulbophyllum nyassanum Schltr. in Bot. Jahrb. Syst. **53**: 571 (1915). Type from Tanzania.
    Bulbophyllum oxypterum var. *mozambicense* (Finet) De Wild., Pl. Bequaert. **1**: 94 (1921).

Robust epiphytic herb. Pseudobulbs set 2–10 cm apart on a woody rhizome, 2–3(4)-leaved, 4–10 × 1–3 cm, very variable in size and shape, but usually ovoid or oblong, 4–5-angled, sometimes with knobs. Leaves 4 × 20 × 1.5–5.5 cm, oblong, obtuse, emarginate, coriaceous, thick. Inflorescence 15–90 cm long, several to many-flowered. Rhachis flattened, 1–5 cm wide, sometimes leaf-like and yellow, straight or undulate, the edge entire or crenate. Flowers distichous, set slightly off-centre from the median line, only a few open at a time, green or yellow-green with purple-brown marks. Pedicel and ovary 2–5 mm long; bracts c. 4 mm long, reflexed. Dorsal sepal 5–8 × 1–2 mm, ovate-lanceolate, acute, concave; lateral sepals 4–6 × 2–3.5 mm, slightly recurved, ovate. Petals 2–3 × c. 0.5 mm, lanceolate. Lip recurved, up to 3 × 1.5 mm, ovate-oblong, fleshy. Column 1.5–2.2 mm long, including stelidia 0.2 mm long.

Zambia. W: Mwinilunga Distr., Zambezi Rapids, 6 km N of Kalene Hill Mission, fl. 20.ii.1975, *Williamson & Gassner* 2445 (K; SRGH). C: Lusaka Distr., 22 km from Lusaka on Great East Road, fl. & fr. 16.xi.1961, *Morze* 75 (K). **Zimbabwe**. C: Harare Distr., near Harare, 1360 m, st. viii.1927, *Eyles* 5028 (K). E: Nyamkwarara Valley, fl. ii.1935, *Gilliland* 1541 (K). **Malawi**. N: Nkhata Bay Distr., Kavuzi Estate, 640 m, fl. 3.xii.1985, *la Croix* 738 (K). C: Lilongwe Distr., Dzalanyama F.R., 1250 m, fl. 11.xii.1985, *la Croix* 751 (K). S: Thyolo Distr., Bvumbwe, 1150 m, fl. xii.1970, *Westwood* 578 (K). **Mozambique**. N: Ribáuè Mts., 900 m, st. ix.1931, *Gomes e Sousa* 846 (K). MS: Serra de Vumba, Nascenta da Fábrica de Águas, 1050 m, fl. 7.v.1980, *Pettersson* 126 (K).
Widespread in tropical Africa. *Brachystegia* or mixed deciduous woodland and riverine forest; 600–1600 m.

11. **Bulbophyllum injoloense** De Wild. in Bull. Jard. Bot. État. **5**: 175 (1916). —Vermeulen in Orchid Monogr. **2**: 102 (1987). —Geerinck in Fl. Afr. Centr., Orchidaceae pt. 2: 345 (1992). Type from Zaire.
    Megaclinium injoloense (De Wild.) De Wild. in Bull. Jard. Bot. État. **5**: 175 (1916); in Pl. Bequaert. **1**: 87 (1921).

Subsp. **pseudoxypterum** (J.J. Verm.) J.J. Verm. in Bull. Jard. Bot. Belg. **56**: 232 (1986); Orchid Monogr. **2**: 103, fig. 60 (1987). —Geerinck in Fl. Afr. Centr., Orchidaceae pt. 2: 346 (1992). Type: Zambia, Solwezi, 3.x.1973, *Fanshawe* 12116 (K, holotype).
Bulbophyllum injolense var. *pseudoxypterum* J.J. Verm. in Bull. Jard. Bot. Belg. **54**: 139 (1984).

Pseudobulbs set 2.5–9.5 cm apart on a woody rhizome 5–6 mm in diameter, 2-leaved, 5–8.5 × 2–3 cm, ovoid, (3)4-angled. Leaves 4–13.5 × 1.5–4 cm, lanceolate to linear-lanceolate coriaceous. Inflorescence 24–80 cm long; peduncle 14–45 mm long with tubular, papery white scales 11–25 mm long. Rhachis 16–35 cm long, flattened, 4–13 mm wide, dull chocolate-brown with fine dark hairs. Pedicel and ovary 3–5 mm long, with fine dark hairs; bracts reflexed, 3.5–8 × 3–7 mm, ovate. Flowers yellow or dirty-white marked with deep chocolate-purple. Dorsal sepal 7–8 × 2–2.5 mm, lanceolate, acute; lateral sepals 5–6.5 × 3 mm, falcate, triangular, acute or acuminate. Petals 3.8–5 × 0.6–1 mm, linear-lanceolate or narrowly lanceolate, finely to coarsely papillose. Lip 2–2.5 × 1–1.4 mm, the margins serrulate or lacerate near the base. Column 1.8–2 mm long, with downward facing wings on lower edge.

Zambia. N: Mbala Distr., 1660 m, fl. 5.vii.1941, *Greenway* 6184 (K). W: Mwinilunga Distr., Matonchi Farm, fl. 31.viii.1930, *Milne-Redhead* 996 (K).
Also in Zaire. Epiphytic in *Cryptosepalum* and *Brachystegia* woodland, sometimes lithophytic.
Subspecies *injloense* occurs in Zaire and may be distinguished by having leaves (13.5)19.5–27 cm long, and by flowers more widely spaced along the rhachis (14–20 mm apart about the middle of the mature rhachis).

12. **Bulbophyllum sandersonii** (Hook.f.) Rchb.f. in Flora **61**: 78 (1878). —Bolus in J. Linn. Soc., Bot. **25**: 181 (1889); Icon. Orchid. Austro-Afric. **1**, 1: pl. 3 (1893). —Goodier & Phipps in Kirkia **1**: 54 (1961). —Grosvenor in Excelsa **6**: 79 (1976). —Williamson, Orch. S. Centr. Africa: 134 (1977). —Ball, S. Afr. Epiph. Orch.: 84 (1978). —Stewart et al., Wild Orch. South. Africa: 219 (1982). —la Croix et al., Malawi Orch. **1**: 55 (1983). —Cribb in F.T.E.A., Orchidaceae: 320 (1984). —Vermeulen, Orchid Monogr. **2**: 107 (1987). —la Croix et al., Orch. Malawi: 205 (1991). —Geerinck in Fl. Afr. Centr., Orchidaceae pt. 2: 349 (1992). Type from South Africa.
Megaclinium sandersonii Hook.f. in Bot. Mag. 97: sub pl. 5936 (1871).
Megaclinium melleri Hook.f. in Bot. Mag. 97: sub pl. 5936 (1871). —Rolfe in F.T.A. **7**: 42 (1897). Type: Malawi, Mt. Chiradzulu, Manganja Range, c. 600 m, Sept. 1861, *Meller* s.n. (K).
Bulbophyllum melleri (Hook.f.) Rchb.f. in Flora **61**: 78 (1878).
Bulbophyllum tentaculigerum Rchb.f. in Flora **61**: 77 (1878). Type from Cameroon.
Megaclinium pusillum Rolfe in Bull. Misc. Inform., Kew **1894**: 362 (1894); in F.T.A. **7**: 42 (1897). Type from E Africa.
Bulbophyllum pusillum (Rolfe) De Wild., Pl. Bequaert. **1**: 95 (1921).
Bulbophyllum oxypterum sensu Moriarty, Wild Fls. Malawi: t. 28, 1 (1975).

Subsp. **sandersonii** TAB. **104**.

Epiphytic herb with a stout rhizome 2–3 mm in diameter. Pseudobulbs set 2–6 cm apart on a woody rhizome, 2-leaved, variable in size and shape, 1.5–7 × 1–2.5 cm, ovoid to ellipsoid, usually 3–6-angled, sometimes with knobs. Leaves 6–10 × 1–2 cm, lanceolate to linear, apex obtuse and emarginate, coriaceous. Inflorescence erect, several to many-flowered, 7–24 cm long; rhachis swollen, flattened, papillose, usually irregularly dentate on the margin, purple; bracts recurved or reflexed, 3.5–7 × 2–5 mm, ovate. Flowers distichous, usually borne off-centre of the median line of the rhachis, 3–15 mm apart, several open at a time. Flowers basically green, but so heavily mottled with purple that the effect is purple; rarely clear yellow. Pedicel and ovary glabrous, 2.5–4.5 mm long. Dorsal sepal 6.5–11 × 1–2.5 mm, linear-lanceolate; lateral sepals 3.5–6 × 2.5–4 mm, basal part ± orbicular, then with a curved apiculus pointing towards the centre. Petals erect, 5–10 mm long, less than 1 mm wide, almost filiform and slightly club-shaped at the tip. Lip about 2 × 1 mm, tongue-like, with 2 low, rounded keels towards the base of the upper surface. Column 2–3.5 mm long, including triangular stelidia 0.3–0.4 mm long.

Zambia. N: NW corner of Lake Mweru, c. 1400 m, fl. in hort. 14.i.1941, *Bredo* in *Moreau* 713 (EAH; K). C: Kundalila Falls, fl. i.1967, *G. Williamson* 20 (K). E: Danger Hill, fl. ii.1969, *Williamson* 1398 (K). **Zimbabwe**. E: Mutare, 1200 m, fl. 4.xii.1961, *Wild & Chase* 5545 (K;

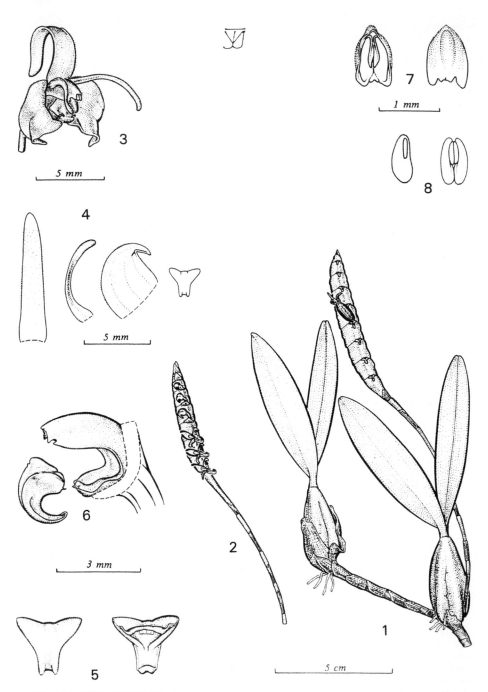

Tab. 104. BULBOPHYLLUM SANDERSONII subsp. SANDERSONII. 1, habit, from *Polhill & Paulo* 1684; 2, inflorescence; 3, flower; 4, flower analysis, from left to right: dorsal sepal, petal, lateral sepal, lip; 5, lip, left: adaxially, right: abaxially; 6, column and lip, lateral view, 2–6 from *Gassner* 16; 7, anther, left: adaxially, right: abaxially; 8, pollinia, left: single pair, right: both pairs, 7 & 8 from *Wubben* cult. s.n. Drawn by J.J. Vermeulen. From Orchid Monographs.

SRGH). **Malawi**. N: South Viphya, base of Nkhalapya, 1600 m, fl. 2.xii.1986, *la Croix* 895 (K). C: Ntchisi For. Res., 1400–1590 m, fl. 26.iii.1970, *Brummitt* 9408 (K). S: Thyolo Distr., Mpeni Estate, 1150 m, fl. 19.ix.1981, *la Croix* 201 (K).

Widespread in tropical and South Africa. Epiphytic in woodland, sometimes riverine forest, occasionally lithophytic; 1000–1600 m.

Subsp. *stenopetalum* (Kraenzl.) J.J. Verm., occurring in W Africa and Zaire, differs from the typical subspecies in having floral bracts as wide as or wider than the fully developed rhachis, flowers medial on the rhachis, lip with 2 prominent adaxial keels close to the margins, sepals white or green or orange and not or faintly marked with purple.

13. **Bulbophyllum scaberulum** (Rolfe) Bolus in J. Linn. Soc., Bot. **25**: 181 (1889). —Summerhayes in Bot. Not. **1937**: 192 (1937). —Grosvenor in Excelsa **6**: 79 (1976). —Ball, S. Afr. Epiph. Orch.: 86 (1978). —Stewart et al., Wild Orch. South. Africa: 220 (1982). —la Croix et al., Malawi Orch. **1**: 56 (1983). —Cribb in F.T.E.A., Orchidaceae: 320 (1984). —Vermeulen in Orchid Monogr. **2**: 116 (1987). —la Croix et al., Orch. Malawi: 206 (1991). —Geerinck in Fl. Afr. Centr., Orchidaceae pt. 2: 352 (1992). Type from South Africa.

Megaclinium scaberulum Rolfe in Gard. Chron., ser. 3, **4**, nr. 80: 6 (1888).

Megaclinium clarkei Rolfe in Bull. Misc. Inform., Kew **1891**: 198 (1891). Type from W Africa.

Bulbophyllum clarkei (Rolfe) Schltr. in Bot. Jahrb. Syst. **38**: 13 (1905).

Bulbophyllum congolanum Schltr., Westafr. Kauschuk-Exp.:281 (1900); in Bot. Jahrb. Syst. **38**: 14 (1905). —Williamson, Orch. S. Centr. Africa: 113, t. 115 (1977). Types from Gabon and Cameroon.

Bulbophyllum chevalieri De Wild., Pl. Bequaert. **1**: 80 (1921). Type from Guinée.

Medium to large epiphytic or lithophytic herb. Rhizome creeping, 2–5 mm in diameter. Pseudobulbs 2-leaved, to 9 × 3 cm, orbicular, ovoid, ellipsoid or oblong, 3–5-angled, the edges often wavy or knobbly, yellow or green. Leaves stiff, 5–17 × 1–4 cm, linear to lanceolate, or ovate-elliptic. Inflorescence erect, up to 24 cm long, the peduncle shorter than the rhachis. Rhachis fleshy, scabrid, bilaterally flattened, to 1 cm wide, green, wavy on the upper margins, up to c. 30-flowered. Bracts reflexed, 2–7 × 2–4 mm, triangular-ovate, placed near the lower edge of the rhachis. Pedicel and ovary 3 mm long. Flowers fleshy, scabrid, mottled with brown or maroon, several to many opening together. Dorsal sepal fleshy, 6–10 × 1.5–2 mm, ovate-lanceolate, curved, subacute; lateral sepals 4–8 × 2.5 mm, ovate, acuminate, the apex curved downwards. Petals 2.5–6 × 0.5–1 mm, lanceolate, acute. Lip 2–6 × 2–3 mm, ovate, deflexed, recurved, the edges entire. Column 1.5 mm long, stelidia short, acute.

Zambia. N: Mbala Distr., Kambole, High Escarpment, 1800 m, fl. 12.ix.1960, *Richards* 13237 (K). W: Kabompo, fl. iv.1969, *G. Williamson* 15 (K; SRGH). C: Kabwe and Mulungushi Distr., fl. x.1960, *Morze* 33 (K). **Zimbabwe**. E: Chimanimani Mts., 900 m, fl. 6.xi.1950, *Wild* 3573 (K; SRGH). S: Chivi Distr., fl. 12.xi.1962, *Leach* 11276 (K; SRGH). **Malawi**. N: Misuku Hills, edge of Mugesse (Mughese) Forest, 1750 m, fl. in hort. 11.ix.1984, *la Croix* 664 (K). S: Thyolo Distr., Nkwadzi River, 900 m, fr. *Morris* 45 (BM; K).

Widespread in tropical and South Africa. Epiphytic or lithophytic in woodland and evergreen forest; 900–1900 m.

This species shows great vegetative variation, and specimens with oblong pseudobulbs and linear-lanceolate leaves closely resemble *B. sandersonii* when sterile, while plants with squat, ± orbicular pseudobulbs and short, ovate leaves are easily recognisable. Vermeulen (1987) distinguished 3 varieties of this species, but only var. *scaberulum* is known from the Flora Zambesiaca area.

14. **Bulbophyllum ballii** P.J. Cribb in Kew Bull. **32**: 159 (1977). —Ball, S. Afr. Epiph. Orch.: 78 (1978). —Vermeulen in Orchid Monogr. **2**: 134, fig. 83 (1987). TAB. **105**. Type: Zimbabwe, Haroni Gorge, near Welgelegen, fl. 7.iii.1954, *Ball* 258 (K, holotype; SRGH).

Bulbophyllum sp. no. 2 (*Ball* 258), Grosvenor in Excelsa **6**: 79 (1976).

Small epiphytic herb. Pseudobulbs 3–8 mm apart on a creeping rhizome, 1(2)-leaved, 8–14 × 6–10 mm, pyriform to ovoid. Leaves 3–7 × c. 1 cm, ligulate or lanceolate, acute. Inflorescence c. 12 cm long; peduncle c. 5 cm long, slightly swollen; rhachis 7 cm long, up to 18-flowered. Sepals yellow-green, marked with purple-red, or purple; petals white and purple; lip deep red-purple. Pedicel and ovary c. 1 mm long, with scattered papillae and dark hairs; bracts recurved, 2–3.5 mm long, with scattered papillae. Sepals with scattered papillae and dark hairs on the outside; dorsal sepal 3 × 1 mm, oblong, acute, concave; lateral sepals c. 3.5 × 1.8 mm,

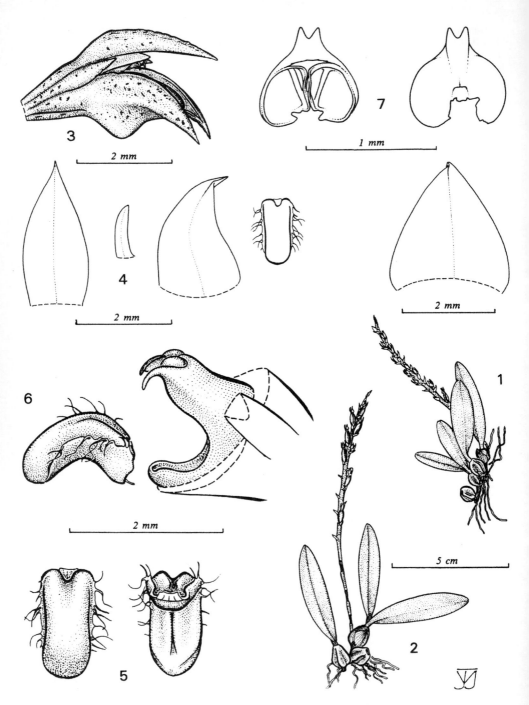

Tab. 105. BULBOPHYLLUM BALLII. 1, habit, flowering, from *Ball* 258; 2, plant, fruiting, from *Ball* 1323; 3, flower; 4, flower analysis, from left to right: dorsal sepal, petal, lateral sepal, lip; 5, lip, left: adaxially, right: abaxially; 6, column and lip, lateral view; 7, anther, left: adaxially, right: abaxially, 3–7 from *Ball* 258. Drawn by J.J. Vermeulen. From Orchid Monographs.

recurved oblong, obliquely lanceolate, acuminate, the edges finely lacerate. Petals to 1.8 × 0.8 mm, lanceolate, slightly falcate. Lip recurved, 1.5 × 0.8 mm, oblong, the edges finely lacerate. Column 1.4–1.8 mm long, including stelidia 0.8–1 mm long; column foot 2 mm long.

Zimbabwe. E: Chimanimani Mts., Haroni Gorge, near Welgelegen, fl. 7.iii.1954, *Ball* 258 (K; SRGH); Chimanimani Distr., Mutzingazi Forest, 915 m, 1956, *Ball* 881 (SRGH).
Known only from the Flora Zambesiaca area. Evergreen forest; c. 1000 m.
G. Williamson 141 (K), from the Zambezi Rapids, Kalene Hill, in Zambia may also belong to this species.

15. **Bulbophyllum elliotii** Rolfe in J. Linn. Soc., Bot. **29**: 51 (1891). —H. Perrier in Fl. Madag., Orch. 1: 420 (1939). —Vermeulen in Orchid Monogr. **2**: 135 (1987). —la Croix et al., Orch. Malawi: 197 (1991). —Geerinck in Fl. Afr. Centr., Orchidaceae pt. 2: 360 (1992). Type from Madagascar.
Bulbophyllum malawiense Morris in Proc. Linn. Soc. **179**: 63 (1968); Epiphyt. Orch. Malawi: 45 (1970). —Grosvenor in Excelsa **6**: 79 (1976). —Williamson, Orch. S. Centr. Africa: 136, t. 118 (1977). —Ball, S. Afr. Epiph. Orch.: 90 (1978). —Stewart et al., Wild Orch. South. Africa: 219 (1982). —la Croix et al., Malawi Orch. 1: 52 (1983). —Cribb in F.T.E.A., Orchidaceae: 312 (1984). Type: Malawi, Thyolo, near Nyasa Church, 900 m, fl. ix.1965, *Morris* 172 (K, holotype).

Small epiphytic herb; rhizome creeping, c. 2 mm in diameter. Pseudobulbs set close together (0.5–1.5 cm apart), forming clumps, 2-leaved, 15–25 × 18–25 mm, orbicular to ovoid, squat, green to reddish. Leaves fleshy, spreading, 25–40 × 10–22 mm, elliptic, rounded at apex, dark green, sometimes tinged with purple. Inflorescence pendulous, 4–15 cm long; peduncle ± equal to the rhachis in length, with several papery sheaths; rhachis fleshy, swollen, densely several- to many-flowered. Flowers fleshy, papillose on outside, red-purple, or green heavily blotched with purple. Pedicel and ovary 1.5–2.5 mm long. Dorsal sepal 3.5–5 × 1.5–2 mm, ovate, concave; lateral sepals 3.5–4.5 × 1.5–2.5 mm, lanceolate, acuminate, slightly recurved. Petals 2–3 × 0.2–1 mm, linear. Lip 2 × 1 mm, tongue-like, fleshy, recurved, the margins long-ciliate. Column 1.5–2.2 mm long, including falcate stelidia 0.6–1 mm long; anther cap with V-shaped ridge on the apex. Pollinia 4, 2 much smaller than the others.

Zambia. N: Mbala Distr., headwaters of Chimeshi, 1660 m, fl. & fr. 30.vii.1962, *Morze* 34B (K). W: Mwinilunga, Zambezi Rapids, 6 km N of Kalene Hill Mission, fl. 21.ii.1975, *Williamson & Gassner* 2455 (K; SRGH). C: Kabwe & Mulungushi Distr., fr. viii/ix.1961, *Morze* 34A (K; SRGH). E: Danger Hill, fl. ii.1969, *G. Williamson* 1367A (K). **Zimbabwe**. E: Vumba Mts., 1500 m, fl. & fr. iii.1959, *Wild* 2808 (K; SRGH). S: Chivi Distr., Nyoni Hills overlooking Tokwe River, fl. 4.ii.1973, *Grosvenor* 809 (K; SRGH). **Malawi**. N: Mzimba Distr., Mzuzu, Marymount Secondary School, 1320 m, fr. 23.v.1970, *Brummitt & Pawek* 11097 (K). S: Thyolo Distr., near Nyasa Church, 900 m, fl. ix.1965, *Morris* 172 (K).
Also in Zaire, Burundi, Tanzania, South Africa and Madagascar. High rainfall *Brachystegia* woodland; 600–1600 m.

16. **Bulbophyllum lupulinum** Lindl. in J. Linn. Soc., Bot. **6**: 126 (1862). —Rolfe in F.T.A. **7**: 28 (1897). —Summerhayes in Kew Bull. **6**: 470 (1952); in F.W.T.A. ed. 2, **3**: 239 (1968). —Vermeulen in Orchid Monogr. **2**: 138 (1987). —Geerinck in Fl. Afr. Centr., Orchidaceae pt. 2: 361 (1992). Type from Cameroon.
Bulbophyllum urbanianum Kraenzl. in Bot. Jahrb. Syst. **28**: 163 (1900). Type from Cameroon.
Bulbophyllum ituriense De Wild. in Rev. Zool. Bot. Africaines **9**: 29 (1921). Type from Zaire.

Epiphytic (lithophytic) herb. Rhizome 2–5 mm in diameter. Pseudobulbs 3–13 cm apart on the rhizome, 2-leaved, sometimes 1-leafed, 3–7.5 × 1–3 cm, ovoid or ellipsoid, somewhat flattened, sharply 4-angled. Leaves up to 25 × 4 cm, lanceolate or linear-lanceolate, obtusely emarginate at the apex. Inflorescence up to 30 cm long; peduncle 10–12 cm long, covered with c. 10 scarious sheaths; rhachis erect, slightly flattened, 4-angled in cross section, with tufts of fine dark hairs, many-flowered; bracts large, scarious, imbricate, 9.5–16 × 7–13 mm, triangular, recurved, grey-purple with some fine dark hairs on the outer surface. Flowers distichous, several opening together, not opening widely, yellow-spotted and streaked with dark red or purple.

Dorsal sepal 4–5.2 × 1.5–3 mm, triangular, acute, not recurved; lateral sepals free, sometimes slightly recurved, 3.8–5.2 × 1.5–3 mm, triangular; petals 2.5–3.2 × 0.25–0.5 mm, spathulate to linear, slightly falcate. Lip 1.5–2.4 × 0.8–1.7 mm, rectangular, reflexed at the apex, denticulate on the margins. Column 1.8–2.6 mm long, including deltoid acute stelidia 0.3–0.5 mm long. Pollinia 4, ellipsoid, the larger ones more than twice as long as smaller.

Zambia. W: Mwinilunga Distr., Zambezi Rapids at Kalene Hill, fl. 29.viii.1960, *Holmes* 0254 (K; SRGH).
Also in W Africa, Zaire and Ethiopia. Riverine forest epiphyte.
The only specimen known from the Flora Zambesiaca area, cited above, is described as having 1-leafed pseudobulbs.

17. **Bulbophyllum fuscum** Lindl. in Bot. Reg. **25**, Misc. Not.: 3 (1839). —Rolfe in F.T.A. **7**: 24 (1897) pro parte. —Summerhayes in F.W.T.A. ed. 2, **3**: 236 (1968). —Vermeulen in Orchid Monogr. **2**: 145 (1987). —la Croix et al., Orch. Malawi: 200 (1991). —Geerinck in Fl. Afr. Centr., Orchidaceae pt. 2: 368 (1992). Type from Sierra Leone.

Var. **melinostachyum** (Schltr.) J.J. Verm. in Bull. Jard. Bot. Belg. **56**: 240 (1986); in Orchid Monogr. **2**: 148 (1987). —Geerinck in Fl. Afr. Centr., Orchidaceae pt. 2: 369 (1992). Type: Mozambique, "25 Miles Station," *Schlechter* 12250 (B†, holotype; K).
  Bulbophyllum melinostachyum Schltr. in Bot. Jahrb. Syst. **26**: 342 (1899).
  Bulbophyllum obanense Rendle, Cat. Talb. S. Nig. Pl.: 101, 146 (1913). Type from Nigeria.
  Bulbophyllum oreonastes auct. non Rchb.f. —Summerhayes in F.W.T.A. ed. 2, **3**: 236 (1968) pro parte. —Morris, Epiphyt. Orch. Malawi: 43 (1970). —Grosvenor in Excelsa **6**: 79 (1976). —Williamson, Orch. S. Centr. Africa: 137 (1977). —Ball, S. Afr. Epiph. Orch.: 88 (1978). —la Croix et al., Malawi Orch. **1**: 53 (1983). —Cribb in F.T.E.A., Orchidaceae: 312 (1984).

Epiphytic or lithophytic herb. Rhizome 1–2 mm in diameter, woody, creeping. Pseudobulbs 1–6 cm apart on the rhizome, 2-leaved, 1–5 × 0.5–1.5 cm, narrowly ovoid or ellipsoid, usually somewhat 4-angled, pale green or yellow. Leaves 2–6(11) × 1–1.5 cm, linear-lanceolate to elliptic, suberect, coriaceous, pale green. Inflorescence erect, 10- to many-flowered; peduncle 2.5–10 cm long; rhachis slightly enlarged and flattened, the flowers borne in channels on opposite sides. Flowers yellow-green to orange, the lip orange to vermilion. Ovary and pedicel c. 1.5 mm long. Bracts scarious, 2.5–6 mm long, ovate, acute, spreading or recurved, partly enclosing the flower. Dorsal sepal 2.5–5 × 1–2 mm, ovate, acute; lateral sepals 2.5–6 × 1.5–2.5 mm, obliquely ovate, acute. Petals 1.5–4 mm long, less than 0.5 mm wide, linear. Lip 1–3 × 0.5–1.5 mm, fleshy, entire, glabrous, channelled at the base, without lateral lobes. Column 0.6–1.4 mm long, including deltoid stelidia c. 0.3 mm long; anther with slightly concave rounded protrusions near the top overtopping the front margin. Pollinia 4, ellipsoid, the larger about twice as long as the smaller.

Zambia. N: c. 7 km NE of Kasama, fl. 17.iii.1957, *Savory* 159 (K; SRGH). W: Ndola Distr., Mwekera River Falls, 1225 m, fl. 9.iii.1973, *Kornas* 3454 (K). **Zimbabwe**. E: Mutare Distr., Vumba Mts., 1220 m, fl. 11.iii.1981, *Philcox, Leppard, Duri & Urayai* 8957 (K). S: Bikita Distr., Turgwe R. Gorge, c. 900 m, fl. & fr. 6.v.1969, *Pope* 108 (K; SRGH). **Malawi**. N: Nkhata Bay Distr., Mzenga Estate, 640 m, fl. 16.iv.1986, *la Croix* 835 (K). C: Lilongwe Distr., Dzalanyama Forest Reserve, by Choulongwe (Chaulongwe) Falls, 1280 m, fl. 28.iii.1970, *Brummitt* 9480 (K). S: Mulanje Distr., Likhubula Stream, 900 m, fl. 23.ii.1965, *Morris* 111 (K). **Mozambique**. MS: Beira Distr., "25 miles Station," 60 m, fl. 11.iv.1898, *Schlechter* 12250 (K; Z).

Also in W Africa, Zaire, Uganda and Tanzania.
Woodland and riverine forest epiphyte or lithophyte; 60–1350 m.
This species has often been confused with *Bulbophyllum oreonastes* Rchb.f. The latter species is much less common in the Flora Zambesiaca area, and may be distinguished by its differently shaped lip and by the flowers standing clear of the rhachis (in *B. fuscum* they appear to be sunk within it).
Var. *fuscum* occurs in W Africa, the Central African Republic, Zaire and Angola. It has a lip with distinct, usually denticulate, lateral lobes near it's base (the lip in var. *melinostachyum* lacks lateral lobes).

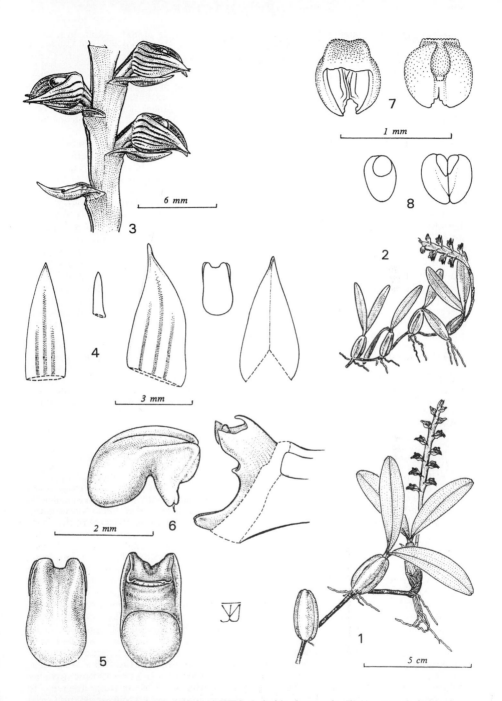

Tab. 106. BULBOPHYLLUM OREONASTES. 1, habit, from cult. *Wubben* s.n.; 2, habit, from *Milne-Redhead* 4349; 3, part of inflorescence, from cult *Wubben.* s.n.; 4, flower analysis, from left to right: dorsal sepal, petal, lateral sepal, lip, floral bract; 5, lip, left: adaxially, right: abaxially; 6, column and lip, lateral view, 4–6 from *Harrington* 546; 7, anther, left: adaxially, right: abaxially; 8, pollinia, left: single pair, right: both pairs, 7 & 8 from *Johansson* 760. Drawn by J.J. Vermeulen. From Orchid Monographs.

18. **Bulbophyllum oreonastes** Rchb.f., Otia Bot. Hamburg. **2**: 118 (1881). —Rolfe in F.T.A. **7**: 24 (1897). —Summerhayes in Kew Bull. **6**: 471 (1952); in F.W.T.A. ed. 2, **3**: 236 (1968) pro parte. —Vermeulen in Orchid Monogr. **2**: 149 (1987). —Geerinck in Fl. Afr. Centr., Orchidaceae pt. 2: 370 (1992). TAB. **106**. Type from Cameroon.
   Bulbophyllum hookerianum Kraenzl. in Bot. Jahrb. Syst. **17**: 49 (1893). Type from Cameroon.
   Bulbophyllum zenkerianum Kraenzl. in Bot. Jahrb. Syst. **48**: 391 (1912). —Summerhayes in Kew Bull. **12**: 116 (1957); in F.W.T.A. ed. 2, **3**: 239 (1968). —Williamson, Orch. S. Centr. Africa: 136 (1977). —Cribb in F.T.E.A., Orchidaceae: 313 (1984). Type from Cameroon.
   Bulbophyllum rhopalochilum Kraenzl. in Bot. Jahrb. Syst. **51**: 384 (1914). Type from Zaire.

Epiphytic or lithophytic herb. Rhizome 0.8–3 mm in diameter. Pseudobulbs 0.7–4 cm apart on the rhizome, 2-leaved, 0.4–3.5 × 0.4–1.2 cm, ovoid, ellipsoid or suborbicular, somewhat 4-angled. Leaves with a petiole-like base up to 8 mm long; lamina 0.6–8.2 × 0.4–2 cm, linear-lanceolate to elliptic. Inflorescence 1.5–17.5 cm long, 5–36-flowered; peduncle 0.7–6.5 cm long, glabrous, 2–5-sheathed; rhachis up to 16 cm long, often somewhat nodding, enlarged and flattened, 4-angled in cross section, with the flowers borne in channels on opposite sides. Flowers distichous, spreading, many open at a time, not opening widely; sepals and petals yellow or orange, usually with red-purple stripes, lip deep orange or purple. Pedicel and ovary 1.8–3 mm long; bracts 3–7 mm long, triangular, spreading, scarious. Dorsal sepal 3–6.2 × 1–1.8 mm, ovate, not recurved; lateral sepals 3–7 × 1.5–3.2 mm, triangular, sometimes slightly recurved. Petals 1–3 × 0.2–0.6 mm, obliquely linear-lanceolate. Lip 1.4–2.5 × 0.6–1.2 mm, elliptic or oblong, recurved about half-way, lip underside swollen at base and apex, the 2 parts separated by a traverse furrow. Anther with a slightly concave, truncate or retuse protrusion which overtops the front margin; pollinia 4, ± orbicular but flattened, the larger about twice as long as the smaller.

Zambia. W: Mwinilunga Distr., Luakera Falls, fl. 25.i.1938, *Milne-Redhead* 4349 (K); Kafue River N of Chingola, fl. 28.xii.1969, *Williamson & Simon* 1843 (K; SRGH). **Malawi**. S: Mangochi Distr., Namizimu Forest Reserve, Mt. Uzuzu, c. 1000 m, fl. 30.iii.1986, *Jenkins* 19 (K).
   Also in W Africa, the Central African Republic, Zaire, Rwanda and Uganda. Riverine forest or *Brachystegia* woodland epiphyte, also lithophytic on mossy rocks in shade of evergreen vegetation; 1000–1200 m.
   See note after *B. fuscum*.

19. **Bulbophyllum encephalodes** Summerh. in Bot. Mus. Leafl., Harv. Univ. **14**: 228 (1951); in Mem. N.Y. Bot. Gard. **9**, 1: 82 (1954); in Hooker's Icon. Pl. **36**: t. 3546 (1956). —Grosvenor in Excelsa **6**: 79 (1976). —Morris, Epiphyt. Orch. Malawi: 44 (1970). —Williamson, Orch. S. Centr. Africa: 139 (1977). —Ball, S. Afr. Epiph. Orch.: 70 (1978). —la Croix el al., Malawi Orch. **1**: 44 (1983). —Cribb in F.T.E.A., Orchidaceae: 307 (1984). —Vermeulen in Orchid Monogr. **2**: 151 (1987). —la Croix et al., Orch. Malawi: 198 (1991). —Geerinck in Fl. Afr. Centr., Orchidaceae pt. 2: 371 (1992). TAB. **107**. Type from Tanzania.

Epiphytic herb with a stout creeping rhizome. Rhizome woody, 2–3.5 mm in diameter. Pseudobulbs 1.5–8 cm apart on the rhizome, 1-leafed, 1.5–3.5 × 1–1.5 cm, ovoid or ellipsoid, 4-angled, green with brown streaks between the angles. Leaf erect, fleshy, with petiole to 6 mm long; lamina 3–14.5 × 1.2–3.2 cm, lanceolate or oblong, emarginate at the apex. Inflorescence 13–43 cm long, many-flowered; peduncle to 32 cm long, rather wiry; rhachis up to 11 cm long, slightly swollen and flattened, distinctly nodding so that it forms an angle with the peduncle. Flowers distichous, fleshy, green with purple striations, lip purple-black, or whitish with purple marks. Ovary and pedicel 1.7–3 mm long; bracts scarious, spreading, 3–5 × 2.6–4 mm, triangular, acuminate. Dorsal sepal 4–6 × 2–3 mm long, ovate, not recurved; lateral sepals similar but acuminate. Petals 2.2–3 × 0.6–1 mm, oblong. Lip 2–3 × 1–1.8 mm, very fleshy, with large, irregular warts towards the apex on upper and lower surfaces. Column 1.5–2 mm long, including triangular stelidia 0.2–0.3 mm long; anther with a retuse protrusion which overtops the front margin; pollinia 4, ellipsoid, the larger pair about twice as long as the smaller.

Zambia. W: Mwinilunga Distr., near Mavundu on Kabompo Road, fl. 20.v.1959, *Holmes* 016 (K). **Zimbabwe**. E: Nyanga (Inyanga), Pungwe Valley near Mozambique border, 700 m, fl. 9.v.1957, *Densen* 18 (K; SRGH). **Malawi**. S: Thyolo Distr., Bvumbwe, Mandimwe Estate, 1100 m, fl. 9.ix.1981, *la Croix* 0139 (MAL).

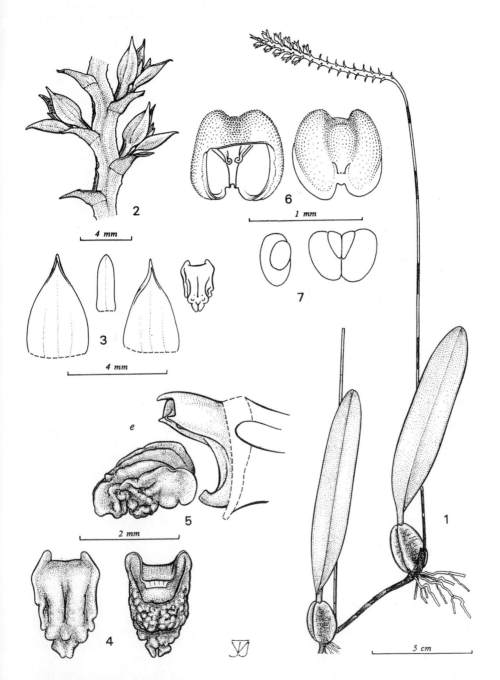

Tab. 107. BULBOPHYLLUM ENCEPHALODES. 1, habit; 2, part of inflorescence; 3, flower analysis, from left to right: dorsal sepal, petal, lateral sepal, lip; 4, lip, left: adaxially, right: abaxially; 5, column and lip, lateral view, 1–5 from *Cribb & Grey-Wilson* 10255; 6, anther, left: adaxially, right: abaxially; 7, pollinia, left: single pair, right: both pairs, 6 & 7 from *Bain* s.n. Drawn by J.J. Vermeulen. From Orchid Monographs.

Also in Cameroon, Zaire, Burundi, Uganda, Kenya and Tanzania. Evergreen forest, often near rivers on *Syzygium cordatum*, occasionally in high rainfall *Brachystegia* woodland, almost always on trunks of trees, very local; 900–1150 m.

20. **Bulbophyllum unifoliatum** De Wild. in Rev. Zool. Bot. Africaines **9**: 34 (1921). —Morris, Epiphyt. Orch. Malawi: 47 (1970). —Geerinck in Bull. Soc. R. Bot. Belg. **109**: 176 (1976). —Williamson, Orch. S. Centr. Africa: 138 (1977). —Vermeulen in Orchid Monogr. **2**: 153 (1987). —la Croix et al., Orch. Malawi: 201 (1991). —Geerinck in Fl. Afr. Centr., Orchidaceae pt. 2: 372 (1992). Type from Zaire.

Epiphytic herb. Pseudobulbs 0.5–9 cm apart on the rhizome, 1-leafed, 1.2–4 × 0.5–1.5 cm, ellipsoid. Leaf coriaceous, 1.8–17 × 0.6–1.5 cm, oblong to linear, rounded or acute at the apex. Inflorescence 5.5–20 cm long, 5–40-flowered. Peduncle 3–14 cm long; rhachis erect, arching or bent down, slightly swollen and flattened, 1.5–8.5 cm long, glabrous or slightly papillose. Flowers distichous, many open at the same time. Dorsal sepal 2.9–5.8 × 1.3–2.5 mm, ovate, acute or acuminate; lateral sepals free, recurved, 2.8–6 × 1.5–2.8 mm, obliquely ovate. Petals 1.5–3 × 0.25–0.9 mm, linear-lanceolate, oblique. Lip sometimes recurved, thick-textured, 1.4–2.6 × 0.8–1.7 mm, elliptic, obtuse, glabrous or papillose, with 2 ridges on the upper surface, margins entire. Column 0.8–1.5 mm long, including rudimentary stelidia.

1. Lip glabrous or very finely papillose · subsp. *infracarinatum*
– Lip coarsely papillose, especially the ridges on the upper surface · · · · · · · · · · · · · · · 2
2. Outer surface of sepals glabrous to very finely papillose · · · · · · · · · · · subsp. *unifoliatum*
– Outer surface of sepals coarsely papillose, especially towards the apex · · · · subsp. *flectens*

Subsp. **unifoliatum** —Vermeulen in Orchid Monogr. **2**: 154 (1987).

Pseudobulbs up to 4 cm long, greenish. Leaves up to 17 × 1.4 cm. Rhachis glabrous. Lip coarsely papillose on the upper surface, except for a glabrous strip down the mid-line. Sepals and petals yellowish or almost orange-brown; lip orange-red, purple towards the base. Dorsal sepal up to 5.2 × 2.4 mm; lateral sepals up to 6 × 2.8 mm; petals up to 3 × 0.9 mm; lip up to 2.5 × 1.6 mm.

Zambia. W: Chingola Distr., iv.1969, *G. Williamson* 21 (SRGH).
Also in Zaire, Rwanda, Angola and Tanzania. Evergreen forest.

Subsp. **infracarinatum** (G. Will.) J.J. Verm. in Orchid Monogr. **2**: 155, fig. 97 (1987). Type: Mozambique, Tsetserra, *Lady Drewe* 33 (SRGH, holotype).
    Bulbophyllum carinatum G. Will. in J. S. African Bot. **46**: 333 (1980) nom. illegit. Type as above.
    Bulbophyllum infracarinatum G. Will. in J. S. African Bot. **47**: 133 (1981). —la Croix et al., Malawi Orch. **1**: 41 (1983). Type as above.
    Bulbophyllum sp. no. 1 (*Ball* 986), Grosvenor in Excelsa **6**: 79 (1976). —Ball, S. Afr. Epiph. Orch.: 74 (1978).
    Bulbophyllum unifoliatum sensu Morris, Epiphyt. Orch. Malawi: 47 (1970) non De Wild.

Pseudobulbs 1.2–2.6 cm long; leaves up to 10 × 1.5 cm. Rhachis glabrous, green. Flowers slightly smaller, with narrower parts, than subsp. *unifoliatum*. Lip almost glabrous on upper surface, being only very finely papillose towards the margins. Sepals yellow-green or pale purple, blotched with purple; petals cream with purple dots; lip purple.

Zimbabwe. E: Chimanimani Mts., Musapa Gap, fl. i.1962, *Ball* 986 (K). **Malawi**. S: Mulanje Mt., Ruo Gorge, near Lujeri Power Station, st. *Morris* 206 (K). **Mozambique**. MS: Tsetserra, fl. i.1959, *Lady Drewe* 33 (SRGH).
Only known from the Flora Zambesiaca area. Submontane evergreen forest and riverine forest; 900–1100 m.

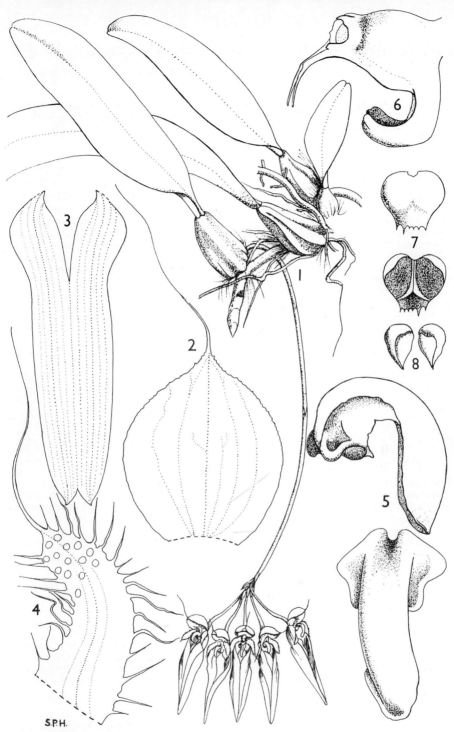

S.P.H.

Tab. 108. BULBOPHYLLUM LONGIFLORUM. 1, habit (×⅔); 2, dorsal sepal (×6); 3, lateral sepal (×2); 4, petal (×8); 5, lip, from side and above (×12); 6, column (×8); 7, anther cap, back and front views (×12); 8, pollinia (×8). Drawn by Susan Hillier, from a plant cultivated at Kew. From F.T.E.A.

Subsp. **flectens** (P.J. Cribb & P. Taylor) J.J. Verm. in Orchid Monogr. **2**: 156, fig. 98 (1987). Type from Tanzania.

 Bulbophyllum flectens P.J. Cribb & P. Taylor in Kew Bull. **35**: 436 (1980). —la Croix et al., Malawi Orch. **1**: 44 (1983). —Cribb in F.T.E.A., Orchidaceae: 307 (1984).

Pseudobulbs up to 3.3 cm long; leaves up to 9 × 1.3 cm. Rhachis glabrous or slightly papillose, purple-brown. Flowers deep purple. Lip coarsely papillose along margins on the upper side, puberulous elsewhere except for a glabrous mid-line. Flowers of similar size to subsp. *unifoliatum*.

 Malawi. C: Lilongwe Distr., Dzalanyama For. Res., 1500–1600 m, fl. 17.xii.1982, *Johnston-Stewart* in *la Croix* 230 (K).
 Also in Tanzania. Riverine and evergreen forest; 1500–1600 m.

21. **Bulbophyllum longiflorum** Thouars, Hist. Orchid. [Fl. Iles Austr. Afr.]: prem. tab. espec. & tab. 98 (1822). —Seidenfaden in Dansk. Bot. Arkiv. **29**, 1: 126 (1973). —la Croix et al., Malawi Orch. **1**: 46 (1983). —Cribb in F.T.E.A., Orchidaceae: 304 (1984). —Vermeulen in Orchid Monogr. **2**: 158 (1987). —la Croix et al., Orch. Malawi: 186 (1991). —Geerinck in Fl. Afr. Centr., Orchidaceae pt. 2: 373 (1992). TAB. **108**. Type from Madagascar.

 Epidendrum umbellatum G. Forst., Fl. Ins. Austr.: 60 (1786), non *Bulbophyllum umbellatum* Lindl. (1830). Type from Society Is.
 Cirrhopetalum thouarsii Lindl., Gen. Sp. Orchid. Pl.: 58 (1830); Bot. Reg. **10**: sub t. 832 (1824). Type as for *B. longiflorum* Thouars.
 Cirrhopetalum umbellatum (G. Forst.) Hook. & Arn., Bot. Beechey Voy.: 71 (1832). — Morris, Epiphyt. Orch. Malawi: 53 (1970). —Grosvenor in Excelsa **6**: 79 (1976). —Ball, S. Afr. Epiph. Orch.: 96 (1978).
 Cirrhopetalum africanum Schltr. in Bot. Jahrb. Syst. **53**: 573 (1915). Type from Tanzania.

Creeping epiphytic herb with a stout woody rhizome 2–5 mm in diameter. Pseudobulbs 2–9 cm apart on the rhizome, 1-leafed, 1.5–4.5 × 0.5–1.5 cm, obliquely ovoid, sometimes slightly 4-angled, green. Leaf erect, coriaceous, 7–20 × 1.5–4 cm, lanceolate to oblong, emarginate at the apex; petiole up to 2 cm long. Inflorescence 5–21 cm long, subumbellate, 2–7-flowered; peduncle with several sheaths. Ovary and pedicel 12–18 mm long; bracts spreading, 4–8 mm long. Flowers light purple, blotched with darker purple, blotchy bronze or clear yellow. Dorsal sepal 9–10 × 8–9 mm, elliptic, concave, with a dark filiform tip 9–10 mm long. Lateral sepals 25–28 × 4–7 mm, twisted so that the outer edges meet becoming connate in the centre. Petals 5–8 × 2–2.5 mm, lanceolate, acute, ciliate on the margin and with a filiform tip c. 5 mm long. Lip fleshy, 6–7.5 × 1.5–2.5 mm, ligulate, recurved, ridged below. Column 5.5–7 mm long, including slender stelidia 3–4.5 mm long. Pollinia 4, obovoid, of equal length.

 Zimbabwe. E: Chimanimani Distr., c. 4 km upstream from Haroni/Lusito junction, c. 340 m, fl. 23.xi.1981, *Pope & Müller* 2040 (K; SRGH). **Malawi**. N: Nkhata Bay Distr., 8 km E of Mzuzu, 1200 m, fr. 7.iii.1977, *Grosvenor & Renz* 1066 (K; SRGH). S: Thyolo Distr., Nchima Tea Estate, 750 m, fl. 27.xi.1984, *Pettersson* 315 (K; UPS).
 Also in Zaire, Uganda, Tanzania, Madagascar and eastwards to the Pacific Islands. Evergreen forest, often riverine, usually on trunks of trees; 800–1450 m.

INDEX TO BOTANICAL NAMES